THE IMPACT OF AIR POLLUTION ON HEALTH, ECONOMY, ENVIRONMENT AND AGRICULTURAL SOURCES

Edited by **Mohamed K. Khallaf**

The Impact of Air Pollution on Health, Economy, Environment and Agricultural Sources
Edited by Mohamed K. Khallaf

Published by InTech

Janeza Trdine 9, 51000 Rijeka, Croatia

Publishing Process Manager Ivana Lorkovic

Technical Editor Teodora Smiljanic

Cover Designer InTech Design Team

Additional hard copies can be obtained from orders@intechopen.com

The Impact of Air Pollution on Health, Economy, Environment and Agricultural Sources, Edited by Mohamed K. Khallaf
p. cm.
ISBN 978-953-307-528-0

For EU product safety concerns:
IN TECH d.o.o., Prolaz Marije Krucifikse Kozulić 3, 51000 Rijeka, Croatia,
info@intechopen.com or visit our website at intechopen.com.

Meet the editor

Jeong-Woo Park was born in Seoul, Republic of Korea. He received the B. S. , M. S., and Ph. D degrees from Seoul National University, Korea, in 1992, 1994, and 1999, respectively. He was a Senior Engineer at memory department, Samsung, Korea, where he worked on digital and analog circuit design, from 1999 to 2001. In 2001 he joined Optical Devices Department, Electronics and Telecommunications Research Institute (ETRI), Korea. Now he is in THz photonics creative team, ETRI, as a Principal member of Engineer. His research interests include silicon based optical devices such as silicon based light source, photodiode, modulator, and MUX/DEMUX. Also, he is working on THz photomixer to find out new applications in an unknown THz area.

Contents

Preface

The problem of air pollution has affects on human health, environment and many areas of life. Modern technology has led to the occurrence of air pollution and its harmful effects, but at the same time tries to push the world community to exert maximum efforts to overcome the harmful effects of it. Atmospheric pollutants have been classified according to their source, chemical composition, size and mode of release into indoor or outdoor environments. Sources of indoor pollutants are cooking and combustion, particle resuspension, building materials, air condition, consumer products, smoking, heating and biologic agents while, sources of outdoor pollutants are industrial, commercial, mobile, urban, regional, agricultural and natural. Air Pollution has an impact on the human body, such as respiratory system and heart disease. Air pollution also effects plants and agricultural products, economy and development rates in many countries all over the world. After they have been invited by the publisher, several authors from various countries, most of whom do not know each other; have collected roughly two dozen chapters which cover a wide area of Air Pollution. This book aims to strengthen the knowledge base dealing with Air Pollution. The book consists of 21 chapters dealing with Air Pollution and its effects in the field of Health, Environment, Economy and Agricultural Sources. This book is divided into four parts. The first one deals with effect of air pollution on health and human body organs, for example the impact of atmospheric pollution in the development of respiratory diseases, air pollution and cardiovascular disease, the impacts of fine particle matter on cardiovascular health, the adverse effects of air pollution on pregnancy, infant death and the association between air pollution, vitamin D status and prevalence of so-called civilization-related diseases. The second part includes the impact of air pollution on plants and agricultural sources and methods of resistance. The chapters deal with the effects of air pollutants on vegetation, current state of phytoremediation of air pollutants, such as particle matters, inorganic and organic pollutants, evaluation of existing air pollution control policies, particularly those targeted at agriculture, improving air quality, as measured by the number of exceeding levels of the CO and NO_2 standards, examining the emission gas concentration and the fundamental thermal decomposition, as well as analyzing the evolved species and the char of wood-based waste processing-materials and in the end community and occupational exposure following off-site Movement of Agricultural Fumigants. The third part includes environmental changes, geographic and climatic conditions due to air pollution. Its chapters deal with formation and reduction pollutants in combustion processes, the mobile monitoring system (MMS), as an example of a new approach for air monitoring along communication lanes, global warming or climate change, chemical pollution prediction models in function of traffic flows, emerging air pollution issues and environmental quality in the context of environmental injustice. The fourth part includes case studies concerning the impact of air pollution in the economy and development goals, such as, indoor air pollution in México, indoor air pollution and millennium development goals in Bangladesh, epidemiological and economic impact of natural gas on indoor air pollution in Colombia and economic growth and air pollution in Iran during development programs. In this book the authors explain the definition of air pollution, the most important pollutants and their different sources and effects on humans and various areas of life. The authors offer different solutions to the problems resulting from air pollution. Finally, this book is an important addition in the subject of air pollution and its impact on various fields.

Mohamed K. Khallaf
Restoration Department,
Faculty of Archaeology,
Fayoum University,
Egypt

Part 1

Effect of Air Pollution on Health and Human Body Organs

1

Impact of Air Pollution on Public Health

E. Marchwinska-Wyrwal[1,2], G. Dziubanek[1], I. Hajok[1], M. Rusin[1], K.Oleksiuk[1] and M.Kubasiak[1]
[1]*Department of Environmental Health, Medical University of Silesia, Katowice*
[2]*University of Ecology and Management, Warsaw*
Poland

1. Introduction

The air we breathe contains emissions from many different sources: industry, motor vehicles, heating and commercial sources, household fuels as well as tobacco smoke. The effects of air pollution on health have been intensively studied in recent years. The results of these studies showed that air pollution harms human health and particularly is harmful for those who are already vulnerable because of their age as children and older people or existing health problems. The epidemiological evidence suggests that adverse health effects are dependent on both exposure concentrations and length of exposure, and that long-term exposures have larger, more persistent cumulative effects than short-term exposures [1]. Ambient air pollution has been associated with a multitude of health effects, including mortality, respiratory and cardiovascular hospitalizations, changes in lung function and asthma attacks. Current scientific evidence indicates that air pollution from the combustion of fossil fuels causes a spectrum of health effects from allergy to death. Recent assessments suggest that the public health impacts may be considerable. Air pollution is associated with a broad spectrum of acute and chronic health effects, the nature of which may vary depending on constituent of the pollutants as well as the group of the population. Current exposure to PM from anthropogenic sources leads to the loss of 8.6 months of life expectancy in Europe - from around 3 months in Finland to more than 13 months in Belgium. The most recent estimates of impacts of PM on mortality, based on PM_{10} and $PM_{2.5}$ monitoring data in 40 European countries, indicate that close to 500 000 deaths per year are accelerated due to exposure to ambient PM in those countries. According to the WHO Health Reports, air pollution at current levels in European cities is responsible for a significant burden of deaths, hospital admissions and exacerbation of symptoms, especially for cardiovascular disease. Because of the tremendous number of people affected, the impact of air pollution on cardiovascular disease represents a serious public health problem. Results from research studies have demonstrated a strong relationship between levels of airborne particles, sulfur dioxide and other fossil fuel emissions and risk of early death from heart disease. People with pre-existing conditions such as high blood pressure, previous heart disease, diabetes, respiratory disease and high cholesterol have been shown to be especially vulnerable. The results of a long-term study on influence of common air pollutants on health of US residents showed that individuals living in the more polluted cities had a higher risk of hospitalization and early death from pulmonary and heart diseases as compared to those living in the less polluted cities. The study focused on the health effects of gaseous pollutants such as sulfur dioxide, which are produced mainly by coal-burning power plants and fine particle air pollution, particles with a diameter of less than 2.5μm, that come from power plant emissions and motor vehicle exhaust. The relationship between air pollution and mortality was much stronger for the fine particle component than for the gaseous pollutants. Exposure to PM is associated with increased hospital admissions and mortality in adults. The risk increases linearly with the concentration of pollution and there is no evidence to suggest a threshold for PM below which no adverse health effects would occur. When inhaled, PM_{10} particles (with a diameter of less than 10μm) penetrate deep into the respiratory system. Finer particles (with a diameter of less than 2.5μm) then go on to penetrate the lungs and pass into the bloodstream and are carried into other body organs. Concerned that these particles cause a wide range of health impacts, WHO has developed guidelines addressing their risks. Knowledge about the links between health and air quality has significantly advanced in the last years. It was determined that short-term exposure to $PM_{2.5}$ significantly increases the risk for cardiovascular and respiratory disease among people over 65 years of age. In the U.S. the National Morbidity, Mortality and Air Pollution Study indicated a 0.41% increase in total mortality in response to a 10-$\mu g/m^3$ increase in PM_{10} in ambient air [2]. The investigators linked $PM_{2.5}$ data to hospital admissions for heart and vascular diseases, heart failure, chronic obstructive pulmonary disease and respiratory infections in an epidemiologic study of over 11.5 million Medicare participants. The study results predict that for each 100 hospital admissions for heart failure, one extra admission will occur for each 10 $\mu m/m^3$ increase in $PM_{2.5}$ [2].

New studies also indicate substantial gains in public health resulting from improvements in air quality. An improvements in air quality over the last 20 years have increased average life expectancy in the U.S. by approximately five months. Researchers at Brigham Young University and the Harvard School of Public Health tracked particulate matter air pollution in 51 major metropolitan areas from 1978 through 2001 and compared those data to death records and census data. On average, life expectancy increased by 2.72 years with about 15% of that increase due to improved air quality. Cities that had the greatest air quality improvements saw the greatest gains in life expectancy. The results shows that a reduction of 10 $\mu g/m^3$ in the ambient air concentration of particulate matter was associated with an estimated increase in average life expectancy of 0.61 years [3]. The magnitude of changes in the health state of population in the polluted part of Poland has been estimated in assuming that the ambient air pollution decrease is 10% [4]. In the case of cardiovascular diseases, a 10% reduction of lead concentration in the air will cause a decrease in the incidence by 17.6 cases per 10 000 people. A 10% reduction of cadmium concentration in the air may result in a decrease in neoplasm mortality by more than 4 cases per 10 000 inhabitants. The reduction of the concentrations of both heavy metals in the air will lower SDR; in the case of cadmium by 24.4, and in the case of lead by 31.6 people. The time scale over which the health effects develop is under investigation. Particulate air pollution is consistently related to the most serious effects, including lung cancer and other cardiopulmonary mortality. Long-term average exposure to PM is associated with both the risks of chronic effects on children's health, such as impaired development of lung function, and the frequency of acute effects, such as the aggravation of asthma or incidence of respiratory symptoms. Children who live in neighborhoods with serious air pollution problems (emissions from the automobile traffic, heavy industry) have lower IQ and score worse memory tests than children from the cleaner environments. The respiratory and cardiovascular effects of air pollution are well documented; however the possible neurodegenerative effects of air pollution have been unexplored and require further intensive research. The research in the US showed that the more heavily exposed children were to black carbon, the lower their scores on several intelligence tests. For example, the average IQ of the most heavily exposed children was 3.4 points less than children with low exposure. When the findings were adjusted for the effects of parents' education, birth weight, and exposure to tobacco smoke, the associations remained. The effects were roughly equivalent to those seen in children whose mothers smoked ten cigarettes per day while pregnant. The researchers assumed that the harmful effects may be caused by the inflammatory and oxidative effects of the black carbon particles [5]. These findings suggest additional research is needed to investigate the effects of air pollution on the development of intelligence in children and on cognitive decline for people of all ages. Globally, the prevalence of asthma and allergies has increased over the last few decades. Asthma has become the commonest chronic disease in children and is one of the major causes of hospitalization for children aged under 15 years. The increasing prevalence of allergic diseases in children throughout Europe is no longer restricted to specific seasons or environments. It has only become fully apparent in the last decade that air pollution, especially of fine particulates, plays a major role in cardiovascular disease. A half of deaths globally arises from cardiovascular disease. Even relatively small increases in the risk of cardiovascular disease will translate into huge numbers of additional people suffering more severely from the disease. There is now substantial evidence concerning the adverse effects of air pollution on pregnancy outcomes and infant death. Evidence reporting associations between maternal exposure to ambient air pollutants and adverse fetal development, in particular growth restriction, pre-term birth, and infant survival due to postnatal respiratory mortality has been growing rapidly in recent years. The association between maternal exposure to ambient air pollution and the risk of congenital anomalies, which are a significant cause of stillbirth and infant mortality has been less well studied. New evidence is also accumulating on the burden of disease due to indoor air pollution. The air pollutants such as asbestos fibers and dioxins, resulting from waste disposal, has been associated with a multitude of health effects. Asbestos fibers are dangerous to health and practically indestructible. Human exposure to asbestos fiber found in inhaled air can lead to diseases such as chronic bronchitis, asbestosis, lung cancer and mesothelioma. The World Health Organization officially recognized asbestos as a carcinogen that pollute the environment globally. It starts a process of gradual elimination of asbestos from the human environment. The building asbestos stripping operations and waste disposal, because of higher emission of asbestos fibers into environment, puts human population at enormous risk. European Union experts estimate that asbestos-related cancers will cause approximately 500 000 deaths up to the year 2030 in Western Europe alone. Dioxin and related compounds always exist in nature as complex mixtures. Dioxins are widely distributed in the environment at low concentrations, primarily as a result of air transport and deposition. Emissions of polychlorinated dibenzodioxin and dibenzofuran (PCDD/F) result from inefficiencies of combustion processes, most typically waste combustion. For uncontrolled combustion, such as open burning of household waste, chlorine content of wastes may play the most significant role in levels of dioxin emissions. Dioxin and related compounds have been shown to be developmental, reproductive, immunological, endocrinological, and cancer hazards, among others in multiple animal species. There is no reason to expect, in general, that humans would not be similarly affected at some dose, and an increasing numbers of data supports this assumption. Tetrachlorodibenzo-*p*-dioxin (TCDD) is best characterized as "carcinogenic to humans." This means that, on the basis of the weight of all of the evidence (human, animal, mode of action), TCDD meets the criteria that allow the scientific community to accept a causal relationship between TCDD exposure and cancer hazard.

2. Human exposure to air pollution: The route of air pollutants to a human organism

Air pollution is a mixture of particulate matter (PM), gases, and vapor-phase molecules [6]. The direct rout of exposure to the air pollution is a respiratory tract. In case of dust pollutants the size of particulate matter is playing an important role in the environmental health risk. PM is categorized by aerodynamic diameter. Particles below 10 µm in diameter are classified as thoracic particles PM_{10}, particles below 2.5 µm in diameter as fine particles, and particles with a diameter < 0.1 µm as ultrafine particles (UFPs) [6]. Particles larger than 10 µm are likely to land in proximal airways, but fine particles reach the lungs and are deposited in the alveoli [7]. Therefore $PM_{2.5}$ may be more harmful than larger ones [8]. Ambient fine particulate pollution was associated with increased risk of cardiovascular diseases [9]. UFPs are deposited deeply into the lungs. The study of Terzano et al. (2010), indicates that the ultrafine particles in contrast to larger-sized particles pass into the bloodstream by different transfer routes and mechanisms and then are distributed into other body organs, including the brain with potential neurotoxic effects [10]. The particulate matter is hazardous to the human health due to absorption on their surface of many harmful contaminants such as: heavy metals (lead, cadmium, mercury and the other), organic compounds (polycyclic aromatic hydrocarbons, PCBs, dioxin and furans). Gaseous pollutants, depending on their solubility in the water, are absorbed in the proximal or the distal parts of the respiratory tract. This is important from the standpoint of the health effects. Sulfur dioxide and formaldehyde are highly water-soluble gases, therefore they do not reach the lungs, and they are irritating the airway epithelium of the upper respiratory tract. For example up to 98% sulfur dioxide may be absorbed in the nasopharynx during nasal breathing [11]. NO_2 is a poorly water-soluble gas, therefore, is deposited far more peripherally in a respiratory tract compared with SO_2, but does not reach the alveoli in any significant quantities [11]. Ozone, in contrast to nitrogen dioxide, does not dissolve in water and in gaseous form reaches the lungs where it begins its malicious activity. Gaseous pollutants can be also absorbed into the body through dermal rout of exposure. However for the general population the role of this route of exposure is insignificant. The indirect rout of exposure to the air pollutants is digestive tract; it follows from the circulation of pollutants in the environment. The food chain is an important pathway of human exposure to polycyclic aromatic hydrocarbons, dioxin, PCBs and heavy metals (cadmium, lead, mercury).

3. Relation between air pollution and life expectancy (LLE)

The World Health Organization has identified ambient air pollution as a high public health priority, indicating the relationship of air pollution with increased mortality and shortened life expectancy [12]. In 2009, life expectancy at birth in twenty seven countries in the European Union [EU-27] was among the highest in the world – almost 76 years for men and 82 years for women [13,14]. In Europe there is a wide variety of life expectancy. In developed countries, located mainly in the western part of Europe and the Nordic countries, people live a few years longer than in the countries of Central and Eastern Europe and these differences are even a dozen years. According to EU experts, current exposure to PM from anthropogenic sources reduces the average life expectancy of 8.6 months in Europe – from around 3 months in Finland and Ireland to more than 13 months in Belgium [15]. It has been estimated that exposure to fine particulate matter in outdoor air leads to 725 000 years of life lost annually in Europe [16,17]. Studies in the USA have shown that people from less polluted cities live longer than those living in more polluted cities. After adjustment for other factors, an association remained between ambient annual average concentrations of fine particles (represented by $PM_{2.5}$) and age-specific risks of mortality, implying shorter life expectancy in more polluted cities [18]. It is estimated that in the European Union an increase of 1 $\mu g/m^3$ of $PM_{2.5}$ for 1 year implies an average Lost of Life Expectancy (LLE) of 0.22 days per person; the number for the United States is similar, but for Russia it is about 40% higher, and for China it is about 25% lower [19]. A study published in 2009, conducted in 51 U.S. metropolitan areas between 1970 to 2000 showed that decrease of 10 $\mu g/m^3$ in the concentration of fine particulate matter was associated with an estimated increase of average life expectancy of 0.61 year. The estimated effect of reduced exposure to pollution on life expectancy was not highly sensitive to adjustment for changes in socioeconomic, demographic, or proxy variables for the prevalence of smoking or to the restriction of observations to relatively large counties [3]. The other studies have shown that reductions in life expectancy of 1.11 years in the Netherlands, 1.37 years in Finland, and 0.80 year in Canada resulting from increases in ambient $PM_{2.5}$ concentrations of 10 $\mu g/m^3$ [20,21]. Long-term exposure to PM is particularly damaging to human health and reduces life expectancy, that is why reducing long-term PM concentrations and exposure is a priority [21,22].

4. Relation between air pollution and mortality

Air pollution is a major environmental risk for health and is estimated to cause approximately 2 million premature deaths worldwide per year. PM air pollution imparts a tremendous burden to the global public health, ranking it as the 13th leading cause of morality [23]. The estimates of health effects of PM exposure in adults are dominated by the increase in the risk of mortality due to long-time exposure to fine PM ($PM_{2.5}$) The total number of premature deaths attributed to exposure amounts to around 348 000 annually in the 25 EU countries. More than half of the burden from air pollution on human health is borne by people in developing countries [15,24]. The short-term health effects of particulate and gaseous air pollutants have been well documented, mainly through time-series studies relating short-term elevations in ambient levels of such pollutants to increases in morbidity and mortality from cardio respiratory conditions. Results of 124 studies of the largest cities in North America and Europe showed an increase in the rate of death from any cause ranging from 0.2 to 0.6% for an increase in ambient PM_{10} concentrations of 10 $\mu g/m^3$ [25]. Long-term epidemiological studies conducted in the U.S. confirm that the adverse effects of fine particulate matter ($PM_{2.5}$) on morbidity and mortality, and indicate that this effect depends on the concentration and time of exposure; long-term exposure gives higher effects than short-term exposure [26]. Long-term exposure to $PM_{2.5}$ increases the risk of no accidental mortality by 6% per a 10 $\mu g/m^3$ increase, independent of age, gender, and geographic region. Exposure to PM was also associated with an increased risk of mortality from lung cancer (range: 15% to 21% per a 10 μg/m^3 increase) and total cardiovascular mortality (range: 12% to 14% per a 10 μg/m^3 increase) [1,27]. The Medicare Cohort Air Pollution Study in the USA has estimated the relative risk of death associated with long-term exposure to $PM_{2.5}$. According to the authors Zeger et al. [28], a 10 mg/m^3 increase in six year average of $PM_{2.5}$ is associated with a 6.8–13.2% increase in mortality. Other studies [27] showed that long-term exposure to $PM_{2.5}$ increases the risk of non-accidental mortality by 6% per a 10 mg/m^3 increase, independent of age, gender, and geographic region. Exposure to $PM_{2.5}$ was also associated with an increased risk of mortality from lung cancer (range: 15–21% per a 10 mg/m^3 increase) and total cardiovascular mortality (range: 12–14% per a 10 mg/m^3 increase) [27]. People with diabetes, heart failure, chronic obstructive pulmonary disease (COPD) and inflammatory diseases such as rheumatoid arthritis are at increased risk of death when they are exposed to particulate air pollution, or soot, for one or more years. Increase of 10 μg/m^3 of PM_{10} over 2 years increased the risk of death by 32% for patients with diabetes, by 28% for patients with COPD, by 27% for patients with congestive heart failure, and by 22% for people with inflammatory diseases such as rheumatoid arthritis or lupus [29]. Significant associations were found between black smoke (BS) and SO_2 concentrations and mortality. The effects were stronger for respiratory illness than other causes of mortality for the most recent exposure periods (shorter latency times) and most recent mortality period (lower pollutant concentrations) [30]. Air pollution has long-term effects on mortality and point to continuing public health risks. They therefore have importance for policies on public health protection through regulation and control of air pollution [30].

5. The impacts of fine particulate matter on cardiovascular health

Cardiovascular disease (CVD) constitute a global problem and is the leading cause of death in the world, especially in highly developed countries. Cardiovascular disease is also a major cause of disability and of reduced quality of life [31,32]. According to forecast, almost 20 million people will die from CVDs, mainly from heart disease and stroke by 2015 [33,34]. Results from many research studies have demonstrated a strong relationship between levels of airborne particles, sulfur dioxide and other air pollutants and risk of early death from heart disease. Air pollutants have been linked with endothelial dysfunction and vasoconstriction, increased blood pressure (BP), prothrombotic and coagulant changes, systemic inflammatory and oxidative stress responses, autonomic imbalance and arrhythmias, and the progression of atherosclerosis [35].

Using data for Chicago area hospitals for years 1988 to 1993 it was found, that an increase in PM_{10} level by 10 μg/m^3 was associated with 1.27%, 1.45%, and 2% increases in hospital admissions for heart disease, chronic obstructive pulmonary disease, and pneumonia, respectively [36]. The Air Pollution and Health: a European and North American Approach (APHENA) project also examined the association between airborne particles and hospital admission for cardiac causes in eight European cities and found that the percentage increases associated with a 10 μg/m^3 elevation in PM_{10} were 0.5% for cardiac admissions in people of all ages and 0.7% for cardiac admissions in people older than 65 years [37,38]. Short-term exposure to $PM_{2.5}$ significantly increases the risk for cardiovascular and respiratory disease among people over 65 years of age. The investigators linked $PM_{2.5}$ data to hospital admissions for heart and vascular diseases, heart failure, chronic obstructive pulmonary disease, and respiratory infections in an epidemiologic study of over 11.5 million

Medicare participants. The study results predict that for each 100 hospital admissions for heart failure, one extra admission will occur for each 10 μg/m^3 increase in $PM_{2.5}$ [2]. People with pre-existing cardiovascular disease, diabetic and elderly individuals are also considered to be more susceptible to air pollution–mediated cardiovascular effects [39].

Long-term exposure to elevated concentrations of ambient $PM_{2.5}$ at levels encountered in the present-day environment (i.e. any increase by 10 μg/m^3) reduces life expectancy within a population probably by several months to a few years [40]. As $PM_{2.5}$ is most strongly associated with cardiovascular deaths in the cohort studies, the reduced life expectancy is most likely predominantly due to excess cardiovascular mortality [40]. It was found that the greater the level of the fine particulate pollution, the greater the risk of cardiovascular disease and death in post-menopausal women, who are considered to be susceptible group within the general population. The increased risk comes from the fine particulate matter typically produced by automobile exhaust. The particles damage arteries in the heart and brain. Even slight elevations in fine particulate matter concentration increased the risk significantly. The risk of dying from heart attack or stroke increased 76% for each ten microgram increase in fine particulate pollution and proved to be about three times higher than previously estimated [41]. The study also indicates that although smoking is a much larger risk factor for cardiovascular disease, exposure to fine particulate combined with smoking imposes additional effects [42]. Additional research is required to establish whether there are independent health effects of the other particulate size fractions beyond those posed by fine particles. Although the focus of the present statement is on PM, it is recognized that other air pollutants may also pose cardiovascular risk alone or in conjunction with fine-particle exposure [40]. There are some evidences that gaseous pollutants may also be a reason for hospitalizations. Hospital admissions for cardiovascular causes, particularly ischemic heart disease, were found to rise in relation to the previous-day and same-day level of SO_2, even after adjustment for PM_{10} levels [40].

Although ozone has been linked to increased cardiopulmonary mortality, strokes, and MIs in some short-term studies, long-term exposure was not associated with cardiovascular mortality after accounting for PM in a recent analysis. The recent finding that small changes in low levels of ambient carbon monoxide concentrations are related to cardiovascular hospitalizations also requires further study [40]. Several secondary aerosols (eg, nitrate and sulfate) are often associated with cardiovascular mortality; however, whether these compounds are directly harmful or are surrogate markers of toxic sources of exposure requires more investigation [40]. The results showed that the daily number of hospitalizations for cardiovascular diseases was significantly associated with daily PM_{10} and NO_2 levels, with stronger associations in the elderly (≥65 years of age) [43]. During the last 15 years air pollution induced cardiovascular toxicity has become the focus of intensive studies among cardiologists and specialists in environmental medicine. They found that long-term particulate matter exposures were most strongly associated with death due to ischemic heart disease, dysrhythmias, heart failure, and cardiac arrest. For these causes of death, a 10 μg/m^3 elevation in particulate matter was associated with 8% to 18% increases in mortality risk. Risks for smokers were comparable or larger than for non-smokers. The researchers conclude that particulate matter exposure is a risk factor for specific cardiovascular disease mortality through mechanisms that likely include pulmonary and systemic inflammation, accelerated atherosclerosis, and changes in cardiac rhythms [42]. According to more recent studies, the ultrafine particles may be translocated into the circulation and directly transported to the vasculature and heart where they can induce cardiac arrhythmias and decrease cardiac contractility and coronary flow [39]. Improving our understanding of the biological mechanisms underlying the acute cardiovascular effects of air pollution is essential to define the best prevention strategies [37]. Cardiovascular disease is very common and, as exposure to air pollution, both in the long and short term, contributes to initiation and exacerbation of disease, it is likely that even modest reductions in exposure will result in significant health gain [43].

6. The adverse effects of air pollution on pregnancy outcomes and infant death

In the last two decades there is observed an increase of the number of scientific reports about a proven influence of air pollution on an occurrence of negative health effects, connected with births. They are: preterm births, stillbirths, intrauterine fetus growth retardation, births of newborns with low birth weight and a risk of newborns' death because of respiratory system disorders [44,45]. Birth weight, gestational age, and fetal growth are important indicators of perinatal health. Low birth weight (LBW), preterm birth, or intrauterine growth retardation (IUGR) are strongly association with infant mortality and morbidity [46]. Long term study shows that low birth weight (LBW) is a risk factor for developing in adulthood coronary health diseases, hypertension and type 2 diabetes [45]. It was observed up to 20% increase in risk of LBW and preterm birth in infants born to women leaving in area with high level of air pollution, specially those exposed to higher levels of motor vehicle exhaust pollution coming from heavy-traffic roadways. Stronger effects were observed for women whose third trimester accounted for months with cold weather, when concentration of air pollutants was the highest because of an activity of local heating sources [47]. In heavily polluted environments the prematurity rate (birth before the 37th week of pregnancy) increases considerably. The study done in the 90's, in the most polluted parts of Poland (Chorzow) shows as high as 14 to 20% of the prematurity rate comparing to 8% rate for Poland this time. The average newborns' birth weight was 515g lower in Chorzow than in the Country [48]. There are more evidence each year which indicate that maternal exposures to air pollutants, including particulate matter (aerodynamic diameter 10 μm and 2.5 μm), sulfur dioxide, nitrogen dioxide and benzopyrene, are associated with adverse pregnancy outcomes [49]. Some of polycyclic aromatic hydrocarbons (PAHs), especially benzopyrene, prove carcinogenic and mutagenic effects and when penetrating through placenta, have a negative influence into fetus. The research showed the newborns, whose mothers were exposed to PAHs during pregnancy, more often born with lower birth weight and smaller head circumference [50, 51, 52]. According to Dejmek et al. (2000), the risk of delivering a growth-retarded infant increases with the level of PAHs in early gestation (first month) [53]. A bond of benzopyrene and DNA in the placenta, has an influence on intrauterine growth retardation - IUGR [54,55]. The exposure to particulate matter (PM) causes increase of risk of occurrence an intrauterine fetus growth retardation [56,57,58]. Children, whose mothers were exposed to high concentrations of PM during pregnancy, more often were born with low birth weight and 10 $\mu g/m^3$ change in PM_{10} accounted for 13.7 g less of weight [59,60,61]. Effect of mothers' exposure to high concentration (above the median 36.3 $\mu g/m^3$) of fine particles ($PM_{2,5}$) was reflected in significantly lower mean weight (128.3g) and length (0.9cm) and lower mean head circumference (0.3) of newborns

[62]. The researchers in the USA found that mothers who lived in areas with the highest levels of $PM_{2.5}$ during their pregnancy delivered slightly smaller babies than their counterparts who lived in areas with lower levels of $PM_{2.5}$ exposure. They also observed association between number of traffic-related pollutants and small for gestational birth weight as well as preterm births (before 37 weeks) [63,64]. Maternal exposure to sulfur dioxide during the first month of pregnancy increased risk of intrauterine growth retardation as well as LBW when preterm birth was associated with exposure to SO_2 during the last month of pregnancy. These results suggest an association between VLBW (below 1500g) and maternal exposures to high levels of sulfur dioxide [46,60]. Increased risk of intrauterine growth retardation was observed also in case of maternal exposure to nitrogen dioxide during the first month of pregnancy [65]. An association between exposure to levels of nitrogen dioxide above 40 $\mu g/m^3$ during the first trimester of pregnancy and a reduction in birth weight was found [66]. Study on CO influence on pregnancy [67] were the basis for estimation that one unit change in mean CO concentration during the last trimester of pregnancy increases the risk of low birth weight by 8%. Furthermore, a one unit change in mean CO concentration during the first 2 weeks after birth increases the risk of infant mortality by 2.5% relative to baseline levels [67].

7. Human's exposure to air pollution resulting from waste disposal

Asbestos is a mineral fiber that due to the unique physical and chemical properties was produced in the past and used in over 3000 products. In the 20th century, asbestos has dominated the building industry, with a maximum global production of 5 million tonnes per year. As a result, the world's asbestos (in products) is currently estimated at around 550 million tonnes. Asbestos fibers are indestructible and dangerous to health. Human exposure to mineral fiber found in inhaled air can lead to diseases such as chronic bronchitis, asbestosis, lung cancer and mesothelioma [68]. International Agency for Research on Cancer (IARC) recognized asbestos (actinolite, amosite, anthophyllite, chrysotile, crocidolite, tremolite) as Group I carcinogen [69]. This category is used when there is sufficient evidence of carcinogenicity in humans. In 1980, the US National Institute of Occupational Safety and Health (NIOSH) and the Occupational Safety and Health Administration (OSHA) working group concluded that there are no levels of exposure to asbestos below which clinical effects did not occur [70]. In the 80's of the past century, the World Health Organization officially recognized asbestos as a carcinogen that pollutes the environment globally. Environmental exposure either in the houses of asbestos workers or in the neighborhood of asbestos mines or factories has been noted in some of the cases [71]. It has been estimated that a third of the mesotheliomas occurring in the USA may be due to nonoccupational exposure [72]. The relationship between asbestos exposure and smoking indicates a synergistic effect of smoking with regard to lung cancer [69]. Further evaluations indicate that this synergistic effect is a multiplicative model [73]. Exposure to asbestos occurs through inhalation of fibers from contaminated air in the working environment, as well as from ambient air in the vicinity of point sources, or indoor air in housing and building containing asbestos materials [74]. Although, in many countries the production and utilization of asbestos-containing materials has been banned, the numerous active environmental sources still exist [70]. Exposure can also occur during installation and use of asbestos-containing products and maintenance of vehicles. Asbestos products are still in place in many buildings and continue to give rise to exposure during use, maintenance, renovation, repairs, removal and demolition [74]. The results of environmental concentrations of respirable asbestos fibers show a wide range of values. The observed discrepancy in the concentrations are dependent on the different environments, specific sampling locations and presence of more than one emission sources [70]. Unfortunately, in recent years, emissions of asbestos fibers into the urban environment has significantly intensified. This follows from the fact that the durability of asbestos-cement building products is estimated for 30 years and the possibility of exploitation of these used in the 70's and 80's of the past century is coming to an end. As is apparent from analysis the deterioration of asbestos-containing construction materials, such as asbestos-cement sheets (AC) used in residential and industrial buildings causes additional contamination of the urban environment. The study performed in a highly urbanized and densely populated town in south part of Poland, revealed that asbestos fibers identified in the air samples near buildings covered with AC panels derived from 2 groups of asbestos minerals, i.e. crocidolite and chrysotile. The observed concentrations of respirable asbestos fibres varied from 0.0010–0.0090 f/cm^3. Significantly higher values were noted in the immediate vicinity of the buildings with asbestos-containing materials, compared to sampling sites located at a distance of 100–500m from such buildings or the sites treated as an asbestos free [70]. Kovalevskiy and Tossavainen, taking measurements near a building with asbestos-containing materials in Moscow, showed that when outdoor concentrations reach the level of 0.009 f/cm^3, at the same time, indoor concentrations approach 0.049 f/cm^3 in residential premises, or even 0.57 f/cm^3 if the building was undergoing renovation [75]. The measurements of respirable fibers in the air on the playgrounds in housing estates, where path was made with admixture of asbestos-containing material showed contamination range from 0.165-0.54 f/m^3 and in apartments adjacent to the playground around 0.01 f/m^3 [76]. The significant increase in concentrations of asbestos also recorded in the immediate vicinity of buildings, at which work is ongoing disassembly of asbestos-cement facades or roofing. The work conducted by a specialized company working according to safety regulations causing dust in the workplace ranged from 1000 to 4000 f/m^3, while the same work done improperly can lead to maximum levels of respirable asbestos fibers in the amount of 80 000 f/m^3 [77]. Starting from the last decade of the 20th century, the world began a process of gradual elimination of asbestos, what in fact results the higher emission of asbestos fibers into the municipal environment. The individual disassembly of utilized asbestos panels, not obeying safe methods of removal, storage, transport and treatment of asbestos waste intensify the environmental exposure of the general population. Since 1980, the number of deaths caused by exposure to asbestos fibers increases gradually, even in countries that have banded the use of asbestos in the early 1990s [78]. Currently about 125 million people in the world are exposed to asbestos at the workplace and at least 90 000 people die each year from asbestos-related lung cancer, mesothelioma and asbestosis resulting from occupational exposures. In addition, it is believed that 7000 of deaths can be attributed to asbestos-related diseases as well as to non-occupational exposures to asbestos [74]. The Report of UN EWG shows that in the U.S. each day 30 people die, which represents 10 000 deaths per year only as a result of diseases caused by exposure to asbestos [79]. Because of long latency periods attached to the diseases, stopping the use of asbestos now will result in a decrease in the number of asbestos-related deaths after a number of decades [74]. It is estimated that in the next 40 years asbestos fibers will cause the death of about 100 000 Americans [79]. European Union experts estimate that the total number of deaths caused by asbestos-related diseases in the UK, Belgium, Germany, Switzerland, Norway, Poland and Estonia is around 15 000 annually; only in Western Europe asbestos-related diseases will caused 500 000 deaths subsequent till the end of 2030 [80]. Considering the numerous health hazards resulting from the inhalation of asbestos dust, there is no safe environmental level for his harmful factor, therefore, the exposure should be kept as low as possible [70].

8. References

[1] Pope C.A. 3rd. *Mortality effects of longer term exposure to fine particulate air pollution: Review of recent epidemiological evidence.* Inhal Toxicol. 2007, 19(Suppl. 1):33–38.

[2] Dominici F., Peng R.D., Bell M.L., Pham L., McDermott A., Zeger S.L., Samet J.M. *Fine particulate air pollution and hospital admission for cardiovascular and respiratory diseases.* JAMA. 2006 Mar 8;295(10):1127-34.

[3] Pope CA 3rd, Ezzati M., Dockery D.W. *Fine-particulate air pollution and life expectancy in the United States.* N Engl J Med. 2009 Jan 22;360(4):376-86.

[4] Marchwinska-Wyrwal E., Dziubanek G., Skrzypek M., Hajok I.. *Study of the health effects of long-term exposure to cadmium and lead in a region of Poland.* Int J Environ Health Res. 2010 Apr;20(2):81-6.

[5] Suglia S.F., Gryparis A., Wright R.O., Schwartz J., Wright R.J.. *Association of black carbon with cognition among children in a prospective birth cohort study.* Am J Epidemiol. 2008 Feb 1;167(3):280-6.

[6] Brook R.D., Franklin B., Cascio W., Hong Y., Howard G., Lipsett M., et al. *Air pollution and cardiovascular disease: a statement for healthcare professionals from the expert panel on population and prevention science of the American Heart Association.* Circulation 2004; 109:2655–71.

[7] Kim, C.S., Fishe.r, D.M., Lutz., D.J., Gerrity, T.R., 1994. *Particle deposition in bifurcating airway models with varying airway geometry.* J. Aerosol. 25, 567–581

[8] Farina F., Sancini G., Mantecca P, Gallinotti D., Camatini M., Palestini P. *The acute toxic effects of particulate matter in mouse lung are related to size and season of collection.* Toxicol Lett. 2011 Mar 1. [Epub ahead of print]

[9] Pope C.A. 3rd,Muhlestein J.B., May H.T., Renlund DG, Anderson J.L., Horne B.D. *Ischemic heart disease events triggered by short-term exposure to fine particulate air pollution.* Circulation. 2006 Dec 5;114(23):2443-8. Epub 2006 Nov 13.

[10] Terzano C., Di Stefano F., Conti V., Graziani E., Petroianni A. *Air pollution ultrafine particles: toxicity beyond the lung.* Eur Rev Med Pharmacol Sci. 2010 Oct;14(10):809-21.

[11] Sandström T. *Respiratory effects of air pollutants: experimental studies in humans.* Eur Respir J, 1995, 8, 976–995

[12] Samet J, Krewski D. *Health effects associated with exposure to ambient air pollution.* J Toxicol Environ Health A. 2007; 1;70(3-4): 227-42.

[13] http://epp.eurostat.ec.europa.eu

[14] *Systematic Review of Health Aspects of Air Pollution in Europe.* Copenhagen, Denmark, WHO Regional Office for Europe 2004

[15] *Exposure of children to air pollution (particulate matter) in outdoor air.* Copenhagen, WHO Regional Office for Europe 2009 (ENHIS fact sheet 3.3)

[16] Cohen A, Anderson H. *The Global Burden of Disease Due to Outdoor Air Pollution.* Journal of Toxicology and Environmental Health. 2005; 68:1301–1307

[17] EEA 2010. The European Environment - State and outlook 2010: Synthesis. http://www.eea.europa.eu/soer/synthesis/synthesis
[18] Dockery, D, Pope C. A, Xu, X. *An association between air pollution and mortality in six US cities*. New England Journal for Medicine 1993; 329: 1753-1759
[19] Leksell I, Rabl A. *Air Pollution and Mortality: Quantification and Valuation of Years of Life lost*. Risk Anal. 2001; 21 (5):843-57
[20] Coyle D, Stieb D, Burnett RT, et al. *Impact of particulate air pollution on quality-adjusted life expectancy in Canada*. J Toxicol Environ Health A 2003; 66:1847-1863
[21] Krewski D. *Evaluating the Effects of Ambient Air Pollution on Life Expectancy*. N Engl J Med 2009; 360:413-415
[22] Craig L, Brook JR, Chiotti Q, et al. *Air Pollution and Public Health: A Guidance Document for Risk Managers*. J Toxicol Environ Health A 2008;71:588-698
[23] Brook RD. *Cardiovascular effects of air pollution*. Clin Sci (Lond). 2008; 115(6): 175-87
[24] http://www.who.int/mediacentre/factsheets/fs313/en/index.html
[25] Samoli E, Peng R, Ramsay T, et al. *Acute effects of ambient particulate matter on mortality in Europe and North America: results from the APHENA study*. Environ Health Perspect 2008;116:1480-6
[26] Puett R, Hart J, Yanosky J et al. *Chronic Fine and Coarse Particulate Exposure, Mortality and Coronary Heart Disease in the Nurses' Health Study*. Environ Health Perspect. 2009;117(11): 1702–1706
[27] Chen H, Goldberg MS, Villeneuve PJ. *A systematic review of the relation between long-term exposure to ambient air pollution and chronic diseases*. Rev Environ Health. 2008;23(4):243-97, 2008
[28] Zeger SL, Dominici F, McDermott A, Samet JM. *Mortality in the Medicare population and chronic exposure to fine particulate air pollution in urban centers (2000-2005)*. Environ Health Perspect. 2008;116(12):1614-9
[29] ATS 2006 International Conference: *Air Pollution Increases Mortality Risk.*
[30] Elliott P, Shaddick G, Wakefield JC, et al. *Long-term associations of outdoor air pollution with mortality in Great Britain*. Thorax 2007; 62(12):1088-94
[31] World Health Organization Regional Office for Europe. European Heart Health Charter. 2007 (http://www.heartcharter.org/)
[32] The European Heart Network. Annual Report 2009 fighting heart disease and stroke (http://www.ehnheart.org/publications/annual-reports.html)
[33] http://www.who.int/mediacentre/factsheets/fs317/en/index.html
[34] http://www.allcountries.org/health/cardiovascular_diseases.html
[35] Brook RD, Rajagopalan S, Pope A, et al. *Particulate Matter air Pollution and Cardiovascular Disease: An Update to the Scientific Statement From the American Heart Association*. Circulation 2010; 121: 2331-2378
[36] Schwartz J. *Is there harvesting association of airborne particles with daily deaths and hospital admissions?* Epidemiology 2001; 12: 55– 61
[37] Franchini M, Mannucci PM. *Short-term effects of air pollution on cardiovascular diseases: outcomes and mechanisms*. J ThrombHaemost 2007;5:2169-2174
[38] Le Tertre A, Medina S, Samoli E, Forsberg B, Michelozzi P, Boumghar A, Vonk JM, Bellini A, Atkinson R, Ayres JG, SunyerJ,Schwartz J, Katsouyanni K. *Short-term effects of particulate air pollution on cardiovascular diseases in eight European cities*. J Epidemiol Community Health 2002; 56: 773–9
[39] Simkhovich B, Kleinman M, Kloner R. *Air Pollution and Cardiovascular Injury*. Journal of the American College of Cardiology 2008; 52 (9): 719-726

[40] Miller KA, Siscovick DS, Sheppard L et al. *Long-term exposure to air pollution and incidence of cardiovascular events in women.* N EnglJ Med 2007 Feb 1;356(5):447-58
[41] Pope CA 3rd, Burnett RT, Thurston GD et al. *Cardiovascular mortality and long-term exposure to particulate air pollution: epidemiological evidence of general pathophysiological pathways of disease.* Circulation 2004 Jan 6;109(1): 71-7. Epub 2003 Dec 15
[42] Larrieu S, Jusot J-F, Blanchard M et al. *Short term effects of air pollution on hospitalizations for cardiovascular diseases in eight French cities: The PSAS program.* Science of the Total Environment 2007, 387: 105-112
[43] EU Department of Health. *Cardiovascular Disease and Air Pollution.* A Report by the Committee on the Medical Effects of Air Pollutants 2006
[44] Rankin J, Chadwick T, Natarajan M, Howel D, Pearce MS, Pless-Mulloli T. *Maternal exposure to ambient air pollutants and risk of congenital anomalies.* Environ Res 2009; 109: 181-187.
[45] Ricciardi C, Guastadisegni C. *Environmental inequities and low birth weight.* Ann Ist Super Sanita 2003; 39(2): 229-234.
[46] Liu S, Krewski D, Shi Y, Chen Y, Burnett RT. *Association between gaseous ambient air pollutants and adverse pregnancy outcomes in Vancouver, Canada.* Environ Health Perspect 2003; 111(14): 1773-1778.
[47] Wilhelm M, Ritz B. *Residential proximity to traffic and adverse birth outcomes in Los Angeles county, California, 1994-1996.* Environ Health Perspect 2003; 111(2): 207-216.
[48] Kasznia-Kocot J, Buszman Z. *The influence of pollution on children's health in Chorzów.* Pol J Environ Stud 1995; 4(2): 29-32.
[49] Salam MT, Millstein J, Li Y, Lurmann FW, Margolis HG, Gilliland FD. *Birth outcomes and prenatal exposure to ozone, carbon monoxide and particulate matter: results from the children's health study.* Environ Health Perspect 2005; 113(11): 1638-1644.
[50] Barakat AO. *PAHs and petroleum biomarkers in the atmospheric environment of Alexandria City, Egypt.* Water Air Soil Pollut 2002; 139: 289– 310.
[51] Perera F.P., Whyatt R.M., Jędrychowski W., et al. *A study of the effects of environmental polycyclic aromatic hydrocarbons on birth outcomes in Poland.* Am J Epidemiol 1998; 147: 309-314.
[52] Perera F.P., Rauh V., Tsai W.Y., et al. *Effects of transplacental exposure to environmental pollutants on birth outcomes in a multi-ethnic population.* Environ Health Perspect 2003; 111: 201–205.
[53] Dejmek J., Solansky I., Benes I., et al. *The impact of polycyclic aromatic hydrocarbons and fine particles on pregnancy outcome.* Environ Health Perspect 2000; 108: 1159–1164.
[54] Sram R.J., Binkova B., Rossner P., et al. *Adverse reproductive outcomes from exposure to environmental mutagens.* Mutat Res 1999; 428: 203-215.
[55] Topinka J., Binkova B., Stavkova Z., et al. *DNA adducts in human placenta as related to air pollution and to GSTM1 genotype.* Mutat Res 1997; 390: 59-68.
[56] Boy E, Bruce N, Delgado H. *Birth weight and exposure to kitchen wood smoke during pregnancy in rural Guatemala.* Environ Health Perspect 2002; 110(1): 109-114.
[57] Dejmek J, Selevan SG, Benes I, Solansky I, Sram RJ. *Fetal growth and maternal exposure to particulate matter during pregnancy.* Environ Health Perspect 1999; 107(6): 475-480.
[58] Chen L, Yang W, Jennison BL, Goodrich A, Omaye ST. *Air pollution and birth weight in Northern Nevada, 1991-1999.* Inhal Toxicol 2002; 14: 141-157.
[59] Wang X, Ding H, Ryan L, Xu X. *Association between air pollution and low birth weight: a community-based study.* Environ Health Perspect 1997; 105(5): 514-520.

[60] Rogers JF, Thompson SJ, Addy CL, McKeown RE, Cowen DJ, Decoufle P. *Association of very low birth weight with exposure to environmental sulphur dioxide and total suspended particulates*. Am J Epidemiol 1999; 151(6): 602-613.
[61] Gouveia N, Bremner SA, Novaes HDM. *Association between ambient air pollution and birth weight in Sao Paulo, Brazil*. J Epidemiol Community Health 2004; 58: 11-17.
[62] Jędrychowski W, Bendkowska I, Flak E, et al. *Estimated risk for altered fetal growth resulting from exposure to fine particles during pregnancy: an epidemiologic prospective cohort study in Poland*. Environ Health Perspect 2004; 112(14): 1398-1402.
[63] Parker JD, Woodruff TJ, Basu R, Schoendorf KC. *Air pollution and birth weight among trem infants in California*. Pediatrics 2005; 115(1): 121-128.
[64] Brauer M, Lencar C, Tamburic L, Koehoorn M, Demers P, Karr C. *A cohort study of traffic-related air pollution impacts on birth outcomes*. Environ Health Perspect 2008; 116(5): 680-686.
[65] Ballester F, Estarlich M, Iniguez C, et al. *Air pollution exposure during pregnancy and reduced birth size: a prospective birth cohort study in Valencia, Spain*. Environ Health 2010; 9(6).
[66] Maroziene L, Grazuleviciene R. *Maternal exposure to low-level air pollution and pregnancy outcomes: a population-based study*. Environ Health 2002; 1(6).
[67] Currie J, Neidell M, Schmieder JF. *Air pollution and infant health: Lessons from New Jersey*. J Health Econ 2009; 28: 688-703.
[68] Szeszenia-Dąbrowska N., Sobala W. *Environmental pollution by asbestos. Health effects*. The Nofer Institute of Occupational Medicine. ISBN 978-83-923517-5-7, Lodz, 2010.
[69] IARC Monographs, Vol. 14, 1977.
[70] Krakowiak E, Górny R, Cembrzyńska J, Sąkol G, Boissier-Draghi M, Anczyk E. *Environmental Exposure to Airborne Asbestos Fibres in a Highly Urbanized City*. Ann Agric Environ Med, 16, 121-128, 2009.
[71] Bignon J., Sebastien P., di Menza L., Nebut M., Payan H. *French registry of mesotheliomas 1965-1978*. Rev Fr Mal Respir, 7, 223-242, 1979.
[72] Enterline P.E. *Cancer produced by nonoccupational asbestos exposure in the United States*. J Air Pollution Control Assoc., 33, 318-322, 1983.
[73] Saracci R. *Asbestos and lung cancer: an analysis of the epidemiological evidence on the asbestos-smoking interaction*. Int J Cancer, 20, 323-331, 1977.
[74] WHO, Special Report: *Elimination of asbestos – related diseases*, 2006.
[75] Currie G, Watt S, Maskell N. *Reconstruction of Moscow: allergy, dust, asbestos and other medical aspects*. BMJ, 339, 2009.
[76] Więcek E. *Asbestos-exposure and health effects*. Work Safety, 2, 2004.
[77] Jawecki B. *Programming asbestos removal at the local level – a proposal guidlines. Infrastructure and Ecology of Rural Areas*. Polish Academy of Sciences, Krakow, 9, 73-83, 2008.
[78] Łuniewski A, Łuniewski S. *Asbestos-historical burden of the 20th century. Publishing house of Environmental and Resources Economists*. ISBN 978-83-61643-28-9, Białystok, 2008.
[79] Kazan-Allen L. *Asbestos and mesothelioma: Worldwide trends*. Lung Cancer, 49, 3-8, 2005.
[80] Vogel L. *Special Report - Asbestos in the world*. HESA, 2005.

2

Impact of Air Pollution on Vitamin D Status and Related Health Consequences

Alina Kurylowicz
Department of Human Epigenetics, Mossakowski Medical Research Centre,
Polish Academy of Sciences
Poland

1. Introduction

During the last decade a growing interest in vitamin D-related problems can be observed and a few thousands of scientific publications are dedicated to this task every year.

The alarming statistical data regarding propagation of vitamin D deficiency in all continents and our developing knowledge about novel, non-calcemic actions of vitamin D seem to be two main causes of this phenomenon. Indeed, recent epidemiological studies suggest that worldwide prevalence of inadequate vitamin D status is higher than expected, however it may result from different factors, including geographical location and cultural background.

Since vitamin D skin synthesis, which takes place under the influence of the sun's ultraviolet B (UVB) radiation, covers as much as 90% of total body needs, inadequate radiation or insufficient cutaneous absorption of UVB are regarded as major causes of vitamin D deficiency.

Air pollution is a chief factor determining the extent of solar UVB that reaches earth surface and several epidemiological data, which represents different populations living in different geographic latitudes, indicate that atmospheric pollution (especially high tropospheric ozone content) may play a significant independent role in the development of vitamin D deficiency. As a result prevalence of D hypovitaminosis among urban residents can be more than twice higher than that of rural inhabitants.

The problem of vitamin D insufficiency has been underestimated for many years and it has been predominantly associated only with bone diseases. Discovery of vitamin D receptor and its identification in a wide number of tissues other than bones led to the designation of novel, so-called "non-calcemic" (e.g. antiproliferative and immunomodulatory), actions of cholecalciferol. Subsequently, there is a growing number of diseases and conditions, development and/or progression of which can be associated with inadequate vitamin D status, including different types of cancers, autoimmune disorders as well as cardiovascular events. Interestingly, prevalence of many of these diseases also positively correlates with the intensity of air pollution. Therefore one can conclude that with the increasing atmospheric pollution grows a number of people in the world who are vitamin D deficient or insufficient and indirectly exposed to several poor health outcomes.

This chapter gives an overview of the literature on this topic and assesses the available data about the association between air pollution, vitamin D status and prevalence of so-called civilization-related diseases. A general information regarding vitamin D metabolism is given first. Next, current definitions of vitamin D status are presented, followed by the alarming statistical data about vitamin D deficiency pandemic and its correlation with the intensity of air pollution and prevalence of common human diseases. To understand possible mechanism linking vitamin D deficiency with these poor health outcomes, first a brief presentation of traditional (calcemic) and novel ("non-calcemic") actions of vitamin D is given. Subsequently, epidemiological and experimental data regarding the association between vitamin D status and prevalence of bone diseases, different types of cancers, autoimmune disorders and cardiovascular diseases is presented. Finally, therapeutic perspectives and general guidelines about vitamin D supplementation and requirements are included. The chapter ends with a short summary which facts about vitamin D can be generally accepted and which still require more solid scientific background.

2. Vitamin D metabolism

In mammals, vitamin D_3 (cholecalciferol) is either produced in the skin by the non-enzymatic conversion (photochemical cleavage) of provitamin D (7-dehydrocholesterol) to previtamin D_3 under the influence of ultraviolet radiation (290-315nm), or delivered from food sources. In the skin previtamin D_3 can also undergo isomerization that results in creation of biologically inactive compounds (lumisterol and tachysterol) and this mechanism prevents from the vitamin D_3 overproduction and intoxication after prolonged exposure to sunlight. In the liver vitamin D_3 undergoes enzymatic hydroxylation to 25-hydroxyvitamin D_3 ($25(OH)D_3$) in the reaction catalyzed by 25-hydroxylase (CYP2R1).

In the tissues that have 1α-hydroxylaze activity (predominantly in kidneys, but also in activated macrophages, colon, prostate, breast, brain as well as in other tissues), $25(OH)D_3$ is converted to the active vitamin D metabolite - $1{,}25(OH)_2\,D_3$ (calcitriol), whereas hydroxylation in position 24 (by 24-hydroxylase - CYP24) initiates degradation of vitamin D metabolites (Horst & Reinhardt, 1997) [Figure 1]. Efficiency of renal vitamin D_3 hydroxylation is regulated by the level of calcium and phosphate ingestion, parathyroid hormone and, in a negative feedback, by circulating levels of $1{,}25(OH)D_3$. The extra-renal hydroxylation is determined by local factors e.g cytokines, growth factors as well as by $25(OH)D_3$ concentration, making it particularly sensitive to vitamin D deficiency (Marques et al., 2010).

The serum $25(OH)D_3$ concentration is the parameter of choice for the assessment of vitamin D status for several reasons. First of all, it reflects total vitamin D derived from dietary intake and sunlight exposure as well as the conversion of vitamin D from adipose stores in the liver. Furthermore, it is relatively stable and it has a long (2-3 weeks) half-life in circulation. Finally, it was shown in several epidemiological studies that $25(OH)D_3$ levels correlated best with several clinical conditions.

3. Definition and prevalence of vitamin D deficiency, insufficiency and adequacy

Definition of vitamin D status has evolved during the last decade. Whereas $25(OH)D_3$ level below 10 ng/μl (25nmol/l) is still, indisputably defined as vitamin D deficiency, the ranges of vitamin D sufficiency has been recently modified. For years, $25(OH)D_3$ levels between 10 ng/μl and 20 ng/μl have been identified as insufficiency (WHO report 2003), however with the recent changes in laboratory reference rages, nowadays vitamin D adequacy is defined as 30 to 76 ng/μl (75nmol/l) and $25(OH)D_3$ concentrations between 10 and 30 ng/μl are described as insufficiency. Vitamin D intoxication which is extremely rare, occurs when $25(OH)D_3$ concentrations exceed 150 ng/μl.

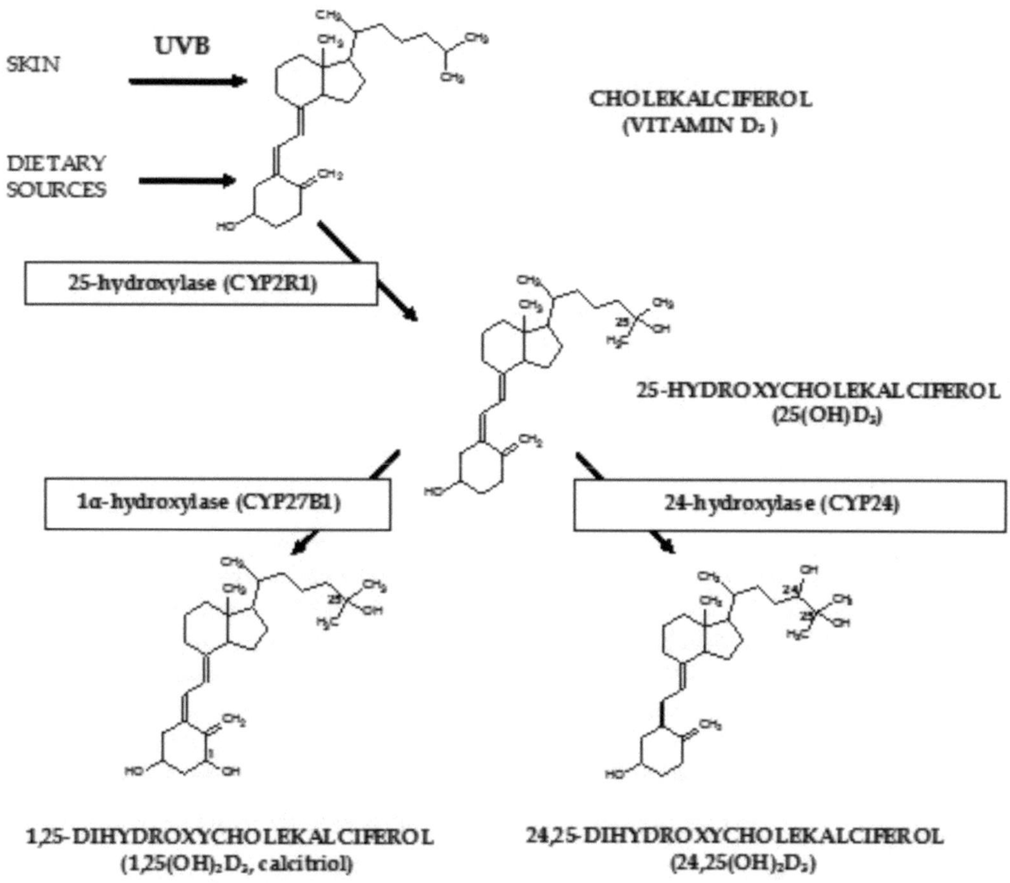

Fig. 1. The simplified scheme of the vitamin D synthesis
Cholekalciferol (synthesized in the skin upon the UV radiation or delivered with food undergoes several hydroxylations, firstly in the position 25 in the liver. The next hydroxylation catalyzed by the 1α-hydroxylase (CYP27B1) leads to the synthesis of the active metabolite – 1,25-dihydroxycholekalciferol. The alternative metabolic pathway of 25-hydroxycholekalciferol leads via hydroxylation in position 24 (the reaction is catalyzed by the vitamin D 24-hydroxylase – CYP24).

There are at least two reasons for setting the low end of the normal range of vitamin D levels at 30 ng/µl. First of all, it is suggested that serum concentration of vitamin D over 30 ng/µl guarantees optimal calcium absorption and below this concentration levels of parathyroid hormone (PTH) rise (Rosen, 2011). Secondly, this concentration provides an adequate amount of substrate for the nonrenal conversion of $25(OH)D_3$ to $1,25(OH)_2D_3$. It is assumed that children have the same requirements as adults, however no comparable studies have been carried out on intestinal calcium absorption and parthyroid hormone levels in children.

During last years vitamin D deficiency and insufficiency has been recognized as a 21st century pandemic and after changing the ranges of vitamin D concentrations defined as normal, it is estimated that this problem concerns as much as 30-50% (according to some sources up to 80%) of the general population (Hollick & Chen, 2008, Ovesen et al., 2003). In a Euronut-Seneca study that compared vitamin D status in elderly citizens in Europe living in latitudes from 35^0 to more than 60^0 the mean $25(OH)D_3$ concentrations ranged from 20 to 60 nmol/l and surprisingly, were higher in the northern countries than in the Mediterranean area (Wielen et al., 1995). This data was confirmed by the results of MORE study conducted in women with osteoporosis (Lips, 2010). In Middle East and in India, vitamin D status correlates with clothing style and is believed to be the lowest among all continents (Arya et al., 2004, Sachan et al., 2005). In South East Asia vitamin D status is generally better (Fraser, 2004). In studies performed in North America several ethnic and life-style related differences in vitamin D status were observed. However, in general, vitamin D deficiency was higher in non-hispanic whites compared to non-hispanic blacks and was in between in Mexican-American (Looker et al., 2008). In Africa, it was reported that the vitamin D status improves from north to south (Prentice et al., 2009) whereas in Australia and Oceania, the trend is opposite (Rockell et al., 2006).

Although studies performed in different continents cannot be exactly compared due to e.g. (i) different methods used to asses $25(OH)D_3$ concentrations (precision and accuracy may vary depending on the kind of the assay) (ii) seasonal fluctuations in the vitamin D dietary intake (iii) seasonal variations in efficacy of cholekalciferol skin synthesis (it is estimated that in northern latitudes serum $25(OH)D_3$ levels decline from late summer to midwinter by 20%), some general patterns suggested by these studies can be recognized. In adults, groups at the highest risk of vitamin D deficiency are elderly (especially those in nursing homes) and obese subjects, and those who due to religious habits are wearing clothes covering most of the body (Lips, 2010).

Based on epidemiological studies it is estimated that an average dietary intake of cholecalciferol is as small as only 200 IU per day, thus the main source of vitamin D remains its skin-derived synthesis that covers as much as 90% of body's total requirements. However it should be emphasized that its efficiency is highly variable and anything that hampers transmission of solar UVB radiation to the earth surface and anything that diminishes the penetration of UVB radiation into the skin will affect efficacy of vitamin D skin synthesis.

Since melanin is extremely efficient in absorbing UVB radiation, individuals with increased skin pigmentation have reduced abilities of vitamin D skin synthesis and are at higher risk of vitamin D deficiency. Appropriate application of sunscreens results in the similar effect. The angle at which the sun reaches the earth has a significant effect on the number of UVB photons that reach the earth surface therefore geographic latitude and season (autumn and winter) as well as agegeographic latitude and season (autumn and winter) as well as age (elderly people have 75% lower concentrations of vitamin D precursor 7-dehydrocholesterol), sex (vitamin D status is usually lower in women) and cultural conditionings (especially practice of purdah that totally prevents exposition of the body to sunlight) are another factors potentially reducing efficacy of vitamin D skin synthesis (Webb, 2006).

Apart from UVB-related vitamin D deficiency, there are several medical or physical conditions which may impair vitamin D status, that include, among others: fat malabsorption, use of anticonvulsant drugs that induce catabolism of vitamin D active metabolites, chronic kidney disease or obesity (fat tissue is known to entrap vitamin D) (Zhang & Naughton, 2010).

Recently the scientists focused their attention on the problem of air pollution as an important factor reducing the amount of UVB radiation reaching the earth surface and therefore correlating with prevalence of vitamin D deficiency.

4. Correlation of air pollution and prevalence of vitamin D deficiency, insufficiency and adequacy

Air pollution is regarded as a dominant factor influencing the extent of solar UVB reaching earth surface. It was proved by both: observational and prospective studies, performed in different populations living in different geographic latitudes that atmospheric pollution (especially high tropospheric ozone content) may play a significant independent role in the development of vitamin D deficiency. Tropospheric ozone can efficiently absorb UVB radiation and decrease the amount of photons reaching ground level. It was proved that the level of air pollution is inversely related to the extent of solar UVB that reaches earth surface. Since industrial areas are those, of the highest intensity of air pollution, prevalence of D hypovitaminosis among urban residents can be more than twice higher compared to rural inhabitants.

To date, there have been published only 3 studies that attempted to correlate the intensity of air pollution with the prevalence of vitamin D deficiency, however their results are unequivocal.

In the tropics, children who live in regions with higher levels of ambient air pollution have been shown to be at increased risk of developing vitamin D-dependent rickets, compared to those living in less polluted areas (Agarwal et al., 2002). Children living in the highly polluted part of Delhi, despite similar types of housing, had significantly lower mean serum concentrations of 25(OH)D_3 compared to those living in the less polluted areas of the city. The prevalence of vitamin D deficiency was correlated with the intensity of air pollution that in turn correlated inversely with the amount of UVB light reaching the ground level. None of the children included into the study used any vitamin D supplementation or consumed vitamin D fortified food, so the differences in the vitamin D status of the children cannot be explained by different dietary habits. However the authors did not collect data regarding time which each of individuals spent on exposition to sunlight that seems to be a chief limitation of this work.

In another study performed in adult European citizens of urban areas above 35^0 N (where cutaneous vitamin D synthesis in winter is nearly negligible), prevalence of vitamin D insufficiency was significantly higher compared to rural residents (38% *vs* 18%) (Manicourt & Devogealer, 2008). The compared groups did not differ in the mean age, body mass index and amount of time spent outdoor. Therefore the authors concluded that the observed difference in vitamin D status must be due to the efficacy of vitamin D cutaneous synthesis. The main reason of this phenomenon was 3 times higher tropospheric ozone concentration in urban compared to rural areas. These results were confirmed by a subsequent study performed in urban and rural areas of Iran (Hosseinpanah et al., 2010), where air pollution (determined as the high tropospheric ozone content) was found to be an independent, significant risk factor of vitamin D deficiency.

For many years the problem of vitamin D insufficiency has been underestimated, associated only with calcium/phosphorus metabolism and regarded as an area of interest reserved only for pediatricians or orthopedists. However, the last years brought a number of epidemiological studies revealing the unexpected connection between the vitamin D hypovitaminosis and prevalence of several diseases that have forced the scientist to revise their opinions on the mechanism of vitamin D action and its role in the maintenance of body homeostasis (Walters, 1992).

5. Vitamin D mechanisms of action

$1,25(OH)_2D_3$ may act in target tissues via both genomic and non-genomic mechanisms. The non-genomic mechanism which is still not fully understood, is associated with stimulation of the enzymatic activity of a nonreceptor protein tyrosine kinase Src that results in activation of the mitogen activated protein kinase (MAP) signaling pathway (Gniadecki, 1998). Much more is known about the interaction of $1,25(OH)_2D_3$ with its nuclear receptor VDR (vitamin D receptor).
VDR together with thyroid hormone receptor (TR) and retinoid-X receptor (RXR) belongs to the class II of nuclear receptors family that act as a transcription factors modulating expression of vitamin D-directed genes. Upon binding with $1,25(OH)_2D_3$, the VDR forms a heterodimer with the retinoid-X receptor and translocates from cytoplasm to nucleus where it interacts with vitamin D responsive elements (VDRE) in promoter regions of target genes and regulates their expression (Dusso & Brown, 1998) [Figure 2].

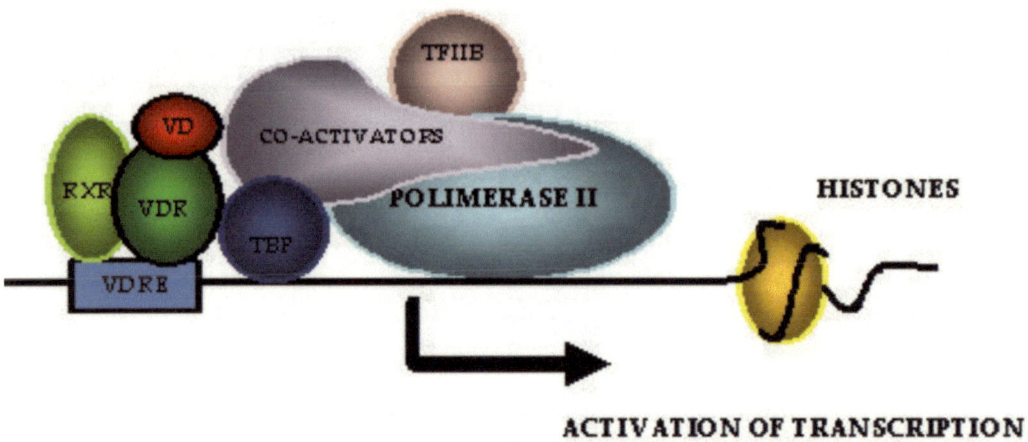

Fig. 2. The simplified scheme of activation of the target gene by the vitamin D receptor. The vitamin D receptor (VDR) with vitamin D (VD) binds as a heterodimer with the retinoid X receptor (RXR) specific sequence in the promoter region of the target gene – the vitamin D responsive element (VDRE). *Via* TATA binding protein (TBP) transcription factor II B (TFIIB), VDR gets in touch with the RNA II polymerase and other transcription factors (not present on these scheme). The transcription level is regulated by the complex of co-activators which bind VDR.

Identification of VDR in the intestine, kidney, bones and parathyroid glands (organs traditionally associated with mineral homeostasis) was not surprising. It is an undisputable fact that vitamin D is essential for maintenance of calcium and phosphorus homeostasis. In response to hypocalcemia-induced secretion of parathyroid hormone, and subsequent increase of *CYP27B1* expression and conversion of $25(OH)D_3$ to an active $1,25(OH)_2D_3$, calcitriol, secreted to circulation, reaches the target cells and, after binding with VDR interacts with VDR-responsive elements in target genes that results in increased calcium and phosphate absorption in gut and in release of calcium and phosphate from the mineral phase of bones. Other actions of vitamin D related to bone metabolism include: inhibition of type 1 collagen synthesis, induction of osteocalcine production, stimulation of monocytes-macrophages differentiation into osteoclasts and production of RANK ligand which mediates in maturation of osteoclast precursors into osteoclasts, responsible for calcium mobilization from bones (Clarke & Kohsla, 2010).

The identification of VDR expression as well as the demonstration of the vitamin D 1α-hydroxylaze (CYP27B1) activity in a wide number of tissues other than bones (including, e.g.: skin, blood cells, prostate, breast, brain and skeletal muscles) started a new era in the understanding of vitamin D action and led to the designation of novel, so-called "non-calcemic" actions of calcitriol.

Nowadays, $1{,}25(OH)_2D_3$ can be defined as both a hormone (when it is synthesized in kidneys and secreted to the circulation) and as a cytokine. As a cytokine calcitriol participates in the regulation of innate immunity. It is synthesized locally by monocytes-macrophages and in an intracrine manner, *via* interaction with VDR, modulates immune response towards microbial agents. *In vitro* studies suggest also crucial role of $1{,}25(OH)_2D_3$ in regulation of differentiation, maturation and function of other antigen presenting cells - dendritic cells. Other functions of vitamin D in the immune system include: regulation of the differentiation and activation of CD4 lymphocytes, increase in the number and function of regulatory T cells (Treg) , reduction in the production Th1-derived cytokines, stimulation of the Th2 helper and natural killer (NK) T cells and probably many others (Marques et al., 2010].

$1{,}25(OH)_2D_3$ has been also shown to have antiproliferative and antiapoptotic properties. Although the exact mechanism by which $1{,}25(OH)_2D_3$ regulates cellular proliferation is not fully understood and may differ between tissues and cell lines, several pathways have been proposed. In *in vitro* studies, $1{,}25(OH)_2D_3$ *via* interaction with VDR, increases expression of cyclin-dependant kinase (CDK) inhibitors - e.g.: proteins p21 and p27. It results in keeping the cell in G1/S phase and in prevention of DNA synthesis (see Figure 3) as it was shown in experiments performed on human prostate adenocarcinoma (lymph node, carcinoma, prostate - LNCaP) and on human leukemia U937 cell lines (Zhuang & Burnstein, 1998; Liu et al., 1996). The inhibition of mitogenic signals transmitted via e.g. epithelial growth factor (EGF) receptor, inhibition of prostaglandins, activation of transforming growth factor β (TGF-β) and proteins binding insulin-like growth factor (IGF-BP3) are examples of other (and not only) mechanisms by which vitamin D is able to regulate cell cycle progression (Desprez et al., 1991, Moreno et al., 2006).

Apoptosis is an example of another cellular process which can be regulated by $1{,}25(OH)_2D_3$, however the exact mechanisms of this phenomenon are still being investigated. Experiments conducted on human breast cancer and on chronic lymphatic leukemia cell lines revealed that $1{,}25(OH)_2D_3$ is able to inhibit expression of the protooncogen *bcl-2* and increase expression of the pro-apoptotic protein Bax (Elstner et al., 1995, Xu et al., 1993). In addition, in breast and prostate cancers cell lines, calcitriol was shown to induce release of cytochrome c in the mechanism that does not depend on kaspases (Spina et al., 2006).

Apart from its engagement in cell proliferation and apoptosis, vitamin D has been also found to be involved in the regulation of cell adhesion and angiogenesis, two other processes important for cancer development and progression. "Anti-invasive" properties of

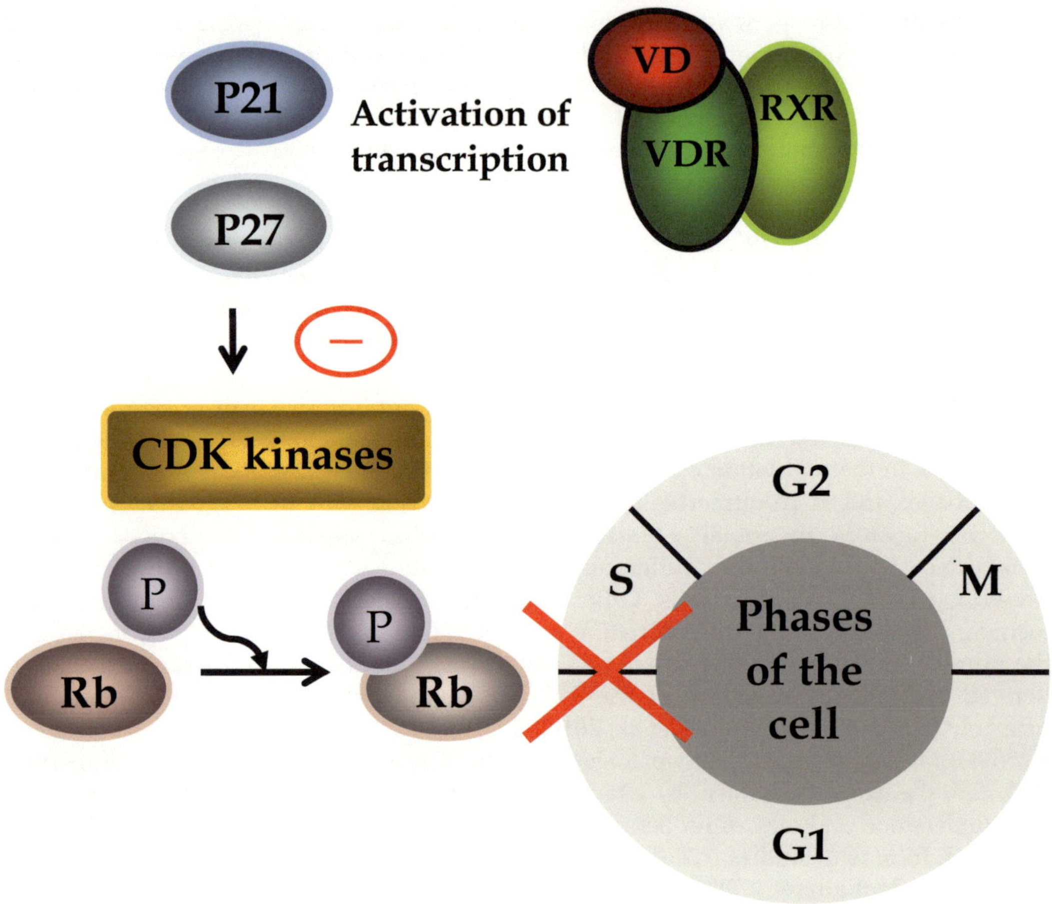

Fig. 3. Inhibition of cell cycle progresion by vitamin D
One of the factors determining the switch between G1 and S phases of the cell cycle is retinoblastoma (Rb) protein phosphorylation that leads to the release of several transcription factors activating genes involved in the cell cycle progression. Rb phosphorylation is catalyzed by the G1 cyclines and cyclin dependent kinases - CDKs. The CDKs activity can be inhibited by the p21 and p27 proteins. Vitamin D with the vitamin D receptor (VDR) binds to the regulatory sequences in the promoters of *p21* and *p27* genes activating their transcription that leads to the inhibition of the CDKs, lack of Rb phosphorylation and inhibition of the cell cycle progression.

$1,25(OH)_2D_3$ has been shown both in *in vitro* experiments (performed on human breast and lung cancer cell lines) as well as on animal models of prostate and bladder cancers, and include:

i. inhibition of metalloproteinases and serine proteases,
ii. up-regulation of cadherin E expression,
iii. down-regulation of integrin α6 and β4 expression (Bao, et al., 2006a, Hansen et al., 1994, Konety et al., 2001).

Inhibition of interleukin 8 gene expression (*via* interaction of $1{,}25(OH)_2D_3$ with p65 subunit of nuclear factor κB) is proposed as one of potential mechanisms by which calcitriol may interfere with the process of angiogenesis (Bao, et al., 2006b).
Knowledge of novel, non-calcemic actions of vitamin D that were proved in studies *in vitro* and on animal models, helps to understand the connection between the vitamin D and several human diseases that was reported by epidemiological studies.

6. Expected and unexpected consequences of vitamin D deficiency

Knowledge of traditional and novel, non-calcemic actions of vitamin D that were proved in studies *in vitro* and on animal models, helps to understand the connection between the vitamin D and several human diseases which was reported by epidemiological studies.

6.1 Musculoskeletal symptoms

For years vitamin D deficiency deficiency has been predominantly associated with impaired bone mineralization and development of rickets in children as well as osteopenia, osteoporosis and fractures in adults.
The consequences of vitamin D deficiency in mothers can be observed in the fetal skeleton just at the beginning of the 19th week of human gestation, resulting in rachitic phenotype, the severity of which is directly associated with a decreasing $25(OH)D_3$ level in the maternal circulation (Mahon et al., 2009). Vitamin D levels in infants correlate with vitamin D levels in their mothers during the first two months of life. Later, vitamin D status in babies, like in adults, depends on their diet and exposure to sunlight.
In rickets caused by vitamin D deficiency (a leading cause of rickets), both intestinal calcium absorption and renal phosphate reabsorption are significantly reduced. The decreased levels of serum calcium and phosphorus levels result in decreased bone mineralization. In addition, phosphorus deficiency contributes to the failure of the expected apoptosis of hypertrophic chondrocytes that results in deorganization of the growth plate. Clinical presentation of nutritional rickets depends on the duration and severity of vitamin D deficiency. At the beginning hypocalcemic symptoms are predominant, whereas skeletal deformities become obvious in later, more advanced stages. Classical symptoms of so-called "blooming" rickets include: craniotabes in infants older than 2-3 months, delayed fontanel closure, wrists enlargement, rachitic rosary, delayed teething, carious teeth, enamel hypoplasia, "O"- or "X"-type leg deforminty, kyphosis and narrow pelvis, chest deformities, costal or lower extremity fractures, caput quadratum, frontal bossing, fractures, brown tumor and extremity pain. Extra-skeletal symptoms include: hypotonia, constipation, proximal myopathy, cardiomyopathy and heart failure, myelofibrosis and pancytopenia, growth retardation, hypocalcemic convultions and benign intracranial hypertension (Levine, 2009).
Although full-symptomatic rickets seems to be nowadays a curiosity, epidemiological data indicate that rickets is not a disease of the past, nor it is limited to developing countries. It is estimated that nowadays 5 per 1,000,000 children aged between 6 months and 5 years have rickets, and the main risk factors for its development are breastfeeding and dark carnation. The peak prevalence of vitamin D-deficient rickets is characteristically between 6 and 18 months of age, with a further smaller peak occurring during adolescence (Nield et al., 2006).
In adults vitamin D deficiency can also cause a skeletal mineralization defect, resulting in osteomalacia demonstrating with isolated or global bone discomfort accompanied by joints

and muscle pain. However some observational studies conducted in adults and concerning the association between skeletal health and 25(OH)D_3 serum levels had conflicting results, suggesting both a strong correlation with the incidence of fractures and falls or only a fair relation (Chung et al., 2009). In general, it is assumed that the decrease of serum 25(OH)D_3 concentration that results in the persistent secondary parathyroidism enhances osteoclastogenesis and subsequent bone resorption.

The critical role of vitamin D in bone mineralization is well established, but our growing knowledge about non-calcemic actions of calcitriol led to the elongation of the list of diseases and conditions which development and/or progression can be associated with inadequate vitamin D status, including: different types of cancers, autoimmune disorders, cardiovascular events and probably many others. Interestingly, prevalence of this diseases frequently correlates with intensity of environmental pollution and many of them are diagnosed more frequently in inhabitants of industrial districts compared to rural areas (Pope, 2002, Pope, 2003, Grant, 2006, Ritz, 2010).

6.2 Vitamin D deficiency and cancer risk

The observational study conducted in 1941 reported that living at higher latitudes (where vitamin D skin synthesis is impaired) is associated with an increased risk of the development and dying of many common cancers including colon, prostate, ovarian and breast cancers (Apperly, 1941).

This observation was followed by the epidemiological studies performed in Europe and in the North America, assessing influence of many potential risk factors for incidence of cancers (e.g. smoking, alcohol consumption, economic burdens and environmental pollution) revealing that morbidity of several cancers (including colon, gastric, lung and breast carcinomas) is inversely associated with exposition to UVB radiation (Grant, 2003, Grant, 2005, Grant & Garland, 2006, Grant, 2009).

Next, on the one hand, another epidemiological prospective, study proved that vitamin D insufficiency (25(OH)D_3 concentrations below 20ng/ml) is associated with 30-50% higher risk of colorectal cancers (Garland et al., 2009). On the other hand, it was shown in a meta-analysis that appropriate (>400 IU) or increased (>1000 IU) vitamin D intake may be associated with a decreased risk of colon, breast, pancreas, esophagus and non-Hodgkin lymphomas (Garland et al., 2006, Giovannucci et al., 2006, Holick & Chen, 2008). It is estimated that the daily intake of 2000 IU of vitamin D would lead to 25% reduction in incidence of breast cancer and 27% reduction in incidence of colorectal cancer (Garland, 2009). Association of vitamin D deficiency with prevalence of breast, skin and prostate carcinomas has been also reported, however the concentrations at which the increased risk has been observed varied. Moreover, there are studies suggesting a U-shaped association between vitamin D concentration and cancers risk, pointing that some individuals may be adversely affected by elevated 25(OH)D_3 concentrations with respect to risk of prostate, breast, pancreas and esophagus cancers (Toner et al., 2010). This inconsistency may be caused by inadequate consideration of modifiers of 25(OH)D_3 concentrations. Therefore, to date, there is not yet sufficient evidence to recommend high-dose vitamin D supplementation for the prevention of cancer.

The hypothesis, about the connection between the vitamin D deficiency and cancer development, based on the observational data, has been confirmed by the *in vitro* studies described in the previous section and by experiments on animal models. For example: in vitamin D deficient Balb/c mice, injected with MC-26 colon cancer cells, the tumor growth

was accelerated, compared to vitamin D sufficient animals. In addition, vitamin D sufficient animals presented higher intra-tumor expression of VDR and CYP27B1, which suggests possible autocrine/paracrine cell growth regulation by vitamin D (Tangpricha et al., 2005). In turn mice with *vdr* knock-out (mutation that eliminates genomic action of $1,25(OH)_2D_3$) have been found to be more susceptible to develop leukemias, breast and skin cancers under the influence of common carcinogens compared to wild-type animals (Welsh et al., 2004).
Discovery of anti-cancer properties of vitamin D (that has no antioxidant properties) required a modification of a classic two-hit model of cancerogenesis, that provides that development of cancer depends on both: activation of proto-oncogenes and deactivation of tumor suppressing genes. Therefore, in context of vitamin D, a novel model of cancer pathogenesis has been proposed - a so-called: the Disjunction-Initiation-Natural Selection-Metastasis-Involution-Transition (DINO-MIT) model. This model is based on the classical concepts of carcinogenesis like initiation and promotion, however it includes also the life cycle of malignancies and provides an explanation of the ability of vitamin D to prevent or arrest the development of cancer (described in detail by Garland et al., 2006).

6.3 Vitamin D deficiency and autoimmune disorders

Similarly to carcinomas, prevalence of some autoimmune disorders such as multiple sclerosis and type 1 diabetes mellitus also correlates with the geographical latitudes. It has been proposed that vitamin D deficiency can disturb the immunological equilibrium and therefore contribute to the development of autoimmunity (e.g. by exacerbation of Th1 immune response) (Cantorna, 2004). In order to understand the complexity of these mechanism, again, insights from animal models of common human autoimmune diseases occurred to be particularly helpful.
Vitamin D deprivation leads to the acceleration of the development of experimental allergic encephalomyelitis (EAE - an animal model of multiple sclerosis) in mice immunized with myelin antigens (e.g. MOG35-55) (Cantorna et al., 1996). In turn, supplementation with active vitamin D metabolite and its analogs may favorably influence the course of autoimmune diseases or even prevent their occurrence. In mice immunization with type 2 collagen leads to the development of collagen induced arthritis (CIA - an animal model of rheumatoid arthritis), however, if the animals are given $1,25(OH)_2D_3$ till the 14th day after immunization, they do not present any symptoms of arthritis at all. If $1,25(OH)_2D_3$ is given to mice with symptomatic artritis, it may alleviate the disease course (Cantorna et al., 1998). Similar evidence comes from the studies performed in nonobese diabetic (NOD) mice that due to the autoimmune inflammation and destruction of pancreatic islets develop insulin-dependent type 1 diabetes. In young subjects, administration of $1,25(OH)_2D_3$ till the third week after birth may completely prevent development of diabetes whereas in mature mice leads to the 50% reduction of inflammatory infiltrates in the pancreatic islets, compared with placebo-treated animals (Gregori et al., 2002). Reduction of T cells infiltration and decrease in the number of demyelinization sites has been also found in spinal cords of mice with experimental allergic encephalomyelitis treated with vitamin D analogs, compared to wild type animals (Mattner et al., 2000).
In humans, epidemiological data revealed that multiple sclerosis morbidity in Caucasians is significantly higher in less insolated areas, and the course of disease correlates with seasonal fluctuation of $25(OH)D_3$ serum concentrations (exacerbations in spring when $25(OH)D_3$ concentrations are the lowest) (Cantorna, 2008). Similarly, in patients with systemic lupus

erythematosus (SLE) severe vitamin D deficiency has been reported by many authors, and it correlates with disease activity (Kamen, 2010).

In contrary, like in epidemiological studies performed in different cancers, adequate consumption of cholecalciferol (400 IU per day) was shown, in a prospective study, to decrease a risk of multiple sclerosis development (relative risk - RR 0.59) (Munger et al., 2004). Vitamin D deficiency during the first year of life was found also to be a severe risk factor of type 1 diabetes whereas its proper administration in early childhood may reduce the risk of the development of this disease by 78% (RR 0.12) (Hypponen et al., 2001). A correlation between the adequate vitamin D intake and lower risk of the development of rheumatoid arthrits (RR=0.67) was reported (Merlino et al., 2004), however it was not confirmed by further studies [Costenbader, 2008]. Inconsistent are also results of the studies regarding the association between vitamin D intake during pregnancy and risk of type 1 diabetes in the offspring as well as with prevalence of other autoimmune disorders.

These discrepancies may be owed to the fact that, like in carcinogenesis, vitamin D deficiency is probably not the chief trigger in the development of autoimmunity. This hypothesis is supported by the *in vivo* experiments, where animals with vitamin D deficiency or with the knock-out of *vdr* gene do not present autoimmune disorders spontaneously and require other stimuli to develop autoimmunity. However these animals may present more severe disease phenotype of autoimmune disorders, as it was shown in mice depleted with IL-10 gene (animal model of inflammatory bowel diseases - IBD) (Froicu & Cantorna, 2007).

6.4 Vitamin D deficiency and cardiovascular health

Cardiovascular diseases are the most common cause of mortality and morbidity worldwide. There is also a growing number of evidence that vitamin D deficiency can be associated with development of several cardiovascular diseases including hypertension, ischemic heart disease and cardiac hypertrophy.

In vitro studies revealed that addition of $1,25(OH)_2D_3$ to the cardiomyocyte cells culture resulted in: inhibition of cell proliferation, enhanced cardiomyocyte formation, decrease of apoptosis and cell-cycle associated genes' expression. In turn animals with *vdr* gene knock-out were shown to develop cardiac hypertrophy, display hypertension as well as increased serum angiotensin and tissue renin levels. These studies suggest that vitamin D *via* its influence on cardiac and vascular structure and function may modulate cardiovascular risk (Artaza et al., 2009).

Observational studies reported a strong association between vitamin D hypovitaminosis and other traditional cardiovascular risk factors. Moreover, according to other reports, vitamin D deficiency is a predictor of all-cause and cardiovascular mortality, whereas vitamin D adequacy is associated with 43% reduction in cardiometabolic disorders (Parker et al., 2010).

The importance of vitamin D status in context of coronary artery disease is well established based on observational studies that reported:

i. inverse correlation of mortality from ischemic heart disease and the exposure to sunlight (Fleck, 1989),
ii. high prevalence of vitamin D deficiency in patients with ischemic heart disease (Kendrick et al., 2009),

iii. increased risk of sudden cardiac death, heart failure and fatal stroke in patients with ischemic heart disease who were vitamin D deficient (Pilz et al., 2008).

A correlation between the time of exposition to UVB radiation and decrease in blood pressure was also observed (Krause et al., 1998) and vitamin D replacement in deficient subjects led to a significant improvement of flow-mediated dilatation of the brachial artery that suggested an influence of vitamin D on vascular smooth cells (Tarcin et al., 2009). Indeed, it was shown on animal models that active vitamin D metabolite inhibits rennin expression in the juxtaglomerular apparatus and may prevent proliferation of vascular smooth muscle cells (Li et al., 2002; Carthy et al., 1989). Both, small retrospective observational studies and a large, cross-sectional study, confirmed that the mean blood pressure vary inversely with serum 25(OH)D_3 concentrations and the association remained significant after adjustment for age, gender, race, ethnicity and physical activity (Scragg et al., 2007, Reddy Vanga et al., 2010).

Additionaly, individuals with vitamin D insufficiency have been found to higher prevalence of peripheral artery disease (Melamed et al., 2008) and worse outcomes in end-stage heart failure (Zittermann et al., 2008).

7. Vitamin D in treatment

Encouraging epidemiological data coming from observational studies, linking vitamin D deficiency with common human diseases, rose a hope that calcitriol and its derivates may be useful in everyday clinical practice (Holick, 2004, Holick, 2007). However, to date bone diseases are the only examples where administration of vitamin D is generally accepted for treatment.

7.1 Rickets and osteoporosis

In order to restore vitamin D reserve in rickets two methods of treatment are proposed. In a low dosage and long-term therapy model, vitamin D is administered 1000-10000 IU/day (dose depends on age) for 2-3 months. After that period 400 IU/day therapy is recommended to maintain the serum vitamin D level. Another regimen, reserved mainly for patients suspected of poor compliance, includes 100 000 - 600 000 IU of vitamin D in a single dose (Wharton & Bishop, 2003).

Prospective, randomized, placebo-controlled trials provided evidence supporting the benefit of vitamin D supplementation in patients with osteoporosis (LaCroix et al., 2009, Meier et al., 2004, Jackson et al., 2006). Data regarding a beneficial effect of calcium plus vitamin D on bone density in postmenopausal women and older men is consistent, however reports on fracture risk are more variable. Reduction in frequency of fractures was reported in some studies (Larsen et al., 2004) whereas large randomized trials have not shown any association between vitamin D supplementation and reduction in fracture risk. Additionally, in many of these trials, it is difficult to differentiate the effect of calcium from that of vitamin D. Nevertheless, in the Women's Health Initiative (the largest of the available trials) subgroup analysis revealed that calcium and vitamin D supplementation was associated with reduced fracture incidence in those subjects who were most compliant (Jackson et al., 2006).

Although the optimal serum concentration of 25(OH)D_3 in patients with osteoporosis is not clear, based upon meta-analyses, one approach to vitamin D supplementation would be to give 400-800 IU daily with a target serum 25(OH)D_3 concentration >20 ng/ml. Older individuals with greater risk may require higher amounts that allow to maintain serum levels of 25(OH)D_3 from 30 to 40 ng/ml (Dawson-Hughes, 2005). Yearly high-dose of vitamin D (e.g. 500,000 IU) for the osteoporosis treatment is not recommended.

7.2 Cancers

In observational studies, sufficient exposition to UVB radiation and vitamin D adequacy have been found to improve prognosis in cancer patients. A study assessing the survival rates for several cancers (including breast, colon, prostate, lung carcinomas and Hodgkin's lymphoma) revealed that patients diagnosed in autumn had approximately 30% higher 18-month survival rate than those diagnosed in winter or spring (Robsahm et al., 2004, Porojnicu et al, 2007).

Results of recent observational studies, reporting that maintenance of 25(OH)D_3 levels about 30ng/ml may be crucial in the prevention of breast, colon and prostate carcinomas (Garland, 2009), together with newly discovered, proapoptotic and anti-proliferative properties of vitamin D stimulate hope that 1,25$(OH)_2D_3$ and its analogues will have practical application in treatment of cancers. Severe hypercalcemia has occurred to be the chief limitation for the common use of vitamin D derivates in oncology and therefore vitamin D analogues are used mainly in combination with other anti-cancer drugs in reduced doses. Both *in vitro* and *in vivo* studies suggest that addition of vitamin D may potentializate action of many drugs commonly used in cancer treatment, including e.g: dexamethasone, docetaxel, paclitaxel, tamoxifen, retinoids, platinium derivatives and others (Gewirtz et.al., 2002). Administration of 1,25$(OH)_2D_3$ may also sensitize cancer cells to radiotherapy as it was shown in prostate cancer line LNCaP or breast cancer line MCF-7 (Dunlap et al., 2003, Polar et al., 2003).

Numerous vitamin D analogs with minimal calcemic activity have been designed with intent to be applied in cancer therapy and some of them have been found to be effective in inhibiting tumor growth (e.g. colon) in animal models. Some of them are tested for treatment (e.g. calcipotriol administered locally in skin metastases of breast cancer) however to date, no active analogs of vitamin D have been proved to be efficacious for the treatment of any human cancer by themselves.

No prospective, randomized, double-blind study has been conducted in order to asses the anticancer potential of vitamin D analogues in humans. However, in a small pilot study performed in 7 patients with prostate cancer, daily administration of 0.5-2.5 µg/24h of 1,25$(OH)_2D_3$ per 6 to 15 months, resulted in significant reduction of prostate specific antigen (PSA) concentration in 6 individuals (Gross et al., 1998). In another study, 37 patients with prostate cancer insensitive to androgens, were given sequentially 0.5 µg of 1,25$(OH)_2D_3$ (day 1) and 36 mg/m^3 of docetaxel (day 2) during 6 following weeks and put on diet containing 400-500 mg of calcium per day. In 30 of them (81%) a significant reduction of PSA concentration was observed after 8 weeks from the beginning of the treatment (Beer et al., 2003).

7.3 Autoimmune diseases

Despite solid clinical and experimental evidence regarding the positive influence of vitamin D derivates on the development and progression of autoimmune disorders, to date vitamin D is not routinely administered in their treatment.

The only disease with partially autoimmune pathogenesis, where vitamin D and its analogs are generally accepted for the treatment is psoriasis. Both systemic and topical administration of vitamin D and its analogs leads to the significant improvement in 70-80% of the treated patients, measured by the Total Severity Score (TSS) or Psoriasis Area Severity Index (PASI) (Ashcroft et al., 2000). Administration of vitamin D for treatment of other autoimmune diseases is still experimental since in humans only small and non-controlled studies have been conducted. For instance, in a an open study performed in 19 patients with rheumatoid arthritis, addition of alphacalcidol ($1\alpha(OH)D_3$) to the traditional disease-modifying anti-rheumatic drugs for three months, resulted in a significant reduction of symptoms in 89% of patients (45% achieved complete remission and 44% had satisfactory results) without side effects (Andjelkovic et al., 1999). There is also one report on a positive influence of $1,25(OH)_2D_3$ on thyroid hormones levels in patients with autoimmune hyperthyroidism, who acquired the treatment in order to improve bone mineral density (Kelman & Lane, 2005).

7.4 Cardiovascular health

To date, studies evaluating influence of vitamin D supplementation on cardiovascular health are few and have inconsistent results. Observational studies reported that individuals who take oral vitamin D supplementation have lower blood pressure and some interventional trials found a correlation between oral vitamin D administration or increased exposure to UVB and decrease of blood pressure. However, this phenomenon was not confirmed by a large prospective studies and a meta-analysis on prevalence of hypertension and vitamin D intake (Forman et al., 2005, Witham et al., 2009). Vitamin D supplementation seemed also to have no effect on the risk of cardiovascular mortality (La Croix et al., 2009).

8. Guidelines regarding vitamin D supplementation

The recognition of vitamin D deficiency as a worldwide problem rises a task of means of its supplementation. Historically, fortification of milk in the 1930s with vitamin D was effective in eradicating rickets in Europe and US. However, after an unexpected outbreak of hypercalcemia in British children in the 1950s caused by the excessive vitamin D consumption from fortified food, fortification of dairy products with vitamin D had been forbidden (Holick, 2010).This situation was probably caused by the imperfection of methods applied to assess the vitamin D concentration in food at that time.

Naturally the main source of vitamin D are wild-caught, not farm-raised oily fish, e.g. salmon, mackerel, herring, and oils from fish. However, since no quality control over vitamin D content in natural products is available, they are not recommended as reliable sources of vitamin D supplementation. In many countries dairy products, juices, bread and other products are fortified with vitamin D.

The recommendations of American Institute of Medicine from 1999 state that all children and adults up to the age of 50 should consume 200 IU of vitamin D daily, whereas individuals aged 51-70 and those who are over 71 years old, 400 IU and 600 IU, respectively. However, these amounts have been insufficient in prevention of osteomalacia and osteoporosis, and therefore many experts now suggest that, in the absence of the adequate sun exposure, as much as 800-1000 IU of vitamin D per day is needed for healthy individuals, regardless of age (Holick and Chen, 2008). In case of diseases leading to vitamin D mal-absorption or sequestration in fat tissue (obesity), higher doses may be required. In aspect of cancer prevention, the National Academy of Sciences-Institute of Medicine (USA) recommends even 2000-4000 IU/day (Garland, 2009). It is estimated that raising the mean population level of 25(OH)D_3 up to 42 ng/ml would result in 18% reduction in all-cause mortality, 25% reduction in prevalence of cancers and cardiovascular diseases, 15% in morbidity of diabetes and a prevalence of multiple sclerosis would be reduced by half (Grant & Schuitemaker, 2010), however approaching this vitamin D status would require supplementation of 2500-4000 IU of cholecalciferol per day.

According to the nutritional guidelines established during the 14th vitamin D workshop in October 2009, appropriate vitamin D status should be achieved rather through supplementation than diet fortification and rather through the use of vitamin D_3 than vitamin D_2. It is suggested that vitamin D_2 is in half less effective than vitamin D_3 in maintaining the 25(OH)D_3 serum levels and therefore it should be administered in higher doses. Moreover, all evidence reported to date on the efficacy of vitamin D in e.g. cancer prevention has been based on vitamin D_3. To verify this theory, a number of prospective vitamin D replacement trials with vitamin D_2 and vitamin D_3 has been conducted (Adams & Hewison, 2010).

In the treatment of evident vitamin D deficiency 50 000 IU of vitamin D is recommended weekly for 8 weeks, resulting usually in 25(OH)D_3 concentration of 30 ng/ml, and after 50 000 IU of vitamin D every 2 weeks to maintain its level. It has been estimated that for every 100 IU of vitamin D ingested, there is an increase in the blood level of 25(OH)D_3 of 1 ng/ml and sensible exposure to sunlight is more effective in raising blood levels of 25(OH)D_3 than 1000 IU vitamin D_3 taken daily for adults of most skin types. However, with the growing intensity of atmospheric pollution, which as it was shown above, is a chief factor limiting the amount of UVB radiation reaching the earth surface, obtaining appropriate amount of vitamin D from skin synthesis is getting more and more challenging.

9. Conclusions

Based on the available observational and experimental data regarding the connection between the air pollution, vitamin D status and related health consequences, following statements can be formulated:

i. the worldwide problem of vitamin D deficiency can be partially explained by the growing intensity of atmospheric pollution, however this observation is based only on the results of relatively small studies and should be replicated in larger scale,
ii. both: vitamin D deficiency and intensity of air pollution correlate with prevalence of common human diseases including different types of cancers, cardiovascular and autoimmune diseases that was confirmed by large, prospective studies and meta-analyses,
iii. association between vitamin D status and extra-skeletal health outcomes can be explained by the molecular mechanism of vitamin D action that exceed bone metabolism,
iv. despite promising results of *in vitro* and *in vivo* studies regarding the favourable influence of vitamin D supplementation on prevalence and course of different diseases, its therapeutic application is still limited only to the treatment of bone diseases,
v. novel definition of vitamin D insufficiency together with the increasing intensity of the atmospheric pollution that impairs vitamin D skin synthesis and with epidemiological data regarding the benefits of vitamin D intake should result soon in the novel guidelines regarding vitamin D supplementation.

10. References

Adams JS & Hewison M. (2010) Update in vitamin D. *Journal of Clinical Endocrinology & Metabolism*, Vol. 95, No. 2, (Febuary 2010), pp. 471-478. PMID: 20133466

Agarwal KS, Mughal MZ, Upadhyay P, Berry JL, Mawer EB & Puliyel JM. (2002) The impact of atmospheric pollution on vitamin D status of infants and toddlers in Delhi, India. *Archives of Disease in Childhood*, Vol. 87, No. 2, (August 2002), pp. 111-113, PMID: 12138058

Andjelkovic Z, Vojinovic J, Pejnovic N, Popovic M, Dujic A, Mitrovic D, Pavlica L & Stefanovic D. (1999) Disease modifying and immunomodulatory effects of high dose 1 alpha (OH) D3 in rheumatoid arthritis patients. *Clinical and Experimental Rheumatology*. Vol. 17, No. 4, (July,August 1999), pp. 453-456, PMID: 10464556

Apperly FL. (1941) The relation of solar radiation to cancer mortality in North America. *Cancer Research*, Vol 1, No. 1, (1941), pp. 191–195.

Artaza JN, Mehrotra R & Norris KC. (2009) Vitamin D and the cardiovascular system. *Clinical Journal of the American Society of Nephrology*, Vol. 4, No. 9, (September 2009), pp. 1515-1522, PMID: 19696220

Arya V, Bhambri R, Godbole MM &Mithal A.(2004) Vitamin D status and its relationship with bone mineral density in healthy Asian Indians. *Osteoporos International*, Vol. 15, No. 1, (January 2004), pp. 56-61. PMID: 13680103

Ashcroft DM, Po AL, Williams HC & Griffiths CE. (2000) Systematic review of comparative efficacy and tolerability of calcipotriol in treating chronic plaque psoriasis. *British Medical Journal*, Vol. 320, No. 7240, (April 2000), pp. 963-967, PMID: 10753146

Bao BY, Yeh SD & Lee YF. (2006a) 1alpha,25-dihydroxyvitamin D3 inhibits prostate cancer cell invasion via modulation of selective proteases. *Carcinogenesis*, Vol. 27, No. 1,(January 2006), pp. 32-42, PMID: 15987715

Bao BY, Yao J & Lee YF. (2006b) 1alpha, 25-dihydroxyvitamin D3 suppresses interleukin-8-mediated prostate cancer cell angiogenesis. *Carcinogenesis*. Vol. 27, No. 9, (September 2006), pp. 1883-1893, PMID: 1662482

Beer TM, Eilers KM, Garzotto M, Egorin MJ, Lowe BA & Henner WD. (2003) Weekly high-dose calcitriol and docetaxel in metastatic androgen-independent prostate cancer. *Journal of Clinical Oncology*, Vol. 21, No. 1, (January 2003), pp. 123-128, PMID: 12506180

Cantorna MT, Hayes CE & DeLuca HF. (1996) 1,25-Dihydroxyvitamin D3 reversibly blocks the progression of relapsing encephalomyelitis, a model of multiple sclerosis. *Proceedings of the National Academy of Sciences of the United States of America*, Vol. 93, No. 15, (July 1996), pp. 7861-7864, PMID: 8755567

Cantorna MT, Hayes CE & DeLuca HF. (1998) 1,25-Dihydroxycholecalciferol inhibits the progression of arthritis in murine models of human arthritis. *The Journal of Nutrition*, Vol. 128, No. 1, (January 1998), pp. 68-72, PMID: 9430604

Cantorna MT & Mahon BD. (2004) Mounting evidence for vitamin D as an environmental factor affecting autoimmune disease prevalence. *Experimental Biology and Medicine (Maywood, N.J.)*, Vol. 229, No. 11, (December 2004), pp. 1136-1142. PMID:15564440

Cantorna MT. (2008) Vitamin D and multiple sclerosis: an update. *Nutrition Reviews*, Vol 66, No. 10, Suppl. 2, (October 2008), pp. S135-138, PMID: 18844840

Carthy EP, Yamashita W, Hsu A & Ooi BS. (1989) 1,25-Dihydroxyvitamin D3 and rat vascular smooth muscle cell growth. *Hypertension*, Vol.13, No. 6, Pt. 2, (June 1989), pp.954-959, PMID: 2786849

Chung M, Balk EM, Brendel M, Ip S, Lau J, Lee J, Lichtenstein A, Patel K, Raman G, Tatsioni A, Terasawa T & Trikalinos TA. (2009) Vitamin D and calcium: a systematic review of health outcomes. *Evidence Report - Technology Assessment*, Vol. 183, pp.1-420, PMID: 20629479

Clarke BL & Khosla S.(2010) Physiology of bone loss. Radiologic Clinic of North America. Vol. 48, No. 3, (May, 2010), pp. 483-495, PMID: 20609887

Costenbader KH, Feskanich D, Holmes M, Karlson EW & Benito-Garcia E. (2008) Vitamin D intake and risks of systemic lupus erythematosus and rheumatoid arthritis in women. *Annals of Rheumatic Diseases*, Vol. 67, No. 4, (April 2008), pp. 530-535, PMID: 17666449

Dawson-Hughes B, Heaney RP, Holick MF, Lips P, Meunier PJ & Vieth R.(2005) Estimates of optimal vitamin D status. *Osteoporos International,* Vol. 16, No. 7, (July 2005), pp. 713-716, PMID: 15776217

Desprez PY, Poujol D, Falette N, Lefebvre MF & Saez S.(1991) 1,25-Dihydroxyvitamin D3 increases epidermal growth factor receptor gene expression in BT-20 breast carcinoma cells. *Biochemical and Biophysical Research Communications*, Vol. 176, No. 1, (April 1991), pp. 1-6, PMID: 167333

Dunlap N, Schwartz GG, Eads D, Cramer SD, Sherk AB, John V & Koumenis C. (2003) 1alpha,25-dihydroxyvitamin D(3) (calcitriol) and its analogue, 19-nor-1alpha,25(OH)(2)D(2), potentiate the effects of ionising radiation on human prostate cancer cells. *British Journal of Cancer*, Vol. 89, No. 4, (August 2003), pp. 746-753, PMID 12915889

Dusso AS & Brown AJ.(1998) Mechanism of vitamin D action and its regulation. *American Journal of Kidney Diseases*, Vol.32, No. 2, Supplement 2, (October 1998), pp. 13-24, PMID: 9808140

Elstner E, Linker-Israeli M, Said J, Umiel T, de Vos S, Shintaku IP, Heber D, Binderup L, Uskokovic M & Koeffler HP.(1995) 20-epi-vitamin D3 analogues: a novel class of potent inhibitors of proliferation and inducers of differentiation of human breast cancer cell lines. *Cancer Research*, Vol. 55, No. 13, (July 1995),pp. 2822-2830, PMID: 7796409

Fleck A.(1989) Latitude and ischaemic heart disease.*Lancet*. Vol.1, No. 8638, (March 1989), pp. 613, PMID: 2564129

Forman JP, Bischoff-Ferrari HA, Willett WC, Stampfer MJ & Curhan GC. (2005) Vitamin D intake and risk of incident hypertension: results from three large prospective cohort studies. *Hypertension*, Vol. 46, No. 4, (October 2005), pp: 676-682, PMID: 16144983

Fraser DR.(2004) Vitamin D-deficiency in Asia. *The Journal of* Steroid Biochemistry *and Molecular* Biology, Vol. 89-90, No..1-5, (May 2004), pp. 491-495, PMID: 15225826

Froicu M & Cantorna MT. (2007) Vitamin D and the vitamin D receptor are critical for control of the innate immune response to colonic injury. *BMC Immunology*, Vol. 8, No. 5, (March 2007), pp. 1-11, PMID: 17397543

Garland CF, Garland FC, Gorham ED, Lipkin M, Newmark H, Mohr SB & Holick MF.(2006) The role of vitamin D in cancer prevention. *American Journal of Public Health*, Vol. 96, No. 2, (February 2006), pp. 252-261, PMID: 16380576

Garland CF, Gorham ED, Mohr SB & Garland FC. (2009) Vitamin D for cancer prevention: global perspective. *Annals of epidemiology*, Vol. 19, No. 7, (July 2009), pp. 468-83, PMID 19523595

Gewirtz DA, Gupta MS & Sundaram S. (2002) Vitamin D3 and vitamin D3 analogues as an adjunct to cancer chemo-therapy and radiotherapy. *Current Medical Chemistry: Anticancer Agents*, Vol. 2, No. 6, (November 2002), pp.683-690, PMID: 12678720

Giovannucci E, Liu Y, Rimm EB, Hollis BW, Fuchs CS, Stampfer MJ & Willett WC. (2006) Prospective study of predictors of vitamin D status and cancer incidence and mortality in men. Journal of the National Cancer Institute, Vol. 98, No. 7, (April 2006), pp. 451-459, PMID: 1659578

Gniadecki R. (1998) Nongenomic signaling by vitamin D: a new face of Src. *Biochemical Pharmacology*, Vol. 56, No. 10 (November 1998), pp. 1273-1277, PMID: 9825725

Grant WB. (2003) Ecologic studies of solar UV-B radiation and cancer mortality rates. *Recent Results in Cancer Research*, Vol. 164, (2003), pp. 371-377. PMID: 12899536.

Grant WB, Garland CF& Holick MF.(2005) Comparisons of estimated economic burdens due to insufficient solar ultraviolet irradiance and vitamin D and excess solar UV irradiance for the United States. *Photochemistry and Photobiology*, Vol. 81, No. 6, (November - December 2005), pp. 1276-1286, PMID: 16159309

Grant WB & Garland CF. (2006) The association of solar ultraviolet B (UVB) with reducing risk of cancer: multifactorial ecologic analysis of geographic variation in age-adjusted cancer mortality rates. *Anticancer Research* Vol. 26, No. 4A, (July – August 2006), pp. 2687-2699, PMID: 16886679

Grant WB. (2006) Epidemiology of disease risks in relation to vitamin D insufficiency. *Progress in Biophysics and Molecular Biology*, Vol. 92, No. 1, (September 2006), pp. 65-79. PMID: 16546242

Grant WB & Mohr SB. (2009) Ecological studies of ultraviolet B, vitamin D and cancer since 2000. *Annals of Epidemiology*, Vol. 19, No. 7, (July 2009), pp. 446-454, PMID: 19269856

Grant WB & Schuitemaker GE. (2010) Health benefits of higher serum 25-hydroxyvitamin D levels in The Netherlands. *The Journal of Steroid Biochemistry and Molecular Biology*, Vol.121, No. 1-2, (July 2010), pp. 456-458, PMID: 20398763

Gregori S, Giarratana N, Smiroldo S, Uskokovic M & Adorini L. (2002) A 1alpha,25-dihydroxyvitamin D(3) analog enhances regulatory T-cells and arrests autoimmune diabetes in NOD mice. *Diabetes*, Vol. 51, No. 5, (May 2002), pp.1367-1374, PMID: 11978632

Gross C, Stamey T, Hancock S & Feldman D. (1998) Treatment of early recurrent prostate cancer with 1,25-dihydroxyvitamin D3 (calcitriol). *The Journal of Urology*, Vol. 159, No. 6, (June 1998), pp. 2035-2039, PMID: 9598513

Hansen CM, Frandsen TL, Brünner N & Binderup L.(1994) 1 alpha,25-Dihydroxyvitamin D3 inhibits the invasive potential of human breast cancer cells in vitro. Clinical & Experimental Metastasis, Vol. 12, No. 3, (May 1994), pp. 195-202, PMID: 819419

Holick MF. (2004) Sunlight and vitamin D for bone health and prevention of autoimmune diseases, cancers, and cardiovascular disease. *The American Journal of Clinical Nutrition*, Vol. 80, No. 6, Supplement, (December 2004), pp. 1678-8168, PMID: 15585788

Holick MF. (2007) Vitamin D deficiency. *The New England Journal of Medicine*, Vol. 357, No. 3, (July 2007), pp. 266-281, PMID:17634462

Holick MF & Chen TC. (2008) Vitamin D deficiency: a worldwide problem with health consequences. *The American Journal of Clinical Nutrition*, Vol. 87, No. 4, (April 2008), pp. 1080S-1086S, PMID: 18400738

Holick MF. (2010) Vitamin D: Evolutionary, Physiological and Health Perspectives. *Current Drugs Targets*, Vol. 12, No. 1, (January 2010), pp. 4-18, PMID: 20795941

Horst RL & Reinhardt TA. (1997) Vitamin D metabolism in *Vitamin D* Feldman D, pp. 13-31, Academic Press, ISBN 0-12-252685-6, San Diego

Hosseinpanah F, Pour SH, Heibatollahi M, Moghbel N, Asefzade S & Azizi F. (2010) The effects of air pollution on vitamin D status in healthy women: a cross sectional study. *BMC Public Health*, Vol. 10, (August 2010), pp. 519-525. PMID: 20799984

Hyppönen E, Läärä E, Reunanen A, Järvelin MR & Virtanen SM. (2001) Intake of vitamin D and risk of type 1 diabetes: a birth-cohort study. *Lancet*, Vol. 358, No. 9292, (November 2001), pp. 1500-1503, PMID: 11705562

Jackson RD, LaCroix AZ, Gass M, Wallace RB, Robbins J, Lewis CE, Bassford T, Beresford SA, Black HR, Blanchette P, Bonds DE, Brunner RL, Brzyski RG, Caan B, Cauley JA, Chlebowski RT, Cummings SR, Granek I, Hays J, Heiss G, Hendrix SL, Howard BV, Hsia J, Hubbell FA, Johnson KC, Judd H, Kotchen JM, Kuller LH, Langer RD, Lasser NL, Limacher MC, Ludlam S, Manson JE, Margolis KL, McGowan J, Ockene JK, O'Sullivan MJ, Phillips L, Prentice RL, Sarto GE, Stefanick ML, Van Horn L, Wactawski-Wende J, Whitlock E, Anderson GL, Assaf AR & Barad D; Women's Health Initiative Investigators. (2006) Calcium plus vitamin D supplementation and the risk of fractures. *The New England Journal of Medicine*, Vol. 354, No. 7, (February 2006), pp. 669-683, PMID: 16481635

Kamen DL. (2010) Vitamin D in Lupus. *Bulletin of the NYU Hospital for Joint Diseases*, Vol. 68, No. 3, (2010), pp. 218-22, PMID: 20969555

Kelman A & Lane NE. (2005) The management of secondary osteoporosis. *Best Practice and Research: Clinical Rheumatology*. Vol. 19, No. 6, (December 2005), pp.1021-1037, PMID: 16301195

Kendrick J, Targher G, Smits G & Chonchol M. 25-Hydroxyvitamin D deficiency is independently associated with cardiovascular disease in the Third National Health and Nutrition Examination Survey. (2009) *Atherosclerosis*, Vol. 205, No. 1, (July 2009), pp. 255-260, PMID: 19091317

Konety BR, Lavelle JP, Pirtskalaishvili G, Dhir R, Meyers SA, Nguyen TS, Hershberger P, Shurin MR, Johnson CS, Trump DL, Zeidel ML & Getzenberg RH.(2001) Effects of vitamin D (calcitriol) on transitional cell carcinoma of the bladder in vitro and in vivo. *The Journal of Urology*, Vol 165, No. 1, (January 2001), pp. 253-258, PMID: 11125420

Krause R, Bühring M, Hopfenmüller W, Holick MF& Sharma AM.(1998) Ultraviolet B and blood pressure. *Lancet*, Vol. 352, No. 9129, (August 1998), pp. 709-710, PMID: 9728997

LaCroix AZ, Kotchen J, Anderson G, Brzyski R, Cauley JA, Cummings SR, Gass M, Johnson KC, Ko M, Larson J, Manson JE, Stefanick ML & Wactawski-Wende J.(2009) Calcium plus vitamin D supplementation and mortality in postmenopausal women: the Women's Health Initiative calcium-vitamin D randomized controlled trial. *Journal of Gerontology: Biological Sciences*, Vol. 64, No. 5, (May 2009), pp. 559-567, PMID: 19221190

Larsen ER, Mosekilde L & Foldspang A. (2004) Vitamin D and calcium supplementation prevents osteoporotic fractures in elderly community dwelling residents: a pragmatic population-based 3-year intervention study. *Journal of Bone and Mineral Research,* Vol. 19, No. 3, (March 2004), pp. 370-378, PMID: 15040824

Levine MA (2009) Common bone and mineral disorders of childchood, in: *Manual of Endocrinology and metabolism,* N Lavin (Fourth Edition), 381-413, Lippincott Williams & Wilkins, a Wolters Kluwer business, ISBN-13: 978-0-7817-6886-3, Baltimore USA

Li YC, Kong J, Wei M, Chen ZF, Liu SQ & Cao LP. (2002) 1,25-Dihydroxyvitamin D(3) is a negative endocrine regulator of the renin-angiotensin system. *The Journal of Clinical Investigation,* Vol. 110, No. 2, (July 2002), pp. 229-38, PMID: 12122115

Lips P. (2010) Worldwide status of vitamin D nutrition. *The Journal of Steroid Biochemistry and Molecular Biology,* Vol. 121, No. 1-2, (July 2010), pp. 297-300, PMID: 20197091

Liu M, Lee MH, Cohen M, Bommakanti M & Freedman LP. (1996) Transcriptional activation of the Cdk inhibitor p21 by vitamin D3 leads to the induced differentiation of the myelomonocytic cell line U937. *Genes & Development,* Vol. 10, No. 2, (January 1996), pp. 142-153, PMID: 856674

Looker AC, Pfeiffer CM, Lacher DA, Schleicher RL, Picciano MF & Yetley EA.(2009) Serum 25-hydroxyvitamin D status of the US population: 1988-1994 compared with 2000-2004. *The American Journal of Clinical Nutrition,* Vol. 88, No. 6, (December 2009), pp. 1519-1527, PMID: 19064511

Mahon P, Harvey N, Crozier S, Inskip H, Robinson S, Arden N, Swaminathan R, Cooper C & Godfrey K; SWS Study Group. (2010) Low maternal vitamin D status and fetal bone development: cohort study. *Journal of Bone and Mineral Research,* Vol. 25, No. 1, (January 2010), pp. 14-19, PMID: 19580464

Manicourt DH & Devogelaer JP. (2008) Urban tropospheric ozone increases the prevalence of vitamin D deficiency among Belgian postmenopausal women with outdoor activities during summer. *The Journal of Clinical Endocrinology & Metabolism,* Vol. 93, No. 10, (October 2008), pp. 3893-3899, PMID: 18628525

Marques CD, Dantas AT, Fragoso TS & Duarte AL. (2010) The importance of vitamin D levels in autoimmune diseases. *Revista Brasileira de Reumatologia,* Vol 50, No. 1, (February 2010), pp.67-80, PMID: 21125142

Mattner F, Smiroldo S, Galbiati F, Muller M, Di Lucia P, Poliani PL, Martino G, Panina-Bordignon P & Adorini L. (2000) Inhibition of Th1 development and treatment of chronic-relapsing experimental allergic encephalomyelitis by a non-hypercalcemic analogue of 1,25-dihydroxyvitamin D(3). *European Journal of Immunology,* Vol. 30, No. 2, (February 2000), pp. 498-508, PMID: 10671205

Meier C, Woitge HW, Witte K, Lemmer B & Seibel MJ.(2004) Supplementation with oral vitamin D3 and calcium during winter prevents seasonal bone loss: a randomized controlled open-label prospective trial. *Journal of Bone and Mineral Research,* Vol. 19, No. 8, (August 2004), pp.1221-1230, PMID: 15231008

Melamed ML, Muntner P, Michos ED, Uribarri J, Weber C, Sharma J & Raggi P. (2008) Serum 25-hydroxyvitamin D levels and the prevalence of peripheral arterial disease: results from NHANES 2001 to 2004. *Arteriosclerosis Thrombosis and Vascular Biology,* Vol. 28, No. 6, (June 2008), pp. 1179-1185, PMID: 18417640

Merewood A, Mehta SD, Grossman X, Chen TC, Mathieu JS, Holick MF & Bauchner H. (2010) Widespread vitamin D deficiency in urban Massachusetts newborns and their mothers. *Pediatrics*, Vol. 125, No. 4, (April 2010), pp. 640-647, PMID: 20308219

Merlino LA, Curtis J, Mikuls TR, Cerhan JR, Criswell LA & Saag KG; Iowa Women's Health Study. (2004) Vitamin D intake is inversely associated with rheumatoid arthritis: results from the Iowa Women's Health Study. *Arthritis and Rheumatism*, Vol. 50, No. 1, (January 2004), pp. 72-77, PMID: 14730601

Moreno J, Krishnan AV, Peehl DM & Feldman D. (2006) Mechanisms of vitamin D-mediated growth inhibition in prostate cancer cells: inhibition of the prostaglandin pathway.*Anticancer Research*, Vol. 26, No. 4A, (July-August 2006), pp. 2525-2530, PMID: 16886660

Munger KL, Zhang SM, O'Reilly E, Hernán MA, Olek MJ, Willett WC & Ascherio A.(2004) Vitamin D intake and incidence of multiple sclerosis. *Neurology*, Vol. 62, No. 1, (January 2004), pp.60-65, PMID: 14718698

Nield LS, Mahajan P, Joshi A & Kamat D. (2006) Rickets: not a disease of the past. *American Family Physician*, Vol. 74, No. 4, (August 2006), pp. 619-626, PMID: 16939184

Ovesen L, Andersen R & Jakobsen J. (2003) Geographical differences in vitamin D status, with particular reference to European countries. *The Proceedings of Nutrition Society*, Vol. 62, No. 4, (November 2003), pp. 813-821, PMID: 15018480

Parker J, Hashmi O, Dutton D, Mavrodaris A, Stranges S, Kandala NB, Clarke A & Franco OH. (2010) Levels of vitamin D and cardiometabolic disorders: systematic review and meta-analysis, *Maturitas*, Vol. 65, No. 3, (March 2010), pp. 225-236, PMID: 20031348

Pilz S, März W, Wellnitz B, Seelhorst U, Fahrleitner-Pammer A, Dimai HP, Boehm BO & Dobnig H (2008) Association of vitamin D deficiency with heart failure and sudden cardiac death in a large cross-sectional study of patients referred for coronary angiography. *Journal of Clinical Endocrinology and Metabolism*, Vol. 93, No. 10, (October 2008), pp. 3927-39-35, PMID: 18682515

Polar MK, Gennings C, Park M, Gupta MS & Gewirtz DA. (2003)Effect of the vitamin D3 analog ILX 23-7553 on apoptosis and sensitivity to fractionated radiation in breast tumor cells and normal human fibroblasts. *Cancer Chemotherapy and Pharmacology*, Vol. 51, No. 5, (May 2003), pp. 415-421, PMID: 12690516

Pope CA 3rd, Burnett RT, Thun MJ, Calle EE, Krewski D, Ito K & Thurston GD. (2002) Lung cancer, cardiopulmonary mortality, and long-term exposure to fine particulate air pollution. *JAMA: the journal of American medical Association*, Vol. 287, No. 9, (March 2002), pp. 1132-1141, PMID: 11879110

Pope CA 3rd, Burnett RT, Thurston GD, Thun MJ, Calle EE, Krewski D & Godleski JJ.(2004) Cardiovascular mortality and long-term exposure to particulate air pollution: epidemiological evidence of general pathophysiological pathways of disease. *Circulation*, Vol. 109, No. 1, (January 2004), pp. 71-77, PMID: 1467614

Porojnicu A, Robsahm TE, Berg JP & Moan J. (2007) Season of diagnosis is a predictor of cancer survival. Sun-induced vitamin D may be involved: a possible role of sun-induced Vitamin D. *The Journal of Steroid Biochemistry and Molecular Biology*, Vol. 103, No. 3-5, (March 2007), pp. 675-678, PMID: 17229569

Prentice A, Jarjou LM, Goldberg GR, Bennett J, Cole TJ & Schoenmakers I. (2009) Maternal plasma 25-hydroxyvitamin D concentration and birthweight, growth and bone mineral accretion of Gambian infants. *Acta Paediatrica*, Vol. 98, No. 8, (August 2009), pp. 1360-1362, PMID: 1959447

Reddy Vanga S, Good M, Howard PA & Vacek JL. (2010) Role of vitamin D in cardiovascular health. *American Journal of Cardiology*, Vol. 106, No. 6, (September 2010), pp. 798-805, PMID: 2081612

Ritz SA. (2010) Air pollution as a potential contributor to the 'epidemic' of autoimmune disease. Medical Hypotheses, Vol. 74, No. 1, (January 2010), pp.110-117, PMID: 19665849

Robsahm TE, Tretli S, Dahlback A & Moan J. (2004) Vitamin D3 from sunlight may improve the prognosis of breast-, colon- and prostate cancer (Norway). Cancer Causes & Control, Vol. 15, No. 2 (March 2004), pp. 149-158. PMID: 15017127

Rockell JE, Skeaff CM, Williams SM & Green TJ. (2006) Serum 25-hydroxyvitamin D concentrations of New Zealanders aged 15 years and older. *Osteoporosis International*, Vol. 17, No. 9, (July 2006), pp. 1382-1389, PMID:16832715

Rosen CJ. (2011) Clinical practice. Vitamin D insufficiency. *The New England Journal of Medicine*, Vol. 364, No. 3, (January 2011), pp.248-254, PMID: 21247315

Sachan A, Gupta R, Das V, Agarwal A, Awasthi PK & Bhatia V. (2005) High prevalence of vitamin D deficiency among pregnant women and their newborns in northern India. *The American Journal of Clinical Nutrition*, Vol. 81, No. 5, (May 2005), pp. 1060-1064, PMID:15883429

Scragg R, Sowers M & Bell C.(2007) Serum 25-hydroxyvitamin D, ethnicity, and blood pressure in the Third National Health and Nutrition Examination Survey. *American Journal of Hypertension*, Vol. 20, No. 7, (July 2007), pp. 713-719, PMID: 17586404

Spina CS, Tangpricha V, Uskokovic M, Adorinic L, Maehr H & Holick MF. (2006) Vitamin D and cancer. *Anticancer Research*, Vol. 26, No. 4A, (July-August 2006), pp. 2515-2524, PMID: 16886659

Tangpricha V, Spina C, Yao M, Chen TC, Wolfe MM & Holick MF. (2005) Vitamin D deficiency enhances the growth of MC-26 colon cancer xenografts in Balb/c mice. *The Journal of Nutrition*, Vol. 135, No. 10, (October 2005), pp. 2350-2354, PMID: 1617719

Tarcin O, Yavuz DG, Ozben B, Telli A, Ogunc AV, Yuksel M, Toprak A, Yazici D, Sancak S, Deyneli O & Akalin S. (2009) Effect of vitamin D deficiency and replacement on endothelial function in asymptomatic subjects. *The Journal of Clinical Endocrinology and Metabolism*, Vol. 94, No. 10, (October 2009), pp. 4023-4030, PMID: 19584181

Toner CD, Davis CD & Milner JA. (2010) The vitamin D and cancer conundrum: aiming at a moving target. *Journal of American Dietetic Association*, Vol. 110, No. 10, (October 2010), pp. 1492-1500. PMID: 20869488

Walters MR. (1992) Newly identified actions of the vitamin D endocrine system. *Endocrine Reviews*, Vol. 13, No. 4, (November 1992), pp. 719-764. PMID: 1333949

Webb AR. (2006) Who, what, where and when-influences on cutaneous vitamin D synthesis. *Progress in Biophysics and Molecular Biology*, Vol. 92, No. 1, (September 2006), pp. 17-25. PMID: 16766240

Welsh J. (2004) Vitamin D and breast cancer: insights from animal models. *The American Journal of Clinical Nutrition*, Vol. 80, No. 6 (Supplement), pp. 1721S-1724S, PMID: 15585794

WHO Scientific Group on the Prevention and Management of Osteoporosis.(2003) Prevention and management of osteoporosis: report of WHO scientific group. *World Health Organ Technical Report Series*, Vol. 921, (2003), pp. 1-164, PMID: 15293701

van der Wielen RP, Löwik MR, van den Berg H, de Groot LC, Haller J, Moreiras O & van Staveren WA. (1995) Serum vitamin D concentrations among elderly people in Europe. *Lancet.*, Vol. 346, No. 8969, (July 1995), pp. 207-210, PMID: 7616799

Willheim M, Thien R, Schrattbauer K, Bajna E, Holub M, Gruber R, Baier K, Pietschmann P, Reinisch W, Scheiner O & Peterlik M. (1999) Regulatory effects of 1alpha,25-dihydroxyvitamin D3 on the cytokine production of human peripheral blood lymphocytes. *The Journal of Clinical Endocrinology and Metabolism*, Vol. 84, No. 10, (October 1999), pp. 3739-3744, PMID: 10523023

Witham MD, Nadir MA &Struthers AD. (2009) Effect of vitamin D on blood pressure: a systematic review and meta-analysis.Journal of Hypertension, Vo. 27, No. 10, (October 2009), pp. 1948-1954, PMID: 19587609

Wharton B & Bishop N. (2003) Rickets. *Lancet*, Vol. 362, No. 9393, (October 2003), pp.1389-1400, PMID: 14585642

Xu HM, Tepper CG, Jones JB, Fernandez CE & Studzinski GP.(1993) 1,25-Dihydroxyvitamin D3 protects HL60 cells against apoptosis but down-regulates the expression of the bcl-2 gene. *Experimental Cell Research*, Vol. 209, No. 2, (December 1993), pp. 367-374, PMID: 8262155

Zhang R & Naughton DP. (2010) Vitamin D in health and disease: current perspectives. *Nutrition Journal*, Vol. 9, No. 65, (December 2010), pp. 1-13, PMID: 21143872

Zhuang SH & Burnstein KL (1998) Antiproliferative effect of 1alpha,25-dihydroxyvitamin D3 in human prostate cancer cell line LNCaP involves reduction of cyclin-dependent kinase 2 activity and persistent G1 accumulation. *Endocrinology*, Vol. 139, No. 3, (March 1998), pp. 1197-1207, PMID: 949205

Zittermann A, Schleithoff SS, Götting C, Dronow O, Fuchs U, Kuhn J, Kleesiek K, Tenderich G, & Koerfer R (2008) .Poor outcome in end-stage heart failure patients with low circulating calcitriol levels. *European Journal of Heart Failure*, Vol. 10, No. 3, (March 2008), pp. 321-327, PMID: 18304873

3

Air Pollution and Its Effects in the Respiratory System

Fortoul, T.I. et al.*
Departamento de Biologia Celular y Tisular, Facultad de Medicina
Universidad Nacional Autonoma de Mexico (UNAM)
Mexico

1. Introduction

1.1 Respiratory system structure and normal function

The respiratory system is the gateway for 15,000 liters of air that enter through the nostrils, and after its appropriate conditioning, arrives to the delicate net structured by the lung parenchyma. At this site gas diffusion takes place, and oxygen diffuses through the Pneumocyte I cytoplasm to reach the erythrocyte's hemoglobin, passing by the endothelial cell.

To make this possible, a well-structured conducting system leads the air (trachea, bronchi, bronchioles, alveoli) from the nostrils to the alveoli. A specialized epithelium blankets the tubes, with some local modifications (Fortoul et al., 2010). This epithelium is constituted by a variety of cells with different functions (Figure 1).

In the submucose under the bronchi, mucous and serous glands are located, and liberate its secretion to the surface of the epithelium (Figure 2). These glands, as well as the goblet cells, produce mucus that in normal situations is Alcian blue/ PAS+ (Davis & Dicker, 2008). The biochemistry of the mucus, produced by the goblet or the submucosal glands is important because they determinate their viscosity, pH, and charge, as well as its stain affinity (Figure 3) (Rose & Voynow, 2006).

With the help of intercellular junctions, this tubes system separates spaces between the epithelium, which is known as interstitial space and it houses connective tissue, fibroblast, lymphocytes, macrophages, and other cells that may migrate from the capillaries.

Along these tubes the air that enters through the nostrils becomes moistened, filtered and temperated, to arrive into the alveoli for gas diffusion. A change in the composition and function of the epithelial components will be associated with a variety of diseases which are associated to air pollutants exposure (Figure 4) (Mussali-Galante & Fortoul, 2008).

* Rojas-Lemus, M.[1], Rodriguez-Lara V.[1], Cano-Gutierrez, G.[1], Gonzalez-Villalva, A.[1], Ustarroz-Cano, M.[1], Garcia-Pelaez, I.[1], Lopez-Valdez, N.[1], Falcon-Rodriguez C.I.[1], Silva-Martinez, J.[1], Gonzalez-Rendon, E.S.[1], Montaño, L.F.[1], Cano-Gutierrez, B.[1], Bizarro-Nevares P.[1], Colin Barenque L.[2]
[1]*Departamento de Biologia Celular y Tisular, Facultad de Medicina, Universidad Nacional Autonoma de Mexico (UNAM), Mexico*
[2]*Departamento de Neurociencias, Iztacala, UNAM, Mexico*

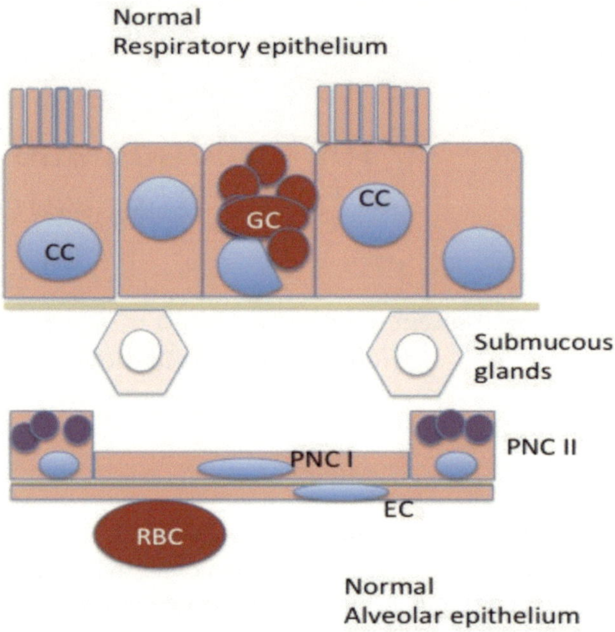

Fig. 1. Respiratory system epithelial cells. Ciliated Cell (CC), Goblet Cell (GC), Pneumocyte I (PNCI), Pneumocyte II (PNCII), Red Blood Cell (RBC), Endothelial Cell (EC).

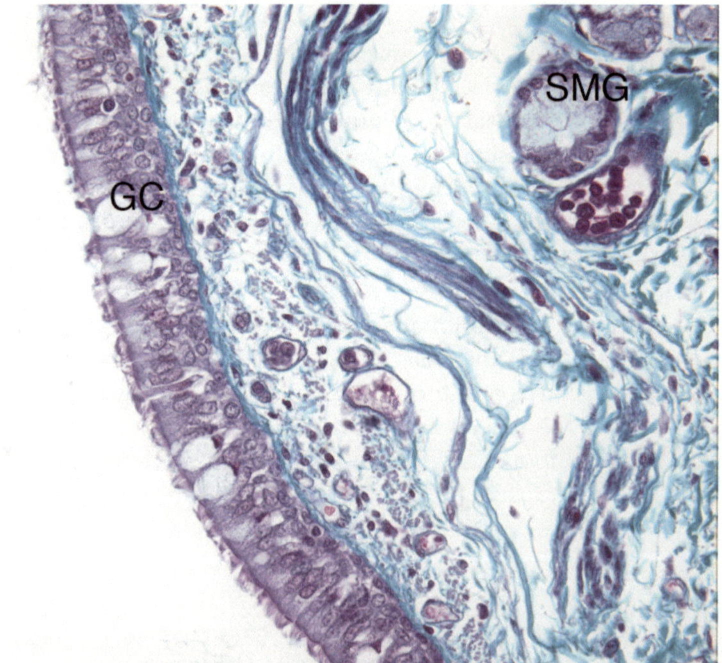

Fig. 2. Bronchial epithelium with Ciliated and Goblet Cells (GC). Submucose glands (SMC) with serous demilunes are also observed.

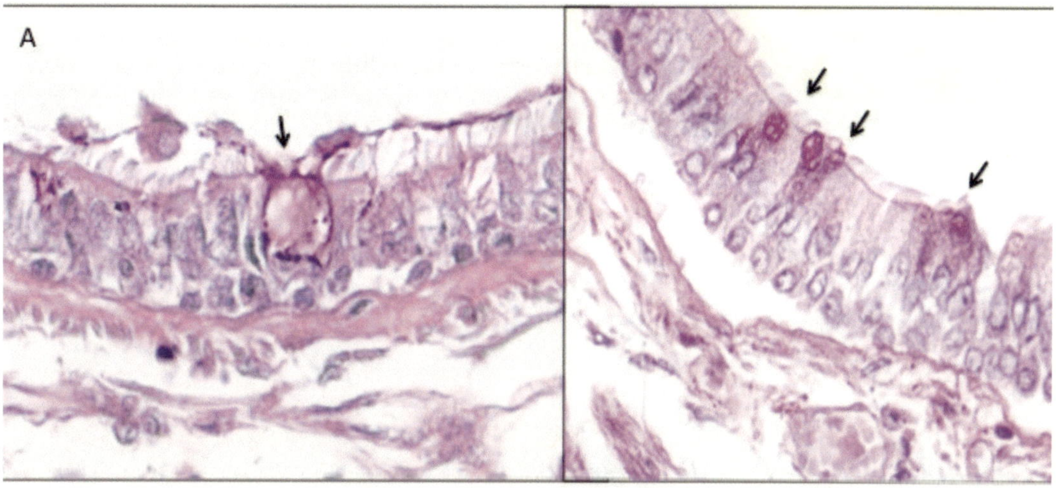

Fig. 3. Respiratory epithelium stained with PAS Schiff. (A) In normal situations, few PAS+ cells are observed (arrow). (B) In disease an increase in these cells is observed (arrows).

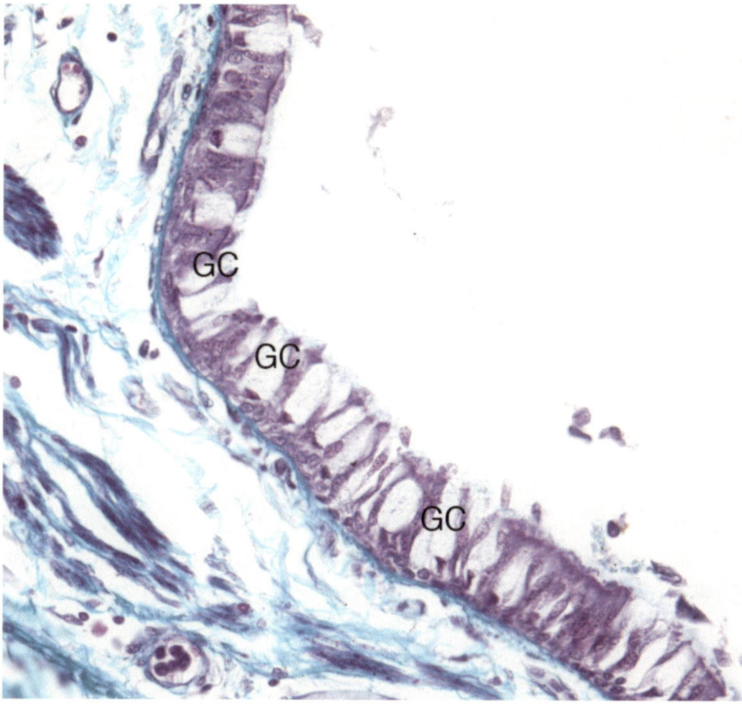

Fig. 4. The increase in the number of Goblet cells is a response of the respiratory epithelium to atmospheric aggressions.

1.2 Diseases associated with air pollutants (inflammation, fibrosis, COPD, asthma, cancer, immunologic modifications)

Air pollutants, described latter in this chapter, induce different reactions and diseases. Inflammation as a consequence of the exposure to irritants such as Ozone, NOx, or particulate matter has been described (Yang & Omaye, 2009). Also, fibrosis as a consequence of the exposure to gases is explored with more detail in a next section (Figure 5).

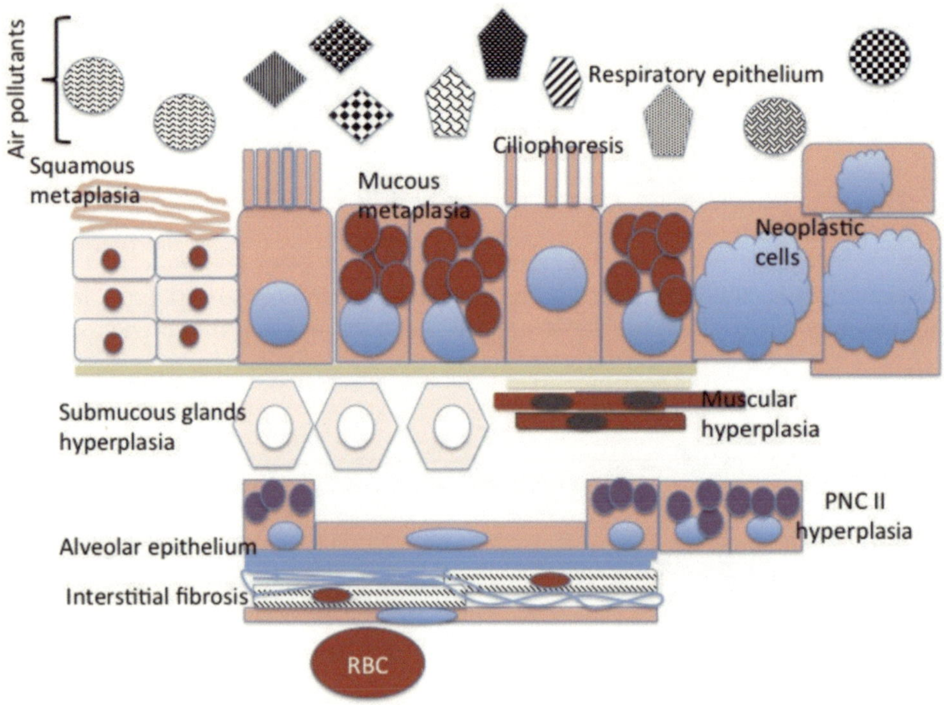

Fig. 5. Schematic representation of the changes that may be found in the respiratory epithelium, after its exposure to air pollutants.

Chronic Obstructive Pulmonary Diseases (COPD) included asthma, increases its frequency in cities with high atmospheric pollution (Olivieri & Scoditti, 2005). The epithelium in these cases may increase the release of IL-6, TNFα, PGE_2, PDGF, TGFβ, VEGF, and a variety of chemokines, and other mediators capable of inducing proliferation of fibroblasts and mucus production (Holgate, 2008).
In addition other components of air pollutants, such as metals, Volatile Organic Compounds (VOCs) are carcinogenic; other pathologies are associated with the exposure (Yang & Omaye, 2009).
In the next sections the different responses of the lung to air pollutants will be explored.

2. Air pollutants

Air Pollutants are natural constituents of the air. Animals produce carbon dioxide as the end result of respiration, volcanic action produces sulfur oxides, and wind movement ensures the presence of suspended particulates. Pollutants are part of our everyday life, and it is difficult to remove them from the respirable air. However, man has caused a severe imbalance, in the natural mechanisms for atmosphere clearance, increasing the discharges of pollutants in the atmosphere, resulting in severe effects on human health (Atash, 2007).

2.1 Classification of air pollutants

Atmospheric pollutants have been classified according to their source; chemical composition, size and release form into indoor or outdoor environments (Bernstein et al., 2004):

A. Primary - secondary pollutants

1. Primary: pollutants emitted directly into the atmosphere
2. Secondary: pollutants that form in the air as a result of chemical reactions with other pollutants and gases.

B. Indoor - outdoor pollutants

1. Indoor pollutants
   1.1 Sources: cooking and combustion, particle resuspension, building materials, air condition, consumer products, smoking, heating, biologic agents
   1.2 Products: combustion products, CO, CO_2, Specific volatile organic compounds, microbial agents and organic dusts, radon, manmade vitreous fibers
2. Outdoor pollutants
   2.1 Sources: industrial, commercial, mobile, urban, regional, agricultural, natural.
   2.2 Products: SO_2, ozone, NO_x, CO, PM, Specific volatile organic compounds

C. Gaseous - particle pollutants

1. Gaseous: SO_2, NO_x, ozone, CO, Specific volatile organic compounds
2. Particle: coarse PM (2.5 - 10 μm; regulatory standard = PM_{10}), fine PM (0.1 - 2.5 μm regulatory standard = $PM_{2.5}$); ultrafine PM (<0.1 μm; no regulated).

2.2 Gases

Problems in air pollution were associated with high concentrations of sulfur dioxide (SO_2) in 20th Century. Through controlled exposure, human studies to SO_2 (0.25 ppm) for only 5 minutes a rapid bronchoconstriction, in both healthy and asthmatic subjects was described. In patients exposed to inhalation of SO_2 a relationship with TNF-α promoter polymorphism was identified, which is know to be associated also with asthma (Winterton et al., 2001; Bernstein et al., 2004).

Ozone is formed in the troposphere through a complex series of reactions involving the action of sunlight on nitrogen dioxide and hydrocarbons. The global concentration of O_3 has increased due to an increase in nitric oxide emissions associated with the switch to fossil fuels during the industrial period. Nitric oxide is rapidly transformed into nitrogen dioxide by atmospheric oxidants such as ozone (Finlayson & Pitts, 1997). Exposure to ozone causes a decreased in forced vital capacity and FEV_1 associated with chest discomfort on inspiration and increased nonspecific airway hyper hyperresponsiveness (Bernstein et al., 2004).

NO_2 is emitted directly into the atmosphere by combustion processes; however, the main source is the oxidation of NO by reactive species. Once NO is converted to NO_2, a variety of reactions can generate nitrate radical (NO_3), dinitrogen pentoxide (N_2O_5) (Finlayson & Pitts, 1997). NO_2 exposure (2 - 6 ppm) induces an inflammatory response in the airways characterized by neutrophil influx and reduced lymphocyte subpopulations, also it might play a more prominent role as sensitizing agent to inhaled allergens (Strand et al., 1997).

Carbon monoxide (CO) is produced by the incomplete combustion of carbon or carbon compounds. This gas can be bound to hemoglobin, forming carboxy hemoglobin and immobilizing hemoglobin function. Another compound of carbon, carbon dioxide is needed in plants' life cycle, in the reaction of photosynthesis. But its rise in the atmosphere in the presence of other gases such as methane and chlorofluorocarbons, is the cause of the greenhouse effect, (Raub et al., 2000).

2.3 Suspended particles

Particle air pollution is a mixture of solid, liquid or solid and liquid particles suspended in the air. The size of suspended particles varies, from a few nm to tens of μm. The PM 10 (thoracic) are particles smaller than 10 μm in diameter that can penetrate into lower respiratory tract; PM 2.5 (respirable) particles smaller than 2.5 μm that can penetrate into gas-exchange region of the lung, and ultrafine particles smaller than 100 nm which have a limited contribution to particle mass, but which in terms of numbers are more abundant than the other sizes, and offer a very large surface area, with increased degrees of lung penetration.

Major natural sources of particles include organic material terrestrial dust caused by winds, sea spray, biogenic emissions, volcanic eruptions and wild fires; the contamination through anthropogenic or technogenic is produced by combustion, industrial waste, nuclear energy, anthropogenic fire, and burning of household waste (Finlayson-Pitts & Pits, 2000). Metal particles (mercury, cadmium, nickel and lead) are part of PM2.5.

3. Lung development modifications by air pollutants

The respiratory system begins its development at the 4th week of pregnancy, in the fetal larynx as a respiratory *primordium*. This structure is covered by endoderm that will differentiate into the respiratory epithelium, and the respiratory glands.

Other structures will be developed from the splanchnic mesoderm that surrounds the endoderm.

The regulation of the respiratory system morphogenesis is coordinated by different structures, in different time periods. It has been reported the expression of molecules such as Fibroblast Growth Factor (FGF), sonic hedgehog, Bone Morphogenic Protein (BPM), retinoic acid and Wnt signaling pathways, as well as various transcription factors as part of this regulatory network (Cardoso & Lü, 2006).

At the end of the 7-week, the lungs are already developed, but its maturation is extended during the pre and postnatal periods. 80% of the alveolar tissue ends its development until the end of adolescence. This large maturation time exposes the lung to suffer damage and modify its development; this also increases the risk for developing pulmonary diseases early in adulthood such as restriction or a decrease in lung function (Wang & Pinkerton, 2008; Rojas-Martinez, et al., 2007).

Prenatal exposure to Environmental Tobacco Smoke (ETS) has been associated with an increased risk of asthma, and during the postnatal period, the exposure increases asthma exacerbations, as well as the risk for respiratory infections (Wang & Pinkerton, 2008). Studies in mice indicate that prenatal exposure to ETS increases allergic responses in postnatal period (Penn et al., 2007; Gern, 2010).

The exposure to outdoor pollutants such as Particulate Matter (PM), carbon monoxide (CO), sulphur (SO_2), nitrogen (NOx), and ozone, decreases lung function during childhood

(Mortimer et al., 2008). Also O_3 exposures increase hyper-reactivity along with an increase in TNF-**a**, IL-1**b**, KC, IL-6, and MCP-1 with non-visible structural lung modifications (Auten et al., 2009).

4. Xenobiotic metabolism

4.1 Lung metabolic active cells

The lung is one of the main sites for xenobiotics metabolism and in some cases, biotransformation (Castell et al., 2005). Biotransformation is the process by which cells modify xenobiotics, with the ultimate goal of facilitating the elimination of lipophilic substances. These reactions are classified in three phases. Phase I: Enzymes encoded by cytochrome P450 (hemoproteins actively involved in the biotransformation of xenobiotics). Phase II: Reactions tend to render more water-soluble products and less active metabolites, and finally, Phase III: Elimination (Choudhary et al., 2005).

More than 40 different types of cells have been described in the lung (Pavek & Dvorak, 2008), and some are metabolically active. Immunohistochemical analysis has evidenced the presence of cytochrome P450 (Cytochrome Proteins) (CYP) in lung cells such as: macrophages (Pavek & Dvorak, 2008), endothelium, alveolar cells types I and II, ciliated cells (Castell et al., 2005) and Non-Ciliated Bronchiolar Cell or Clara cell. Clara cells are the leading cells for xenobiotic metabolism in the lung because its profuse cytochrome P450 mono oxygenase activity (Katavolos et al., 2009).

4.2 CYPs and lung metabolism (CYPs and cell metabolism)

CYP families and subfamilies are responsible of the oxidative metabolism of the majority of xenobiotics such as: drugs, environmental pollutants and carcinogens. The main families are: CYP1, CYP2 and CYP3 and comprise about half of the total CYPs. The subfamilies are classifieds according to the degree of nucleonic and amino acid sequence homology (Castell et al., 2005). All CYPs are localized in the cell's smooth endoplasmic reticulum. CYP 450 enzymes act as mono oxygenases, and use one atom of molecular oxygen to oxidase xenobiotics, requiring the aid of NADPH-cytochrome P450 reductase, to provide the electrons required for the reduction of the second oxygen atom to H_2O_2 (Ioannides, 2008). The lung contains a variety of subfamilies such as: CYP1A1, CYP1B1, CYP2A6, CYP2B6, CYP2E1, CYP3A5 (Castell et al., 2005), CYP2B1, CYP3A1 (Pons et al., 2000), CYP2F1, CYP2F2 (Carlson, 2008) and CYP2S1 (Deb & Bandiera, 2010).

4.3 Air pollution modification of CYPs

Air pollution increases or decreases the CYPs amount in the lung. Exposure to environmental factors such as: dioxins as 2,3,7,8-Tetrachlorodibenzo-p-dioxin (TCDD), polycyclic aromatic hydrocarbons (PAH) as benzo [a] pyrene (BaP) and tobacco smoke (Chang et al., 2006), enhances CYP1A1 activity and has been reported to be a prognostic factor for lung cancer development. (Ioannides, 2008). CYP1B1 and CYP2S1 also increase its activity after the exposure to PAH and BaP (Deb & Bandiera, 2010), also CYP1A1 and CYP1B1 are localized in the Clara cell (Chang et al., 2006). PAH and BaP also bind to arylhidrocarbonyl receptor (AhR) that translocates to the nucleus and acts as a transcription factor that binds to a specific DNA recognition sequence, termed the xenobiotic responsive element (XRE) (Anwar-Mohamed et al., 2009) and all CYPs can generate ROS during NADPH-dependent CYP catalysis (Ioannides, 2008) (Figure 6).

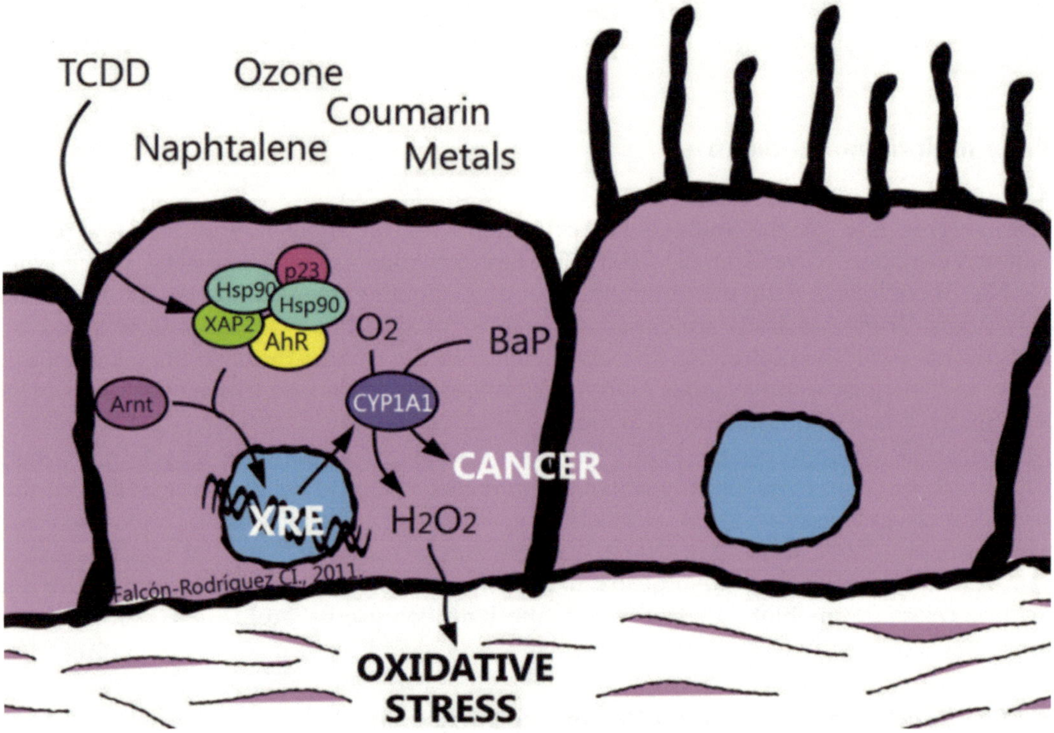

Fig. 6. Molecular mechanism of CYP1A1 activation by AhR signal transduction pathway in Clara cell (Bronchiolar epithelium. A ligand enters to the cell (TCDD) and binds to the cytosolic complex of AhR, chaperones hsp90, co-chaperone p23. Ligand-AhR complex translocates into the nucleus. AhR-ARNT heterodimer then binds to xenobiotic response element (XRE).

Other hydrocarbons as toluene, occurs naturally in crude oil and in the tolu tree. It is also produced in gasoline process, fuels from crude oil, making coke from coal and it is used in paints and paint thinners. Toluene exposure enhances CYP2B1 activity in rat lung (Pons et al., 2000). Naphthalene, derived from petroleum, coal, and tobacco smoke (Morris & Buckpitt, 2009), increases CYP2F2 activity. In mouse lung CYP2F2 the metabolism rate was 107 nmol/min/nmol P450, whereas for human lung CYP2F1 was 0.045 nmol/min/nmol P450 (Carlson, 2008). Other products such as coumarin, a natural product used widely as a fragrance ingredient, and also been used clinically at high dosages in humans for the treatment of high-protein lymphedemas, and as an antineoplastic agent in the treatment of renal cell carcinoma and malignant melanoma. The Coumarin-Epoxidation increases the expression of CYP2F2, and shows a specific metabolism in mouse Clara cell (Born et al., 2002). Furthermore, Clara cell does not contain CYP2E1, but, when rats are exposed to ozone, the CYP2E1 is induced in the bronchiolar epithelium (Ioannides, 2008). In general Particulate Matter induces gene expression of CYP1A1 and CYP2E1 (Abbas et al., 2009). Finally, metals are components of Total Suspended Particles (PST). Scarce information is available on the metal effects on lung CYPs, although, demonstrated recently that V^{5+} was able to decrease the TCDD-mediated induction of CYP1A1 mRNA, protein and catalytic activity (Anwar-Mohamed et al., 2009).

5. Air pollutant effects on respiratory system

There are changes associated with atmospheric pollutants that are evidenced with different patterns and diseases. Changes in the local lymphoid tissue, carcinogenic and genotoxic changes, fibrosis, COPD, are some of the outcomes observed after air pollution. Inflammation is core event in all the changes observed after the contact of the epithelial cells with the air pollutants, so it will be the first described in this section.

5.1 Inflammation and air pollution

The first suggestion of the consequences of air pollution date back to the serious events that occurred in Belgium in 1948 in the so called "killer fog" incident, and in the so called "London fog" that happened in the first week of December 1952 and provoked 4000 more deaths than those expected. After those episodes air pollution control policies were introduced since it became clear that air pollution was associated with an increase in morbidity and mortality in individuals with cardiovascular disease or with chronic obstructive pulmonary disease (Mills et al., 2009). The root of this increased mortality lay in the lung chronic inflammation that affects the pulmonary vascular endothelium, the thrombotic potential and the fibrinolitic balance in exposed individuals. These processes favor atherosclerotic plaque rupture, thrombosis and translocation of particulate matter to the blood, through the pulmonary capillaries; the latter affects directly the body vascular endothelium leading to the loss of endothelium integrity thus initiating a pulmonary and a systemic inflammatory reaction (Mills et al., 2006). A fraction of the translocated particulate matters accumulates in the liver, the spleen, the thymus and others (Nemmar et al., 2002).

Ambient particulate matter (PM) is a mixture of inhalable particles that are considered as serious contaminants. These particles come from the combustion of biomass fuels used for cooking and heating homes, emissions from internal combustion motors and industrial machinery, and forest fires (Torres-Duque et al., 2008). This matter has been grouped in coarse (2.5-10μm) and fine (2.5 μm or less) depending on their diameter.. Both types induce serious health consequences in exposed individuals. Both types of particles penetrate into the lung, however the coarse particles are more dangerous because of their mass (Hetland et al., 2005). Obviously there are many inherent conditions in the exposed individual that increase the risk to develop cardiovascular or pulmonary disease: alfa-1 antitrypsin deficiency, family history of chronic obstructive pulmonary disease or atherosclerosis, personal history of frequent upper respiratory tract infections, hypersensitivity to inhaled irritants, tobacco, asthma, and being female (Ekici et al., 2005). Also, we must consider the enhanced air pollution in overcrowded cities with serious traffic problems. These cities usually have a higher concentration of PMs in the air. An increase of 7 μg/m3 in the PM2.5-10 concentration is associated with a 5% decrease in FEV1 and an odds ratio of 1.33 for chronic obstructive pulmonary disease in women compared with men (Schikowski et al., 2005).

Both types of particles recruit and activate neutrophils, but PM2.5-10 induces a higher proinflammatory activity (Wegesser & Last, 2009). There are important differences in the relative concentration and type of components (metals, organic compounds, ultrafine particles adhered to larger particles) (Donaldson et al., 2005) between coarse and fine particles that depend on the season and geographic site where the sample is collected (Seagrave et al., 2006).

Other constituents of ambient particulate matter are biological materials, especially b-glucans, fungi spores and endotoxins that derive from gram-negative bacteria (Schwarze et al., 2006). It has been shown that PM constituent responsible for the pro-inflammatory activity induced by PM2.5-10 are endotoxins and particularly a soluble fraction known as lipopolysaccharides (LPS) (Schins et al., 2004). The concentration of endotoxin in PM depends on the site and season, as we have already mentioned, but environmental humidity is associated with endotoxin concentration in particulate matter (Spaan et al., 2008). One example of the impact of biological components in PM is the heat shock protein HSP60 derived from *Chlamydia pneumoniae,* detected in PM2.5-10 that promotes lung inflammation and pulmonary dendritic cells activation through the innate immune response receptor TLR4 and the MyD88 pathway. The inflammation induced by HSP is secondary to an increase in the number of immune-related cells in the broncho alveolar lavage (BAL), enhanced recruitment of neutrophils, increased synthesis of IL-6 and over-expression of CD80 and CD86 in BAL dendritic cells (Bulut et al, 2009). HSP60 also activate pulmonary macrophages and endothelial cells through TLR4 in a MyD88-dependent pathway (Bulut, 2002). Interestingly, and independently from its endotoxin content, PM2.5 per se activate macrophages through TLR-2 y TLR-4 (Shoenfelt et al., 2009). Similarly, the respiratory epithelium, the first point of contact for inhaled foreign organism, also express TLR4 and secrete, upon activation, pro-inflammatory cytokines and chemokines that recruit neutrophils and T lymphocytes to the infection site (Parker & Prince, 2011).

Pulmonary inflammation induced by exposition to or inhalation of PM is closely related to particle processing by alveolar macrophages. Once these macrophages become activated a cascade of pro-inflammatory cytokines is initiated leading to endothelial damage. A recent study shows that in the presence of PM2.5-10 the amount of alveolar macrophages is duplicated and the amount of activated macrophages triplicates leading to persistent lung and systemic inflammation that were both associated with vascular endothelial dysfunction (Tamagawa et al., 2008). The increase in IL-6 serum concentration in PM2.5-10 exposed individuals was important during the first two weeks of exposure and was directly related to the amount of activated macrophages; afterwards, the concentration of IL-6 was similar in exposed individuals and controls. Alveolar macrophages activated with PMs secrete TNF-a, granulocyte-monocyte colony stimulating factor, and IL-1b (van Eeden et al., 2001). It is highly probable that these cytokines act not only locally but systemically thus generating a more organized inflammatory response that include the bone marrow. The acute exposure to PMs induces a rapid bone marrow response liberating leukocytes and platelets into the systemic circulation. As far as inflammatory cytokines in the pulmonary inflammation process, it has been shown that IL-6 inhibits directly the expression of eNOS thus diminishing nitric oxide production by the endothelium (Saura et al., 2006). Chemokine secretion is also altered in individuals exposed to particulate matter. Interleukin 8, a chemokine responsible for the recruitment and activation of pulmonary neutrophils in inflammation sites, binds to PMs (Seagrave, 2008). Activated neutrophils and macrophages secrete IL-8 and IL-1b.The excessive recruitment of neutrophils is clearly detected in BAL samples of individuals exposed to particulate matter.

A recent study analyzed the effect of daily changes in particulate matter air pollution upon the inflammatory cells response. The results showed that as the concentration of PM2.5-10 increases in the air the serum concentration of fibrinogen and the expression of E-selectin in exposed individuals increases whereas the concentration of prothrombin and von Willebrand antigen diminishes. The serum concentration of C reactive protein, coagulation factor VII, amyloid A and the soluble ICAM-1 fraction was not modified (Hildebrandt et al., 2009). Fibrinogen is an acute phase protein and a coagulation factor synthetized by the liver in response to elevated IL-6 serum concentration (Gabay & Kushner, 1999). E-selectin reflects the activation of the vascular endothelium and it is well known that the increased expression of this adhesion molecule is associated with the serum concentration of inflammatory cytokines such as TNF-a, IFN-y and IL-6 (Rice & Bevilacqua, 1989) as well as the recruitment of leukocytes and T lymphocytes in the inflammation zone. The presence of neutrophils in an IL-6 enriched environment enhances the expression of E-selectin in endothelial cells but also its apoptotic death thus perpetuating the inflammatory damage (Barnes et al., 2011). It has also been shown that prolonged exposure to PM2.5 is associated with an important plasmatic increase in endothelin-1 concentration and in an increased pulmonary artery pressure (Calderon-Garcidueñas et al., 2007). The contribution of all these processes in the lung keeps the inflammatory process alive.

Heat inactivation of PM2.5-10 diminishes the expression of CD14, CD11b/CR3 y HLA-DR and the phagocytic activity of alveolar macrophages (Alexis et al., 2006) but its influence in neutrophil recruitment into the airways is still controversial. Exposures to viruses, increase the serum concentration of INF-g, previous to the inhalation of particulate matter induce a stronger pulmonary inflammation response that include oxidative damage to the lung. The result of these changes is a loss in the antibacterial function of neutrophils and alveolar macrophages and consequently, an increase in the local content of bacterial endotoxins (Sigaud et al., 2007) that perpetuates the pulmonary inflammation. Sigaud and coworkers (2007) also demonstrated that macrophages exposed to PMs have an enhanced expression of multiple inflammation-related genes: MCP-5, IL-9, IL-17B, IL-1b, MIP-1b, MIP-3b, IL-8R, C10, CCR-1, CCR-2 and MDC.

In summary, exposure to particulate matter induces excessive production of pro-inflammatory cytokines and chemokines by alveolar macrophages and lung dendritic cells, both of which are activated through TLR2 and TLR4 and the MyD88 signaling pathway. The excessive amount of these inflammatory molecules directly affect the pulmonary and systemic vascular endothelium by diminishing its capacity to regulate properly its vascular tone and permeability, triggering abnormal coagulation and fibrinolysis mechanisms, and increasing the adhesion of inflammatory cells to the vascular endothelium.

5.2 Bronchial associated lymphoid tissue (BALT) modifications

Bronchus-associated lymphoid tissue (BALT) is a constitutive mucosal lymphoid tissue adjacent to major airways. BALT is composed by B cells surrounded by a parafollicular region of T cells, dendritic (DCs) and macrophages. As the result of air pollution BALT can acquire antigens, allergens or contaminants from the airways, then complex interactions occur increasing its efficiency. For example, BALT can initiate local immune responses and the amount of BALT increases (Randall, 2010).

The presence of BALT in adult mammals depends on species, antigen stimulation and age. BALT is found in normal lungs of most healthy adult rabbits, rats, guinea pigs and old adult mice. In contrast, the presence and frequency of BALT in normal lungs of healthy adult humans is controversial (Kawamata et al., 2009).

In humans BALT is neither found at birth nor in healthy adults but transiently arises during childhood and adolescence. In both humans and mice, air pollution can induce BALT, data derives from splenectomized lymphotoxin α-deficient mice, which lack all secondary lymphoid organs but do develop BALT. This suggests that BALT can serve as induction sites for adaptive immune responses to contaminants. However, mechanisms that control the development and maintenance of BALT are largely unknown (Halle et al., 2009).

5.3 Carcinogenic and mutagenic effects

Lung cancer is the leading cause of cancer death worldwide. Smoking is the major risk factor for lung cancer. Although some subjects who have never smoked get lung cancer, smoking causes 9 out of 10 cases of this pathology. So there are other factors that promote the carcinogenic process. It has been reported that air pollution exposure may cause lung cancer. In a six US cities study, Dockery and coworkers (1993) found that the greatest effects were for lung cancer and cardiopulmonary disease, between the least and most polluted cities. Pope et al. (1995) found increased long-term effects on cardiopulmonary mortality and lung cancer in a 17-year follow-up (Naess et al., 2007). Also, Cao and coworkers (2011) analysis provides the first prospective evidence in China that air pollution (e.g., SO2) may contribute to the increased risk for lung cancer mortality. In Europe, the proportion of lung cancers attributable to urban air pollution is estimated to be 11% (Molina et al., 2008). Likewise, it have been reported that the type of lung cancer is related to air pollution, for example, in a Spain study, individuals living near industries displayed an excess risk of lung cancer (OR=1.49; 95%CI=0.93-2.39), which attained statistical significance for small cell carcinomas (OR=2.23; 95%CI=1.01-4.92), residents in urban areas showed a statistically significant increased risk for adenocarcinoma (OR=1.92; 95%CI=1.09-3.38). In Aviles a health area, no differences in risk was found (López-Cima et al., 2011).

Urban air, particularly in densely populated urban environments, contains inorganic particulates (arsenic, asbestos, chromium, cadmium, lead and nickel), radionuclides (210Pb, 212Pb and 222Rn), gaseous and particulate organic species (benzene, benzo[α]pyrene, 1,3-butadiene and benzene-soluble organics), oxidants such as ozone and sulfur and nitrogen oxides in particle form. These substances are present as components of complex mixtures proceeding basically from industries emissions, combustion of fossil fuels for power generation or transportation, and all are related with carcinogenesis process (Naess et al., 2007).

Particulate matter (PM), especially fine particles of less than 2.5 μm in diameter (PM2.5) is related with lung cancer (Perez-Padilla et al., 20010). Pope and coworkers (2002) reported that each 10-μg/m3 elevation in long-term average PM2.5 ambient concentrations was associated with approximately a 8% increased risk of lung cancer mortality, although the magnitude of the effect somewhat depended on the time frame of pollution monitoring. Particulate matter contains a lot of compounds that are considered human carcinogens like heavy metals such as cadmium, cobalt, chromium and nickel. Several epidemiological studies have clearly demonstrated that exposure to metals has toxic and carcinogenic affects in animals and humans; some of them have been demonstrated to be lung carcinogenesis promoters (Salnikow & Zhitkovich, 2008). Human cadmium exposure is associated with lung cancer. Also in animal models, cadmium induces lung carcinomas after inhalation (Beyersmann & Hartwig, 2008). At the cellular level, this metal affects cell proliferation, differentiation, apoptosis, and other cellular activities and may cause numerous molecular lesions that would be relevant to carcinogenesis (Mates et al., 2010). Inorganic cobalt compounds, both soluble and particulate forms, caused lung tumors in animal experiments (Beyersmann & Hartwig, 2008). The IARC recently classified the mixture cobalt/tungsten carbide (Co/WC) as carcinogenic to humans (Mates et al., 2010).

Chromium is other carcinogenic metal. Epidemiological studies have consistently shown that the lower respiratory tract is the target organ of Cr(VI) compound exposure, and occupational exposure to these compounds is strongly associated with a higher incidence of lung cancer. Chromium exists in the environment in two major valence states, Cr(VI) and chromium (III) [Cr(III)], and Cr(VI) is actively transported into cells by the anionic transport system. The reduction of Cr(VI) to Cr(III) can lead to the formation of DNA-chromium adducts, DNA-DNA and DNA-protein cross-links, DNA-Cr(III)-amino acid ternary complexes and radical-mediated DNA strand breaks. In addition, it has been reported that lung cancer from workers exposed to Cr(VI) has a high percentage of G to T transversion mutations in the non-transcribed strand of the p53 gene (Feng et al., 2003). Chromium compounds also are capable to induce oxidative stress and the deregulation of cell proliferation (Beyersmann & Hartwig, 2008). On the other hand, several epidemiological studies demonstrated a strong correlation between nickel exposure and risk of lung and nasal cancer, especially in the case of workers at nickel refineries. Evidence from experimental animals has demonstrated the carcinogenicity of metallic nickel, which is also classified as possibly carcinogenic to humans (Group 2B). Nickel carcinogenesis involves epigenetic alterations, disruption of cellular iron homeostasis (by interfering with iron-dependent enzymes), generation of ROS, and activation of the hypoxia-signalling pathway (Salnikow & Zhitkovich, 2008).

Sulfur oxide pollution (as measured by sulfate particles and/or sulfur dioxide) is significantly associated lung cancer mortality. Elevated mortality risks have been associated primarily with measures of fine particulate and sulfur oxide pollution (Naess et al., 2007; Pope et al., 2002). Inhalation exposure to air pollutants, e.g., SO2, has been associated with the DNA damage of multiple organs including the lung, providing a possible biological pathway through which air pollution may affect lung cancer incidence (Cao et al., 2011).

Exposure to nitrogen oxide (NOx and NO2) also is related to lung cancer (Naess et al., 2007; Raaschou-Nielson et al. 2010). Studies have been shown that women had particularly large effects for lung cancer in the young age group, somewhat less so for the old exposed to nitrogen oxide (Naess et al., 2007). Moreover NOx exposure has significant correlations with adenocarcinoma (AC) type cancer incidence rates for both genders (Chen et al., 2009), although women seem to be more susceptible (Liaw et al., 2010). NOx may potentially trigger mutagenic and carcinogenic activity and play significant roles in the metabolism and behavior of AC type lung cancer (Fujimoto et al., 1998). It appears that the higher the NOx concentration, the higher the AC incidence rate. Recent studies on the role of NO in tumor progression suggest that NO is an important bioregulatory and signaling molecule and may play a role in the process of carcinogenesis (Tamir & Tannenbaum, 1996). NO is an endothelial growth factor that specifically mediates tumor vascularization (Jenkins et al., 1995) and tumor blood flow (Tozer et al., 1997). Exposure of cells to high NO concentrations cause DNA damage and apoptosis. Moreover, recent results have shown that NO stimulates p53 accumulation (Forrester et al., 1996). In summary, according to previous literature, NO and its derivatives can cause DNA damages (Wink et al., 1993) and play important roles in human lung AC (Fujimoto et al.,1998) (Liaw et al., 2010).

Airborne polycyclic aromatic hydrocarbons (PAHs) are emitted when organic matter is burned. It has long been known that several PAHs can produce cancers in experimental animals, and epidemiologic studies of exposed workers, especially in coke ovens and aluminum smelters, have shown clear excesses of lung cancer (Bostrom et al., 2002; Ben et al., 2004). PAHs have been suggested as being responsible for the initiation and development of lung cancer. PAHs and their metabolites are involved in mechanisms of carcinogenesis; produce early chromosomal changes, transformation of cells in culture, cytotoxicity and mutagenicity. The benzo(a)pyrene (BAP) for example, directly damages p53. In smoking-related lung cancer, 40% of the p53 gene mutations are G to T transversions, and 90% of this type of mutation can be attributed to the non-transcribed (coding) strand. G to T transversion has been regarded as a hallmark of PAH-induced mutations in smoking-related lung cancer. Activated metabolites of PAHs in cigarette smoke, including benzo[a]pyrene diol epoxide (BPDE), preferentially form DNA adducts at methylated CpG sites along the p53 gene corresponding to the afore mentioned major mutational hotspots in smoking-related lung cancer (Feng et al., 2003).

On the other hand, radon gas is a naturally occurring radioactive gas that can seep out of the soil into buildings, also emanates from uranium-bearing soil and porous rock. Radon is the second biggest cause of lung cancer after smoking. The cancer risk from radon increases the risk from smoking. Radon induces damage to a checkpoint tumor suppressor gene such as Tp53 (which codes for p53) since alpha particle radiation is a key mechanism for radon-related lung cancer (Harley et al., 2008; Bissett & McLaughlin 2010). Some studies suggest a relationship between the AGGARG-ATGMET transversion in codon 249 of P53 from people exposed to high radon concentrations (Ruano-Ravina et al., 2009). The damage done to epithelial cells of the lung occurs when radiation interacts either directly with DNA in the cell nucleus or indirectly through the affect of free radicals (UNSCEAR, 2000). Recently, *in vitro* studies of cells exposed to alpha-particle radiation gave evidence that more cells showed damage than those that were traversed by alpha-particles (Sawant et al., 2001; Alavanja 2002). Radon produces oxidative stress, although some similarities in the increased frequency of p53 mutations at a later stage in this process, DNA damage in the form of sister-chromatid exchange and mutations, have been observed for both smokers and those exposed to radon gas (Alavanja 2002). Other studies provided evidence of a GSTM1 and radon interaction in the increasing risk for lung cancer. Glutathione-S-transferase M1 (GSTM1) conjugates known carcinogens such as epoxides of polycyclic aromatic hydrocarbons (Risch & Plass, 2008). Other studies show that radon exposure in miners induces gene mutations and chromosomal aberrations. Numerous *in vitro* cytogenetic studies demonstrated that radon induces different types of genetic and cytogenetic damage that is likely to play a role in radon lung carcinogenesis (Al-Zoughool & Krewski, 2009).

5.4 Genotoxic effects

Generation of DNA damage is considered an important initial event in carcinogenesis. Multiples assays exist for the detection of different genotoxic effects of compounds in experimental systems, or for exposure investigation for genotoxic agents in environmental or occupational settings (Moller 2005). Cells with DNA damage are more susceptible to develop mutations after exposure to xenobiotics (Olive et al. 2001). For this reason, genotoxic evaluation of environmental pollutants is necessary in the respiratory system, because it is the first contact for inhaled xenobiotics.

Studies on environmental pollution (and their components), genotoxicity and respiratory tract, have been carried out *in vivo* and *in vitro*. Table 1 summarizes the results obtained in the last years.

A number of studies have considered DNA damage as an endpoint for the effects of air pollutants (Vineis & Husgafvel- Pursiainen 2005). In this report, we evidence the results obtained by genotoxic profiles, indicating that air pollutants cause alterations in the genetic material of the tested cells (strand breaks [illustrate in Figure 7], oxidative damage, adducts and micronucleus). DNA damage could provoke mutations in any cell from the respiratory tract and, may facilitate the development of neoplastic events.

	Exposure	System/Tissue	End Point	Results	Reference
In vivo					
	DEP Inhalation	Big Blue rats Male Lungs	Mutant frequency Mutation spectra DNA adducts 8-OHdG levels	+ + + +	Sato et al., 2000
	DEP Inhalation	Mice Female Lungs Bronchioalveolar lavage cells	8-OHdG levels DNA strand breaks (comet assay)	+ +	Risom et al., 2003
	DEP Inhalation	Mice	DNA adducts DNA strand breaks (comet assay) DNA oxidative damage Mutation frequency	+ + + -	Dybdahl et al., 2004
	Environmental Pollution	Human Females/males Nasal epithelial cells	DNA strand breaks (comet assay)	+	Fortoul et al., 2010
In vitro	DEP	A549 cells	DNA strand breaks (comet assay)	+	Dybdahl et al., 2004
	Particulate Matter (PM2.5 and PM 10)	A549 cells	DNA strand breaks (comet assay)	+	Gutierrez-Castillo et al., 2006
	Particulate Matter (PM10)	A549 cells	Micronucleus	+	Roubicek et al., 2007
	Fine Particles	A549 cells	DNA strand breaks (comet assay)	+	Sharma et al., 2007
	Nitrogen dioxide (NO_2)	Nasal epithelial cells	DNA strand breaks (comet assay) Micronucleus	+ -	Koehler et al., 2010

DEP: Diesel Exhaust Particles. A549 cells: human lung epithelial cell line. +Positive results. – Negative results. $PM_{2.5}$: Particulate Matter with aerodynamic diameter ≤2.5 μm. PM_{10}: Particulate Matter with aerodynamic diameter ≤10 μm.

Table 1. Genotoxic damage studies associated with air pollutants.

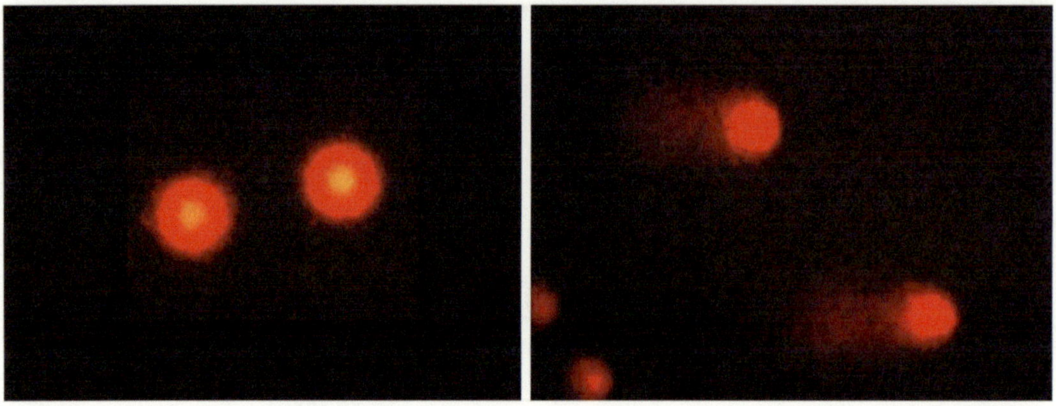

Fig. 7. Comet assay in nasal epithelial cells. Left: Undamaged Cells. Right: Damaged Cells (comets).

5.5 Pulmonary fibrosis and atmospheric pollution

Pulmonary fibrosis is the end result of a diverse group of lung disorders. Although there are multiple initiating agents for pulmonary fibrosis, including toxins, fibres/particles, autoimmune reactions, drugs and radiation, the etiology of the majority of cases of pulmonary fibrosis is unknown and these cases are referred to as Idiopathic Pulmonary Fibrosis (IPF). The harmful effects of environmental pollution on the respiratory system are undeniable. The combustion of fuels and its derivatives are the main cause for pollutant emission through engines and industrial plants (Zuurbier, et al., 2011). Chronic exposure to particulate matter, ozone and cigarette smoke can produce long-term effects on the lungs. This source of pollutants sets the lungs in a situation of constant aggression that result in a state of chronic inflammation, which could lead to pulmonary fibrosis (Churg & Wright, 2002).

5.5.1 Fibrotic lung reactions due to air pollution

The etiology for lung's fibrotic reactions is still unknown and it has been proposed a variety of environmental stimuli, such as metals, cigarette smoke, drugs and infectious agents (Araya & Nishimura, 2010). The degree of fibrosis will depend on the response of each individual, even with similar exposures, indicating that host genetic factors influence the fibrotic response of the patient (Westergren-Thorsson, et al., 2010).

Diesel exhaust particulate matter in polluted environments derived from internal combustion engines, increases the transcription of inflammatory cytokines and antimicrobial peptides, contributing to increased inflammatory response of airways in patients with chronic obstructive pulmonary diseases (COPD) (Nam, et al., 2006). This induces an oxidative state caused by lung macrophages that increase pulmonary responses and could result in irreversible lung fibrosis (Figure 8).

It has now been implicated the Transforming Growth Factor β1 (TGF-β1) as a key factor in the lung fibrotic response (Datta et al., 2011; Koli et al., 2008). The changes caused by TGF-β1 could be observed in repeated injuries of the airways, as occurs in asthma, chronic obstructive pulmonary disease and pulmonary fibrosis (Araya & Nishimura, 2010).

Exposure to cigarette smoke produces a large amount of reactive oxygen species and activates latent TGF-β1. The inhalation of this smoke promotes the recruitment of macrophages and neutrophils that are important sources of reactive oxygen species and also

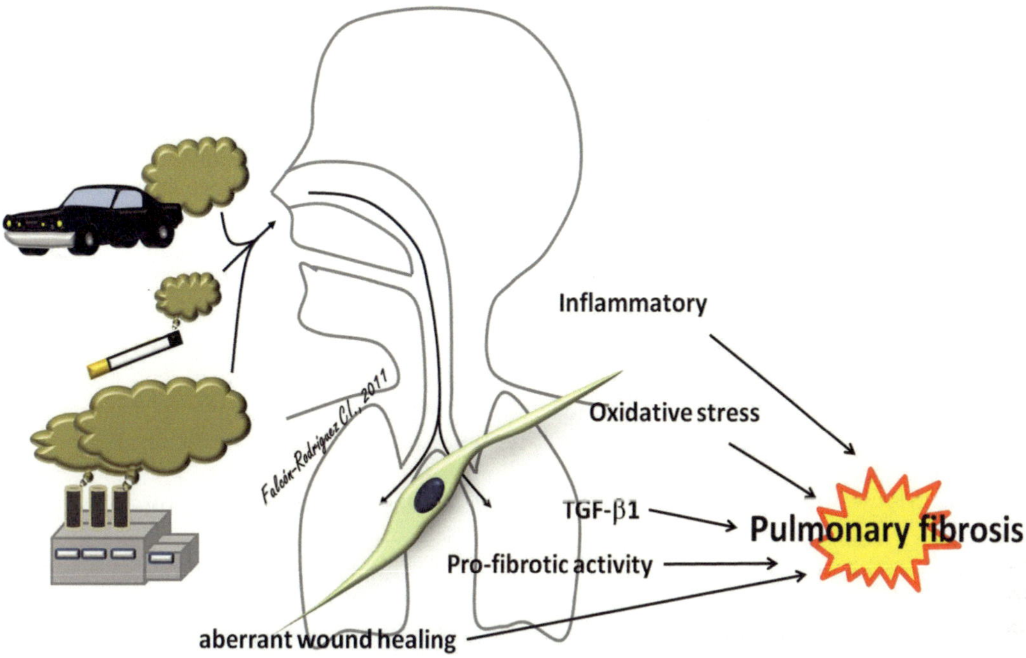

Fig. 8. Environmental pollution from various sources, such as transition metals, particle exposure, noxious drugs, cigarette smoke, induces repetitive cellular injury and inflammation. This leaves the tissue susceptible to increased TGF-β1, inflammation, profibrotic signals and aberrant wound healing, all of which may contribute to the progression of pulmonary fibrosis.

contribute to activate TGF-β1 (Westergren-Thorsson, et al., 2010). On the other hand, TGF-β1 induces transdifferentiation of fibroblasts to myofibroblasts and perpetuates the fibrogenic process, also providing a protective effect against myofibroblasts apoptosis, which are not removed once the lung lesions healed (Song et al., 2011).

Asbestos can induce lung fibrosis in occupational and experimental exposures, (Ross and Murray.2004; Dai & Churg, 2001). Mineral dusts can directly induce fibrosis in the airway wall, and this has been studied *in vivo*. Coexposures to cigarette smoke or ozone increase the fibrogenic effect of mineral dusts (Churg & Wright, 2002; Churg et al., 1996). Asbestosis and mineral dusts exposure increase the gene expression of profibrotic factors: TGF-β 1 and platelet derived growth factor (PDGF) that also increases procollagen. These changes may explain fibrosis progression (Churg and Wright, 2002; Churg et al., 1999). There is also human evidence of small airway remodeling in chronic exposure to high levels of particulate air pollution (Churg et al., 2003).

Oxidative stress, as a consequence of an inflammatory stimulus such as air pollution, plays a critical role in the pathogenesis of IPF (Park et al., 2009). Fibrotic stimuli of unknown origin are thought to create an imbalance between oxidant production and antioxidant protection, resulting in the accumulation of reactive oxygen species (ROS) (Rahman et al., 1999). The precise pathways leading from injury to fibrosis are not well established, but oxidants may contribute to the production of profibrotic factors such as TGF β, and oxidized proteins have been reported in human subjects with IPF. In addition, some studies have reported that various antioxidant enzyme systems protect against lung fibrosis (Gao et al., 2008; Khang et al., 2003).

5.6 Chronic obstructive pulmonary disease (COPD)

Chronic Obstructive Pulmonary Disease (COPD) is a slowly progressive condition characterized by airflow limitation which is usually reversible (MacNee, 2007). Although smoking habit is the main factor associated with COPD, air pollution has been implicated in its pathogenesis, and there is enough evidence to support the association between air pollution and COPD exacerbations and worsening of those with pre-existing COPD. Particulate Matter (PM), Nitrogen Oxides (NOx) are associated with its development (Salvi & Barnes, 2009).
Some of the observed changes are the increase in the presence of goblet cells (mucous metaplasia) as well as submucose glands hyperplasia. This increase in mucus production will result in reduced mucocliliary clearance. Emphysema, the enlargement and destruction of the alveolar spaces, is also part of this entity, because the decrease in the elastic recoil of the lung the air is trapped in the alveoli. Small airways are very important components in this entity. The changes, such as inflammation and fibrosis modify the prognosis: the greater the damage the poorer the prognosis.
The main participant in the progression of the bronchial damage is the inflammatory response. The epithelial cells generate this response when they are exposed to air pollutants or other irritants such as cigarette smoke, then a variety of pro-inflammatory, pro-fibrotic and mitotic factors are liberated. These factors increase endothelial permeability; activate macrophages, CD8+ Lymphocytes, and neutrophils. These cells produce more inflammatory factors, which, in chronic exposure inflammation leads to airway remodeling. (Figure 9) (Roth, 2008).

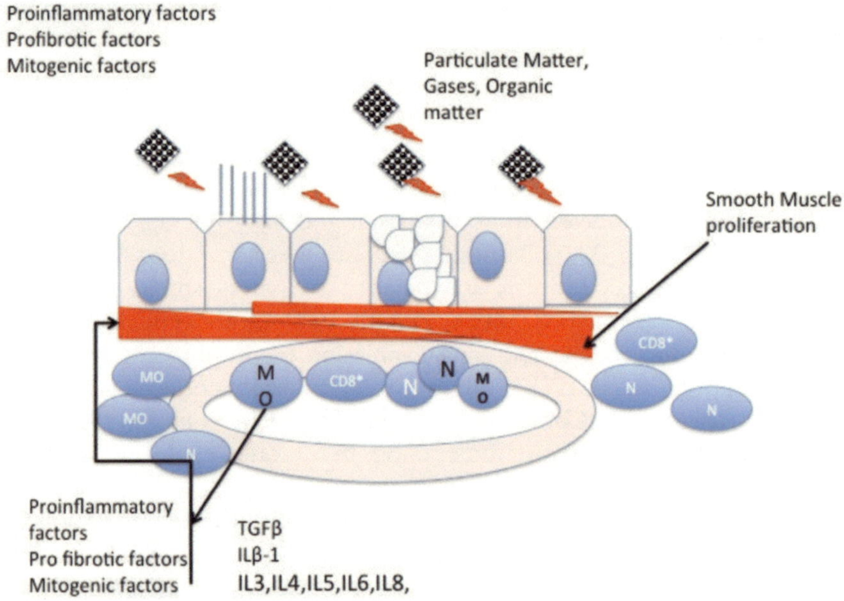

Fig. 9. The respiratory epithelium produces proinflammatory factors that will activate macrophages (MO), will increase neutrophil (N) and CD8* lymphocytes migration (CD8+). Also, smooth muscle proliferation is observed, as well as an alteration of the extracellular matrix.

Remodeling includes smooth muscle and fibroblast proliferation, synthesis of extracellular matrix (MMPs), decrease in proteases synthesis and increase in gelatinases. The sum of all these events will maintain the obstruction observed in these patients.

6. Conclusion

Further studies are needed to assess the impact of atmospheric pollution in the development of respiratory diseases, and to explore new therapeutic approaches to reverse the progression of the chronic changes that currently are observed in asthma, COPD, fibrosis and cancer.

7. Perspectives

More information about the differences in the response to air pollutants is needs, as well as the possible treatments, if any of the biomarkers is found. This drives us to the need for more specific biomarkers to identify the severity of the inflammation, or the type of inflammation. Because air Pollution is a complex mixture of organic and inorganic elements, more detailed information about specific characteristics of each pollutant response would be very helpful.

The lung has its own responses and support for more research in this field of interest, must be encouraged.

8. Acknowledgements

Authors thank to Veronica Rodriguez-Mata for her technical work in the technical work for the histological preparations.

9. References

Abbas, I.; Saint-Georges; F.; Billet, S.; Verdin, A.; Mulliez, P.; Shirali, P. & Garçon, G. (2009). Air Pollution Particulate Matter (PM 2.5)-Induced Gene Expression of Volatile Organic Compound and/or Polycyclic Aromatic Hydrocarbon-metabolizing Enzymes in an In Vitro Coculture Lung Model. *Toxicology in Vitro,* Vol.23, No.1, (February 2009), pp. 37–46, ISSN 0887-2333

Alavanja, M.C.R. (2002). Biologic Damage Resulting from Exposure to Tobacco Smoke and from Radon: Implication for Preventive Interventions. *Oncogene,* Vol.21, No.48, (October 2002), pp.7365 - 7375, ISSN 0950-9232

Alexis, N.E.; Lay J.C.; Zeman, K.; Bennet, W.E; Peden, D.B.; Soukup, J.M.; Devlin, R.B. & Becker, S. (2006). Biological material on inhaled coarse fraction particulate matter activates airway phagocytes in vivo on healthy volunteers. *Journal of Allergy Clinical Immunology,* Vol.117, No. pp. 1367-1373 ISSN 00916749

Al-Zoughool, M.& Krewski, D. (2009). Health Effects of Radon: A Review of the Literature. *International Journal Radiation Biology,* Vol. 85, No. 1, (2009),pp. 57-69 ISSN 0955-3002

Anwar-Mohamed, A.; Elbekai, R.H. & El-Kadi, A.O.S. (2009) Regulation of CYP1A1 by heavy metals and consequences for drug metabolism. *Expert Opinion on Drug Metabolism & Toxicology,* Vol.5, No.5, (May 2009), pp. 501-521, ISSN 1742-5255

Araya, J. & Nishimura, S. L. (2010). Fibrogenic reactions in lung disease. *Annual Review of Pathology,* No.5, pp. 77-98, *ISSN*: 1553-4014,

Atash, F. (2007). The deterioration of urban environments in developing countries: Mitigating the air pollution crisis in Theran, Iran. *Cities,* Vol.24, No.6, (February 2007), pp. 339 - 409, *ISSN* 0264-2751

Auten, R.L.; Potts, E.N.; Mason, S.N.; Fischer, B.; Huang, Y. & Foster, W.M. (2009). Maternal Exposure to Particulate Matter Increases Postnatal Ozone-induced Airway Hyper reactivity in Juvenile Mice. *American Journal Respiratory Critical Care Medicine,* Vol.180, No.12, (December 2009), pp. 1218-1226, ISSN 1073-449X

Barnes, T.C., Spiller, D.G., Anderson, M.E; Edwards, S..W., Moots, R.J.(2011). Endothelial activation and apoptosis mediated by neutrophil-dependent interleukin 6 trans-signaling: a novel target for systemic sclerosis. *Annals of the Rheumatic Diseases* (February, 2011) Vol. 70, No. 2, pp. 366-372. ISNN 1468-2060

Ben A.; Hutchinson E.; Unwin, J.; & Fletcher, T. Lung Cancer Risk after Exposure to Polycyclic Aromatic Hydrocarbons: A Review and Meta-Analysis. *Environmental Health Perspectives,* Vol.112, No. 9, (January 2004), pp. 970-78, ISSN: 0091-6765

Bernstein, J.; Neil, A.; Barnes, Ch.; Bernstein, L.; Bernstein, J.; Nel, A.; Peden, D.; Diaz S.D.; Tarlo, S. & Williams, B. (2004). Health effects of air pollution. *Environmental and Occupational Respiratory Disorders,* Vol.114, No. 5 (August 2004), pp. 1116 -1123

Beyersmann, D. & Hartwig, A. (2008). Carcinogenic Metal Compounds: Recent Insight into Molecular and Cellular Mechanisms. *Archives of Toxicology,* Vol. 82, (May 2008), pp. 493–512, ISSN 0340-5761

Bissett, R.J. & McLaughlin J.R. Radon. *Chronic Disease Canada,* Vol.29, Suppl 1, (2010), pp.38-50, ISSN 0228-8699

Born, S.L.; Caudill D.; Fliter K.L. & Purdon M.P. (2002). Identification of the Cytochromes P450 that Catalyze Coumarin 3,4-epoxidation and 3-hydroxylation. *Drug Metabolism and Disposition,* Vol.30, No.5, (May 2002), pp. 483–487, ISSN 0090-9556

Bostrom, CE.; Gerde, P.; Hanberg, A.; Jernstrom, B.; Johansson, C.; Kyrklund, T.; Rannung, A.; Tornqvist, M.; Victorin, K. & Westerholm R. (2002). Cancer Risk Assessment, Indicators, and Guidelines for Polycyclic Aromatic Hydrocarbons in the Ambient Air. *Environmental Health Perspectives,* Vol.110, (Jun 2002), pp.451–488, ISSN 0091-6765

Bulut, Y.; Shimada, K.; Wong, M.H; Chen, S; Gray, P; Alsabeh, R.; Doherty, T.M.; Crother, T.R; & Arditi, M.(2009). Chlamydial heat shock protein 60 induces acute pulmonary inflammation in mice via the Toll-like receptor 4- and MyD88-dependent pathway. *Infection and Immunity,* Vol. 77, No. 7, (April, 2009) pp. 2683-2690. ISSN:0019-9567

Calderon-Garcidueñas, L.; Vincent, R; Mora-Tiscareño, A; Franco-Lira, M; Henriquez-Roldan, C; Barragan-Mejia, G; Garrido-Garcia, L; Camacho-Reyes, L; Valencia-Salazar, G; Paredes, R; Romero, L.; Osnaya, H.; Villareal-Calderon, R; Torres-Jardon, R; Hazucha, M.J; & Reed, W. (2007). Elevated plasma endothelin-1 and pulmonary arterial pressure in children exposed to air pollution. *Environmental Health Perspectives,* (August, 2007) Vol. 115, No. 8, pp. 1248-1253. ISSN:1552-9924

Cao, J.; Chunxue Yang,; Jianxin Li,; Renjie Chen,; Bingheng Chen,; Dongfeng Gu & Haidong Kan J. (2011). Association Between Long-Term Exposure to Outdoor Air Pollution and Mortality in China: A Cohort Study. *Journal of Hazardous Materials* . Vol.186, No. 2-3, (January 2011), pp.1594–1600. ISSN: 0304-3894

Cardoso, W.V. & Lü, J. (2006). Regulation of Early Lung Morphogenesis: Questions, Facts and Controversies. *Development,* Vol.133, No.9 (May 2006),pp. 1611-1624, ISSN 0950-1991

Carlson, G.R. (2008). Critical Appraisal of the Expression of Cytochrome P450 Enzymes in Human Lung and Evaluation of the Possibility that such Expression Provides Evidence of Potential Styrene Tumorigenicity in Humans. *Toxicology*, Vol.254, No.1-2, (December 2008), pp. 1–10, ISSN 030-483X

Castell, J.V.; Donato, M.T. & Gomez-Lechon, J.M. (2005). Metabolism and Bioactivation of Toxicants in the Lung. The In Vitro Cellular Approach. *Experimental and Toxicologic Pathology*, Vol.57, (July 2005), pp.189–204, ISSN 0940-2993

Chang, H.; Chang, L.W.; Cheng, Y.H.; Tsai, W.T.; Tsai, M.X. & Lin, P.P. (2006). Preferential Induction of CYP1A1 and CYP1B1 in CCSP-Positive Cells. *Toxicological Sciences*, Vol.89, No.1, (January 2006), pp. 205–213, ISSN 1096-6080.

Chen, F.; Jackson, H. & Bina, WF. (2009). Lung Adenocarcinoma Incidence Rates and Their Relation to Motor Vehicle Density. *Cancer Epidemiology Biomarkers Prevention*. Vol.18, No.3, (March 2009), 760–764, ISSN 1055-9965

Choudhary, D.; Jansson, I.; Stoilov, I.; Sarfarazi, M. & Schenkman J.B. (2005). Expression Patterns of Mouse and Human CYP Orthologs (Families 1–4) During Development and in Different Adult Tissues. *Archives of Biochemistry and Biophysics*, Vol.436, No.1, (April 2005), pp. 50–61, ISSN 0003-9861

Churg, A. & Wright, J.L. (2002). Airway wall remodeling induced by occupational mineral dusts and air pollutant particles. *Chest* ; Vol.122, No.6 suppl (December 2002), pp. 306S-309S, ISSN 0012-3692

Churg, A.; Brauer M. & Keeling, B. (1996). Ozone enhances the uptake of mineral particles by tracheobronchial epithelial cells in organ culture. *American Journal of Respiratory and Critical Care of Medicine, No.*153, No.4, pp.1230–1234, *ISSN* 1073-449X

Churg, A.; Brauer, M.; Avila-Casado, M.C.; Fortoul, T.I. & Wright, J.L. (2003). Chronic exposure to high levels of particulate air pollution and small airway eemodeling. *EnvironmentalHealth Perspectives*, Vol.111, No.5, (May 2003) pp. 714–718, *ISSN*: *0091-6765*.

Churg, A.; Gilks, B. & Dai, J. (1999). Induction of fibrogenic mediators by fine and ultrafine titanium dioxide in rat tracheal explants. *American Journal of Physiology*, Vol.277, No. 5 pt 1, pp. L975–L982, *ISSN*: 0363-6143;

Dai, J. & Churg, A. (2001). Relationship of fiber surface iron and active oxygen species to expression of procollagen, PDGF-A, and TGF 1 in tracheal explants exposed to amosite asbestos. *American Journal of Respiratory Cell and Molecular Biology*, Vol.24, No.4, (April 2001)pp. 427–435, *ISSN*: 1535-4989

Datta, A.; Scotton, C. J. & Chambers, R. C. (2011). Novel Therapeutic Approaches for Pulmonary Fibrosis. *British Journal of Pharmacology*. Epub ahead of print, ISSN *0007-1188*.

Davis, C.W. & Dicker, B.E. (2008). Regulated Airway Goblet Cell Mucin Secretion. *Annual Review of Physiology*, Vol.70, (March 2008), pp.487-512, ISSN 0066-4278

Deb, S. & Bandiera, S.M. (2010). Characterization of a New Cytochrome P450 Enzyme, CYP2S1, in Rats: Its Regulation by Aryl Hydrocarbon Receptor Agonists. *Toxicology*, Vol.267, No.1-3, (January 2010), pp. 91–98, ISSN 0300-483X

Dockery, DW.; Pope, CA; Xu, X.; Spengler, JD.; Ware, JH,; Fay, ME,; Ferris, BG. Jr. & Speizer, FE. (1993). An Association Between Air Pollution and Mortality in Six U.S. Cities. *New England Journal Medicine*, Vol.329, (December 1993), pp.1753–9, ISSN 0028-4793

Donaldson, K; Mills, N; MacNee, W.; Robinson, S; & Newby, D. (2005). Role of inflammation in cardiopulmonary health effects of PM. *Toxicology and Applied Pharmacology*, Vol. 207, No. 2 (Sept, 1, 2005) pp. 483-488. ISSN:0041-008X

Dybdahl, M.; Risom, L.; Bornholdt, J.; Autrup, H.; Loft, S. & Wallin, H. (2004). Inflammatory and Genotoxic Effects of Diesel Particles In Vitro and In Vivo. *Mutation Research*, Vol.562, No.1-2, (August 2004), pp.119-131, ISSN 1383-5718

Ekici, A.; Ekici, M., Kurtipek, E., Akin, A., Arsian, M., Kara, T., Apaydin, Z., Demir, S. (2005) Obstructive airway diseases in women exposed to biomass smoke. *Environmental Research*, Vol. 99, No.1, (September, 2005) pp. 93-98. ISSN:0013-935

Feng, Z.; Hu, WW.; Rom, WN.; Costa, M. & Tang, MS. (2003). Chromium(VI) exposure enhances polycyclic aromatic hydrocarbon DNA binding at the p53 gene in human lung cells. *Carcinogenesis*, Vol.24, No.4, (April 2003), pp.771-778, ISSN 0143-3334.

Finlayson-Pitts, B. J. & Pitts, J. N. (1997). Tropospheric air pollution: ozone, airborne toxics, polycyclic aromatic hydrocarbons and particles. *Science*, Vol.276, No. 5315, (May 1997), pp.1045-1052, *ISSN 0036-8075*

Finlayson-Pitts, B.J. & Pitts, J.N. (2000). Chemistry of the Upper and Lower Atmosphere, Academic Press, San Diego, California, xxii+969 pp., ISBN 0-12-257060xMills, N.L. Donaldson, D., Hadoke, P.W., Boon, N.A., MacNee, W., Cassee, F.R., Sandström, T., Blomberg, A. & Newby D.E. (2009). Adverse cardiovascular effects of air pollution. Nature Clinical Practice Cardiovascular Medicine, Vol. 6, No.1, (November, 2009), pp. 36-44, ISSN: 1759-5002

Forrester, K.; Ambs, S.; Lupold, SE.; Kapust, RB.; Spillare. EP.; Weinberg, WC.; Felley-Bosco, E.; Wang, XW.; Geller, DA.;Tzeng, E. & Billar, TR. (1996). Harris CC. Nitric Oxide-Induced p53 Accumulation and Regulation of Inducible Nitric Oxide Synthase Expression by Wild-type p53. *Proceedings of the National Academy Science USA*. Vol. 93, (March1996), pp. 2442–2447, ISSN 0027-8424

Fortoul, T.I.; Rodriguez-Lara, V.; Lopez-Valdez, N. & Falcon-Rodriguez, C.I. (2010). Sistema Respiratorio, In: *Histologia y Biologia Celular*. T.I. Fortoul van der Goes, (Ed.), 187-195, McGraw-Hill, ISBN: 978-607-15-0340-4, China

Fortoul, T.I.; Rojas-Lemus, M.; Avila-Casado M.C.; Rodriguez-Lara V.; Montaño, L.F.; Muñoz-Comonfort, A. & Lopez-Zepeda, S. (2010), Endogenous Antioxidants and Nasal Human Epithelium Response to Air Pollutants: Genotoxic and Inmmunocytochemical Evaluation. *Journal of Applied Toxicology*, Vol.30, No.7, (October, 2010), pp.661-665, ISSN 0260-437X

Fujimoto, H.; Sasaki, J.; Matsumoto, M.; Suga, M.; Ando, Y.; Iggo, R.; Tuda, M.; Saya, H. & Anas, M. (1998). Significant Correlation of Nitric Oxide Synthase Activity and p53 Gene Mutation in Stage I Lung Adenocarcinoma. *Japanese Journal of Cancer Research*.Vol. 89, (July 1998), pp. 696–702, ISSN 0910-5050

Gabay, C., & Kushner, I. (1999) Acute-phase proteins and other systemic responses to inflammation. *New England Journal of Medicine*, (February, 11, 1999) Vol. 340, No.6, pp.448-454. ISSN:0028-4793

Gao, F.; Kinnula, V.L.; Myllarniemi, M. & Oury, T.D. (2008). Extracellular superoxide dismutase in pulmonary fibrosis. *Antioxidants &. Redox Signal*, Vol.10, No. 2, (February 2008), pp.343–54, *ISSN*: 1523-0864

Gern, J.E. (2010). The Urban Environment and Childhood Asthma Study. *Journal of Allergy and Clinical Immunology*, Vol.125, No.3, (March 2010), pp. 545-549, ISSN 0091-6749

Gutierrez-Castillo, M.E.; Roubicek, D.A.; Cebrian-García M.E.; Vizcaya-Ruiz, A.D.; Sordo-Cedeno, M. & Ostrosky-Wegman, P. (2006). Effect of Chemical Composition on the Induction of DNA Damage by Urban Airbone Particulate Matter. *Environmental and Molecular Mutagenesis,* Vol.47, No.3, (April 2006), pp. 199-211, ISSN 0893-6692

Halle, S.; Dujardin, H.; Bakocevic, N.; Fleige, H.; Danzer, H.; Willenzon, S.; Suezer, Y.; Hammerling, G.; Garbi, N.; Sutter, G.; Worbs, T. & Forster, R. (2009). Induced bronchus-associated lymphoid tissue serves as a general priming site for T cells and is maintained by dendritic cells. *Journal of Experimental Medicine,* Vol.206, No.12, (November 2009), pp. 2593-2601, *ISSN: 0022-1007*

Harley, NH.; Chittaporn, P.; Heikkinen, MSA.; Meyers, OA. & Robbins, ES. (2008). Radon Carcinogenesis: Risk Data and Cellular Hits. *Radiation Protection Dosimetry.* Vol.130, No.1, (June 2008), pp.107–109, ISSN 0144-8420

Hetland, R.B; Cassee, F.R; Lag, M., Refsnes, M; Dybing, E; & Schwaze, P.E. (2005) Cytokine release from alveolar macrophages exposed to ambient particulate matter: heterogeneity in relation to size, city and season. *Particle and Fibre Toxicology,* Vol. 17, No.2, (August, 2005) pp. 1-15, SSN:1743-8977

Hildebrandt, K; Ruckerl, R.; Koeening, W.; Schneider, A.; Pitz, M.; Heinrich, J; Marder, V.; Frampton, M.; Oberdorster, G; Wichmann, G; & Peters, A. (2009). Short-term effects of air pollution: a panel study of blood markers in patients with chronic pulmonary disease. *Particle and Fibre Toxicology* (September, 2009). Vol. 26, No.25, pp. 1-13. ISSN:1743-8977

Holgate, S.T. (2008). The Airway Epithelium is Central to the Pathogenesis of Asthma. *Allergology International,* Vol.57, No.1, (March 2008), pp. 1-10, ISSN 1323-8930

Ioannides, C. (2008). *Cytochromes P450. Role in the metabolism and toxicity of drugs and others xenobiotics,* RSC Publishing, ISBN: 978-0-85404-274-6, Cambrigde, UK

Jenkins, DC,; Charles, IG.; Thomsen, LL.; Moss, DW.; Holmes, LS.; Baylis, SA.; Rhodes, P.; Westmore, K.; Emson, PC. & Moncada, S. (1995). Roles of Nitric Oxide in Tumor Growth. *Proceedings of the National Academy Science USA,* Vol.92, (May 1995), pp.4392–6, ISSN 0027-8424

Kang, S.K.; Rabbani, Z.N.; Folz, R.J.; Golson, M.L.; Huang, H.; Samulski; T.S.; Dewhirst, M.W.; Anscher, M.S. & Vujaskovic, Z. (2003). Overexpression of extracellular superoxide dismutase protects mice from radiation-induced lung injury. *International Journal of Radiation Oncology Biology Physics,* Vol.15, No. 57, (November 2003), pp. 1056–66, *ISSN*: 0360-3016.

Katavolos, P.; Ackerley, C.A.; Viel, L.; Clark, M.E.; Wen, X. & Bienzle, D. (2009). Clara Cell Secretory Protein Is Reduced in Equine Recurrent Airway Obstruction. *Veterinary Pathology,* Vol.46, No.4, (July 2009), pp. 604–613, ISSN 0300-9858

Kawamata, N.; Xu, B.; Nishijima, H.; Aoyama, K.; Kusumoto, M.; Takeuchi, T.; Tei, Ch.; Michie, A. & Matsuyama, T. (2009). Expression of endothelia and lymphocyte adhesion molecules in bronchus-associated lymphoid tissue (BALT) in adult human lung. *Respiratory Research,* Vol.10, No.97, (October 2009), pp. 1-11, *ISSN*: 1465-9921

Koehler, C.; Ginzkey, C.; Friehs, G.; Hackenberg, S.; Froelich, K.; Scherzed, A.; Burghartz, M.; Kessler, M. & Kleinsasser, N. (2010). Aspects of Nitrogen Dioxide Toxicity in Environmental Urban Concentrations in Human Nasal Epithelium. *Toxicology and Applied Pharmacology,* Vol.245, No.2, (June 2010), pp. 219-225, ISSN 0041-008X

Koli, K.; Myllarniemi, M.; Keski-Oja, J. & Kinnula, V.L. (2008). Transforming growth factor-ß activation in the lung: focus on fibrosis and reactive oxygen species. *Antioxidants & Redox Signal*, Vol.10, No.2, (February 2008), pp. 333–42, *ISSN*: 1523-0864

Liaw, YP.; Ting, TF.; Ho, CC. & Chiou, ZY. (2010). Cell Type Specificity of Lung Cancer Associated with Nitric Oxide. *Science of the Total Environment*, Vol.408, No.21, (October 2010), pp.4931-4, ISSN 0048-9697

López-Cima, MF.; García-Pérez, J.; Pérez-Gómez, B.; Aragonés, N.; López-Abente G.; Tardón, A. & Pollán, M. (2011). Lung Cancer Risk and Pollution in an Industrial Region of Northern Spain: a Hospital-based Case-Control Study. *International Journal Health Geographics*, Vol.25, No.10, (January 2011), pp.10-23, ISSN 1572-9893

MacNee, W. (2007) Pathogenesis of Chronic Obstructive Pulmonary disease. *Clinics in Chest Medicine, Vol.2, No.4, (November 2005)*, pp. 479-513, *ISSN*: 0272-5231

Mates, JM. ; Segura, JA.; Alonso, FJ. & Marquez, J. (2010). Roles of Dioxins and Heavy Metals in Cancer and Neurological Diseases Using ROS-Mediated Mechanisms. *Free Radical Biology & Medicine*, Vol. 49, (November 2010), pp. 1328–1341. ISSN: 0891-5849

Mills, N.L; Amin, N; Robinson, S.D; Anand, A; Davies, J; Patel, D; de la Fuente, J.M; Cassee, F.R; Boon, R.A; MacNee, W; Millar, A.M; Donaldson, K. & Newby D.E. (2006). Do Inhaled Carbon Nanoparticles Translocate Directly into the Circulation in Humans? *American Journal of Respiratory and Critical Care Medicine* Vol. 173, No. 4, (February, 2006) pp. 426–431, ISSN 1535-4970

Molina, JR.; Yang, P.; Cassivi, SD.; Schild, SE. & Adjei, AA. (2008). Non-Small Cell Lung Cancer: Epidemiology, Risk Factors, Treatment, and Survivorship. *Mayo Clinic Proceedings*, Vol.83, No.5, (May 2008), pp.584-594, ISSN 0025-6196

Moller, P. (2005). Genotoxicity of Environmental Agents Assessed by the Alkaline Comet Assay. *Basic & Clinical Pharmacology and Toxicology*, Vol.96, Suppl1, (January 2005), pp.1-42, ISSN I742-7835

Morris, J.B. & Buckpitt, A.R. (2009). Upper Respiratory Tract Uptake of Naphthalene. *Toxicological Sciences*, Vol.111, No.2, (October 2009), pp. 383–391, ISSN 1096-6080

Mortimer, K.; Neugebauer, R.; Lurmann, F.; Alcorn, S.; Balmes, J. & Tager, I. (2008). Air Pollution and Pulmonary Function in Asthmatic Children: Effects of Prenatal and Lifetime Exposures. *Epidemiology*, Vol.19, No.4, (July 2008), pp. 550-562, ISSN 1044-3983

Mussali-Galante, P. & Fortoul, T.I. (2008). Atmospheric pollution, In: *Environmental Research Progress*, F.P. Maes, (Ed.), 147-160, Nova Science Publishers, ISBN: 978-1-60456-080-0, New York, USA

Næss Øyvind; Per Nafstad; Geir Aamodt; Bjørgulf Claussen & Pa Rosland. (2007). Relation Between Concentration of Air Pollution and Cause-Specific Mortality:Four-Year Exposures to Nitrogen Dioxide and Particulate Matter Pollutants in 470 Neighborhoods in Oslo, Norway. *American Journal of Epidemiology*. Vol.165, No.4, (February 2007), pp. 435–443, ISSN 0002-9262

Nam, H. Y., Ahn, E. K., Kim, H. J., Lim, Y., Lee, C. B., Lee, K. Y. & Vallyathan,V. (2006). Diesel exhaust particles increase IL-1beta-induced human beta-defensin expression via NF-kappaB-mediated pathway in human lung epithelial cells. *Particle & Fibre Toxicology*, Vol.3, No.9 (May 2006) (available on line), *ISSN*: 17438977

Nemmar, A; Vanbilloen, H; Hoylaerts, M.F; Hoet, P.H; Verbruggen, A.; & Nemery, B. (2002). Passage of intratracheally instilled ultrafine particles from the lung into the systemic circulation in hamster. *American Journal of Respiratory and Critical Care Medicine,* Vol. 165, No.12, (June, 2002) pp. 1671–1672, ISSN 1535-4970

Olive, P.L.; Durand, R.E.; Banath, J.P. & Johnston, P.J. (2001). Analysis of DNA Damage in Individual Cells. *Methods in Cell Biology,* Vol.64, (2001), pp.235- 249, ISSN 0091-679X

Olivieri, D. & Scoditti E. (2005). Impact of Environmental Factors on Lung Defences. *European Respiratory Review,* Vol.14, No.95, (December 2005), pp. 51-56, ISSN 0905-9180

Park, H.S.; Kim, S.R. & Lee, Y.L. (2009). Impact of oxidative stress on lung diseases. *Respirology, Vol.*14, No.1, (January 2009), pp. 27- 38, *ISSN*: 13237799

Parker,D., & Prince, A.(2011) Innate Immunity in the Respiratory Epithelium. *American Journal of Respiratory and Critical Care Medicine,* (Februarry, 2011)(Epublish ahead of print) ISSN:1044-1549

Pavek, P. & Dvorak, Z. (2008). Xenobiotic-Induced Transcriptional Regulation of Xenobiotic Metabolizing Enzymes of the Cytochrome P450 Superfamily in Human Extrahepatic Tissues. *Current Drug Metabolism,* Vol.9, No.2, (February 2008), pp. 129-143, ISSN 1389-2002

Penn, A.L.; Rouse, R.L.; Horohov, D.W.; Kearney, M.T.; Paulsen, D.B. & Lomax, L. (2007). In Utero Exposure to Environmental Tobacco Smoke Potentiates Adult Responses to Allergen in BALB/c Mice. *Environmental Health Perspectives,* Vol.115, No.4, (April 2007), pp. 548-555, ISSN 0091-6765

Perez-Padilla, R.; Schilmann, A. & Riojas-Rodriguez, H. (2010). Respiratory Health Effects of Indoor Air Pollution. *International Journal Tuberculosis Lung Disease,* Vol. 14, No. 9, (September 2010), pp. 1079–1086, ISSN 1081-1206

Pons, F.; Haag, M.; Corcos, L.; Bonnet, P.; Guillouzo, A.; Lugnier, A. & Frossard, N. (2000). Inhalation of Toluene Diisocyanate Affects Cytochrome P4502B1 Expression in Rat Lung. *Archives of Toxicology,* Vol.74, No.7, (September 2000), pp. 397-403, ISSN 0340-5761

Pope, CA.; Burnett, RT.; Thun, MJ.; Calle, EE.; Krewski, D.; Ito, K. & Thurston, GD. (2002). Lung Cancer, Cardiopulmonary Mortality, and Long-term Exposure to Fine Particulate Air Pollution. *Journal of the American Medical Association,* Vol. 287, No. 9, (March 2002), pp. 1132-1141, ISSN 0098-7484

Pope, CA.; Thun, MJ.; Namboodiri, MM.; Dockery, DW.; Evans, JS.; Speizer, FE. & Heath, CW. (1995). Particulate Air Pollution as a Predictor of Mortality in a Prospective Study of U.S. Adults. *American Journal Respiratory Critcal Care Medicine,* 151, (March 1995), pp.669–74, ISSN 1073-449X

Raaschou-Nielson, O.; Bak, H.; Sorensen, M.; Jensen, SS.; Ketzel, M.; Huiberg, M.; Schnohr, P.; Tjonneland, A. Overvad, K. & Loft, S. (2010). Air Pollution from Traffic and Risk for Lung Cancer in Three Danish Cohorts. *Cancer Epidemiology Biomarkers and Prevention.* Vol. 19, No.5, (May 2010), pp. 1284-91, ISSN1055-9965

Rahman, I.; Skwarska, E.; Henry, M.; Davis, M.; O'Connor, C.M.; FitzGerald, M.X.; Greening, A. & MacNee, W. (1999). Systemic and pulmonary oxidative stress in idiopathic pulmonary fibrosis. *Free Radical Biology* & *Medicine,* Vol.27, No.1-2, (July 1999), pp. 60–8, *ISSN: 0891-5849.*

Randall, T.D. (2010). Bronchus-associated lymphoid tissue (BALT) structure and function. *Advances in Immunology,* No. 107, (November 2010), pp. 187-241, *ISSN* 0065-2776

Raub, J.; MathieuNolf, M.; Hampson N. & Thom S. (2000). Carbon monoxide poisoning - a public health perspective. *Toxicology*, Vol.145, No.1, (April 2000), pp 1- 14, ISSN 0300-483X

Rice, G.E., & Bevilacqua, M.P. (1989) An inducible endothelial cell surface glycoprotein mediates melanoma adhesion. *Science*, (December,8, 1989) Vol .246, No. 4935, pp.1303-1306. ISSN:0193-4511

Risch, A. & Plass, C. (2008). Lung Cancer Epigenetics and Genetics. *International Journal of Cancer*, Vol. 123, No. 1, (July 2008), pp. 1–7, ISSN 0020-7136

Risom, L.; Dybdahl, M.; Bornholdt, J.; Vogel, U.; Wallin, H.; Moller, P. & Loft, S. (2003). Oxidative DNA Damage and Defence Gene Expression in the Mouse Lung after Short-Term Exposure to Diesel Exhaust Particles by Inhalation. *Carcinogenesis*, Vol.24, No.11, (November 2003), pp.1847-1852, ISSN 0143-3334

Rojas-Martinez, R.; Perez-Padilla, R.; Olaiz-Fernandez G.; Mendoza-Alvarado, L.; Moreno-Macias, H.; Fortoul, T.; McDonell, W.; Loomis, D. & Romieu, I. (2007). Lung Function Growth in Children With Long-term Exposure to Air Pollutants in Mexico City. *American Journal of Respiratory and Critical Care of Medicine, No.*176, pp. 377-384, *ISSN* 1073-449X

Rose, M.C. & Voynow, J.A. (2006). Respiratory Tract Mucin Genes and Mucin Glycoproteins in Health and Disease. *Physiological Reviews*, Vol.86, No.1, (January 2006), pp. 245-278, ISSN 0031-9333

Ross, M.H. & Murray, J. (2004). Occupational Respiratory disease in mining. *Occupational Medicine*, Vol.54, No.5, pp. 304–310, *ISSN*: *0962-7480*.

Roth, M. (2008) Pathogenesis of COPD. Part II. Inflammation in COPD. The *International Journal of Tuberculosis and Lung Disease*, Vol.12, No.4 (April 2008), pp. 375-380, *ISSN* 1027-3719

Roubicek, D.A.; Gutierrez-Castillo, M.; Sordo, M.; Cebrian-García M.E. & Ostrosky-Wegman, P. (2007). Micronuclei Induced by Airbone Particulate Matter from Mexico City. *Mutation Research*, Vol.631, No 1, (July 2007), pp.9-15, ISSN 1383-5718

Ruano-Ravina, A.; Faraldo-Valles, MJ. & Barros-Dios, JM. (2009). Is there a Specific Mutation of p53 Gene Due to Radon Exposure? A Systematic Review. *International Journal Radiation Biology*. Vol.85, No. 7, pp.614–621, ISSN 0955-3002

Salnikow, K. & Zhitkovich, A. (2008). Genetic and Epigenetic Mechanisms in Metal Carcinogenesis and Cocarcinogenesis: Nickel, Arsenic, and Chromium, *Chemical Research Toxicology*, Vol. 21, (January 2008), pp. 28–44, ISSN 0893-228X

Salvi, S.S., & Barnes, P. (2009) Chronic Obstructive Pulmonary Disease in Non-smokers. *The Lancet*, Vol.374, No. 9691 (August 2009), pp. 733-743, *ISSN 0140-6736*

Sato, H.; Sone, H.; Sagai, M.; Suzuki, KT. & Aoki, Y. (2000). Increase in Mutation Frequency in Lung of Big Blue(R) Rats by Exposure to Diesel Exhaust. *Carcinogenesis*, Vol.21, No.4, (April 2000), pp. 653-661, ISSN 0143-3334

Saura, M; Zaragoza, C; Bao, C.; Rodriguez-Puyol, M; & Lowenstein, C.J. (2006). Stat3 mediates interleukin-6 [correction of interelukin-6] inhibition of human endothelial nitric-oxide synthase expression. *Journal of Biological Chemistry*, (October,1, 2006). Vol.281, No.40, pp. 30057-30062. ISSN: 1083-351X

Sawant, S.G.; Randers-Pherson, G.; Geard, CR.; Brenner, D.J. & Hall EJ. (2001). The Bystander Effect in Radiation Oncogenesis: I. Transformation in C3H 10T1/2 Cells in vitro can be initiated in the Unirradiated Neighbors of Irradiated Cells. *Radiation Research*, Vol.155, No.3, (March 2001), pp. 397-401, ISSN 0033-7587

Schikowski, T; Sugiri, D; Ranft, U; Gehring, U; Heinrich, J; Wichmann, H.E.; & Kramer, U. (2005) Long-term air pollution exposure and living close to busy roads are associated with COPD in women. *Respiratory Research*, Vol. 22, No. 6 (December, 2005) pp. 152-161, ISSN:1465-9921

Schins, R.P; Lightbody, J.H; Borm, P.J; Shi, T; Donaldson, K.; & Stone, V. (2004) Inflammatory effects of coarse and fine particulate matter in relation to chemical and biological constituents. *Toxicology and Applied Pharmacology*, Vol. 195, No. 1 (Feb, 2004) pp. 1-11. ISSN:0041-008X

Schwarze, P.E; Oyreyik, J; Lag, M; Refsnes, M.; Nafstad, P.; Hetland, R.B.; & Dybing, E. (2006) Particulate matter properties and health effects: consistency of epidemiological and toxicological studies. *Human and Experimental Toxicology*, Vol. 25, No. 10 (October, 2006) pp.559-579. ISSN:0960-3271

Seagrave, J. (2008). Mechanisms and implications of air pollution particle associations with chemokines. *Toxicology and Applied Pharmacology*, (November, 1, 2008).Vol. 232, No. 3, pp. 469-477. ISSN: 1096-0333

Seagrave, J; McDonald, J.D; Bedrick, E; Edgerton, E.S; Gigliotti, A.P; Jansen, J.J; Ke, L.; Naeher, L.P; Seilkop, S.K; Zheng, M; & Mauderly, J.L. (2006) Lung toxicity of ambient particulate matter from southeastern U.S. sites with different contributing sources: relationships between composition and effects. *Environmental Health Perspectives*, Vol. 114, No. 9, (September, 2006) pp. 1387-1393. ISSN:0091-6765

Sharma, A.K; Jensen, K.A; Rank, J; White, P.A; Lundstedt, S; Gagne, R; Jacobsen, N.R; Kristiansen, J; Vogel, U. & Wallin, H. (2007). Genotoxicity, Inflammation and Physic-Chemical Properties of Fine Particle Samples from an Incineration Energy Plant and Urban Air. *Mutation Research-Genetic Toxicology and Environmental Mutagenesis*, Vol.633, No.2, (October 2007), pp. 95-111, ISSN 1385-5718

Shoenfelt, J.; Mitkus, R. J.; Zeisler, R.; Spatz, R. O.; Powell, J.; Fenton, M. J.; Squibb, K. A.; Medvedev, A. E. (2009). Involvement of TLR2 and TLR4 in inflammatory immune responses induced by fine and coarse ambient air particulate matter. *Journal of Leukocyte Biology*, Vol. 86 No.2, (August, 2009) pp. 303-312. ISSN:0741-5400

Sigaud, S.; Goldsmith, C.A.; Zhou, H.; Yang, Z.; Fedulov, A.; Imrich, A.; & Kobzik, L. (2007). Air pollution particles diminish bacterial clearance in the primed lungs of mice. *Toxicology and Applied Pharmacology*, (August, 1, 2007).Vol. 223, No. 1, pp. 1-9. ISSN: 1096-0333

Song, J. W, Hong, S. B, Lim, C.M, Koh, Y. & Kim, D. S. (2011). Acute exacerbation of idiopathic pulmonary fibrosis: incidence, risk factors and outcome. *European Respiratory Journal*, Vol. 37, No.2, (February 2011) pp. 356-363, *ISSN*: 0903-1936

Spaan, S.; Schinkel, J.; Wouters, I.M.; Preller, L.; Tielemans, E.; Nij, E.T.; & Heederik, D. (2008). Variability in endotoxin exposure levels and consequences for exposure assessment. *Annals of Occupational Hygiene*, Vol.52, No.5 (July 2008) pp. 303-316. ISSN:0003-4878

Strand, V, Rak, S, Svartengren, M. & Bylin, G. (1997). Nitrogen dioxide exposure enhances asthmatic reaction to inhaled allergen in subjects with asthma. *American Journal of Respiratory and Critical Care Medicine*, Vol.155, No. 3, (March 1997), pp 881 -887, ISSN 1073-449XWinterton, DL.; Kaufman, J, Keener CV, Quigley, S, Farin, FM.; Williams PV. & Koenid, JQ. (2001). Genetic Polymorphism as a Biomarkers of Sensitivity to Inhaled Sulfur Dioxide in Subjects with Asthma. (2001). *Annals of Allergy Asthma & Immunology*, Vol.86, No. 2, (February 2001), pp 232-238, ISSN 1073-449X

Tamagawa, E.; Bai, N.; Morimot, K.; Gray, C.; Mui, T.; Yatera, K.; Zhang, X.; Xing, L.; Laher, I.; Sin, D.D.; Man, S.F.; & van Eeden, S.F. (2008). Particulate matter exposure induces persistent lung inflammation and endothelial dysfunction. *American Journal of Physiology-Lung Cellular and Molecular Physiology*. (July, 2008) Vol. 295, No. 1, pp. L79-L85. ISSN:1040-0605

Tamir, S. & Tannenbaum, S.R. (1996). The Role of Nitric Oxide (NO.) In the Carcinogenic Process. *Biochimica et Biophysica Acta*. Vol.1288, No.2, (October 1996), pp. F31–F36, ISSN 0304-419X

Torres-Duque, C.; Maldonado, D.; Perez-Padilla, R.; Ezzati, M.; & Viegi, G. (2008). Biomass fuels and respiratory diseases: a review of the evidence. *Proceedings of the American Thoracic Society*, Vol. 5, No. 5 (August, 2008) pp. 577-590, ISSN:1546-3222

Tozer, G.M.; Prise, V.E.; & Chaplin, D.J. (1997). Inhibition of Nitric Oxide Synthase Induces a Selective Reduction in Tumor Blood Flow That is Reversible with L-arginine. *Cancer Research*, Vol. 57, (March 1997), pp.948–955, ISSN: 0008-5472

United Nations Scientific Committee on the Effects of Atomic Radiation (UNSCEAR). (2000). Sources and Effects of Ionizing Radiation. New York: United Nations. Available from http://www.unscear.org/unscear/publications/2000_1.html

van Eeden, S.F.; Tan, W.C.; Suwa, T.; Mukae, H.; Terashima, T.; Fujii, T.; Oui, D.; Vincent, R.; Hogg, J.C. (2001). Cytokines involved in the systemic inflammatory response induced by exposure to particulate matter air pollutants (PM(10). *American Journal of Respiratory and Critical Care Medicine* (September, 1, 2001) Vol. 164, No. 5, pp.826-830. ISSN: 1535-4970

Vineis, P. & Husgafvel-Pursiainen, K. (2005). Air Pollution and Cancer: Biomarker Studies in Human Populations. *Carcinogenesis*, Vol.26, No.11, (November 2005), pp.1846-1855, ISSN 0143-3334

Wang, L. & Pinkerton KE. (2008). Detrimental Effects of Tobacco Smoke Exposure During Development on Postnatal Lung Function and Asthma. *Birth Defects Research Part C: Embryo Today*, Vol.84, No.1, (March 2008), pp. 54-60, ISSN 1542-975X

Wegesser, T.C. & Last, J.A. (2009) Mouse lung inflammation after instillation of particulate matter collected from a working dairy barn. *Toxicology and Applied Pharmacology*, Vol. 236, No. 3 (March, 2009) pp. 348-357. ISSN:0041-008X

Westergren-Thorsson, G.; Larsen, K.; Nihlberg, K.; Andersson-Sjoland, A.; Hallgren, O.; Marko-Varga, G. & Bjermer, L. (2010). Pathological airway remodeling in inflammation. *The Clinical Respiratory Journal*, Vol.4, No.1 suppl, (May 2010), pp. 1-8, *ISSN*: 1752-6981

Wink, DA.; Darbyshire, JF.; Nims, RW.; Saavedra, JE. & Ford, PC. (1993). Reactions of the Bioregulatory Agent Nitric Oxide in Oxygenated Aqueous Media: Determination of the Kinetics for Oxidation and Nitrosation by Intermediates Generated in the NO/O2 Reaction. *Chemical Research Toxicology*, Vol. 6, (Jan-Feb 1993), pp.23–7, ISSN 0893-228X

Yang, W. & Omaye, S.T. (2009). Air Pollutants, Oxidative Stress and Human Health. *Mutation Research/ Genetic Toxicology and Environm Mutagenesis*, Vol.674, No.1-2, (March 2009), pp. 45-54, ISSN 1383-5718

Zuurbier, M.; Hoek, G.; Oldenwening, M.; Meliefste, K.; van den Hazel, P., & Brunekreef, B. (2011). Respiratory Effects of Commuters' Exposure to Air Pollution in Traffic. *Epidemiology*. Vol. 22, No.2, (March 2011), pp. 219-27, *ISSN*: 1044-3983

4

Air Pollution and Cardiovascular Disease

Jan Emmerechts[1], Lotte Jacobs[2] and Marc F. Hoylaerts[1]
[1]*Center for Molecular and Vascular Biology*
[2]*Occupational & Environmental Medicine, Unit of lung toxicology*
University of Leuven
Belgium

1. Introduction

Numerous epidemiological studies report consistent associations between exposure to urban air pollution and cardio-respiratory morbidity and mortality. One of the important discoveries of these epidemiological studies during the last decade was that the increased mortality associated with enhanced air pollution exposure was not due only to pulmonary diseases, but mainly to *cardiovascular* diseases. (Zanobetti et al. 2003, Samet et al. 2000, Dockery et al. 1993, Jerrett et al. 2005, Pope et al. 2004a, Pope et al. 2002, Simkhovich, Kleinman and Kloner 2008, Nawrot, Nemmar and Nemery 2006, Hoek et al. 2002, Katsouyanni et al. 2001, Dominici et al. 2003).

The focus in the initial epidemiological research was directed towards the association between both short-term and long-term exposure to air pollution and arterial cardiovascular effects, such as myocardial infarction. These landmark studies, in the beginning of the 90's, were quickly followed by experimental studies in humans and in rodents, to unravel the underlying pathophysiological mechanisms. The number of publications in this field increased exponentially, so that by the beginning of 2011, a search through PubMed using the MeSH terms 'air pollution' and 'cardiovascular disease' retrieved almost 1300 hits.

Ambient environmental air pollutants include gaseous (carbon monoxide, nitrogen oxides, sulfur dioxide, ozone) and particulate components. The particulate component, particulate matter (PM), is subdivided based on size ranges into 'thoracic particles' (PM_{10}, with a mean aerodynamic diameter <10 μm), 'coarse particles' (>2.5 μm and <10 μm), 'fine particles' ($PM_{2.5}$, <2.5 μm), and ultra-fine particles (UFP, <0.1 μm). Although exposure to some gaseous components has been linked to cardiovascular events, the larger body of evidence points towards the deleterious effects of the particulates in air pollution. Therefore, this chapter will focus mainly on the cardiovascular morbidity induced by PM exposure.

Active cigarette smoking has been established as a major independent cause of cardiovascular disease (HHS 2004). The inhaled dose of fine particles from ambient air pollution, as from secondhand cigarette smoke, is extremely small compared with that from active cigarette smoking. Accordingly, the estimated relative risks from active smoking, even at relatively light smoking levels, are substantially larger than the risks from ambient air pollution or secondhand smoke. However, the risks induced by these latter 2 types of exposure are higher than would be expected from a simple linear extrapolation based on the amount of inhaled PM from active smoking (Pope et al. 2009), and have important public health implications (Nawrot et al. 2011).

Arterial and venous thrombosis share common risk factors (Lowe 2008). The role of air pollution exposure as a risk factor for *arterial* events now being beyond discussion, a few years ago, epidemiologists started investigating a possible association with *venous* thrombotic events. Thus, in 2008, Baccarelli et al. demonstrated a link between chronic exposure to elevated levels of air pollution and deep vein thrombosis (DVT) for the first time. To understand the pathophysiological mechanisms underlying the observed link between air pollution and cardiovascular morbidity, one should take into account the complex interplay of prohemostatic and antihemostatic mechanisms, with different protagonists for the arterial and the venous vasculature. The human cardiovascular system consists of a functional vascular network for blood distribution, subdivided in a systemic and pulmonary circulatory system. The systemic circulation transports oxygenated blood through the arteries from the left heart to the organs and returns oxygen-depleted blood through the veins to the lungs. The pulmonary circulation subsequently transports the oxygen-depleted blood from the heart to the lungs, where it is oxygenated and returned to the heart.

Vascular integrity throughout the vascular tree is maintained by the vessel wall itself, as well as by a complex hemostatic mechanism involving blood platelets and coagulation factors.

The critical need to rapidly form a stable, localized clot in response to vascular injury (='hemostasis') must be balanced with the need to maintain blood flow within the vessels. Different antihemostatic mechanisms prevent clot formation under resting physiological conditions, and limit clot growth to the site of vascular injury. When prohemostatic tendencies proceed beyond the physiological need to maintain vascular integrity, a pathological thrombus may form, obstructing the normal blood flow (='thrombosis'). In the arterial system, thrombus formation induces oxygen-deprivation (ischemia) of the downstream tissues, such as myocardial infarction and cerebral ischemia. The formation of an arterial thrombus largely depends on the activation of blood platelets, and is most often triggered by the rupture of an atherosclerotic plaque. Indeed, the chronic localized deposition of lipids into the arterial vessel wall (atherosclerosis) leads to the formation of plaques that can rupture when unstable, hereby exposing their procoagulant contents to the circulation (Ross 1999). Hence, while often being asymptomatic in itself over many years, atherosclerosis formation may cumulate into an acute burst of symptomatic arterial thrombus formation.

In the venous system, thrombus formation results from a decrease in blood flow, in conjunction with a hypercoagulable state and endothelial dysfunction (Virchow's triad), and most often affects the deep veins of the legs (deep vein thrombosis, DVT). The most serious complication of DVT is the embolisation of clot dislodgements to the lungs (pulmonary embolism, PE).

The following paragraphs will describe how air pollutants affect arterial and venous functionality and lead to pathophysiological manifestations.

2. Particle triggered pathophysiological mechanisms

Inhaled particles deposit in various segments of the human respiratory tract. While the larger PM_{10} particles impact to a large extent in the nasopharyngeal and tracheal region, the smaller $PM_{2.5}$ particles penetrate deeper into the bronchi and bronchioli, whereas the UFP reach the alveolar regions. Inhaled particles are believed to affect the cardiovascular system through 3 different pathways: interference with the autonomic nervous system, direct translocation of UFP into the systemic circulation and pulmonary inflammation.

PM inhalation deranges the autonomic nervous control of the heart (Brook et al. 2004). Numerous studies (e.g. (Park et al. 2010, Pope et al. 1999)) have shown that, by reducing the heart rate variability, PM may increase the risk for cardiac arrhytmias and sudden death . In addition, elevations in air pollution have been associated with ST-segment depression (Pekkanen et al. 2002, Mills et al. 2007), an impaired cardiac deceleration capacity (Schneider et al. 2010), hypertension (Ibald-Mulli et al. 2001) and increased diastolic blood pressure (Urch et al. 2005). The exact underlying mechanisms remain to be elucidated, but stimulation of irritant receptors in the airways and subsequent reflex activation of the nervous system as well as direct effects of pollutants on cardiac ion channels have been suggested (Brook et al. 2004, Pope et al. 2004b).

A second mechanism of action comprises the translocation of inhaled particles into the systemic circulation. Direct effects may occur via UFP that readily cross the pulmonary epithelial barrier, along with soluble constituents released from the larger particles (e.g. transition metals). Systemic translocation of particles was demonstrated in experimental animal models (Nemmar et al. 2001) (Oberdorster et al. 2002). Although evidence of systemic translocation from human studies is less clear, with both positive (Nemmar et al. 2002, Pery et al. 2009) and negative (Mills et al. 2006) findings, it is likely that this pathway also exists in humans, given the deep penetration of UFP into the alveoli and the close apposition of the alveolar wall and the capillary network. Radioactivity in the systemic circulation was already detected 1 minute after the inhalation of radioactively labelled carbon particles in humans, with peak radioactivity levels between 10 and 20 minutes (Nemmar et al. 2002). When measured in rats under resting conditions, only a small fraction (1.6-2.5%) of intratracheally instilled UFP translocated into the circulation. However, this fraction increased to 4.7% following pretreatment of the lungs with lipopolysaccharides, suggesting a role for pulmonary inflammation in enhancing the extrapulmonary translocation of particles (Chen et al. 2006). Different translocation mechanisms, ranging from endocytosis by alveolar type I and endothelial cells, over phagocytosis by macrophages to passage through widened tight junctions are recognized and depend on the particle surface chemistry (Oberdorster, Oberdorster and Oberdorster 2005). However, a detailed description of the different pathways is beyond the scope of this article. Once UFP have translocated to the blood circulation, they can be distributed throughout the body, and interact with the vascular endothelium or circulating cells, such as blood platelets and leukocytes.

Inhaled PM executes its deleterious effects also via a third, more chronic mechanism, namely pulmonary inflammation and oxidative stress. Exposure to PM induces a proinflammatory response in human lungs (Ghio, Kim and Devlin 2000), consistent with observations in *in vivo* animal models (Nemmar et al. 2003c, Emmerechts et al. 2010) and in vitro cellular models (Mitschik et al. 2008, Alfaro-Moreno et al. 2008). The presence of soluble transition metals in PM enhances the inflammatory responses via increased oxidative stress (Jiang et al. 2000). The PM-induced pulmonary inflammation is followed by the release of inflammatory cytokines, such as interleukin (IL)-1β, IL-6 and granulocyte macrophage colony-stimulating factor (van Eeden et al. 2001) in the circulation, resulting in the release of bone-marrow derived neutrophils and monocytes (Tan et al. 2000).

The generation of a systemic inflammatory response, mostly demonstrated by increases in C-reactive protein (CRP) (Peters et al. 2001b, Hertel et al. 2010), is of major importance in the pathogenesis of cardiovascular events. Upon PM exposure, IL-6 translocates from the lung into the systemic circulation (Kido et al. 2011) and is directly involved in the regulation of the synthesis of CRP in the liver. Elevated concentrations of IL-6 are associated with an increased risk of cardiovascular events (Ridker et al. 2000, Lindmark et al. 2001) and mortality (Volpato et al. 2001). Knock-out mice that lacked IL-6 were protected against the prothrombotic effects of PM exposure (Mutlu et al. 2007). Increasing evidence points to an extensive cross-talk between inflammation and hemostasis, whereby inflammation leads to activation of blood platelets and of coagulation, and activated blood platelets and coagulation factors also considerably contribute to the inflammatory action (Levi and van der Poll 2010).

In the following paragraphs, the deleterious effects of PM exposure on arterial and venous parameters will be discussed. By virtue of their respective protagonist roles, blood platelet activation will mainly be discussed in the paragraph on arterial events, while coagulation activation will mainly be discussed in the paragraph on venous events. Formally, arterial thrombosis, the basis for myocardial infarction, is the result of vessel wall injury and formation of a platelet-rich thrombus. Venous thrombosis, the basis for VTE (venous thromboembolism) results from coagulation activation and formation of a fibrin-rich thrombus. It should be noted, however, that both blood platelet and coagulation activation intervene in arterial and venous thrombosis, and that both systems highly interact with each other (Prandoni 2009).

3. Air pollution and arterial events

Over the last 2 decades, a vast number of epidemiological studies (reviewed in (Maitre et al. 2006)) have provided convincing evidence to conclude that chronic exposure to PM enhances atherosclerosis and that acute exposure increases the risk of atherosclerotic plaque rupture, triggering arterial thrombosis, myocardial infarction and cardiovascular mortality. Relative risk levels for cardiovascular disease may differ between different studies, due to differences in study design. Short-term effects have been most often studied in time-series and case-crossover studies, while long-term effects have been studied in case-control and cohort studies. Relative risk levels are generally lower in time series studies than in other epidemiological designs. Nevertheless, the associations between cardiovascular disease and PM exposure are consistent, whatever the type of method used (Maitre et al. 2006).

An initial landmark report was that of the Harvard Six Cities study (Dockery et al. 1993), in which a cohort of 8111 adults were followed up for 14 to 16 years. The adjusted overall mortality rate for the most polluted city vs. the least polluted was 1.26 (95%CI 1.08-1.47). Cardiovascular deaths accounted for the largest single category of increased mortality. Each 10 $\mu g/m^3$ increase in long-term levels of $PM_{2.5}$ has been associated with a 8 to 18% increase in cardiovascular mortality (Pope et al. 2004a). An association with mortality was also found for traffic-related air pollution and several traffic exposure variables, although relative risks were small (Beelen et al. 2008). The effects of long-term PM exposure on cardiovascular mortality have been shown elegantly by the demonstration of a parallelism between air quality improvement and reduction in cardiovascular events on a population-based level (Laden et al. 2006, Boldo et al. 2011). A potential benefit in general mortality can be expected within 2 years after the reduction of PM exposure (Schwartz et al. 2008).

The magnitude of these associations appeared to be more pronounced for the smaller $PM_{2.5}$ fraction than for the larger PM_{10} fraction (Puett et al. 2009). Considering a large body of evidence, a recent updated version of the American Heart Association scientific statement on 'Air Pollution and Cardiovascular Disease' (Brook et al. 2010) concluded that per 10 $\mu g/m^3$ increase in long-term levels of $PM_{2.5}$, all-cause mortality increased by an approximate 10%. The mortality risk specifically related to cardiovascular disease appears to be elevated to a similar, or perhaps even greater extent, ranging from 3 to 76% over different studies.

3.1 Chronic PM exposure and atherosclerosis

What etiological agent can explain the link between chronic air pollution exposure and cardiovascular mortality? Künzli et al. provided the first epidemiological evidence for an association with atherosclerosis: living in the areas of Los Angeles with highest annual mean concentrations of ambient $PM_{2.5}$ was associated with an increased intima-media thickness of the carotid artery (Kunzli et al. 2005).

Distance from the residence to a major road correlated with the degree of coronary artery calcification, a measure for atherosclerosis (Hoffmann et al. 2007).

Another study in 5172 adults investigated 20-year PM exposure and found an association, although weaker than in the previous studies, with carotid intima media thickness, but not with other measures of atherosclerosis i.e. coronary calcium and ankle brachial index (Diez Roux et al. 2008).

A recent study demonstrates that long-term PM exposure is not only related to the degree, but also to a faster progression rate of atherosclerosis (Kunzli et al. 2010).

Along with this epidemiological evidence, experimental research also established a link between exposure to PM and the development of atherosclerosis. Repeated exposure to PM_{10} in rabbits was associated with both systemic inflammation and the progression of the atherosclerotic process, the extent of which correlated with the extent of PM_{10} phagocytosed by alveolar macrophages (Suwa et al. 2002).

Exposing genetically susceptible apolipoprotein E-null mice for 6 months to an equivalent concentration of 15.2 $\mu g/m^3$ $PM_{2.5}$ over a lifetime, enhanced abdominal aortic plaque formation as compared to mice exposed to filtered air (Sun et al. 2005). Interestingly, ultrafine (<0.18 μm) particle-exposed mice exhibited significantly larger atheroslerotic lesions than mice exposed to fine (<2.5 μm) particles or filtered air (Araujo et al. 2008).

Atherosclerosis is now considered an inflammatory disease with low density lipoprotein (LDL) cholesterol accumulation in the arteries as the primary risk factor (Ross 1999). However, up to 50% of the patients who develop atherosclerosis do not have high cholesterol (Braunwald 1997). Therefore, it is the relationship between the accumulated lipids and other harmful components of inflammation in the arterial vessel wall that is of concern. LDL infiltration of the arterial vessel wall is followed by oxidative modification to oxidized LDL (ox-LDL) in the subendothelial space and chemotaxis of monocytes. These monocytes differentiate into macrophages and the subsequent phagocytosis of ox-LDL leads to the formation of foam cells and the release of inflammatory mediators, inducing a vicious cycle of inflammation. Further stages include smooth muscle cell proliferation, formation of a fibrous cap with necrotic core and calcification (Ross 1999). Thickening of the vessel wall and obliteration of the vascular lumen induces downstream ischemia of the tissues.

PM exposure can induce atherosclerosis via different pathways: systemically translocated UFP or their chemical constituents induce activation of proatherogenic molecular pathways in endothelial cells, by oxidative stress. Inflammatory mediators released from the lungs may promote chemotaxis of monocytes into the vessel wall. PM induces high-density lipoprotein (HDL) dysfunction with loss of its anti-inflammatory properties (Araujo and Nel 2009).

Oxidative transformation of LDL into ox-LDL is a key step in the initiation and progression of atherosclerosis (Stocker and Keaney 2004), and circulating levels of ox-LDL are therefore an early marker, and a risk factor for the disease (Wallenfeldt et al. 2004). The correlation between PM exposure and circulating levels of ox-LDL on an individual level was shown by Jacobs et al., demonstrating a dose-dependent association between this parameter and the carbon load of airway macrophages, a personal marker for chronic exposure to fossil fuel derived ultrafine particles (Jacobs et al. 2011).

It has been previously shown that particles can induce oxidative stress both in vitro (Jimenez et al. 2000, Carter et al. 1997) and in exposed animals (Costa and Dreher 1997, Kadiiska et al. 1997, Tao, Gonzalez-Flecha and Kobzik 2003, Araujo et al. 2008).

In agreement with epidemiological findings (Puett et al. 2009), experimental studies suggest that the smaller particles are more pathogenic, as a result of their greater propensity to induce systemic prooxidant and proinflammatory effects (Araujo et al. 2008). Indeed, ambient UFP trigger the induction of the antioxidant gene heme oxygenase 1 (HO-1) to a higher degree than ambient $PM_{2.5}$ or coarse particles, both *in vitro* (Li et al. 2004) and *in vivo* (Araujo et al. 2008, Araujo and Nel 2009). Several mechanisms contribute to the greater proatherogenic potential of UFP: because of their small size, particles < 0.1-0.2 μm contribute very little to overall $PM_{2.5}$ mass. However, they represent >85-90% of the total $PM_{2.5}$ particle number (Sioutas, Delfino and Singh 2005). The high number of UFP, in conjunction with a large surface-to-mass ratio increases the bioavailability of the pro-oxidant chemicals (polycyclic aromatic hydrocarbons, transition metals etc.) present on the UFP's surface. The number of chemicals that are displayed on the surface of particles increases exponentially as the size shrinks below 100 nm (Oberdorster et al. 2005). Deep penetration in the lung and subsequent translocation of UFP into the circulation make these pro-oxidant chemicals more bioavailable at the contact sites of the particles with cells and tissues.

3.2 Acute PM exposure and arterial thrombosis

Not only chronic, but also short-term PM exposure has been linked to cardiovascular mortality: Both the American NMMAPS (National Morbidity, Mortality, and Air Pollution Study (Dominici et al. 2003)) and the European APHEA2 (Air Pollution and Health: A European Approach (Katsouyanni et al. 2001, Zanobetti et al. 2003)) studies (approximately 50 million and 43 million persons included respectively) demonstrated small increases in cardiovascular mortality with increasing PM exposure. In an attempt to evaluate the coherence of studies across continents, the APHENA (A Combined European and North American Approach) analyzed data of these 2 aforementioned studies and Canadian studies (Samoli et al. 2008). The combined effect on all-cause mortality ranged from 0.2% to 0.6% for a 10 μg/m^3 increase in daily levels of ambient PM_{10}, with greater effects for the elderly (>75 years) and the unemployed. An extensive review of studies investigating a link between short-term PM exposure and cardiovascular mortality is provided in (Brook et al. 2010).

Peters et al. (Peters et al. 2001a) demonstrated an increased risk of myocardial infarction in association with elevated concentrations of fine $PM_{2.5}$, both in the previous 2-hours period and the day before the onset. Likewise, the onset of myocardial infarction was linked to participation in traffic, as soon as 1 h afterward (odds ratio 2.92, 95%CI 2.22-3.83) (Peters et al. 2004).

Exposure to ambient $PM_{2.5}$ is associated with short-term increases in hospital admission rates for cerebrovascular, peripheral and cardiac ischemic disease, heart rhythm and heart failure, with the strongest association for heart failure (1.28 % 95%CI 0.78-1.78% increase in risk per 10 $\mu g/m^3$ increase in same-day $PM_{2.5}$) (Dominici et al. 2006).

The risk of mortality from coronary heart disease related to PM exposure appears to be higher in women (RR 1.42, 95%CI 1.06-1.90) than in men (RR 0.90, 95%CI 0.76-1.05 per 10 $\mu g/m^3$ increase in $PM_{2.5}$)(Chen et al. 2005). In a study of 65893 postmenopausal women with a median follow-up of 6 years, each increase in long-term levels of $PM_{2.5}$ of 10 $\mu g/m^3$, measured at the women's residence, was associated with a 24% (95%CI 09-41%) increase in the risk of a cardiovascular event, and a 76% (95%CI 25-147%) increase in the risk of death of cardiovascular disease (Miller et al. 2007).

Although the magnitude of the risk on myocardial infarction induced by short-term PM exposure is rather small on a personal level, it is of major importance on a population level, by virtue of the large number of people exposed. Taking into account both risk magnitude and risk prevalence by measurement of the population attributable fraction (PAF), Nawrot et al. showed that a short-term increase in air pollution exposure is an important trigger for myocardial infarction, of similar magnitude (PAF 5-7%) as other well accepted triggers such as physical exertion, alcohol and coffee (Nawrot et al. 2011).

Epidemiological studies suggest an association between short-term increases in PM exposure and atherosclerotic plaque rupture, causing arterial thrombosis and myocardial infarction. In contrast to the growing number of mechanistic studies investigating the role of chronic PM exposure on atherogenesis, the precise mechanisms explaining the role of short-term PM exposure in acute plaque rupture largely remain to be elucidated. However, several epidemiological and mechanistic studies demonstrated that, in parallel to atherosclerotic plaque rupture, direct or indirect activation of circulating blood platelets by PM contributes to the arterial thrombosis risk. Indeed, the extent to which a growing thrombus occludes the vascular lumen may in part depend on platelet hyperactivity.

Under physiological circumstances, the high blood pressure generated on the arterial side of the circulation requires a powerful, almost instantaneous prohemostatic response in order to minimize blood loss from sites of vascular injury. Blood platelets play a critical role in this response. Upon damage of the endothelial cell layer covering the luminal side of blood vessels, circulating blood platelets adhere to the exposed subendothelial matrix through the binding of the glycoprotein (GP) Ib-IX-V receptor to exposed von Willebrand factor (vWF). Blood platelet adhesion is further enhanced by the binding of different GP receptors to other subendothelial matrix proteins, such as collagen and fibrin(ogen). Upon adhesion and activation of the blood platelets by various agonists, vWF and fibrinogen molecules cross-link different platelets, resulting in blood platelet aggregation and the formation of an initial platelet plug which covers the site of endothelial lesion. The simultaneous activation of the coagulation cascade leads to the formation of a network of insoluble fibrin strands that further stabilize the initial platelet plug.

Air pollution exposure can induce an inappropriate activation of blood platelets beyond the physiological need to restore vessel damage, resulting in arterial thrombosis (Fig. 1).

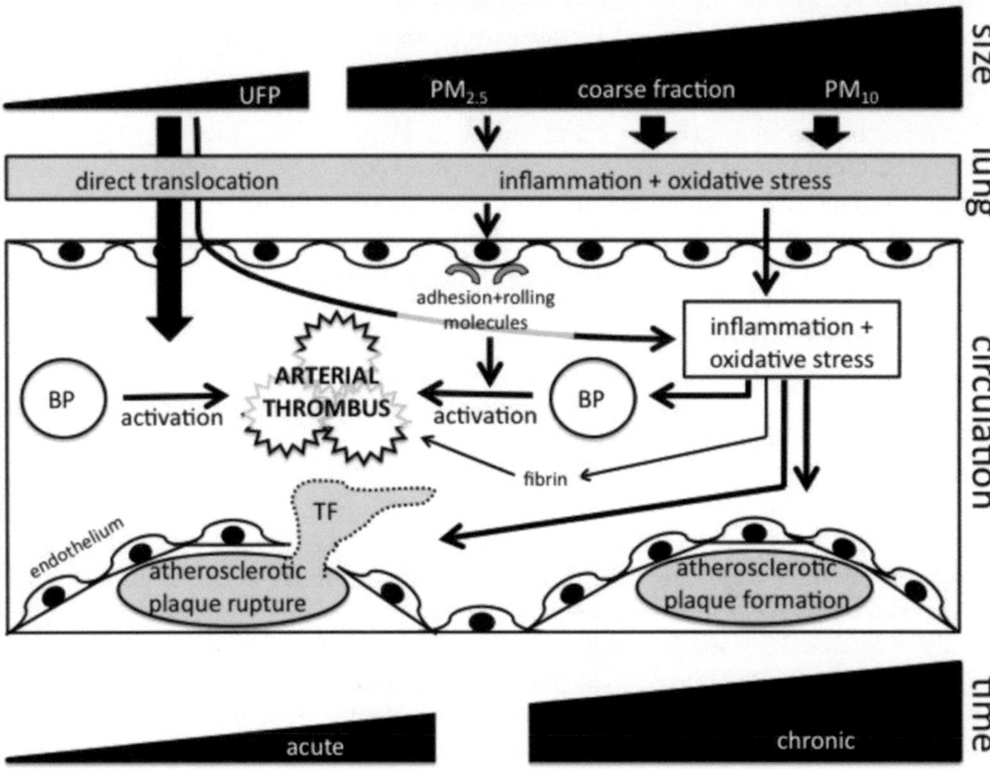

BP: blood platelet, PM: particulate matter, TF: tissue factor, UFP: ultra-fine particles

Fig. 1. Biological pathways linking PM exposure and arterial thrombosis

By exposing healthy volunteers to diluted diesel exhaust, Lucking et al. showed an association with enhanced platelet activity and thrombus formation in an *ex vivo* perfusion chamber, 2 hours and 6 hours after exposure, in conjunction with increased numbers of platelet-neutrophil (+52%) and platelet-monocyte (+30%) conjugates (Lucking et al. 2008).

Short-term, but not long-term PM exposure was found to enhance platelet function, as measured *ex vivo* by a shortening of the closure time of the Platelet Function Analyzer (PFA-100, Siemens Healthcare Diagnostics), in patients with diabetes (Jacobs et al. 2009). In this study, an interquartile range (39.2 $\mu g/m^3$) increase in the PM_{10} concentration, measured 2 hours before the clinical investigation at the entrance of the hospital, was associated with a decrease of 21.1 sec (95%CI -35.3 to -6.8) in the PFA-100 closure time. Platelet function was not correlated with leukocyte counts, suggesting that short-term PM exposure may have effects on platelet function independently of systemic inflammation, as was also shown in experimental animal models (Nemmar et al. 2003c).

Ambient PM_{10} levels have also been associated with augmented platelet aggregation 24 to 96 hours after exposure in healthy adults, in the absence of increased CRP or fibrinogen (Rudez et al. 2009). In patients with coronary heart disease, mean concentrations over 24 hours of ambient UFP, but not $PM_{2.5}$ or PM_{10} were positively associated with the levels of soluble CD40 ligand, a marker for platelet activation. No assocations were found with longer time frames, up to 5 days (Ruckerl et al. 2007b).

In experimental conditions using DEP, Nemmar et al. demonstrated a prothrombotic tendency and activation of circulating blood platelets (confirmed by PFA-100), as well as lung inflammation, which persisted up to 24 hr after intratracheal instillation of DEP in hamsters (Nemmar et al. 2003a, Nemmar et al. 2003c).

However, different pathophysiological mechanisms seem to be responsible for the observed prothrombotic risk at different time points. Pretreatment of hamsters with a histamine H1-receptor antagonist, an anti-inflammatory drug, abolished pulmonary inflammation at all time points and reduced DEP-induced thrombosis at 6 and 24 hours post-instillation, indicating a crucial role for inflammation in thrombogenicity at these time points. Likewise, the administration of other anti-inflammatory drugs, such as dexamethasone and selective inhibitors of basophils, macrophages and neutrophils, also significantly reduced the PM-induced prothrombogenicity at 24 hours (Nemmar et al. 2004, Nemmar et al. 2005).

In contrast, pretreatment with the histamine H1-receptor antagonist did not reduce thrombosis as soon as 1 hour after DEP exposure (Nemmar et al. 2003c). Therefore, the early prothrombotic tendency appears not to result from pulmonary inflammation, but possibly from direct effects of systemically translocated particles on the blood platelets and/or the (pulmonary) vessel wall (Nemmar et al. 2003c). The direct activating effect of PM on blood platelets was shown by the addition of as little as 0.5 μg/mL DEPs to untreated hamster blood, significantly shortenening the PAF-100 closure time (Nemmar et al. 2003a), as well as by a dose-dependent (0.1-1 μg/mL) effect of PM on in vitro platelet aggregation in rat blood (Nemmar et al. 2010), although similar experiments in human blood were negative (Rudez et al. 2009).

In agreement with these results, 1 hour after intratracheal instillation, well-defined positively charged ultrafine (60 nm) polystyrene particles significantly enhanced platelet-rich thrombus formation, while 400 nm particles, incapable of systemic translocation, did not affect thrombus formation, despite similar increases in neutrophils, lactate dehydrogenase and histamine levels in the bronchoalveolar lavage fluid (Nemmar et al. 2003b).

Pulmonary instillation of carbon nanotubes elevated platelet-leukocyte conjugates at 6 hours and increased the peripheral thrombotic potential at 24 hours after exposure. Inhibition of P-selectin abrogated these responses (Nemmar et al. 2007). P-selectin is found in storage Weibel-Palade bodies of endothelial cells and in α-granules of platelets, from where it can be expressed on the outer membrane upon activation. Surface expression of P-selectin initiates capture and rolling of circulating leukocytes over stimulated endothelium (Theilmeier et al. 2002) and the formation of platelet-leukocyte conjugates (Yokoyama et al. 2005). Increased levels of platelet-leukocyte conjugates have been demonstrated in Indian women who used biomass as cooking fuel, producing higher levels of PM, as compared to women cooking with a cleaner fuel (liquefied petroleum gas) (Ray et al. 2006). In a panel study of 60 elderly subjects with coronary artery disease, Delfino et al. demonstrated associations between soluble P-selectin levels and the mean 1 to 5-day concentrations of ambient finer particles ($PM_{0.25}$ and $PM_{2.5}$), but not the bigger PM_{10} (Delfino et al. 2009). Taken together, these studies suggest that the release of pulmonary cell-derived mediators (eg. histamine) and the expression of endothelial and platelet surface proteins (eg. P-selectin) after several hours, along with the more rapid activation of circulating platelets by direct contact with UFP may mediate peripheral prothrombotic effects.

4. Air pollution and venous thromboembolism

4.1 Epidemiology

In addition to the well-recognized PM-related adverse effects on the arterial vascular system, recent epidemiological evidence also suggests an association between exposure to PM and venous thromboembolism (VTE). Baccarelli et al. reported a 70% increase in the risk of deep vein thrombosis (DVT) for each 10 μg/m^3 increase of the annual mean level of PM_{10} in the areas of residence of the study subjects (OR 1.70, 95%CI 1.30-2.23) (Baccarelli et al. 2008). The observed exposure-response relationship was approximately linear over the observed PM_{10} range, so that PM_{10} at the higher concentrations within the international limits can still increase the risk of DVT, as compared to the lowest concentration measured. These authors found, in the same study subjects, that living near major traffic roads was also associated with an increased risk of DVT, even after controlling for the community-level PM pollution (Baccarelli et al. 2009). Very recently, exposure to PM has also been associated with hospital admission for VTE in Chile. Both for DVT and for PE, pooled estimates of relative risk of hospitalization were 1.05 (95%CI 1.03-1.06) for a 20.02 μg/m^3 increase in $PM_{2.5}$ (Dales, Cakmak and Vidal 2010).

However, these initial epidemiological reports on the association between air pollution exposure and venous thrombosis were followed by a number of prospective cohort studies that failed to demonstrate an association: 26,450 post-menopausal women, enrolled in the Women's Health Initiative (WHI) Hormone Therapy trials, were randomized to treatment with either hormone therapy or placebo. Regardless of the treatment category, no evidence was found of an association between short- or long-term (up to 1 year) PM exposure and VTE (Shih et al. 2010). Of note, the aforementioned study of Baccarelli et al. also observed lower PM-induced VTE risk among women compared to men (Baccarelli et al. 2008). A prospective study in 13,134 middle-aged persons, including men and women, also provided evidence against an association between VTE and long-term air pollution exposure, as assessed by residential distance to a major road (Kan et al. 2011).

Hence, in contrast to the well-accepted and documented deleterious effects of air pollution exposure on arterial events, data are scarce and the link with venous thrombosis is less straightforward, prompting further epidemiological investigation.

4.2 Pathophysiology

At lower rates of shear found in the venous circulation, the contribution of blood platelets to clot formation is of lesser importance than in the arterial circulation, leaving a protagonist role for the coagulation cascade in venous hemostasis. Activation of the coagulation cascade is initiated by activation of coagulation factor VII (FVII) by binding to tissue factor (TF), expressed on subendothelial cells such as fibroblasts and vascular smooth muscle cells. The complex of TF and activated FVII (FVIIa) initiates a cascade of subsequent coagulation factor activations, resulting in the generation of thrombin. Thrombin (FII) is a key enzyme, converting fibrinogen monomers to fibrin polymers that clot into a fibrin plug, and amplifying the coagulation cascade through activation of FV, FVIII and FXI.

The mechanisms underlying the observed increase in venous thrombosis in association with exposure to air pollution remain largely unknown, and published results with regard to markers of secondary hemostasis activation are conflicting. Although some epidemiological and controlled exposure studies demonstrated an association between PM exposure and shortening of the prothrombin time (PT) or increased levels of fibrinogen and vWF, others failed to demonstrate positive associations with these or other markers of coagulation, in humans (Table 1). In fact, disappointingly few studies reported on PM-induced coagulatory changes that could form the basis for the observed link between air pollution and DVT. How can this conundrum of PM-induced DVT in the absence of a procoagulant phenotype be explained?

One explanation for the lack of positive associations between PM exposure and measurement of parameters of coagulation might be found in the short observation time frame that was used in most studies. While short-term PM exposure enhances blood platelet activation, a more chronically sustained exposure appears to be necessary to induce significant changes in the coagulatory cascade.

This hypothesis is corroborated by epidemiological findings in which the risk for DVT was only associated with the mean PM concentration over a one year period, and not with any shorter time-point (Baccarelli et al. 2008). This was confirmed by animal studies in which short- term exposure of healthy mice to intratracheally instilled DEP or UPM enhanced arterial, but not venous thrombosis (Emmerechts et al. 2010). In this study, even very high doses of PM (up to 200 µg/mouse), given as a single dose, induced only mild increases in the levels of FVII, FVIII and fibrinogen. Likewise, exposure of rats to concentrated PM from New York City air did not alter levels of fibrinogen, FVII or thrombin-antithrombin complexes (TAT) (Nadziejko et al. 2002).

Significant increases in the level of fibrinogen, or decreases in the levels of the anticoagulant proteins activated protein C or tissue factor pathway inhibitor (TFPI) upon short-term PM exposure have been observed in rodents, but at doses of 100 µg or higher per mouse (Cozzi et al. 2007, Inoue et al. 2006). One study stands out among other studies on procoagulant changes and PM exposure: Mutlu et al. observed a pronounced prothrombotic phenotype in mice upon a single intratracheal instillation of as few as 10 µg of PM_{10}, characterized by shortenings in bleeding time, PT and aPTT, and relatively high increases in the levels of circulating blood platelets, FVII, FVIII, FX and fibrinogen (Mutlu et al. 2007). The reason for the discrepancy between this and other studies being unclear, this study is of value since it demonstrated the absence of a PM-induced prothrombotic phenotype in interleukin-6 (IL-6) knock-out mice, recognizing a major role of inflammatory factors in the induction of procoagulant changes following PM exposure.

Indeed, although some studies suggest a short-term effect of directly translocated UFP through the activation of the coagulation cascade via contact activation, as demonstrated *in* vitro (Kilinc et al. 2010), evidence seems to favor a more prominent role for inflammatory changes related to chronic PM exposure. In this context, it is of interest that the only coagulation factor for which the associations with air pollution were consistent over different studies in humans is fibrinogen (Table 1), an acute phase protein that is upregulated during inflammatory processes.

However, although considered to be a (minor) risk factor, elevated levels of fibrinogen seem unlikely to be solely responsible for the PM-induced increased risk of DVT.

Through the expression of procoagulant proteins and lipids on their surface, microvesicles (also called microparticles, a term we prefer to avoid in the context of pollution by particles) could offer an alternative explanation. Microvesicles are circulating vesicles released from stimulated or apoptotic cells in the vasculature, or during thrombogenesis in the bone marrow, with a mean diameter smaller than 1 µm. Through their surface expression of

reference	subjects					exposure		coagulatory changes	
author + year	n	type	mean age (SD or range)	gender (% male)	controlled exposure	type of air pollution	exposure time	significant changes	no significant changes
(Seaton et al. 1999)	112	NA	70 (7)	?	no	ambient PM_{10}	3 days	FVII (-), fbg (-)	
(Ghio et al. 2000)	38	healthy subjects	26 (0.7)	95	yes	concentrated $PM_{2.5}$	2h	fbg	
(Pekkanen et al. 2000)	7205	healthy subjects	NA	69	no	ambient PM_{10}	1-3 days	fbg	
(Ghio et al. 2003)	20	healthy subjects	25 (0.8)	70	yes	concentrated $PM_{2.5}$	2h	fbg	D-dim, PC, vWF
(Riediker et al. 2004)	9	healthy subjects	27 (23-30)	100	no	in-vehicle $PM_{2.5}$	9h	vWF	
(Beckett et al. 2005)	12	healthy subjects	35 (23-52)	50	yes	zinc oxide particles	2h		fbg, FVII, vWF
(Blomberg et al. 2005)	15	COPD patients	66 (56-72)	NA	yes	diesel exhaust	1h		fbg, D-dim, vWF
(Barregard et al. 2006)	13	healthy subjects	34 (20-56)	46	yes	wood smoke	4h	FVIII	fbg, FVII, D-dim, vWF
(Ruckerl et al. 2006)	57	CHD patients	66 (6)	100	no	ambient $PM_{2.5}$ and PM_{10}	1-5 days	FVII (-), vWF	fbg, D-dim
(Baccarelli et al. 2007)	1218	healthy subjects	44 (11-84)	40	no	ambient PM_{10}	t0 - 30days	PT	aPTT, fbg, AT, PC, PS
(Carlsten et al. 2007)	13	healthy subjects	25 (20-42)	85	yes	diesel exhaust	2h		D-dim, vWF
(Chuang et al. 2007)	76	healthy students	21 (18-25)	65	no	ambient $PM_{2.5}$ and PM_{10}	1-3 days	fbg	
(Ruckerl et al. 2007a)	1003	MI survivors	65 (45-78)	69	no	ambient $PM_{2.5}$ and PM_{10}	1-4 days	fbg	
(Scharrer et al. 2007)	20	healthy subjects	29 (8)	60	yes	welding fume	1h		FVIII, vWF, AT
(Brauner et al. 2008)	41	healthy subjects	67 (60-75)	51	yes	indoor $PM_{2.5}$ and PM_{10}	2 days		fbg, FII+VII+X
(Lucking et al. 2008)	20	healthy subjects	26 (21-44)	NA	yes	diesel exhaust	1-2h		PT, aPTT
(Rudez et al. 2009)	40	healthy subjects	41 (15)	35	no	ambient PM_{10}	6h-4days		fbg,TG
(Samet et al. 2009)	19	healthy subjects	18-35	53	yes	concentrated ambient UFP	2h	D-dim	fbg, FIX, FXII, vWF
(Bonzini et al. 2010)	37	steel plant workers	42 (7)	100	no	occupational PM_{10}	1-3 days	PT,TG	aPTT, D-dim
(Stewart et al. 2010)	19	T2DM patients	48 (9)	47	yes	carbon UFP	2h		FVII, FIX, D-dim, TF
(Thompson et al. 2010)	45	healthy subjects	27 (19-48)	49	no	ambient $PM_{2.5}$	t0 - 7 days		fbg
(Jacobs et al. 2011)	70	DM patients	57 (14)	53	no	carbon load in alveolar Mφ	NA		vWF

COPD: chronic obstructive pulmonary disease, MI: myocardial infarction, CHD: coronary heart disease, DM: diabetes mellitus, T2DM: type 2 diabetes mellitus, PM: particulate matter, UFP: ultra-fine particles, Mφ: macrophages, PT: prothrombin time, aPTT: activated partial prothrombin time, AT: antithrombin, PC: protein C, PS: protein S, F: factor, fbg: fibrinogen, D-dim: D-dimers, vWF: von Willebrand factor, TG: thrombin

Table 1. Associations between PM exposure and coagulatory changes according to different studies.

negatively charged phospholipids and of tissue factor (TF), they create a procoagulant surface on which coagulation factors can bind and be activated (Morel et al. 2006). Indeed, the initial concept that TF presence is limited to a hemostatic envelope surrounding blood vessels has been challenged by the identification of 'blood borne' TF, either on circulating white blood cells or microvesicles, as a soluble protein, or possibly on stimulated endothelial cells (Pawlinski et al. 2010).

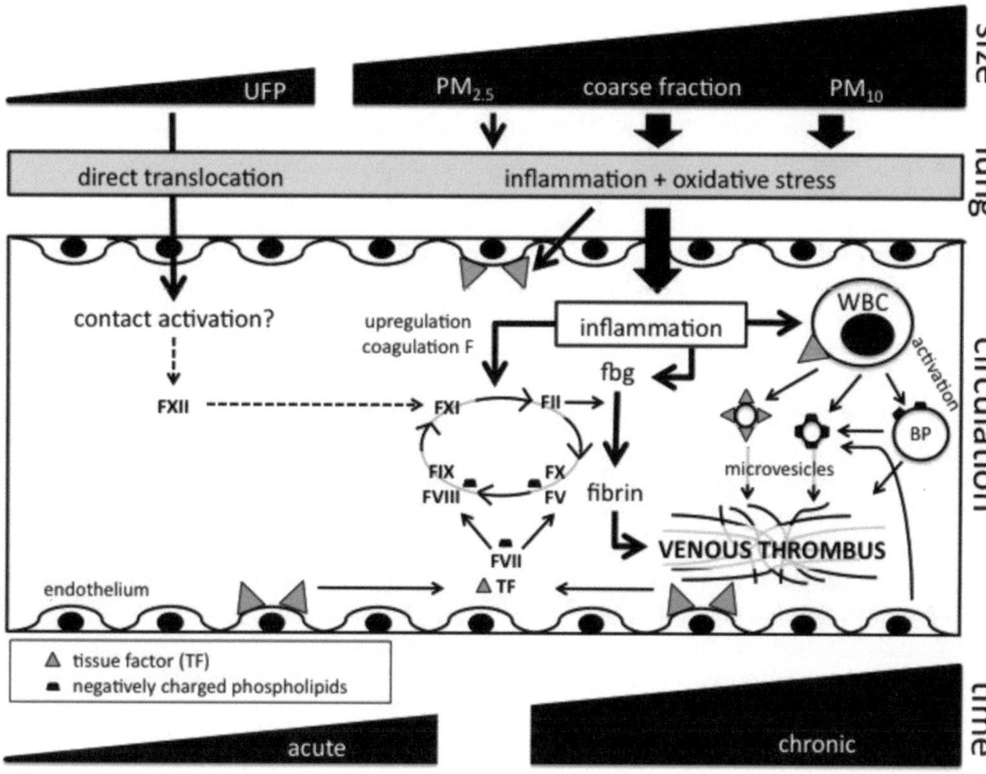

BP: blood platelet, F: factor, fbg: fibrinogen, PM: particulate matter, TF: tissue factor, UFP: ultra-fine particles, WBC: white blood cell

Fig. 2. Biological pathways linking PM exposure and venous thrombosis

A role for microvesicles has been suggested by the work of Bonzini et al., investigating blood samples collected in steel-production plant workers. Besides shortening the PT, elevated PM exposure also enhanced thrombin generation, but only when measured in an assay without the exogenous addition of a coagulation trigger or negatively charged phospholipids (Bonzini et al. 2010). These findings suggest that PM exposure may induce the release of small amounts of endogenous TF and/or negatively charged phospholipids that may function as triggers of thrombin generation in the assay system. Circulating microvesicles might well be the source of these triggers. This hypothesis is corroborated by animal studies demonstrating elevated numbers of procoagulant microvesicles, 24 hours after intratracheal instillation of carbon nanotubes in mice (Nemmar et al. 2007). Likewise, when stimulated *ex vivo*, blood platelets from mice exposed to concentrated ambient PM for 2 weeks released more microvesicles relative to platelets from ambient air-exposed control animals (Wilson et al. 2010). However, observational or controlled exposure studies in humans are needed for further confirmation. Figure 2 summarizes the possible pathophysiological pathways linking PM exposure and venous thrombogenicity.

5. Endothelial function and fibrinolysis

The effects of air pollution on the endothelial function and the fibrinolytic system have mainly been investigated in controlled exposure studies by 2 research groups who joined forces. The groups of Newby and Blomberg used exposure chambers to expose healthy and compromised volunteers to the diluted exhaust of an iddling diesel engine for several hours in randomized cross-over studies. They demonstrated an impaired bradykinin-induced endothelial release of tissue plasminogen activator (t-PA) upon diesel exhaust inhalation (estimated reduction of net t-PA release of 34%) (Mills et al. 2005, Mills et al. 2007), in addition to an attenuated agonist-induced increase in blood flow at 6 hours post-inhalation, in the absence of inflammatory changes (Mills et al. 2005). At 24 hours post-inhalation, endothelium-dependent vasodilatation (induced by acetylcholine and bradykinin) remained impaired, while endothelium-independent vasodilatation (using sodium nitroprusside and verapamil) and t-PA release were unaffected, in the presence of mild sytemic inflammation (Tornqvist et al. 2007).

These and other (Bonzini et al. 2010, Chuang et al. 2007, Ghio et al. 2003, Samet et al. 2009) studies did not demonstrate an assocation between PM exposure and baseline levels (not bradykinin-induced) of t-PA.

While studies, based on controlled exposure to diluted diesel exhaust (Mills et al. 2007, Tornqvist et al. 2007, Carlsten et al. 2007) or concentrated ambient particles (Ghio et al. 2003), did not observe increases in the levels of plasminogen activator inhibitor-1 (PAI-1), some epidemiological or animal studies, focussing on urban PM, did: a study in 76 young healthy students demonstrated elevated PAI-1 concentrations in association with the mean $PM_{2.5}$ or PM_{10} concentration at their university's campus over 1 to 3 days (Chuang et al. 2007). Likewise, urban PM upregulated PAI-1 levels, 24 hours after intratracheal instillation in mice (Cozzi et al. 2007).

PM exposure could also impair the endothelial repair mechanisms by reducing the number of endothelial progenitor cells, as demonstrated by a recent report (O'Toole et al. 2010).

Taken together, these studies indicate a potential deleterious effect of PM inhalation on the endothelial and fibrinolytic function that may aggravate the prothrombotic phenotype induced by blood platelet and coagulation activation.

6. Conclusions

A wide array of epidemiological and experimental studies have provided persuasive evidence that air pollutants, the PM fraction in particular, contribute to cardiovascular morbidity and mortality. By virtue of the heterogeneity in both study design and the composition of the PM considered by these studies, it is not surprising that not all findings have been consistent. However, considering the overall weight of scientific evidence, some general conclusions can be drawn: through the induction of inflammation and oxidative stress, the inhalation of particulates, especially the finest fractions ($PM_{2.5}$ and UFP), over longer time periods contributes to atherosclerotic plaque formation. At shorter time points (<24 h), these particles may induce plaque rupture and activate blood platelets, leading to acute peripheral arterial events such as myocardial infarction. Blood platelet activation within the first few hours is inflammation-independent, most probably resulting from direct contact with systemically translocated particles and/or activated endothelium. Thereafter, inflammatory changes are responsible for further platelet activation.

Although evidence linking PM exposure with venous thromboembolic events is less established than with arterial events and warrants further investigation, recent findings suggest that chronic air pollution exposure is also a risk factor for venous thrombosis. Inflammatory changes, along with the generation of circulating procoagulant microvesicles might be of larger importance than coagulation factor upregulation, favoring a role for the larger particles (PM_{10}) with higher pro-inflammatory endotoxin content on their surface.

Air pollution exposure may not be the highest risk factor for arterial or venous thrombosis on an individual level. However, because of the huge number of persons exposed, on a global scale it is a major, and more importantly, a modifiable risk factor for cardiovascular disease and mortality.

7. References

Alfaro-Moreno, E., T. S. Nawrot, B. M. Vanaudenaerde, M. F. Hoylaerts, J. A. Vanoirbeek, B. Nemery & P. H. Hoet (2008) Co-cultures of multiple cell types mimic pulmonary cell communication in response to urban PM10. *Eur Respir J,* 32, 1184-94.

Araujo, J. A., B. Barajas, M. Kleinman, X. Wang, B. J. Bennett, K. W. Gong, M. Navab, J. Harkema, C. Sioutas, A. J. Lusis & A. E. Nel (2008) Ambient particulate pollutants in the ultrafine range promote early atherosclerosis and systemic oxidative stress. *Circ Res,* 102, 589-96.

Araujo, J. A. & A. E. Nel (2009) Particulate matter and atherosclerosis: role of particle size, composition and oxidative stress. *Part Fibre Toxicol,* 6, 24.

Baccarelli, A., I. Martinelli, V. Pegoraro, S. Melly, P. Grillo, A. Zanobetti, L. Hou, P. A. Bertazzi, P. M. Mannucci & J. Schwartz (2009) Living near major traffic roads and risk of deep vein thrombosis. *Circulation,* 119, 3118-24.

Baccarelli, A., I. Martinelli, A. Zanobetti, P. Grillo, L. F. Hou, P. A. Bertazzi, P. M. Mannucci & J. Schwartz (2008) Exposure to particulate air pollution and risk of deep vein thrombosis. *Arch Intern Med,* 168, 920-7.

Baccarelli, A., A. Zanobetti, I. Martinelli, P. Grillo, L. Hou, S. Giacomini, M. Bonzini, G. Lanzani, P. M. Mannucci, P. A. Bertazzi & J. Schwartz (2007) Effects of exposure to air pollution on blood coagulation. *J Thromb Haemost,* 5, 252-60.

Barregard, L., G. Sallsten, P. Gustafson, L. Andersson, L. Johansson, S. Basu & L. Stigendal (2006) Experimental exposure to wood-smoke particles in healthy humans: effects on markers of inflammation, coagulation, and lipid peroxidation. *Inhal Toxicol,* 18, 845-53.

Beckett, W. S., D. F. Chalupa, A. Pauly-Brown, D. M. Speers, J. C. Stewart, M. W. Frampton, M. J. Utell, L. S. Huang, C. Cox, W. Zareba & G. Oberdorster (2005) Comparing inhaled ultrafine versus fine zinc oxide particles in healthy adults: a human inhalation study. *Am J Respir Crit Care Med,* 171, 1129-35.

Beelen, R., G. Hoek, P. A. van den Brandt, R. A. Goldbohm, P. Fischer, L. J. Schouten, M. Jerrett, E. Hughes, B. Armstrong & B. Brunekreef (2008) Long-term effects of traffic-related air pollution on mortality in a Dutch cohort (NLCS-AIR study). *Environ Health Perspect,* 116, 196-202.

Blomberg, A., H. Tornqvist, L. Desmyter, V. Deneys & C. Hermans (2005) Exposure to diesel exhaust nanoparticles does not induce blood hypercoagulability in an at-risk population. *J Thromb Haemost,* 3, 2103-5.

Boldo, E., C. Linares, J. Lumbreras, R. Borge, A. Narros, J. Garcia-Perez, P. Fernandez-Navarro, B. Perez-Gomez, N. Aragones, R. Ramis, M. Pollan, T. Moreno, A. Karanasiou & G. Lopez-Abente (2011) Health impact assessment of a reduction in ambient PM(2.5) levels in Spain. *Environ Int,* 37, 342-8.

Bonzini, M., A. Tripodi, A. Artoni, L. Tarantini, B. Marinelli, P. A. Bertazzi, P. Apostoli & A. Baccarelli (2010) Effects of inhalable particulate matter on blood coagulation. *J Thromb Haemost,* 8, 662-8.

Brauner, E. V., L. Forchhammer, P. Moller, L. Barregard, L. Gunnarsen, A. Afshari, P. Wahlin, M. Glasius, L. O. Dragsted, S. Basu, O. Raaschou-Nielsen & S. Loft (2008) Indoor particles affect vascular function in the aged: an air filtration-based intervention study. *Am J Respir Crit Care Med,* 177, 419-25.

Braunwald, E. (1997) Shattuck lecture--cardiovascular medicine at the turn of the millennium: triumphs, concerns, and opportunities. *N Engl J Med,* 337, 1360-9.

Brook, R. D., B. Franklin, W. Cascio, Y. Hong, G. Howard, M. Lipsett, R. Luepker, M. Mittleman, J. Samet, S. C. Smith, Jr. & I. Tager (2004) Air pollution and cardiovascular disease: a statement for healthcare professionals from the Expert Panel on Population and Prevention Science of the American Heart Association. *Circulation,* 109, 2655-71.

Brook, R. D., S. Rajagopalan, C. A. Pope, 3rd, J. R. Brook, A. Bhatnagar, A. V. Diez-Roux, F. Holguin, Y. Hong, R. V. Luepker, M. A. Mittleman, A. Peters, D. Siscovick, S. C. Smith, Jr., L. Whitsel & J. D. Kaufman (2010) Particulate matter air pollution and cardiovascular disease: An update to the scientific statement from the American Heart Association. *Circulation,* 121, 2331-78.

Carlsten, C., J. D. Kaufman, A. Peretz, C. A. Trenga, L. Sheppard & J. H. Sullivan (2007) Coagulation markers in healthy human subjects exposed to diesel exhaust. *Thromb Res,* 120, 849-55.

Carter, J. D., A. J. Ghio, J. M. Samet & R. B. Devlin (1997) Cytokine production by human airway epithelial cells after exposure to an air pollution particle is metal-dependent. *Toxicol Appl Pharmacol,* 146, 180-8.

Chen, J., M. Tan, A. Nemmar, W. Song, M. Dong, G. Zhang & Y. Li (2006) Quantification of extrapulmonary translocation of intratracheal-instilled particles in vivo in rats: effect of lipopolysaccharide. *Toxicology,* 222, 195-201.

Chen, L. H., S. F. Knutsen, D. Shavlik, W. L. Beeson, F. Petersen, M. Ghamsary & D. Abbey (2005) The association between fatal coronary heart disease and ambient particulate air pollution: Are females at greater risk? *Environ Health Perspect,* 113, 1723-9.

Chuang, K. J., C. C. Chan, T. C. Su, C. T. Lee & C. S. Tang (2007) The effect of urban air pollution on inflammation, oxidative stress, coagulation, and autonomic dysfunction in young adults. *Am J Respir Crit Care Med,* 176, 370-6.

Costa, D. L. & K. L. Dreher (1997) Bioavailable transition metals in particulate matter mediate cardiopulmonary injury in healthy and compromised animal models. *Environ Health Perspect,* 105 Suppl 5, 1053-60.

Cozzi, E., C. J. Wingard, W. E. Cascio, R. B. Devlin, J. J. Miles, A. R. Bofferding, R. M. Lust, M. R. Van Scott & R. A. Henriksen (2007) Effect of ambient particulate matter exposure on hemostasis. *Transl Res,* 149, 324-32.

Dales, R. E., S. Cakmak & C. B. Vidal (2010) Air Pollution and hospitalization for venous thromboembolic disease in Chile. *J Thromb Haemost.*

Delfino, R. J., N. Staimer, T. Tjoa, D. L. Gillen, A. Polidori, M. Arhami, M. T. Kleinman, N. D. Vaziri, J. Longhurst & C. Sioutas (2009) Air pollution exposures and circulating biomarkers of effect in a susceptible population: clues to potential causal component mixtures and mechanisms. *Environ Health Perspect,* 117, 1232-8.

Diez Roux, A. V., A. H. Auchincloss, T. G. Franklin, T. Raghunathan, R. G. Barr, J. Kaufman, B. Astor & J. Keeler (2008) Long-term exposure to ambient particulate matter and prevalence of subclinical atherosclerosis in the Multi-Ethnic Study of Atherosclerosis. *Am J Epidemiol,* 167, 667-75.

Dockery, D. W., C. A. Pope, 3rd, X. Xu, J. D. Spengler, J. H. Ware, M. E. Fay, B. G. Ferris, Jr. & F. E. Speizer (1993) An association between air pollution and mortality in six U.S. cities. *N Engl J Med,* 329, 1753-9.

Dominici, F., A. McDermott, M. Daniels, S. L. Zeger & J. M. Samet (2003) Mortality among residents of 90 cities. In Revised Analyses of Time-Series Studies of Air Pollution and Health. *Boston, MA: Health Effects Institute,* 9-24.

Dominici, F., R. D. Peng, M. L. Bell, L. Pham, A. McDermott, S. L. Zeger & J. M. Samet (2006) Fine particulate air pollution and hospital admission for cardiovascular and respiratory diseases. *JAMA,* 295, 1127-34.

Emmerechts, J., E. Alfaro-Moreno, B. M. Vanaudenaerde, B. Nemery & M. F. Hoylaerts (2010) Short-term exposure to particulate matter induces arterial but not venous thrombosis in healthy mice. *J Thromb Haemost,* 8, 2651-61.

Ghio, A. J., A. Hall, M. A. Bassett, W. E. Cascio & R. B. Devlin (2003) Exposure to concentrated ambient air particles alters hematologic indices in humans. *Inhal Toxicol,* 15, 1465-78.

Ghio, A. J., C. Kim & R. B. Devlin (2000) Concentrated ambient air particles induce mild pulmonary inflammation in healthy human volunteers. *Am J Respir Crit Care Med,* 162, 981-8.

Hertel, S., A. Viehmann, S. Moebus, K. Mann, M. Brocker-Preuss, S. Mohlenkamp, M. Nonnemacher, R. Erbel, H. Jakobs, M. Memmesheimer, K. H. Jockel & B. Hoffmann (2010) Influence of short-term exposure to ultrafine and fine particles on systemic inflammation. *Eur J Epidemiol,* 25, 581-92.

HHS (2004) The Health Consequences of Smoking: A Report of the Surgeon General. *US Department of Health and Human Services, Centers for Disease Control and Prevention, National Center for Chronic Disease Prevention and Health Promotion, Office of Smoking and Health.*

Hoek, G., B. Brunekreef, S. Goldbohm, P. Fischer & P. A. van den Brandt (2002) Association between mortality and indicators of traffic-related air pollution in the Netherlands: a cohort study. *Lancet,* 360, 1203-9.

Hoffmann, B., S. Moebus, S. Mohlenkamp, A. Stang, N. Lehmann, N. Dragano, A. Schmermund, M. Memmesheimer, K. Mann, R. Erbel & K. H. Jockel (2007) Residential exposure to traffic is associated with coronary atherosclerosis. *Circulation,* 116, 489-96.

Ibald-Mulli, A., J. Stieber, H. E. Wichmann, W. Koenig & A. Peters (2001) Effects of air pollution on blood pressure: a population-based approach. *Am J Public Health,* 91, 571-7.

Inoue, K., H. Takano, M. Sakurai, T. Oda, H. Tamura, R. Yanagisawa, A. Shimada & T. Yoshikawa (2006) Pulmonary exposure to diesel exhaust particles enhances coagulatory disturbance with endothelial damage and systemic inflammation related to lung inflammation. *Exp Biol Med (Maywood),* 231, 1626-32.

Jacobs, L., J. Emmerechts, M. F. Hoylaerts, C. Mathieu, P. H. Hoet, B. Nemery & T. S. Nawrot (2011) Traffic Air Pollution and Oxidized LDL. *PLoS One,* 6.

Jacobs, L., J. Emmerechts, C. Mathieu, M. F. Hoylaerts, F. Fierens, P. H. Hoet, B. Nemery & T. S. Nawrot (2009) Air pollution related prothrombotic changes in persons with diabetes. *Environ Health Perspect,* 118, 191-6.

Jerrett, M., R. T. Burnett, R. Ma, C. A. Pope, 3rd, D. Krewski, K. B. Newbold, G. Thurston, Y. Shi, N. Finkelstein, E. E. Calle & M. J. Thun (2005) Spatial analysis of air pollution and mortality in Los Angeles. *Epidemiology,* 16, 727-36.

Jiang, N., K. L. Dreher, J. A. Dye, Y. Li, J. H. Richards, L. D. Martin & K. B. Adler (2000) Residual oil fly ash induces cytotoxicity and mucin secretion by guinea pig tracheal epithelial cells via an oxidant-mediated mechanism. *Toxicol Appl Pharmacol,* 163, 221-30.

Jimenez, L. A., J. Thompson, D. A. Brown, I. Rahman, F. Antonicelli, R. Duffin, E. M. Drost, R. T. Hay, K. Donaldson & W. MacNee (2000) Activation of NF-kappaB by PM(10) occurs via an iron-mediated mechanism in the absence of IkappaB degradation. *Toxicol Appl Pharmacol,* 166, 101-10.

Kadiiska, M. B., R. P. Mason, K. L. Dreher, D. L. Costa & A. J. Ghio (1997) In vivo evidence of free radical formation in the rat lung after exposure to an emission source air pollution particle. *Chem Res Toxicol,* 10, 1104-8.

Kan, H., A. R. Folsom, M. Cushman, K. M. Rose, W. D. Rosamond, D. Liao, F. Lurmann & S. J. London (2011) Traffic exposure and incident venous thromboembolism in the atherosclerosis risk in communities (ARIC) study. *J Thromb Haemost.*

Katsouyanni, K., G. Touloumi, E. Samoli, A. Gryparis, A. Le Tertre, Y. Monopolis, G. Rossi, D. Zmirou, F. Ballester, A. Boumghar, H. R. Anderson, B. Wojtyniak, A. Paldy, R. Braunstein, J. Pekkanen, C. Schindler & J. Schwartz (2001) Confounding and effect modification in the short-term effects of ambient particles on total mortality: results from 29 European cities within the APHEA2 project. *Epidemiology,* 12, 521-31.

Kido, T., E. Tamagawa, N. Bai, K. Suda, H. H. Yang, Y. Li, G. Chiang, K. Yatera, H. Mukae, D. D. Sin & S. F. Van Eeden (2011) Particulate matter induces translocation of IL-6 from the lung to the systemic circulation. *Am J Respir Cell Mol Biol,* 44, 197-204.

Kilinc E, van Oerle R, Borissoff JI, Oschatz C, Gerlofs-Nijland ME, Janssen NA, Cassee FR, Sandstrom T, Renne T, Ten Cate H, Spronk HM. Factor XII Activation is Essential to Sustain the Procoagulant Effects of Particulate Matter. *J Thromb Haemost.* 2011.

Kunzli, N., M. Jerrett, R. Garcia-Esteban, X. Basagana, B. Beckermann, F. Gilliland, M. Medina, J. Peters, H. N. Hodis & W. J. Mack (2010) Ambient air pollution and the progression of atherosclerosis in adults. *PLoS One,* 5, e9096.

Kunzli, N., M. Jerrett, W. J. Mack, B. Beckerman, L. LaBree, F. Gilliland, D. Thomas, J. Peters & H. N. Hodis (2005) Ambient air pollution and atherosclerosis in Los Angeles. *Environ Health Perspect,* 113, 201-6.

Laden, F., J. Schwartz, F. E. Speizer & D. W. Dockery (2006) Reduction in fine particulate air pollution and mortality: Extended follow-up of the Harvard Six Cities study. *Am J Respir Crit Care Med,* 173, 667-72.

Levi, M. & T. van der Poll (2010) Inflammation and coagulation. *Crit Care Med,* 38, S26-34.

Li, N., J. Alam, M. I. Venkatesan, A. Eiguren-Fernandez, D. Schmitz, E. Di Stefano, N. Slaughter, E. Killeen, X. Wang, A. Huang, M. Wang, A. H. Miguel, A. Cho, C. Sioutas & A. E. Nel (2004) Nrf2 is a key transcription factor that regulates antioxidant defense in macrophages and epithelial cells: protecting against the proinflammatory and oxidizing effects of diesel exhaust chemicals. *J Immunol,* 173, 3467-81.

Lindmark, E., E. Diderholm, L. Wallentin & A. Siegbahn (2001) Relationship between interleukin 6 and mortality in patients with unstable coronary artery disease: effects of an early invasive or noninvasive strategy. *JAMA,* 286, 2107-13.

Lowe, G. D. (2008) Common risk factors for both arterial and venous thrombosis. *Br J Haematol,* 140, 488-95.

Lucking, A. J., M. Lundback, N. L. Mills, D. Faratian, S. L. Barath, J. Pourazar, F. R. Cassee, K. Donaldson, N. A. Boon, J. J. Badimon, T. Sandstrom, A. Blomberg & D. E. Newby (2008) Diesel exhaust inhalation increases thrombus formation in man. *Eur Heart J,* 29, 3043-51.

Maitre, A., V. Bonneterre, L. Huillard, P. Sabatier & R. de Gaudemaris (2006) Impact of urban atmospheric pollution on coronary disease. *Eur Heart J,* 27, 2275-84.

Miller, K. A., D. S. Siscovick, L. Sheppard, K. Shepherd, J. H. Sullivan, G. L. Anderson & J. D. Kaufman (2007) Long-term exposure to air pollution and incidence of cardiovascular events in women. *N Engl J Med,* 356, 447-58.

Mills, N. L., N. Amin, S. D. Robinson, A. Anand, J. Davies, D. Patel, J. M. de la Fuente, F. R. Cassee, N. A. Boon, W. Macnee, A. M. Millar, K. Donaldson & D. E. Newby (2006) Do inhaled carbon nanoparticles translocate directly into the circulation in humans? *Am J Respir Crit Care Med,* 173, 426-31.

Mills, N. L., H. Tornqvist, M. C. Gonzalez, E. Vink, S. D. Robinson, S. Soderberg, N. A. Boon, K. Donaldson, T. Sandstrom, A. Blomberg & D. E. Newby (2007) Ischemic and thrombotic effects of dilute diesel-exhaust inhalation in men with coronary heart disease. *N Engl J Med,* 357, 1075-82.

Mills, N. L., H. Tornqvist, S. D. Robinson, M. Gonzalez, K. Darnley, W. MacNee, N. A. Boon, K. Donaldson, A. Blomberg, T. Sandstrom & D. E. Newby (2005) Diesel exhaust inhalation causes vascular dysfunction and impaired endogenous fibrinolysis. *Circulation,* 112, 3930-6.

Mitschik, S., R. Schierl, D. Nowak & R. A. Jorres (2008) Effects of particulate matter on cytokine production in vitro: a comparative analysis of published studies. *Inhal Toxicol,* 20, 399-414.

Morel, O., F. Toti, B. Hugel, B. Bakouboula, L. Camoin-Jau, F. Dignat-George & J. M. Freyssinet (2006) Procoagulant microparticles: disrupting the vascular homeostasis equation? *Arterioscler Thromb Vasc Biol,* 26, 2594-604.

Mutlu, G. M., D. Green, A. Bellmeyer, C. M. Baker, Z. Burgess, N. Rajamannan, J. W. Christman, N. Foiles, D. W. Kamp, A. J. Ghio, N. S. Chandel, D. A. Dean, J. I. Sznajder & G. R. Budinger (2007) Ambient particulate matter accelerates coagulation via an IL-6-dependent pathway. *J Clin Invest,* 117, 2952-61.

Nadziejko, C., K. Fang, L. C. Chen, B. Cohen, M. Karpatkin & A. Nadas (2002) Effect of concentrated ambient particulate matter on blood coagulation parameters in rats. *Res Rep Health Eff Inst*, 7-29; discussion 31-8.

Nawrot, T. S., A. Nemmar & B. Nemery (2006) Air pollution: To the heart of the matter. *Eur Heart J*, 27, 2269-71.

Nawrot, T. S., L. Perez, N. Kunzli, E. Munters & B. Nemery (2011) Public health importance of triggers of myocardial infarction: a comparative risk assessment. *Lancet*, 377, 732-40.

Nemmar, A., S. Al-Salam, S. Zia, S. Dhanasekaran, M. Shudadevi & B. H. Ali (2010) Time-course effects of systemically administered diesel exhaust particles in rats. *Toxicol Lett*, 194, 58-65.

Nemmar, A., P. H. Hoet, D. Dinsdale, J. Vermylen, M. F. Hoylaerts & B. Nemery (2003a) Diesel exhaust particles in lung acutely enhance experimental peripheral thrombosis. *Circulation*, 107, 1202-8.

Nemmar, A., P. H. Hoet, P. Vandervoort, D. Dinsdale, B. Nemery & M. F. Hoylaerts (2007) Enhanced peripheral thrombogenicity after lung inflammation is mediated by platelet-leukocyte activation: role of P-selectin. *J Thromb Haemost*, 5, 1217-26.

Nemmar, A., P. H. Hoet, B. Vanquickenborne, D. Dinsdale, M. Thomeer, M. F. Hoylaerts, H. Vanbilloen, L. Mortelmans & B. Nemery (2002) Passage of inhaled particles into the blood circulation in humans. *Circulation*, 105, 411-4.

Nemmar, A., P. H. Hoet, J. Vermylen, B. Nemery & M. F. Hoylaerts (2004) Pharmacological stabilization of mast cells abrogates late thrombotic events induced by diesel exhaust particles in hamsters. *Circulation*, 110, 1670-7.

Nemmar, A., M. F. Hoylaerts, P. H. Hoet, J. Vermylen & B. Nemery (2003b) Size effect of intratracheally instilled particles on pulmonary inflammation and vascular thrombosis. *Toxicol Appl Pharmacol*, 186, 38-45.

Nemmar, A., B. Nemery, P. H. Hoet, N. Van Rooijen & M. F. Hoylaerts (2005) Silica particles enhance peripheral thrombosis: key role of lung macrophage-neutrophil cross-talk. *Am J Respir Crit Care Med*, 171, 872-9.

Nemmar, A., B. Nemery, P. H. Hoet, J. Vermylen & M. F. Hoylaerts (2003c) Pulmonary inflammation and thrombogenicity caused by diesel particles in hamsters: role of histamine. *Am J Respir Crit Care Med*, 168, 1366-72.

Nemmar, A., H. Vanbilloen, M. F. Hoylaerts, P. H. Hoet, A. Verbruggen & B. Nemery (2001) Passage of intratracheally instilled ultrafine particles from the lung into the systemic circulation in hamster. *Am J Respir Crit Care Med*, 164, 1665-8.

O'Toole, T. E., J. Hellmann, L. Wheat, P. Haberzettl, J. Lee, D. J. Conklin, A. Bhatnagar & C. A. Pope, 3rd (2010) Episodic exposure to fine particulate air pollution decreases circulating levels of endothelial progenitor cells. *Circ Res*, 107, 200-3.

Oberdorster, G., E. Oberdorster & J. Oberdorster (2005) Nanotoxicology: an emerging discipline evolving from studies of ultrafine particles. *Environ Health Perspect*, 113, 823-39.

Oberdorster, G., Z. Sharp, V. Atudorei, A. Elder, R. Gelein, A. Lunts, W. Kreyling & C. Cox (2002) Extrapulmonary translocation of ultrafine carbon particles following whole-body inhalation exposure of rats. *J Toxicol Environ Health A*, 65, 1531-43.

Park, S. K., A. H. Auchincloss, M. S. O'Neill, R. Prineas, J. C. Correa, J. Keeler, R. G. Barr, J. D. Kaufman & A. V. Diez Roux (2010) Particulate air pollution, metabolic syndrome, and heart rate variability: the multi-ethnic study of atherosclerosis (MESA). *Environ Health Perspect,* 118, 1406-11.

Pawlinski, R., J. G. Wang, A. P. Owens, 3rd, J. Williams, S. Antoniak, M. Tencati, T. Luther, J. W. Rowley, E. N. Low, A. S. Weyrich & N. Mackman (2010) Hematopoietic and nonhematopoietic cell tissue factor activates the coagulation cascade in endotoxemic mice. *Blood,* 116, 806-14.

Pekkanen, J., E. J. Brunner, H. R. Anderson, P. Tiittanen & R. W. Atkinson (2000) Daily concentrations of air pollution and plasma fibrinogen in London. *Occup Environ Med,* 57, 818-22.

Pekkanen, J., A. Peters, G. Hoek, P. Tiittanen, B. Brunekreef, J. de Hartog, J. Heinrich, A. Ibald-Mulli, W. G. Kreyling, T. Lanki, K. L. Timonen & E. Vanninen (2002) Particulate air pollution and risk of ST-segment depression during repeated submaximal exercise tests among subjects with coronary heart disease: the Exposure and Risk Assessment for Fine and Ultrafine Particles in Ambient Air (ULTRA) study. *Circulation,* 106, 933-8.

Pery, A. R., C. Brochot, P. H. Hoet, A. Nemmar & F. Y. Bois (2009) Development of a physiologically based kinetic model for 99m-technetium-labelled carbon nanoparticles inhaled by humans. *Inhal Toxicol,* 21, 1099-107.

Peters, A., D. W. Dockery, J. E. Muller & M. A. Mittleman (2001a) Increased particulate air pollution and the triggering of myocardial infarction. *Circulation,* 103, 2810-5.

Peters, A., M. Frohlich, A. Doring, T. Immervoll, H. E. Wichmann, W. L. Hutchinson, M. B. Pepys & W. Koenig (2001b) Particulate air pollution is associated with an acute phase response in men; results from the MONICA-Augsburg Study. *Eur Heart J,* 22, 1198-204.

Peters, A., S. von Klot, M. Heier, I. Trentinaglia, A. Hormann, H. E. Wichmann & H. Lowel (2004) Exposure to traffic and the onset of myocardial infarction. *N Engl J Med,* 351, 1721-30.

Pope, C. A., 3rd, R. T. Burnett, D. Krewski, M. Jerrett, Y. Shi, E. E. Calle & M. J. Thun (2009) Cardiovascular mortality and exposure to airborne fine particulate matter and cigarette smoke: shape of the exposure-response relationship. *Circulation,* 120, 941-8.

Pope, C. A., 3rd, R. T. Burnett, M. J. Thun, E. E. Calle, D. Krewski, K. Ito & G. D. Thurston (2002) Lung cancer, cardiopulmonary mortality, and long-term exposure to fine particulate air pollution. *JAMA,* 287, 1132-41.

Pope, C. A., 3rd, R. T. Burnett, G. D. Thurston, M. J. Thun, E. E. Calle, D. Krewski & J. J. Godleski (2004a) Cardiovascular mortality and long-term exposure to particulate air pollution: epidemiological evidence of general pathophysiological pathways of disease. *Circulation,* 109, 71-7.

Pope, C. A., 3rd, M. L. Hansen, R. W. Long, K. R. Nielsen, N. L. Eatough, W. E. Wilson & D. J. Eatough (2004b) Ambient particulate air pollution, heart rate variability, and blood markers of inflammation in a panel of elderly subjects. *Environ Health Perspect,* 112, 339-45.

Pope, C. A., 3rd, R. L. Verrier, E. G. Lovett, A. C. Larson, M. E. Raizenne, R. E. Kanner, J. Schwartz, G. M. Villegas, D. R. Gold & D. W. Dockery (1999) Heart rate variability associated with particulate air pollution. *Am Heart J,* 138, 890-9.

Prandoni, P. (2009) Venous and arterial thrombosis: Two aspects of the same disease? *Clin Epidemiol,* 1, 1-6.

Puett, R. C., J. E. Hart, J. D. Yanosky, C. Paciorek, J. Schwartz, H. Suh, F. E. Speizer & F. Laden (2009) Chronic fine and coarse particulate exposure, mortality, and coronary heart disease in the Nurses' Health Study. *Environ Health Perspect,* 117, 1697-701.

Ray, M. R., S. Mukherjee, S. Roychoudhury, P. Bhattacharya, M. Banerjee, S. Siddique, S. Chakraborty & T. Lahiri (2006) Platelet activation, upregulation of CD11b/ CD18 expression on leukocytes and increase in circulating leukocyte-platelet aggregates in Indian women chronically exposed to biomass smoke. *Hum Exp Toxicol,* 25, 627-35.

Ridker, P. M., N. Rifai, M. J. Stampfer & C. H. Hennekens (2000) Plasma concentration of interleukin-6 and the risk of future myocardial infarction among apparently healthy men. *Circulation,* 101, 1767-72.

Riediker, M., W. E. Cascio, T. R. Griggs, M. C. Herbst, P. A. Bromberg, L. Neas, R. W. Williams & R. B. Devlin (2004) Particulate matter exposure in cars is associated with cardiovascular effects in healthy young men. *Am J Respir Crit Care Med,* 169, 934-40.

Ross, R. (1999) Atherosclerosis--an inflammatory disease. *N Engl J Med,* 340, 115-26.

Ruckerl, R., S. Greven, P. Ljungman, P. Aalto, C. Antoniades, T. Bellander, N. Berglind, C. Chrysohoou, F. Forastiere, B. Jacquemin, S. von Klot, W. Koenig, H. Kuchenhoff, T. Lanki, J. Pekkanen, C. A. Perucci, A. Schneider, J. Sunyer & A. Peters (2007a) Air pollution and inflammation (interleukin-6, C-reactive protein, fibrinogen) in myocardial infarction survivors. *Environ Health Perspect,* 115, 1072-80.

Ruckerl, R., A. Ibald-Mulli, W. Koenig, A. Schneider, G. Woelke, J. Cyrys, J. Heinrich, V. Marder, M. Frampton, H. E. Wichmann & A. Peters (2006) Air pollution and markers of inflammation and coagulation in patients with coronary heart disease. *Am J Respir Crit Care Med,* 173, 432-41.

Ruckerl, R., R. P. Phipps, A. Schneider, M. Frampton, J. Cyrys, G. Oberdorster, H. E. Wichmann & A. Peters (2007b) Ultrafine particles and platelet activation in patients with coronary heart disease--results from a prospective panel study. *Part Fibre Toxicol,* 4, 1.

Rudez, G., N. A. Janssen, E. Kilinc, F. W. Leebeek, M. E. Gerlofs-Nijland, H. M. Spronk, H. ten Cate, F. R. Cassee & M. P. de Maat (2009) Effects of ambient air pollution on hemostasis and inflammation. *Environ Health Perspect,* 117, 995-1001.

Samet, J. M., F. Dominici, F. C. Curriero, I. Coursac & S. L. Zeger (2000) Fine particulate air pollution and mortality in 20 U.S. cities, 1987-1994. *N Engl J Med,* 343, 1742-9.

Samet, J. M., A. Rappold, D. Graff, W. E. Cascio, J. H. Berntsen, Y. C. Huang, M. Herbst, M. Bassett, T. Montilla, M. J. Hazucha, P. A. Bromberg & R. B. Devlin (2009) Concentrated ambient ultrafine particle exposure induces cardiac changes in young healthy volunteers. *Am J Respir Crit Care Med,* 179, 1034-42.

Samoli, E., R. Peng, T. Ramsay, M. Pipikou, G. Touloumi, F. Dominici, R. Burnett, A. Cohen, D. Krewski, J. Samet & K. Katsouyanni (2008) Acute effects of ambient particulate matter on mortality in Europe and North America: results from the APHENA study. *Environ Health Perspect,* 116, 1480-6.

Scharrer, E., H. Hessel, A. Kronseder, W. Guth, B. Rolinski, R. A. Jorres, K. Radon, R. Schierl, P. Angerer & D. Nowak (2007) Heart rate variability, hemostatic and acute inflammatory blood parameters in healthy adults after short-term exposure to welding fume. *Int Arch Occup Environ Health,* 80, 265-72.

Schneider, A., R. Hampel, A. Ibald-Mulli, W. Zareba, G. Schmidt, R. Schneider, R. Ruckerl, J. P. Couderc, B. Mykins, G. Oberdorster, G. Wolke, M. Pitz, H. E. Wichmann & A. Peters (2010) Changes in deceleration capacity of heart rate and heart rate variability induced by ambient air pollution in individuals with coronary artery disease. *Part Fibre Toxicol,* 7, 29.

Schwartz, J., B. Coull, F. Laden & L. Ryan (2008) The effect of dose and timing of dose on the association between airborne particles and survival. *Environ Health Perspect,* 116, 64-9.

Seaton, A., A. Soutar, V. Crawford, R. Elton, S. McNerlan, J. Cherrie, M. Watt, R. Agius & R. Stout (1999) Particulate air pollution and the blood. *Thorax,* 54, 1027-32.

Shih, R. A., B. A. Griffin, N. Salkowski, A. Jewell, C. Eibner, C. E. Bird, D. Liao, M. Cushman, H. G. Margolis, C. B. Eaton & E. A. Whitsel (2010) Ambient Particulate Matter Air Pollution and Venous Thromboembolism in the Women's Health Initiative Hormone Therapy Trials. *Environ Health Perspect.*

Simkhovich, B. Z., M. T. Kleinman & R. A. Kloner (2008) Air pollution and cardiovascular injury epidemiology, toxicology, and mechanisms. *J Am Coll Cardiol,* 52, 719-26.

Sioutas, C., R. J. Delfino & M. Singh (2005) Exposure assessment for atmospheric ultrafine particles (UFPs) and implications in epidemiologic research. *Environ Health Perspect,* 113, 947-55.

Stewart, J. C., D. C. Chalupa, R. B. Devlin, L. M. Frasier, L. S. Huang, E. L. Little, S. M. Lee, R. P. Phipps, A. P. Pietropaoli, M. B. Taubman, M. J. Utell & M. W. Frampton (2010) Vascular effects of ultrafine particles in persons with type 2 diabetes. *Environ Health Perspect,* 118, 1692-8.

Stocker, R. & J. F. Keaney, Jr. (2004) Role of oxidative modifications in atherosclerosis. *Physiol Rev,* 84, 1381-478.

Sun, Q., A. Wang, X. Jin, A. Natanzon, D. Duquaine, R. D. Brook, J. G. Aguinaldo, Z. A. Fayad, V. Fuster, M. Lippmann, L. C. Chen & S. Rajagopalan (2005) Long-term air pollution exposure and acceleration of atherosclerosis and vascular inflammation in an animal model. *JAMA,* 294, 3003-10.

Suwa, T., J. C. Hogg, K. B. Quinlan, A. Ohgami, R. Vincent & S. F. van Eeden (2002) Particulate air pollution induces progression of atherosclerosis. *J Am Coll Cardiol,* 39, 935-42.

Tan, W. C., D. Qiu, B. L. Liam, T. P. Ng, S. H. Lee, S. F. van Eeden, Y. D'Yachkova & J. C. Hogg (2000) The human bone marrow response to acute air pollution caused by forest fires. *Am J Respir Crit Care Med,* 161, 1213-7.

Tao, F., B. Gonzalez-Flecha & L. Kobzik (2003) Reactive oxygen species in pulmonary inflammation by ambient particulates. *Free Radic Biol Med,* 35, 327-40.

Theilmeier, G., C. Michiels, E. Spaepen, I. Vreys, D. Collen, J. Vermylen & M. F. Hoylaerts (2002) Endothelial von Willebrand factor recruits platelets to atherosclerosis-prone sites in response to hypercholesterolemia. *Blood,* 99, 4486-93.

Thompson, A. M., A. Zanobetti, F. Silverman, J. Schwartz, B. Coull, B. Urch, M. Speck, J. R. Brook, M. Manno & D. R. Gold (2010) Baseline repeated measures from controlled human exposure studies: associations between ambient air pollution exposure and the systemic inflammatory biomarkers IL-6 and fibrinogen. *Environ Health Perspect,* 118, 120-4.

Tornqvist, H., N. L. Mills, M. Gonzalez, M. R. Miller, S. D. Robinson, I. L. Megson, W. Macnee, K. Donaldson, S. Soderberg, D. E. Newby, T. Sandstrom & A. Blomberg

(2007) Persistent endothelial dysfunction in humans after diesel exhaust inhalation. *Am J Respir Crit Care Med,* 176**,** 395-400.

Urch, B., F. Silverman, P. Corey, J. R. Brook, K. Z. Lukic, S. Rajagopalan & R. D. Brook (2005) Acute blood pressure responses in healthy adults during controlled air pollution exposures. *Environ Health Perspect,* 113**,** 1052-5.

van Eeden, S. F., W. C. Tan, T. Suwa, H. Mukae, T. Terashima, T. Fujii, D. Qui, R. Vincent & J. C. Hogg (2001) Cytokines involved in the systemic inflammatory response induced by exposure to particulate matter air pollutants (PM(10)). *Am J Respir Crit Care Med,* 164**,** 826-30.

Volpato, S., J. M. Guralnik, L. Ferrucci, J. Balfour, P. Chaves, L. P. Fried & T. B. Harris (2001) Cardiovascular disease, interleukin-6, and risk of mortality in older women: the women's health and aging study. *Circulation,* 103**,** 947-53.

Wallenfeldt, K., B. Fagerberg, J. Wikstrand & J. Hulthe (2004) Oxidized low-density lipoprotein in plasma is a prognostic marker of subclinical atherosclerosis development in clinically healthy men. *J Intern Med,* 256**,** 413-20.

Wilson, D. W., H. H. Aung, M. W. Lame, L. Plummer, K. E. Pinkerton, W. Ham, M. Kleeman, J. W. Norris & F. Tablin (2010) Exposure of mice to concentrated ambient particulate matter results in platelet and systemic cytokine activation. *Inhal Toxicol,* 22**,** 267-76.

Yokoyama, S., H. Ikeda, N. Haramaki, H. Yasukawa, T. Murohara & T. Imaizumi (2005) Platelet P-selectin plays an important role in arterial thrombogenesis by forming large stable platelet-leukocyte aggregates. *J Am Coll Cardiol,* 45**,** 1280-6.

Zanobetti, A., J. Schwartz, E. Samoli, A. Gryparis, G. Touloumi, J. Peacock, R. H. Anderson, A. Le Tertre, J. Bobros, M. Celko, A. Goren, B. Forsberg, P. Michelozzi, D. Rabczenko, S. P. Hoyos, H. E. Wichmann & K. Katsouyanni (2003) The temporal pattern of respiratory and heart disease mortality in response to air pollution. *Environ Health Perspect,* 111**,** 1188-93.

5

Ambient Air Pollution and Reproductive Health

Xiaohui Xu[1], Haidong Kan[2] and Sandie Ha[1]
[1]*Department of Epidemiology, College of Public Health and Health Professions, University of Florida,*
[2]*Department of Environmental Health, School of Public Health, Fudan University,*
[1]*USA*
[2]*China*

1. Introduction

Adverse health effects of ambient air pollution on mortality and morbidity in adults and children have been extensively studied across the world (Barnett et al. 2006; Bell et al. 2004; Dominici et al. 2003; Gold et al. 1999; Jerrett et al. 2005; Middleton et al. 2008; Pope 1999; Pope et al. 1991; Pope and Kanner 1993; Samoli et al. 2007; Wietlisbach et al. 1996). Considerable consistency across studies has been observed for many health endpoints including total mortality, cardiopulmonary mortality and morbidity. Moreover, air pollution studies suggest that the opposite ends of the age spectrum are more susceptible than the general population (Dockery and Pope 1994; Saldiva et al. 1995; Schwartz et al. 1994). Therefore, fetuses are thought to be a vulnerable subgroup of the population who could be most endangered by the effects of air pollution (Pope 2000). Early studies had shown that maternal active and passive smoking could impair reproductive outcomes. Thus, there is a strong belief that prenatal exposure to air contaminants, which is similar to the effects of maternal smoking, can also lead to some adverse pregnancy outcomes. The number of studies linking air pollution with adverse pregnancy outcomes has recently grown steadily since the late 1990s. The adverse effects of air pollutants including particulate matter (PM), nitrogen oxide (NOx), sulfur dioxide (SO_2), carbon monoxide (CO), and ozone (O_3) on measures of fetal size, gestational duration and other reproductive outcomes have been studied.

Adverse reproductive outcomes including low birth weight (LBW: birth weight <2,500g) or preterm delivery (PTD: birth at <37 weeks of gestation) have arisen fairly consistently in recent years. During 1990-2006, the rates have risen 21% for PTD and 19% for LBW in the United States, respectively (Martin et al. 2008). Studies have suggested that LBW or PTD has been associated with not only childhood mortality and morbidity but also the risk of diseases in adulthood such as heart diseases and diabetes (Clapp Iii and Lopez 2007; Osmond and Barker 2000; Rinaudo and Lamb 2008; Thompson 2007). The prevention of adverse pregnancy outcomes is a renewed national and international priority in maternal and child health (Damus 2008). There is an emerging need to identify the etiological factors of adverse birth outcomes such as environmental exposures, which could be modifiable in order to help reverse the increasing rates.

In this chapter, we review the health effects of ambient air pollution on pregnancy outcomes such as LBW, very low birth weight (VLBW: <1,500g), small for gestational age or intrauterine growth retardation (SGA or IUGR, i.e. birth weight below the 10th percentile for that gestational age), PTD, and birth defects. Further, we discuss the potential biological mechanisms underlying the associations between air pollution and adverse reproductive outcomes. Finally, we briefly introduce the methodology including study designs, exposure measurements, windows of exposure, confounders, and future directions for study of air pollution and pregnancy outcomes.

2. Health effects of air pollution on birth outcomes

Birth outcomes, including outcomes of fetal size such as LBW, VLBW, SGA, and IUGR, outcomes of gestational age such as PTD, and birth defects, have frequently been studied in epidemiological research on health effects of air pollution. The associations between air pollution and these birth outcomes may suggest different etiologic and pathogenic mechanisms. Moreover, air pollution is a mixture of pollutants, which also vary in nature and possible health effects. Therefore, existing evidence of health effects of air pollution on birth outcomes are briefly examined by types of birth outcome and/or criteria air pollutants including PM, NO_x, SO_2, and CO as below. Although the associations between O_3 exposure and birth outcomes have also been investigated by several studies, available evidence seems only to support an association with birth defects (Shah and Balkhair 2011). Therefore, the health effects of O3 are merely discussed for birth defects.

2.1 Health effects on fetal size - LBW, VLBW, SGA and IUGR

The effects of ambient air pollutants on fetal size, specifically LBW have been well studied during the past decades. Overall, there were some positive findings, which suggest that air pollution exposure, through all periods during pregnancy, is associated with adverse health effects on fetal size including LBW, VLBW, SGA and IUGR. However, air pollution consists of many components, making it difficult to pinpoint which pollutant can actually affect fetal size. Therefore, we present below the results from some selected studies by criteria air pollutants individually.

2.1.1. Particulate matter (PM)

PM generally attracts much more attention in epidemiological research of air pollution because of its physical and chemical characteristics of consisting of organic and inorganic components. The effects of PM on birth outcomes have been reported to be more pronounced compared to other pollutants. Several exposure indices of PM such as total suspend particles (TSP), particulate matter smaller than 10μm aerodynamic diameter (PM_{10}) and smaller than 2.5μm aerodynamic diameter ($PM_{2.5}$) have been used in studies. Wang et al firstly reported that exposure to TSP in the third trimester of pregnancy was significantly associated with LBW (Wang et al. 1997). After that, many studies have been conducted to report associations between particulate air pollution and LBW, SGA, as well as IUGR. Bobak et al found that annual concentration of TSP during the year of birth was associated with an increased risk of LBW (Bobak and Leon 1999). Rogers et al also found that annual TSP in the year of birth is associated with the risk of LBW (Rogers et al. 2000). Consistent evidence on adverse health effects of PM on birth weight has been reported when PM_{10} and $PM_{2.5}$ have been used as measures of exposure. Mannes et al found that each 1 $\mu g/m^3$

increase in PM_{10} is associated with on average a birth weight reduction of 4 grams (95% confidence interval (CI): 3g to 6g) after adjusting for other important covariates (Mannes et al. 2005). Dugandzic et al also found that PM_{10} exposure in the highest quartile during the first trimester increased the risk of LBW by 33% (OR= 1.33, 95% CI: 1.02 to 1.74) compared to exposure in the lowest quartile (Dugandzic et al. 2006). Similar results of health effects of PM_{10} on LBW were also observed in Xu et al's study (Xu et al. 2010). In addition, $PM_{2.5}$ is found to be associated with a reduction in birth weight by Parker et al in California in 2005. They found that exposure to $PM_{2.5}$ in the highest quartile during pregnancy is associated with a reduction in infant birth weight of -36.1 g (95% CI: -16.5 g to -55.8 g) compared to exposure in the lowest quartile of <11.9 $\mu g/m^3$ after adjusting for other covariates (Parker et al. 2005). Another study reported by Rich et al also found that every 4 $\mu g/m^3$ increase in $PM_{2.5}$ exposure during the first and the third trimester increases the risk for SGA by 4.5% (95% CI: 0.5-8.7) and 4.1% (95% CI: 0.3-8.0), respectively (Rich et al. 2009). In addition, Dejmek et al 1999 also found that exposure to a high tertile of $PM_{2.5}$ during the first gestational month is associated with an increased risk of IUGR (OR=2.11, 95% CI: 1.20-3.70) among northern Bohemian women (Dejmek et al. 1999). Liu et al (2007) reported that every 10 $\mu g/m^3$ increase in $PM_{2.5}$ is associated with roughly six to seven percent increase (95% CI: 3-10%) in the risk of IUGR after adjusting for other risk factors (Liu et al. 2007). However, other studies have not found a significant association between particulate air pollution and fetal size (Maisonet et al. 2001; Salam et al. 2005).

2.1.2 NO_x

Health effects of nitric oxide (NO) and nitrogen dioxide (NO_2) on birth weight have also been studied. Studies on health effects of NO on fetal size is limited (Bobak 2000; Bobak and Leon 1999; Landgren 1996) and no existing evidence shows adverse health effects of NO on fetal size. Several studies show the associations between NO_2 exposure and LBW (Ballester et al. 2010; Bell et al. 2007; Ha et al. 2001; Lee et al. 2003). Some other studies also show that exposure to NO_2 is associated with increased risks of SGA. For example, an earlier report also found that each 10 parts per billion (ppb) increase in NO_2 exposure during the first month of pregnancy increases the risk of IUGR by 5% after adjusting for important maternal, infant and other environmental factors (Liu et al. 2003). Similar results were found by Mannes et al in 2005. This study indicated that each 1 ppb increase in NO_2 exposure during pregnancy is associated with 1 to 34 grams reductions in birth weight among infants in Sydney, Australia (Mannes et al. 2005). Liu et al (2007) also reported that every 20ppb increase in NO_2 exposure during the first, second, and third trimesters is associated with increased risks of IUGR by 16%, 14%, and 16%, respectively (Liu et al. 2007). However, several recent studies did not find a significant association between NO_2 exposure and fetal size (Gehring et al. 2010; Madsen et al. 2010).

2.1.3 SO_2

SO_2 is another important component of air pollution, and has been found to be associated with fetal size. Liu et al 2003 found that every 5.0 ppb increase in SO_2 exposure during the first month of pregnancy increases the risk of IUGR by approximately 7% (OR = 1.07, 95% CI: 1.01–1.13) among Canadian women (Liu et al. 2003). Several other studies have also linked SO_2 exposure with LBW. Wang et al have found that each 100 $\mu g/m^3$ increase in SO_2 exposure during pregnancy is associated with 11% increase (OR= 1.11, 95% CI: 1.06-1.16) in the risk of LBW and a 7.3 gram reduction in birth weight among Chinese women (Wang et al. 1997). Bobak and Leon also found that every 50 µg/m^3 increase in SO_2 exposure during the whole pregnancy period is associated with a 10 % increase (OR=1.10, 95% CI: 1.02 -1.17) in the risk of having a LBW infant (Bobak and Leon 1999). Dugandzic et al 2006, in their cohort study in Canada, found that SO_2 exposure in the highest quartile during the first trimester is associated with a 36% increase in the risk of having a LBW infant (OR=1.36, 95% CI:1.04 to 1.78) (Dugandzic et al. 2006).

2.1.4 CO

Carbon monoxide is found to be associated with decreased birth weight. The earliest study reported by Alderman suggests a potential link between maternal CO exposure and risk of LBW (Alderman et al. 1987). Later, Ritz and Yu reported a significant association between CO exposure and risk of LBW in 1999 (Ritz and Yu 1999). Liu et al (2003) also found that each 1 ppb increase in CO exposure during the first month of pregnancy is associated with a 6% increase in the risk of IUGR (OR = 1.06, 95% CI:1.01–1.10) among Canadian women (Liu et al. 2003). Another later report by Liu et al in 2007 found that after adjusting for important maternal, infant and environmental factors, every 1ppb increase in CO exposure during the first, second, and third trimester of pregnancy is associated with a 18%, 15%, and 19% increase in the risk of IUGR, respectively (Liu et al. 2007). Significant findings were also reported in other studies (Lee et al. 2003; Maisonet et al. 2001). However, the evidence of health effects of CO exposure on fetal size is less consistent. Some studies have not shown a significant association (Huynh et al. 2006; Salam et al. 2005). Moreover, one other study found a protective effect of CO exposure on LBW (Lin et al. 2004).

2.2 Health effects on gestational age - PTD

PTD has also been extensively studied in the past decades. Studies have reported the effects of various air pollutants such as PM, NO_x, SO_2, and CO on PTD. The effect size and time of exposure of a specific air pollutant are inconsistent across studies because of the differences in study design, exposure assessment, and confounders being controlled for.

2.2.1 PM: TSP, PM_{10} and $PM_{2.5}$

Two studies had used TSP as a measure of exposure. One study reported that a decrease in the gestation age of 0.042 weeks was associated with each 100 µg/m^3 increase in TSP with a 7-day lag (Xu et al. 1995) while the other showed no significant association between TSP and PTD (Bobak 2000). Meanwhile, several studies have reported a significant association between PM_{10} exposure and PTD. A cohort study showed 16% and 20% increased risks of preterm birth for a 50 µg/m^3 increase in TSP exposure during the first month of pregnancy and during the 6 weeks before birth, respectively (Ritz et al. 2000). A study conducted in Australia suggested that exposure to PM_{10} during the first trimester was associated with a 15% increase in the risk of PTD (Hansen et al. 2006). A time-series analysis conducted in Pennsylvania also suggested a significant association between PTD and PM_{10} exposure with 2-day as well as 5-day lag (Sagiv et al. 2005). Another time-series study conducted in Shanghai also shows that a 4.4% increase in the risk of PTD was PTD was observed for each 10 µg/m^3 increase in PM_{10} exposure during the 8 weeks before pregnancy (Jiang et al. 2007). However, some other studies did not observe a significant effect of PM_{10} on PTD (Brauer et al. 2008; Kim et al. 2007; Lee et al. 2008). The effect of exposure to $PM_{2.5}$ was also assessed by several studies. Huynh et al (2006) found that exposure to to the highest quartile of $PM_{2.5}$ (> 22.1µg/m^3) during pregnancy is associated with a 15% increase in the risk of PTD compared to the lowest quartile of exposure. In the same study, similar results were found for exposure during the first month (OR=1.21, 95%CI: 1.12, 1.30) and last two weeks of pregnancy (OR=1.17 , 95%CI: 1.09, 1.27) (Huynh et al. 2006). A two-stage design study being conducted in Los Angeles reported that exposure to $PM_{2.5}$ > 21.4 μg/m³ during the first trimester increased the risk of PTD by about 10% (OR=1.10, 95%CI: 1.01, 1.20).

2.2.2 NO_x

The association of NO_2 exposure and PTD was explored in several studies. Bobak et al reported that 10% and 11% increased risks of PTD were observed for each 50 μg/m³ increase of NO_x during the first and third trimesters, respectively (Bobak 2000). NO_2 is also found to be associated with PTD among residents in Lithuania (Maroziene and Grazuleviciene 2002). Each 10 μg/m³ increase in NO_2 exposure during pregnancy period and first trimester is associated with 25% (OR = 1.25, 95%CI: 1.07-1.46) and 67% (OR = 1.67, 95% CI: 1.28-2.18) increase in the risk of PTD, respectively. A time series analysis indicated that daily PTD rate is associated with average exposure to NO_2 during the 6 weeks preceding delivery (Darrow et al. 2009). A recent cohort study conducted in Spain reported that the risk of PTD was shown to be significant when women were exposed to NO_2 levels >46.2 μg/m³ during the second (OR=1.11, 95% CI: 1.03-1.25) and third trimesters (OR=1.10, 95% CI: 1.00-1.21) as well as throughout the entire pregnancy (OR=1.29, 95% CI: 1.13-1.46) (Llop et al. 2010). However, evidence on the effect of NO_2 on PTD is inconsistent. Several other studies did not show a significant association (Gehring et al. 2010; Liu et al. 2003; Ritz et al. 2007).

2.2.3 SO_2

The earliest study conducted in China reported a significant effect of SO_2 on PTD. It shows that each 100 μg/m³ increase in SO_2 during the 7 days preceding delivery is associated with .075 wk (12.6h) decrease in duration of gestation (Xu et al. 1995). Bobak et al reported that exposure to SO_2 during all trimesters was significantly associated with the risk of PTD (Bobak 2000). Liu et al in 2003 also reported the significant effects of SO_2 on PTD among women living in Vancouver, Canada. This study indicated that each 5.0 ppb increase in SO_2 exposure during the last month of pregnancy is associated with a 9% increase in risk of PTD (Liu et al. 2003). A time series analysis also revealed that a 15% increased risk for PTD was observed for each 15 ppb increase in average SO_2 in the 6 weeks before birth in Pennsylvania (Sagiv et al. 2005). Similar findings from another time-series analysis conducted in Shanghai were observed (Jiang et al. 2007). Leem et al found that exposure to the highest quartile of SO_2 during first trimester increased the risk of PTD by 21% compared to the lowest quartile of exposure (OR= 1.21, 95% CI: 1.04-1.42) in Korea (Leem et al. 2006). A study in Australia also revealed that SO_2 levels in the first and third trimester of pregnancy were a significant predictor of preterm birth (ORs between 2.30 to 3.15) (Jalaludin et al. 2007). However, no significant effect of SO_2 was observed in other studies (Brauer et al. 2008; Darrow et al. 2009; Landgren 1996).

2.2.4 CO

CO is among one of the air pollutants that have been demonstrated to have negative effects on PTD. Ritz et al reported that exposure to CO during the first month of pregnancy and 6 weeks before birth was significantly associated with an increased risk of PTD in the inland regions of Southern California (Ritz et al. 2000). A study reported by Liu et al showed that an 8% increase (95% CI: 1%-15%) in the risk of PTD was observed for each 1.0 parts per million (ppm) increase in exposure to CO during the last month of pregnancy (Liu et al. 2003). Leem et al found that exposure to the highest quartiles of CO during first trimester increased the risk of PTD by 26% (OR= 1.26, 95% CI: 1.11-1.44) in Korea (Leem et al. 2006). Similar results reported by Wilhelm and Ritz in 2005 suggest that exposure to the highest quartile of CO during the first trimester increases the risk of PTD by 27% (95% CI: 7%-50%) (Wilhelm and Ritz 2005). Another cohort study found that exposure to high level of CO during the first trimester of pregnancy (> 1.25 ppb) is associated with 1.25 (95% CI: 1.12, 1.38) times of the risk of PTD compared to low level of CO exposure (< 0.59ppb) after adjusting for important maternal, infant, and environmental characteristics (Ritz et al. 2007). However, several other studies including a recent study did not observe significant effects of CO on PTD (Darrow et al. 2009; Huynh et al. 2006; Rudra et al. 2011).

2.3 Health effects on birth defects

Birth defects are another group of adverse birth outcomes that have recently been found to be associated with air pollution in several studies. Types of birth defects that have been reported to be associated with air pollution include heart defects and cleft palate. Ritz et al examined the effects of air pollutants including CO, NO_2, O_3, and PM_{10} on cardiac and orofacial defects in southern California. They found that second-month CO and O_3 exposures were significantly associated with cardiac defects after adjusting for maternal age, ethnicity, education, access to prenatal care, infant gender, decade of infant birth, parity, time since last pregnancy, birth type, and other air pollutants (Ritz et al. 2002).

Gilboa et al in 2005 investigated the effects of air pollutants such as CO, PM_{10} and SO_2 on different types of heart defects (e.g tetralogy of Fallot, and atrial and ventricular septal defects) among Texas women (Gilboa et al. 2005). Specifically, this study found that exposure to CO in the highest quartile (>0.7ppm) during pregnancy was associated with 2.04 times of the risk of tetralogy of Fallot compared to the lowest quartile after adjusting for infant sex, plurality, maternal education, maternal race, and season of conception (Gilboa et al. 2005). In the same study, PM_{10} and SO_2 were also found to be significantly associated with isolated atrial septal defects (OR = 2.27, 95% CI: 1.43-3.60) and isolated ventricular septal defects (OR =2.16, 95% CI: 1.51-3.09), respectively.

Marshall et al in 2010 examined the effects of criteria air pollutants including CO, O_3, NO_2, SO_2, PM_{10} and $PM_{2.5}$ on oral cleft defects in New Jersey. This study suggested that exposure to the highest quartile (>0.033ppm) of O_3 during the 3-8th month of pregnancy is associated with 2.2 times (95% CI: 1.0-4.9) greater risk of cleft palate compared to exposure to the lowest quartile (<0.015ppm). However, this study found no significant effects of other air pollutants on cleft defect. On the contrary, CO was surprisingly found to be a protective factor of cleft palate defect (OR=0.4, 95% CI: 0.2-0.7) (Marshall et al. 2010).

Another study conducted in the northeast of England had examined the effects of black smoke and SO_2 on congenital heart defects. This analysis found a weak association between maternal exposure to black smoke and heart defects but not for SO_2 exposure (Dadvand et al. 2011).

In summary, criteria air pollutants have been linked with adverse birth outcomes of fetal size, gestational age and birth defects. Several recent publications had also systematically reviewed epidemiologic evidence on the associations between maternal air pollution exposure and birth outcomes (Bosetti et al. 2010; Glinianaia et al. 2004; Maisonet et al. 2004;

Shah and Balkhair 2011). Although the heterogeneity and/or absence of associations between individual air pollutant and birth outcomes exists in published studies, current epidemiological evidence suggests that maternal exposure to PM and SO_2 is associated with several adverse birth outcomes. Reviewing the biologically plausible mechanistic pathways underlying the impacts of air pollution on birth outcomes would greatly contribute to our understanding of these associations.

3. Biologically plausible mechanisms

Although the specific mechanisms through which air pollution affects birth outcomes remain to be fully understood, with emerging bio-molecular technologies, an increasing number of studies show evidence towards potential mechanisms linking air pollution with birth outcomes. Air pollution has been frequently shown to affect the respiratory, cardiovascular, circulatory, and nervous system through multiple potential pathways, some of which might be appropriate for linking air pollution with birth outcomes (Block and Calderon-Garciduenas 2009; Brook 2008; Kunzli and Tager 2005). Several publications have described the biologically plausible mechanistic pathways through which air pollution impacts on birth outcomes (Kannan et al. 2006; Slama et al. 2008). We also used a figure from Slama et al's publication to show the possible biological mechanisms (see figure 1). These mechanisms are described in the following section.

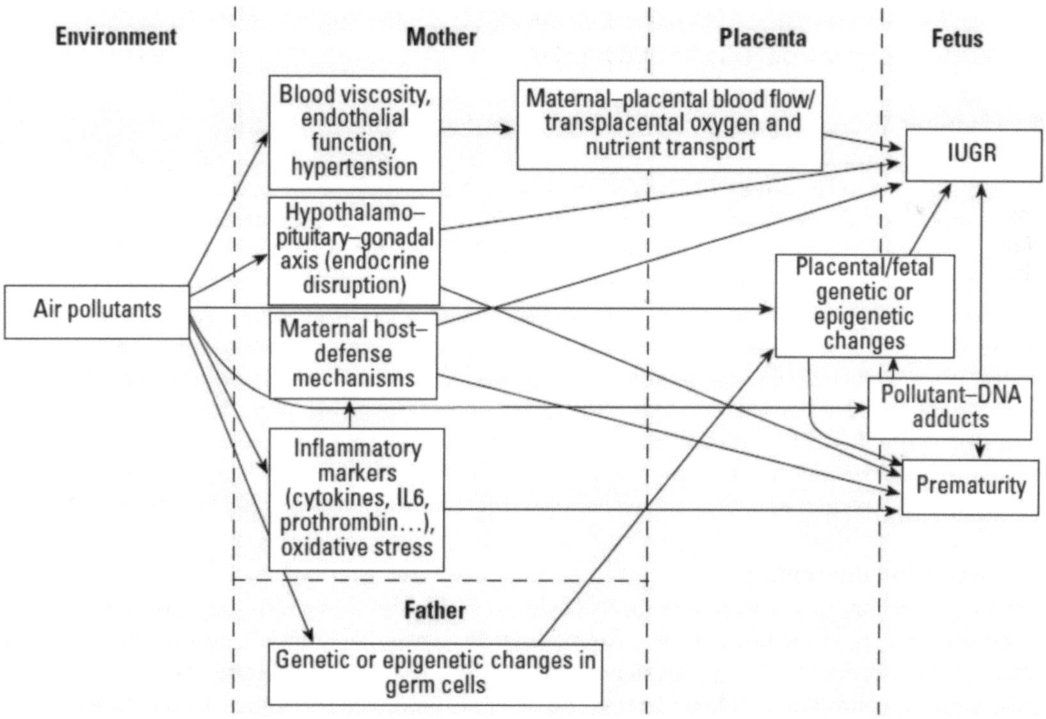

Source: Slama et al. Envrionmental Health Perspective, 2008, 116(6):795

Fig. 1. Possible biological mechanisms by which air pollutants may affect birth outcomes

3.1 Oxidative stress and inflammation

It has been believed that oxidative stress and inflammation play central roles in adverse health effects of air pollution (Donaldson et al. 2001; Tao et al. 2003). Oxidative stress is a condition in which the body encounters more reactive oxygen species than its ability to properly remove them, resulting in excess peroxides and free radicals that can ultimately damage cell components and biological processes (Leeuwenburgh and Heinecke 2001). For example, several studies reported that transition-metal constituents in PM can induce oxidative stress (Kadiiska et al. 1997; Prahalad et al. 2001). PM can also activate inflammatory cells to generate of oxidative stressors such as reactive oxygen species and reactive nitrogen species (Tao et al. 2003). Consequently, oxidative stress can directly lead to DNA damage and may affect the embryo at its earlier stage of growth (Mohorovic 2004; Risom et al. 2005). Moreover, oxidative stress may also influence sperm motility and concentration, which is relevant for male reproductive health (Agarwal et al. 2006).

Air pollution-induced inflammation is considered as another potential biological mechanism through which air pollution cause adverse effects in organs such as the lung (Kunzli and Tager 2005). Inflammation could be a direct result of oxidation or up-regulation of pro-inflammatory mediators induced by air pollutants (Risom et al. 2005). Several studies have shown that air pollutants can enter the blood stream from lung and get deposited onto various body organs via active transport or passive diffusion (Chen et al. 2008; Peters et al. 2006). If air pollutants reach the placenta, they will also induce acute placental inflammation, which subsequently results in impaired transplacetal nutrient exchanges (Bobak 2000). In addition, inflammation could change maternal immunity, reduce host defense, thus increase the maternal risk of infections, which may in turn increase the risks of adverse birth outcomes (Wilhelm and Ritz 2005).

3.2 Changes in rheological factors of blood and endothelial function

Changes of blood coagulability and viscosity as a result of exposure to air pollutants have been suggested in the studies of cardiovascular health effects of air pollution (Coppola et al. 1989; Peters et al. 1997). In addition to the changes in rheological factors of blood, air pollution exposure could also influence endothelial functions. A study found that plasma concentrations of asymmetric dimethyl arginine, which suggests an impaired vascular function, were increased after exposure to $PM_{2.5}$ (Valkonen et al. 2001). Another study also found that exposure to air pollution could cause conduit arterial vasoconstriction in healthy adults (Brook et al. 2002). These studies suggest that exposure to air pollution may trigger endothelial dysfunction, which leads to vasoconstriction. If air pollution could cause the changes in rheological factors of blood such as blood viscosity and blood coagulability, and artery vasoconstriction among pregnancy women, it in turn could influence the transplacental oxygen and nutrient transport which further impacts fetal development.

3.3 Endocrine disruption

Air pollution has also been linked to endocrine dysfunctions that can potentially have negative impacts of birth outcomes. Air pollutants, particularly PM, may interfere with the endocrine system and affect progesterone production (Furuta et al. 2004; Takeda et al. 2004). Although this pathway has been less extensively studied, the possibility of the effects of air pollution on the endocrine system was suggested in recent investigations. Growing evidence suggests that inhalation of environmental tobacco smoke (ETS) contributes to disruptions in thyroid function (Carrillo et al. 2009). Air pollution, similar to ETS, could also interfere with thyroid function. Disruption of thyroid function is associated with fetal development (Blazer et al. 2003; Patel et al. 2011). In addition, polycyclic aromatic hydrocarbon (PAH) was also found to have anti-estrogenic activity and can disrupt endocrine functions (Tran et al. 1996). Studies also found that endocrine disruption among pregnant women might cause IUGR (Kanaka-Gantenbein et al. 2003). Therefore, the link between air pollution and IUGR might involve this pathway. Moreover, a study found that air organic compounds such as PAH can also directly affect epidermal growth factor (EGF) and insulin-like growth factor type I and II (IGF-1 and IGF-2) receptors, leading to inhibition of placental cell growth and proliferation (Dejmek et al. 2000). This could cause a decreased fetal-placental exchange of oxygen and nutrients, which are the critical factors regulating fetal growth.

3.4 Hemodymanic changes

Air pollution exposure has been linked to hemodynamic responses such as changes in blood pressure, heart rate and rhythm, and cardiac autonomic tone (Ibald-Mulli et al. 2004). Specifically, PM exposure is associated with increase in heart rate, heart rate variability, and the frequency of ectopic beats. It is also associated with increase in systolic and diastolic blood pressure (Linn et al. 1999; van den Hooven et al. 2011). The effects of air pollution on hemodymanic changes can suggest a potential mechanism through which it can affect birth outcomes. The hemodymanic changes discussed are all risk factors for hypertension and other cardiovascular diseases. Therefore, exposure to air pollution can increase risks of health conditions (e.g. hypertension) that can ultimately increase the risk of adverse birth outcomes. For example, exposure to PM_{10} has been found to increase the risk of pregnancy-induced hypertension (van den Hooven et al 2011), which is found to be associated with elevated risk of preterm birth (OR=3.30), low birth weight (OR=4.68), fetal growth restriction (OR=2.94), and low Apgar scores 1 minute (OR=2.99) and 5 minutes (OR=2.08) (Olusanya and Solanke 2011).

3.5 Germ-line mutations

Although most of the mechanisms suggested are maternally related, recent studies suggest that paternal factors, especially germ-line mutation resulted from pollution exposure can also greatly influence birth outcomes. In animal studies, it was found that, germline mutation frequencies in wild herring gulls nesting At a polluted site were 2.8 to 8.5 times higher than birds that nested in non-polluted sites. This is consistent with what was discussed earlier—air pollution can adversely impact DNA replication process (Somers and Cooper 2009). In mice, similar results are found where there is an increased frequency of expanded simple tandem repeats among mice that are exposed to air pollution (Somers and Cooper 2009). Moreover, Somers et al 2002 also found evidence that air pollution is capable of inducing heritable DNA mutations (Somers et al. 2002). Since the sentinel animals that were used to study the association of air pollution and DNA mutations were from areas populated with humans, it is very likely that air pollution can also lead to germline mutations among humans, which ultimately can affect birth outcomes. Moreover, several studies have been conducted to support the hypotheses that air pollution is associated with elevated levels of DNA fragmentation in human sperm (Jafarabadi 2007; Rubes et al. 2005).

In conclusion, research on potentially biological mechanisms through which air pollution can affect birth outcomes is still limited. We review several potential mechanisms above which have been suggested in previous studies. Meanwhile, it is important to recognize the need of further research to increase our understanding about the biological mechanisms underlying the impact of various air pollutants.

4. Methodology

As we reviewed above, evidence on the associations between air pollution and birth outcomes are not consistent in all cases. The heterogeneity and/or absence of association could be due to the differences in study design, exposure measurement, identification of windows of exposure, and selection of confounders. We discussed some common methods in these areas below.

4.1 Study designs

To assess whether air pollution exposure has an effect on pregnancy outcomes, the investigators need to compare the pregnancy outcome occurrence of individuals who have been subject to different exposure levels and, in particular, by comparing the occurrence or prevalence rate of pregnancy outcomes of exposed persons with those of unexposed or less exposed persons (Strickland et al. 2009). Specifically, descriptive studies, ecological studies, case-control studies, and cohort studies can be used to examine the relationship between air pollution and pregnancy outcomes.

The first step in investigating the pregnancy outcomes associated with air pollution exposure could be a **descriptive study**. Descriptive studies examine the distribution of pregnancy outcomes in a defined population. They are helpful in assessing the possibility that an association exists, and identifying hypotheses to be evaluated in analytical studies. Any type of pregnancy outcomes can be used for conducting descriptive studies. Descriptive data are commonly applied to examine patterns of pregnancy outcomes by place, time and person. *Geographic comparisons* based on standardized morbidity rates can be made among different geographic regions. The variations between these regions concerning occurrence of pregnancy outcomes contribute to the basis of causal hypotheses. *Temporal trends* in pregnancy outcomes rates can also be valuable to indicate the possible effects of air pollution.

Ecological studies are studies in which the investigators analyze hypothesized associations between air pollution and pregnancy outcomes using groups of people, rather than individuals, as the unit of analysis. It compares aggregate measures of exposure, such as average exposure or proportion of population exposed, with aggregate measures of pregnancy outcomes rates, for the same population. A traditional approach is to use *geographical areas* (Bobak and Leon 1999) as the basis for defining the study groups, and then correlations between aggregate measures of exposure and pregnancy outcomes at the same geographical location are analyzed. The *time series* design is a special and emerging type of ecological study with widespread application in epidemiological studies of short-term exposure to air pollution and pregnant outcomes (Darrow et al. 2009; Jiang et al. 2007; Lee et al. 2008; Sagiv et al. 2005; Zhao et al.). It investigates the relationship between air pollution levels and pregnancy outcomes - each measured and aggregated over the same time units (e.g., days, weeks) during a specified time period. Modeling time-series data is challenging because the relatively small effect of air pollutants is hard to be identified and quantified in the presence of strong confounding. Typically, one of the most widely-used statistical approaches for time series analysis of air pollution and health is the generalized additive model (GAM). A strength of time-series analyses is the inherent control of individual-level risk factors that do not vary temporally. However, in studies of adverse pregnancy outcomes, risk factors considered time-invariant at the individual level may vary seasonally when aggregated into a pregnancy risk set. Therefore, time-series investigations of seasonally-varying exposures and adverse pregnancy outcomes should consider the potential for bias due to seasonal heterogeneity in the risk set. Moreover, this study design must be used with caution due to the important but hardly predictable ecological bias for which group-level associations do not accurately reflect individual-level associations.

A **case-control study** examines associations between air pollution exposures and adverse pregnancy outcomes by comparing cases, or individuals who developed the outcome with controls who are a sample of the source population from which the cases were identified. Controls are usually individuals who are similar to the cases in terms of risk characteristics, but who have not developed the pregnancy outcomes. Having selected cases and controls, the investigators then determine the prior air pollution exposure of the cases and controls. Hospital-based cases are usually selected for their high accessibility and cooperative attitude but this method is subject to bias. A population-based case control study (Rogers et al. 2000) is the principal alternative, for which all incident cases of the outcome in a defined geographic area are included as cases. The selection of controls can be more challenging. The most commonly used control groups are a random sample of the source population from which the cases are selected or persons seeking medical care at the same institutions as the cases for conditions believed to be unrelated to the health outcome of interest. Individual or frequency matching is another efficient strategy to select controls so that the distributions of some a priori selected risk factors are identical or nearly the same for the controls as for the cases (Dadvand et al. 2011; Hansen et al. 2009).

A cohort study selects subjects who are at the risk of developing a particular pregnancy outcome, and then are divided into groups according to their air pollution exposure status. The study groups are then followed over time to determine the subsequent incidence of the pregnancy outcome within each group. This study design enables investigators to measure incidence rates and to estimate all effect measures, such as rate ratio (RR) and rate difference. Cohort studies have been an effective method to assess the long-term health effects of acute or chronic exposure to air pollution. A fixed cohort consists of a group of individuals who are identified at a point or interval of tine and then followed over time. A dynamic cohort allows the inclusion of members over time as they fit the selection criteria. A historical or retrospective cohort study is conceptually identical, but less expensive and easier to conduct, to a prospective cohort study except that the study takes place after the causal events have unfolded. Most studies of air pollution and birth outcomes have used a population-based cohort study to examine their associations by linking information from birth certificate records with data from air monitoring (Aguilera et al. 2008; Brauer et al. 2008; Kim et al. 2007; Ritz et al. 2007). This approach allows conducting a study with a large study sample size at very low cost because of the utilization of existing datasets. However, it also easily suffers from exposure misclassification and inadequate controlling of confounders.

Two-phase (stage) designs nesting a sample within a cohort, for which both outcome and some exposure information are available, have a long history in epidemiology. In this approach, some additional information at the individual level is obtained for a sample nested within a cohort. Furthermore, the additional information can be used to assess potential confounding effects on the estimated relationships between exposures of interest and outcomes within the context of the larger cohort. Therefore, this is a special case control design and can better control confounding by use of survey data and minimization of any selection or response bias. The two-phase study design has been recently applied to investigate the associations between air pollution and birth outcomes (Ritz et al. 2007). This approach holds promising because it provides a better capability of controlling for confounding than a classical population-based cohort study of air pollution and birth outcomes.

4.2 Methods of exposure assessment

Air pollution exposure assessment in most studies of air pollution and birth outcomes relies on existing networks of ambient monitoring stations. This approach of exposure assessment is mostly to assign exposure to maternal residential location based on the nearest monitor site or the monitor sites within the same administrative unit of maternal address (e.g. county) (Alderman et al. 1987; Dejmek et al. 1999; Xu et al. 1995; Zhao et al. 2011). However, the method is more likely to result in exposure misclassification because the monitoring networks are relatively sparse and insufficient for capturing all spatial variations in exposure and individual time-activity pattern is also not considered. Another approach for maternal exposure assessment is to develop surrogate measures, i.e. proximity methods , for example, the distance to sources of air pollution for estimating the level of air pollution exposure (Kashima et al. 2011; Yorifuji et al. 2011). Although proximity methods are straightforward, they have largely been thought as a form of exploratory analysis in traffic-related air pollution studies because they have considerable limitations such as invalid assumption of isotropic dispersion of pollution and inadequate consideration of other factors of meteorological condition and topography. Recently, interpolation methods have been applied to estimate maternal air pollution exposure, such as the inverse distance weighting (Xu et al. 2010) and Kriging method (Leem et al. 2006; Seo et al. 2010). Interpolation models produce a continuous surface of pollution concentration, and then extract individual exposure from the surface based on residential locations. This approach is limited by issues regarding exaggerated variation of predictions, spatial coverage and representatives of the existing monitoring sites, and incapability of taking other possible predictors into accounts. Furthermore, dispersion models have been developed to address a number of limitations for air pollution exposure assessment (Batterman et al. 2010; Hoffmann et al. 2009a; Hoffmann et al. 2009b). These models have been regarded as a more realistic representation of the problem by making use of data on emissions, meteorological conditions, and topography in conjunction with information from empirical monitoring systems in predicting spatial exposure estimates of air pollution concentration. Some studies have applied this method for maternal air pollution exposure assessments (Ihrig et al. 1998; Madsen et al. 2010; Wu et al. 2009). However, the features of dispersion models such as assumption of Gaussian dispersion, relatively costly data input, and complicated models, validation, impede their wide applications in estimating air pollution exposure. Land Use Regression (LUR) Models are the methods which have been developed to predict air pollutant concentrations at a given site based on the surrounding land use and traffic characteristics (Briggs et al. 1997; Briggs et al. 2000; Clougherty et al. 2008). This approach establishes a multivariate regression model, where monitored air pollution data serves as a dependent variable and traffic information, land use/cover, and other geographic information are considered as independent variables (Ryan and LeMasters 2007). The model can be expanded using GIS approaches, and thus is capable of predicting the level of air pollution at any location of an area. The advance in GIS technology and availability of detailed geo-reference data further increase the feasibility of implementing LUR models and reduce the costs of modeling. Although many studies have demonstrated their success in predicting air pollution levels, a number of limitations remain. First, the LUR models have been primarily utilized for urban areas, but few of them have been expanded to suburban or rural areas because of lack of information. Second, LUR models are limited to explaining spatial variation of air pollutants, but few of them have addressed temporal variations. This is because the model predictors are temporally stable, such as land use, elevation, and population density (Molter et al. 2010; Rose et al. 2011). A recent review has highlighted the potential for improving the LUR approach by including a spatial and a temporal component in the models (Hoek et al. 2008). Several studies have used this method for maternal exposure assessments (Gehring et al. 2011). The main characteristics of the exposure methods described above are summarized in Table 1.

Although several exposure assessment models have been developed to improve maternal air pollution exposure, personal air pollution exposure remains challenge in the field because limited studies have considered microenvironment in personal exposure estimate. Some methods such as personal dosimetry and biomarkers may need to be considered in future studies.

Methods	Characteristics	Strengths	Limitations
Air monitoring networks	Using existing air data for individual exposure estimation	Easy to access the data; Low cost;	Lack of space coverage; increase likelihoods of misclassification
Proximity methods	Developing surrogate measures such as distance to the road as exposure index	Straightforward; Requires basic GIS techniques;	Invalid assumption of isotropic dispersion of pollution; inadequate consideration
Geo-statistical methods	Using geo-statistical methods to created predicted surfaces of air pollutants	Continuous space coverage; Require few other spatial data;	Exaggerated variation of predictions; required extensive monitoring network; incapability of controlling for other covariates
Dispersion models	Making use of data on emissions, meteorological conditions, topography and information from empirical monitoring systems for spatial prediction	Continuous space coverage; More realistic representation of the problem;	Invalid assumption of Gaussian dispersion; relatively costly data input; Complicated model validation
Land Use Regression (LUR) Models	Predicting air pollutant concentrations at a given site based on the surrounding land use and traffic characteristics	Continuous space coverage; Simple regression model;	Requires temporal variation of predictor; Unmatched of existing spatial data;

Table 1. Characteristics of common methods of exposure assessment in studies of air pollution on birth outcomes

4.3 Windows of exposure

It is believed that the fetus at some particular period of pregnancy is more sensitive to adverse health effects of air pollution. While some studies attempted to determine the effects of air pollution exposure during specific phases of pregnancy, evidence currently available does not support any firm conclusions. Most studies of air pollution and birth outcomes have examined windows of exposure such as the whole pregnancy, each trimester and/or each month of pregnancy. This inconsistency in defining windows of exposure across studies makes it extremely difficult to determine a specific critical period of exposure for each birth outcome. However, according to previous research, negative effects of various air pollutants can be seen as early as the first month of pregnancy. For example, the effects of SO_2, $PM_{2.5}$, and PM_{10} exposure are seen as soon as during the first gestation month for low birth weight, preterm delivery and intrauterine growth restriction (Dejmek et al. 1999; Huynh et al. 2006; Liu et al. 2003; Ritz et al. 2000). For studies which considered all three pregnancy trimesters, exposures during the first and third trimesters appeared to be significantly associated with some birth outcomes. For example, Bukowski et al 2007 demonstrated in their study that variation in birth weight might be determined by fetal size during the first 12 weeks after conception (Bukowski et al. 2007).

Most of the studies reviewed had only examined exposure during pregnancy. However, pre-pregnancy exposure to air pollutants which may affect genetic or epigenetic component in parents see **Germ-line mutations)**, might in turn impact birth outcomes. Therefore, future studies might also consider examining pre-pregnancy windows of exposure.

4.4 Confounders

Confounding threats the validity of observational research. This concern is particularly pertinent in studies of air pollution and birth outcomes as the effects of air pollution on birth outcomes are small. Strickland et al had thoroughly discussed the issue of confounding in epidemiological studies of air pollution and birth outcomes (Strickland et al. 2009). Under the counterfactual disease model, the effect of a factor on a disease is determined by comparing the risk of disease in a population exposure to the factor with one in the same population without exposure to the factor (i.e. under a counterfactual condition). As Strickland et al suggested, risk in the population under a counterfactual condition cannot be observed. A substitute population requires representing the risk in the population under a counterfactual condition. Confounding occurs if the risk in the substitute population is different from the risk in the population under a counterfactual condition.

Birth certificates have been used as the primary data source for information of confounders in most of studies of air pollution and birth outcomes. The birth records usually have information on maternal and infantile characteristics such as maternal age, race, educational attainment, pregnancy complications, and infant gender, etc. Confounding in the studies may arise from imperfect classification of existing information, i.e. residual confounding and unmeasured other variables such as socioeconomic status and maternal nutrition. Residual confounding refers to the confounding due to an imperfect classification of the groups of variables, e.g. maternal age groups and educational groups. Within the group, the confounding of the variable may still exist. For confounding effects of unmeasured factors in birth records, the two-phase study, as we described above, may provide a way to deal with the issue. However, selecting potential confounders requires caution. Some factors which are the intermediate factors in the pathway of air pollution and birth outcomes should not be selected as confounders. For example, some pregnancy complications such as preeclampsia might be caused by air pollution and also risk factors of adverse birth outcomes (Wu et al. 2009).

5. Recommendations of future studies

Several limitations exist in previous air pollution studies on birth outcomes. A few papers have already discussed these issues and provide directions for future studies (Slama et al. 2008; Woodruff et al. 2009). Specifically, the following—but not limited to—issues are of major concern: 1) Accurate individual air pollution exposure including identification of windows of exposure remains a major challenge in the field; 2) The effects of specific components of PM have rarely been investigated; 3) Research of air pollution on fetal health rarely considers the concurrent effects of multiple air pollutants; 4) Little has been done to examine the potential effect modifiers of the relationship between air pollution and birth outcomes; 5) Research including experimental and epidemiological studies on biologically plausible mechanisms is limited. These issues might need to be adequately addressed in future.

1. Development of accurate air pollution exposure assessment remains an important work. The exposure models are in need of improvement in terms of spatial and temporal resolution by incorporating some new technologies such as remote sensing. Individual time-activity patterns in combination with the information from the exposure models might be used together for personal exposure assessment. Some methods of personal monitoring and biomarkers might be considered in prospective cohort studies.
2. Few studies have considered PM composition in relation to birth outcomes (Bell et al. 2010). Different sources of PM may have different chemical constituents and thus have different toxicity. The studies on PM composition might provide clues to explain the heterogeneity of health effects of PM across studies. Further, it would provide information on target chemicals or sources for prevention efforts.
3. Individual is simultaneously exposed to multiple air pollutants. As we reviewed, several criteria air pollutants have adverse effects on birth outcomes. It may be more reasonable to assume that there is a mixture of pollutants that is considered harmful to birth outcomes. However, few studies have investigated the adverse health effects of an air pollution mixture on birth outcomes. It is an important area of ongoing research of air pollution on fetal health. Some sophisticated models need to be developed to examine the effect of a mixture of pollutants because of high collinearity between air pollutants.
4. Air pollution studies on mortality and morbidity have suggested that people with different personal characteristics might respond to air pollution in different ways (Dubowsky et al. 2006; Zanobetti et al. 2000; Zeka et al. 2006). The National Research Council has identified the potential effect modifiers as a key data gap and has emphasized the need for continued study of the most susceptible populations related to air pollution (NRC 2004). However, little has been done to examine the potential modifiers of the health effects of air pollution on birth outcomes. Limited evidence suggests that the effect sizes of air pollution on adverse pregnancy outcomes may be different by fetal gender (Ghosh et al. 2007; Jedrychowski et al. 2009), pregnancy complication (Rich et al. 2009), and maternal nutrition (Jedrychowski et al. 2007; Kannan et al. 2006). Much remains to be learned about these relationships and what else could cause fetuses to be especially sensitive to the adverse health effects of air pollution.

5. Research to identify relevant biological mechanisms is needed to augment our understanding about the relationships. Epidemiological studies could consider other perinatal end points such as preeclampsia and placental size in future studies. The findings from these studies would provide important supporting evidence on the potential biological mechanisms as discussed above.

6. Acknowledgements

Publication of this chapter was funded in part by the University of Florida Open-Access Publishing Fund.

7. References

Agarwal A, Sharma RK, Nallella KP, Thomas AJ, Jr., Alvarez JG, and Sikka SC. 2006. Reactive oxygen species as an independent marker of male factor infertility. Fertil Steril 86, 878-885.

Aguilera I, Sunyer J, Fernandez-Patier R, Hoek G, Aguirre-Alfaro A, Meliefste K, et al. 2008. Estimation of outdoor NO(x), NO(2), and BTEX exposure in a cohort of pregnant women using land use regression modeling. Environ Sci Technol 42, 815-821.

Alderman BW, Baron AE, and Savitz DA. 1987. Maternal exposure to neighborhood carbon monoxide and risk of low infant birth weight. Public Health Rep 102, 410-414.

Ballester F, Estarlich M, Iniguez C, Llop S, Ramon R, Esplugues A, et al. 2010. Air pollution exposure during pregnancy and reduced birth size: a prospective birth cohort study in Valencia, Spain. Environ Health 9, 6.

Barnett AG, Williams GM, Schwartz J, Best TL, Neller AH, Petroeschevsky AL, et al. 2006. The effects of air pollution on hospitalizations for cardiovascular disease in elderly people in Australian and New Zealand cities. Environ Health Perspect 114, 1018-1023.

Batterman SA, Zhang K, and Kononowech R. 2010. Prediction and analysis of near-road concentrations using a reduced-form emission/dispersion model. Environ Health 9, 29.

Bell ML, Belanger K, Ebisu K, Gent JF, Lee HJ, Koutrakis P, et al. 2010. Prenatal exposure to fine particulate matter and birth weight: variations by particulate constituents and sources. Epidemiology 21, 884-891.

Bell ML, Ebisu K, and Belanger K. 2007. Ambient air pollution and low birth weight in Connecticut and Massachusetts. Environ Health Perspect 115, 1118-1124.

Bell ML, McDermott A, Zeger SL, Samet JM, and Dominici F. 2004. Ozone and short-term mortality in 95 US urban communities, 1987-2000. Jama 292, 2372-2378.

Blazer S, Moreh-Waterman Y, Miller-Lotan R, Tamir A, and Hochberg Z. 2003. Maternal hypothyroidism may affect fetal growth and neonatal thyroid function. Obstet Gynecol 102, 232-241.

Block ML, and Calderon-Garciduenas L. 2009. Air pollution: mechanisms of neuroinflammation and CNS disease. Trends Neurosci 32, 506-516.

Bobak M. 2000. Outdoor air pollution, low birth weight, and prematurity. Environ Health Perspect 108, 173-176.

Bobak M, and Leon DA. 1999. Pregnancy outcomes and outdoor air pollution: an ecological study in districts of the Czech Republic 1986-8. Occup Environ Med 56, 539-543.

Bosetti C, Nieuwenhuijsen MJ, Gallus S, Cipriani S, La Vecchia C, and Parazzini F. 2010. Ambient particulate matter and preterm birth or birth weight: a review of the literature. Arch Toxicol 84, 447-460.

Brauer M, Lencar C, Tamburic L, Koehoorn M, Demers P, and Karr C. 2008. A cohort study of traffic-related air pollution impacts on birth outcomes. Environ Health Perspect 116, 680-686.

Briggs DJ, Collins S, Elliott P, Fischer P, Kingham S, Lebret E, et al. 1997. Mapping urban air pollution using GIS: a regression-based approach. International Journal of Geographical Information Science 11, 699-718.

Briggs DJ, de Hoogh C, Gulliver J, Wills J, Elliott P, Kingham S, et al. 2000. A regression-based method for mapping traffic-related air pollution: application and testing in four contrasting urban environments. Sci Total Environ 253, 151-167.

Brook RD. 2008. Cardiovascular effects of air pollution. Clin Sci (Lond) 115, 175-187.

Brook RD, Brook JR, Urch B, Vincent R, Rajagopalan S, and Silverman F. 2002. Inhalation of fine particulate air pollution and ozone causes acute arterial vasoconstriction in healthy adults. Circulation 105, 1534-1536.

Bukowski R, Smith GC, Malone FD, Ball RH, Nyberg DA, Comstock CH, et al. 2007. Fetal growth in early pregnancy and risk of delivering low birth weight infant: prospective cohort study. Bmj 334, 836.

Carrillo AE, Metsios GS, and Flouris AD. 2009. Effects of secondhand smoke on thyroid function. Inflamm Allergy Drug Targets 8, 359-363.

Chen L, Yokel RA, Hennig B, and Toborek M. 2008. Manufactured aluminum oxide nanoparticles decrease expression of tight junction proteins in brain vasculature. J Neuroimmune Pharmacol 3, 286-295.

Clapp Iii JF, and Lopez B. 2007. Size at Birth, Obesity and Blood Pressure at Age Five. Metab Syndr Relat Disord 5, 116-126.

Clougherty JE, Wright RJ, Baxter LK, and Levy JI. 2008. Land use regression modeling of intra-urban residential variability in multiple traffic-related air pollutants. Environ Health 7, 17.

Coppola L, Giunta R, Grassia A, Misso L, Verrazzo G, Violano PF, et al. 1989. Air pollution by gasoline exhaust fumes: effect on platelet function and blood viscosity. Med Lav 80, 187-191.

Dadvand P, Rankin J, Rushton S, and Pless-Mulloli T. 2011. Association between maternal exposure to ambient air pollution and congenital heart disease: A register-based spatiotemporal analysis. Am J Epidemiol 173, 171-182.

Damus K. 2008. Prevention of preterm birth: a renewed national priority. Curr Opin Obstet Gynecol 20, 590-596.

Darrow LA, Klein M, Flanders WD, Waller LA, Correa A, Marcus M, et al. 2009. Ambient Air Pollution and Preterm Birth: A Time-series Analysis. Epidemiology.

Dejmek J, Jelinek R, Solansky I, Benes I, and Sram RJ. 2000. Fecundability and parental exposure to ambient sulfur dioxide. Environ Health Perspect 108, 647-654.

Dejmek J, Selevan SG, Benes I, Solansky I, and Sram RJ. 1999. Fetal growth and maternal exposure to particulate matter during pregnancy. Environ Health Perspect 107, 475-480.

Dockery DW, and Pope CA, 3rd. 1994. Acute respiratory effects of particulate air pollution. Annu Rev Public Health 15, 107-132.

Dominici F, McDermott A, Zeger SL, and Samet JM. 2003. Airborne particulate matter and mortality: timescale effects in four US cities. Am J Epidemiol 157, 1055-1065.

Donaldson K, Stone V, Seaton A, and MacNee W. 2001. Ambient particle inhalation and the cardiovascular system: potential mechanisms. Environ Health Perspect 109 Suppl 4, 523-527.

Dubowsky SD, Suh H, Schwartz J, Coull BA, and Gold DR. 2006. Diabetes, obesity, and hypertension may enhance associations between air pollution and markers of systemic inflammation. Environ Health Perspect 114, 992-998.

Dugandzic R, Dodds L, Stieb D, and Smith-Doiron M. 2006. The association between low level exposures to ambient air pollution and term low birth weight: a retrospective cohort study. Environ Health 5, 3.

Furuta C, Suzuki AK, Taneda S, Kamata K, Hayashi H, Mori Y, et al. 2004. Estrogenic activities of nitrophenols in diesel exhaust particles. Biol Reprod 70, 1527-1533.

Gehring U, Wijga AH, Fischer P, de Jongste JC, Kerkhof M, Koppelman GH, et al. 2010. Traffic-related air pollution, preterm birth and term birth weight in the PIAMA birth cohort study. Environ Res 111, 125-135.

Gehring U, Wijga AH, Fischer P, de Jongste JC, Kerkhof M, Koppelman GH, et al. 2011. Traffic-related air pollution, preterm birth and term birth weight in the PIAMA birth cohort study. Environ Res 111, 125-135.

Ghosh R, Rankin J, Pless-Mulloli T, and Glinianaia S. 2007. Does the effect of air pollution on pregnancy outcomes differ by gender? A systematic review. Environ Res 105, 400-408.

Gilboa SM, Mendola P, Olshan AF, Langlois PH, Savitz DA, Loomis D, et al. 2005. Relation between ambient air quality and selected birth defects, seven county study, Texas, 1997-2000. Am J Epidemiol 162, 238-252.

Glinianaia SV, Rankin J, Bell R, Pless-Mulloli T, and Howel D. 2004. Particulate air pollution and fetal health: a systematic review of the epidemiologic evidence. Epidemiology 15, 36-45.

Gold DR, Damokosh AI, Pope CA, 3rd, Dockery DW, McDonnell WF, Serrano P, et al. 1999. Particulate and ozone pollutant effects on the respiratory function of children in southwest Mexico City. Epidemiology 10, 8-16.

Ha EH, Hong YC, Lee BE, Woo BH, Schwartz J, and Christiani DC. 2001. Is air pollution a risk factor for low birth weight in Seoul? Epidemiology 12, 643-648.

Hansen C, Neller A, Williams G, and Simpson R. 2006. Maternal exposure to low levels of ambient air pollution and preterm birth in Brisbane, Australia. Bjog 113, 935-941.

Hansen CA, Barnett AG, Jalaludin BB, and Morgan GG. 2009. Ambient air pollution and birth defects in brisbane, australia. PLoS One 4, e5408.

Hoek G, Beelen R, de Hoogh K, Vienneau D, Gulliver J, Fischer P, et al. 2008. A review of land-use regression models to assess spatial variation of outdoor air pollution. Atmospheric Environment 42, 7561-7578.

Hoffmann B, Moebus S, Dragano N, Stang A, Mohlenkamp S, Schmermund A, et al. 2009a. Chronic residential exposure to particulate matter air pollution and systemic inflammatory markers. Environ Health Perspect 117, 1302-1308.

Hoffmann B, Moebus S, Kroger K, Stang A, Mohlenkamp S, Dragano N, et al. 2009b. Residential exposure to urban air pollution, ankle-brachial index, and peripheral arterial disease. Epidemiology 20, 280-288.

Huynh M, Woodruff TJ, Parker JD, and Schoendorf KC. 2006. Relationships between air pollution and preterm birth in California. Paediatr Perinat Epidemiol 20, 454-461.

Ibald-Mulli A, Timonen KL, Peters A, Heinrich J, Wolke G, Lanki T, et al. 2004. Effects of particulate air pollution on blood pressure and heart rate in subjects with cardiovascular disease: a multicenter approach. Environ Health Perspect 112, 369-377.

Ihrig MM, Shalat SL, and Baynes C. 1998. A hospital-based case-control study of stillbirths and environmental exposure to arsenic using an atmospheric dispersion model linked to a geographical information system. Epidemiology 9, 290-294.

Jafarabadi M. 2007. Episodic air pollution is associated with increased DNA fragmentation in human sperm without other changes in semen quality. Hum Reprod 22, 3263; author reply 3264.

Jalaludin B, Mannes T, Morgan G, Lincoln D, Sheppeard V, and Corbett S. 2007. Impact of ambient air pollution on gestational age is modified by season in Sydney, Australia. Environ Health 6, 16.

Jedrychowski W, Masters E, Choi H, Sochacka E, Flak E, Mroz E, et al. 2007. Pre-pregnancy dietary vitamin A intake may alleviate the adverse birth outcomes associated with prenatal pollutant exposure: epidemiologic cohort study in Poland. Int J Occup Environ Health 13, 175-180.

Jedrychowski W, Perera F, Mrozek-Budzyn D, Mroz E, Flak E, Spengler JD, et al. 2009. Gender differences in fetal growth of newborns exposed prenatally to airborne fine particulate matter. Environ Res 109, 447-456.

Jerrett M, Burnett RT, Ma R, Pope CA, 3rd, Krewski D, Newbold KB, et al. 2005. Spatial analysis of air pollution and mortality in Los Angeles. Epidemiology 16, 727-736.

Jiang LL, Zhang YH, Song GX, Chen GH, Chen BH, Zhao NQ, et al. 2007. A time series analysis of outdoor air pollution and preterm birth in Shanghai, China. Biomed Environ Sci 20, 426-431.

Kadiiska MB, Mason RP, Dreher KL, Costa DL, and Ghio AJ. 1997. In vivo evidence of free radical formation in the rat lung after exposure to an emission source air pollution particle. Chem Res Toxicol 10, 1104-1108.

Kanaka-Gantenbein C, Mastorakos G, and Chrousos GP. 2003. Endocrine-related causes and consequences of intrauterine growth retardation. Ann N Y Acad Sci 997, 150-157.

Kannan S, Misra DP, Dvonch JT, and Krishnakumar A. 2006. Exposures to airborne particulate matter and adverse perinatal outcomes: a biologically plausible mechanistic framework for exploring potential effect modification by nutrition. Environ Health Perspect 114, 1636-1642.

Kashima S, Naruse H, Yorifuji T, Ohki S, Murakoshi T, Takao S, et al. 2011. Residential proximity to heavy traffic and birth weight in Shizuoka, Japan. Environ Res.

Kim OJ, Ha EH, Kim BM, Seo JH, Park HS, Jung WJ, et al. 2007. PM10 and pregnancy outcomes: a hospital-based cohort study of pregnant women in Seoul. J Occup Environ Med 49, 1394-1402.

Kunzli N, and Tager IB. 2005. Air pollution: from lung to heart. Swiss Med Wkly 135, 697-702.

Landgren O. 1996. Environmental pollution and delivery outcome in southern Sweden: a study with central registries. Acta Paediatr 85, 1361-1364.

Lee BE, Ha EH, Park HS, Kim YJ, Hong YC, Kim H, et al. 2003. Exposure to air pollution during different gestational phases contributes to risks of low birth weight. Hum Reprod 18, 638-643.

Lee SJ, Hajat S, Steer PJ, and Filippi V. 2008. A time-series analysis of any short-term effects of meteorological and air pollution factors on preterm births in London, UK. Environ Res 106, 185-194.

Leem JH, Kaplan BM, Shim YK, Pohl HR, Gotway CA, Bullard SM, et al. 2006. Exposures to air pollutants during pregnancy and preterm delivery. Environ Health Perspect 114, 905-910.

Leeuwenburgh C, and Heinecke JW. 2001. Oxidative stress and antioxidants in exercise. Curr Med Chem 8, 829-838.

Lin CM, Li CY, and Mao IF. 2004. Increased risks of term low-birth-weight infants in a petrochemical industrial city with high air pollution levels. Arch Environ Health 59, 663-668.

Linn WS, Gong H, Jr., Clark KW, and Anderson KR. 1999. Day-to-day particulate exposures and health changes in Los Angeles area residents with severe lung disease. J Air Waste Manag Assoc 49, 108-115.

Liu S, Krewski D, Shi Y, Chen Y, and Burnett RT. 2003. Association between gaseous ambient air pollutants and adverse pregnancy outcomes in Vancouver, Canada. Environ Health Perspect 111, 1773-1778.

Liu S, Krewski D, Shi Y, Chen Y, and Burnett RT. 2007. Association between maternal exposure to ambient air pollutants during pregnancy and fetal growth restriction. J Expo Sci Environ Epidemiol 17, 426-432.

Llop S, Ballester F, Estarlich M, Esplugues A, Rebagliato M, and Iniguez C. 2010. Preterm birth and exposure to air pollutants during pregnancy. Environ Res 110, 778-785.

Madsen C, Gehring U, Walker SE, Brunekreef B, Stigum H, Naess O, et al. 2010. Ambient air pollution exposure, residential mobility and term birth weight in Oslo, Norway. Environ Res 110, 363-371.

Maisonet M, Bush TJ, Correa A, and Jaakkola JJ. 2001. Relation between ambient air pollution and low birth weight in the Northeastern United States. Environ Health Perspect 109 Suppl 3, 351-356.

Maisonet M, Correa A, Misra D, and Jaakkola JJ. 2004. A review of the literature on the effects of ambient air pollution on fetal growth. Environ Res 95, 106-115.

Mannes T, Jalaludin B, Morgan G, Lincoln D, Sheppeard V, and Corbett S. 2005. Impact of ambient air pollution on birth weight in Sydney, Australia. Occup Environ Med 62, 524-530.

Maroziene L, and Grazuleviciene R. 2002. Maternal exposure to low-level air pollution and pregnancy outcomes: a population-based study. Environ Health 1, 6.

Marshall EG, Harris G, and Wartenberg D. 2010. Oral cleft defects and maternal exposure to ambient air pollutants in New Jersey. Birth Defects Res A Clin Mol Teratol 88, 205-215.

Martin JA, Kung HC, Mathews TJ, Hoyert DL, Strobino DM, Guyer B, et al. 2008. Annual summary of vital statistics: 2006. Pediatrics 121, 788-801.

Middleton N, Yiallouros P, Kleanthous S, Kolokotroni O, Schwartz J, Dockery DW, et al. 2008. A 10-year time-series analysis of respiratory and cardiovascular morbidity in

Nicosia, Cyprus: the effect of short-term changes in air pollution and dust storms. Environ Health 7, 39.

Mohorovic L. 2004. First two months of pregnancy--critical time for preterm delivery and low birthweight caused by adverse effects of coal combustion toxics. Early Hum Dev 80, 115-123.

Molter A, Lindley S, de Vocht F, Simpson A, and Agius R. 2010. Modelling air pollution for epidemiologic research--part II: predicting temporal variation through land use regression. Sci Total Environ 409, 211-217.

NRC. 2004. National Research Council. Research priorities for airborne particulate matter. IV. Continuing research progress. Washington, DC: The National Academies Press, 2004.

Olusanya BO, and Solanke OA. 2011. Perinatal Outcomes Associated with Maternal Hypertensive Disorders of Pregnancy in a Developing Country. Hypertens Pregnancy.

Osmond C, and Barker DJ. 2000. Fetal, infant, and childhood growth are predictors of coronary heart disease, diabetes, and hypertension in adult men and women. Environ Health Perspect 108 Suppl 3, 545-553.

Parker JD, Woodruff TJ, Basu R, and Schoendorf KC. 2005. Air pollution and birth weight among term infants in California. Pediatrics 115, 121-128.

Patel J, Landers K, Li H, Mortimer RH, and Richard K. 2011. Thyroid hormones and fetal neurological development. J Endocrinol 2011.

Peters A, Doring A, Wichmann HE, and Koenig W. 1997. Increased plasma viscosity during an air pollution episode: a link to mortality? Lancet 349, 1582-1587.

Peters A, Veronesi B, Calderon-Garciduenas L, Gehr P, Chen LC, Geiser M, et al. 2006. Translocation and potential neurological effects of fine and ultrafine particles a critical update. Part Fibre Toxicol 3, 13.

Pope CA, 3rd. 1999. Mortality and air pollution: associations persist with continued advances in research methodology. Environ Health Perspect 107, 613-614.

Pope CA, 3rd. 2000. Epidemiology of fine particulate air pollution and human health: biologic mechanisms and who's at risk? Environ Health Perspect 108 Suppl 4, 713-723.

Pope CA, 3rd, Dockery DW, Spengler JD, and Raizenne ME. 1991. Respiratory health and PM10 pollution. A daily time series analysis. Am Rev Respir Dis 144, 668-674.

Pope CA, 3rd, and Kanner RE. 1993. Acute effects of PM10 pollution on pulmonary function of smokers with mild to moderate chronic obstructive pulmonary disease. Am Rev Respir Dis 147, 1336-1340.

Prahalad AK, Inmon J, Dailey LA, Madden MC, Ghio AJ, and Gallagher JE. 2001. Air pollution particles mediated oxidative DNA base damage in a cell free system and in human airway epithelial cells in relation to particulate metal content and bioreactivity. Chem Res Toxicol 14, 879-887.

Rich DQ, Demissie K, Lu SE, Kamat L, Wartenberg D, and Rhoads GG. 2009. Ambient air pollutant concentrations during pregnancy and the risk of fetal growth restriction. J Epidemiol Community Health 63, 488-496.

Rinaudo PF, and Lamb J. 2008. Fetal origins of perinatal morbidity and/or adult disease. Semin Reprod Med 26, 436-445.

Risom L, Moller P, and Loft S. 2005. Oxidative stress-induced DNA damage by particulate air pollution. Mutat Res 592, 119-137.

Ritz B, Wilhelm M, Hoggatt KJ, and Ghosh JK. 2007. Ambient air pollution and preterm birth in the environment and pregnancy outcomes study at the University of California, Los Angeles. Am J Epidemiol 166, 1045-1052.

Ritz B, and Yu F. 1999. The effect of ambient carbon monoxide on low birth weight among children born in southern California between 1989 and 1993. Environ Health Perspect 107, 17-25.

Ritz B, Yu F, Chapa G, and Fruin S. 2000. Effect of air pollution on preterm birth among children born in Southern California between 1989 and 1993. Epidemiology 11, 502-511.

Ritz B, Yu F, Fruin S, Chapa G, Shaw GM, and Harris JA. 2002. Ambient air pollution and risk of birth defects in Southern California. Am J Epidemiol 155, 17-25.

Rogers JF, Thompson SJ, Addy CL, McKeown RE, Cowen DJ, and Decoufle P. 2000. Association of very low birth weight with exposures to environmental sulfur dioxide and total suspended particulates. Am J Epidemiol 151, 602-613.

Rose N, Cowie C, Gillett R, and Marks GB. 2011. Validation of a spatiotemporal land use regression model incorporating fixed site monitors. Environ Sci Technol 45, 294-299.

Rubes J, Selevan SG, Evenson DP, Zudova D, Vozdova M, Zudova Z, et al. 2005. Episodic air pollution is associated with increased DNA fragmentation in human sperm without other changes in semen quality. Hum Reprod 20, 2776-2783.

Rudra CB, Williams MA, Sheppard L, Koenig JQ, and Schiff MA. 2011. Ambient Carbon Monoxide and Fine Particulate Matter in Relation to Preeclampsia and Preterm Delivery in Western Washington State. Environ Health Perspect.

Ryan PH, and LeMasters GK. 2007. A review of land-use regression models for characterizing intraurban air pollution exposure. Inhal Toxicol 19 Suppl 1, 127-133.

Sagiv SK, Mendola P, Loomis D, Herring AH, Neas LM, Savitz DA, et al. 2005. A time-series analysis of air pollution and preterm birth in Pennsylvania, 1997-2001. Environ Health Perspect 113, 602-606.

Salam MT, Millstein J, Li YF, Lurmann FW, Margolis HG, and Gilliland FD. 2005. Birth outcomes and prenatal exposure to ozone, carbon monoxide, and particulate matter: results from the Children's Health Study. Environ Health Perspect 113, 1638-1644.

Saldiva PH, Pope CA, 3rd, Schwartz J, Dockery DW, Lichtenfels AJ, Salge JM, et al. 1995. Air pollution and mortality in elderly people: a time-series study in Sao Paulo, Brazil. Arch Environ Health 50, 159-163.

Samoli E, Touloumi G, Schwartz J, Anderson HR, Schindler C, Forsberg B, et al. 2007. Short-term effects of carbon monoxide on mortality: an analysis within the APHEA project. Environ Health Perspect 115, 1578-1583.

Schwartz J, Dockery DW, Neas LM, Wypij D, Ware JH, Spengler JD, et al. 1994. Acute effects of summer air pollution on respiratory symptom reporting in children. Am J Respir Crit Care Med 150, 1234-1242.

Seo JH, Leem JH, Ha EH, Kim OJ, Kim BM, Lee JY, et al. 2010. Population-attributable risk of low birthweight related to PM10 pollution in seven Korean cities. Paediatr Perinat Epidemiol 24, 140-148.

Shah PS, and Balkhair T. 2011. Air pollution and birth outcomes: a systematic review. Environ Int 37, 498-516.

Slama R, Darrow L, Parker J, Woodruff TJ, Strickland M, Nieuwenhuijsen M, et al. 2008. Meeting report: atmospheric pollution and human reproduction. Environ Health Perspect 116, 791-798.

Somers CM, and Cooper DN. 2009. Air pollution and mutations in the germline: are humans at risk? Hum Genet 125, 119-130.

Somers CM, Yauk CL, White PA, Parfett CL, and Quinn JS. 2002. Air pollution induces heritable DNA mutations. Proc Natl Acad Sci U S A 99, 15904-15907.

Strickland MJ, Klein M, Darrow LA, Flanders WD, Correa A, Marcus M, et al. 2009. The issue of confounding in epidemiological studies of ambient air pollution and pregnancy outcomes. J Epidemiol Community Health 63, 500-504.

Takeda K, Tsukue N, and Yoshida S. 2004. Endocrine-disrupting activity of chemicals in diesel exhaust and diesel exhaust particles. Environ Sci 11, 33-45.

Tao F, Gonzalez-Flecha B, and Kobzik L. 2003. Reactive oxygen species in pulmonary inflammation by ambient particulates. Free Radic Biol Med 35, 327-340.

Thompson JN. 2007. Fetal nutrition and adult hypertension, diabetes, obesity, and coronary artery disease. Neonatal Netw 26, 235-240.

Tran DQ, Ide CF, McLachlan JA, and Arnold SF. 1996. The anti-estrogenic activity of selected polynuclear aromatic hydrocarbons in yeast expressing human estrogen receptor. Biochem Biophys Res Commun 229, 101-108.

Valkonen VP, Paiva H, Salonen JT, Lakka TA, Lehtimaki T, Laakso J, et al. 2001. Risk of acute coronary events and serum concentration of asymmetrical dimethylarginine. Lancet 358, 2127-2128.

van den Hooven EH, de Kluizenaar Y, Pierik FH, Hofman A, van Ratingen SW, Zandveld PY, et al. 2011. Air pollution, blood pressure, and the risk of hypertensive complications during pregnancy: the generation R study. Hypertension 57, 406-412.

Wang X, Ding H, Ryan L, and Xu X. 1997. Association between air pollution and low birth weight: a community-based study. Environ Health Perspect 105, 514-520.

Wietlisbach V, Pope CA, 3rd, and Ackermann-Liebrich U. 1996. Air pollution and daily mortality in three Swiss urban areas. Soz Praventivmed 41, 107-115.

Wilhelm M, and Ritz B. 2005. Local variations in CO and particulate air pollution and adverse birth outcomes in Los Angeles County, California, USA. Environ Health Perspect 113, 1212-1221.

Woodruff TJ, Parker JD, Darrow LA, Slama R, Bell ML, Choi H, et al. 2009. Methodological issues in studies of air pollution and reproductive health. Environ Res 109, 311-320.

Wu J, Ren C, Delfino RJ, Chung J, Wilhelm M, and Ritz B. 2009. Association between local traffic-generated air pollution and preeclampsia and preterm delivery in the south coast air basin of California. Environ Health Perspect 117, 1773-1779.

Xu X, Ding H, and Wang X. 1995. Acute effects of total suspended particles and sulfur dioxides on preterm delivery: a community-based cohort study. Arch Environ Health 50, 407-415.

Xu X, Sharma RK, Talbott EO, Zborowski JV, Rager J, Arena VC, et al. 2010. PM10 air pollution exposure during pregnancy and term low birth weight in Allegheny County, PA, 1994-2000. Int Arch Occup Environ Health 84, 251-257.

Yorifuji T, Naruse H, Kashima S, Ohki S, Murakoshi T, Takao S, et al. 2011. Residential proximity to major roads and preterm births. Epidemiology 22, 74-80.

Zanobetti A, Schwartz J, and Gold D. 2000. Are there sensitive subgroups for the effects of airborne particles? Environ Health Perspect 108, 841-845.

Zeka A, Zanobetti A, and Schwartz J. 2006. Individual-level modifiers of the effects of particulate matter on daily mortality. Am J Epidemiol 163, 849-859.

Zhao Q, Liang Z, Tao S, Zhu J, and Du Y. 2011. Effects of air pollution on neonatal prematurity in Guangzhou of China: a time-series study. Environ Health 10, 2.

Part 2

Environmental Changes, Geographic and Climatic Conditions due to Air Pollution

6

Reduction of Air Pollution by Combustion Processes

Aurel Gaba and Stefania Felicia Iordache
Valahia University of Targoviste
Romania

1. Introduction

Among human activities, those in which fuel combustion processes intervene are those who contaminate the atmosphere greatly.

The combustion process is a process of rapid oxidation, followed by light phenomena and the release of large amounts of energy, able to maintain it at high temperatures. Compared with slow oxidation processes, it is characteristic to the combustion process sudden acceleration of the reaction rate to achieve theoretically infinite values. This applies, for example, to the stoichiometric mixture of methane - oxygen heated to a temperature of 560°C in a sealed container. Heating the same mixture to a temperature of only 200°C, result in a slow oxidation process, which produces methanol, formic acid, formaldehyde, carbon monoxide and carbon dioxide gas, with an overall response rate with an evolution with measured values up to a maximum, after which rate value decreases with the depletion of reagents. In everyday life we encounter slow oxidation processes at every step. Thus, minerals are subject to slow oxidation process which occurs at ambient temperature by consumption of oxygen from atmospheric air, with production of oxides in a state of maximum stability. Such a process is carbon steel corrosion under the action of atmospheric oxygen at ambient temperatures, which is transformed first into ferrous oxide (FeO) and then in a more stable substance, ferric oxide (Fe_2O_3). Also, living organisms consume oxygen in the atmosphere, at room temperature to oxidize nutrients over a slow but very complex process. In both examples above, as in any oxidation process, there are necessary two substances: the oxidant, which has the ability to quickly combine with the substance subject to oxidation, respectively, the substance that is oxidized, called fuel.

The transformation of chemical energy of fossil fuels in forms of energy directly useable, primarily mechanical energy, electrical energy and heat energy, is practically done only by means of combustion. In the production of electrical and heat energy, are consumed by burning, at present, 87% fossil fuels, the remainder being nuclear energy and regenerative energy (hydraulic, wind energy, solar, geothermal and marine) 6%, respectively 7%. At this consumption of the fossil fuels the consumption to produce mechanical energy in transports and the technological consumption, e.g. consumption of coal to produce metallurgical coke and for injection in blast furnaces, is added.

In 2008, according to the World Energy Outlook (2010), world consumption of fuels was 12 300 million toe (tons oil equivalent), of which 30 660 million barrels of oil and 3100 billion cubic meters of natural gas. For the year 2035, according to the script "new scenario" of the

World Energy Outlook (2010), total consumption will be increased up to16 700 million toe, of which 36 135 million barrels of oil and 4500 billion cubic meters of natural gas. These significant increases in world consumption of fuels results in conditions that anticipate an increase in the share of nuclear energy to 8% and regenerative energy to 14% in electrical and heat energy production. As a result, World Energy Outlook (2010), consider that the regulations of Copenhagen Accord (The United Nations Climate Change Conference in Copenhagen, 7-19 December 2009) to reduce CO2 emissions are not sufficient to halt the temperature increase with up to 2°C, worldwide.

In the fuel combustion processes the maximum release of energy is obtained with the complete combustion with, minimum excess of air, closer to stoichiometric combustion, which generally leads to the exclusive formation of CO_2, H_2O, SO_2 and N_2 gases, respectively O_2 and N_2 of the air excess. In practice, combustion processes slide away more or less from this ideal, both due to the complex structure of used fuels, such as coals and inferior oils, also due to imperfection of the combustion installations. Thus are formed unburned substances as solid particles, rich in carbon, which in most cases are accompanied by unburned gases. Unburned substances discharged into the atmosphere with the combustion gases, lead to heat energy loss, which leads to a decrease of the thermal efficiency of the process but at the same time, they have polluting action (Ghia & Gaba, 2000).

Sources of air pollution from burning fuels are classified as follows: stationary sources, mainly consisting of boilers, furnaces and gas turbines and mobile sources consisting of transport means, whose internal combustion engines are diesel (diesel-consuming, also called compression-ignition engines) and spark ignition engines, consuming gasoline. Lately there are used as stationary sources in heat energy and electricity production in cogeneration and internal combustion engines. Classification of stationary sources is done by industry type:

- Energy Industry (heat energy and electricity production, petroleum refining, manufacture of solid fuel and other industries);
- Manufacturing Industry and construction (iron and steel, chemicals, pulp, paper and printing, food processing, beverage and tobacco), called the EEA Industry (energy) sector. Also in EEA (European Environment Agency) publications mobile sources are classified in road transport and other transport.

Wastes scattered in the atmosphere by fuel combustion processes are called emissions (some are returning back on the ground forming immissions) and form two major categories: gases and solid particles.

Gases released into the atmosphere by fuel combustion process are: CO_2, SO_2, H_2O, N_2, O_2, CO, NO_x, PAHs (polycyclic aromatic hydrocarbons), VOCs (volatile organic compounds), dioxins, furans. Lately it was started the separate monitoring of CH_4 methane from volatile organic compounds, these remaining without CH_4 becoming NMVOCs (non-methane volatile organic compounds). With the exception of H_2O, N_2 and O_2, all other elements pollute the atmosphere directly and some by induced forms (usually synergistic, their effect being amplified, more than their cumulative effect) such as the greenhouse effect, smog and acid rain (acid deposition more accurately).

The greenhouse effect is due to more than 50% to CO_2 and in very small proportion to nitrous oxide N_2O, mainly as gases from combustion processes. 65% of CO_2 in the atmosphere comes from burning fuels. Carbon dioxide recently considered as pollutant, prevents the heat energy discharge from the surface of the Earth which is received by radiation from the sun. With the passage of time, even a few degrees increase in the Earth temperature could lead to reduction of the ice cap, the spread of deserts, rising the levels of oceans and sea, as important changes of the climate. In consequence, inside United Nations it was negotiated a convention on climate change which was signed at the Rio Conference in 1992, with the objective of stabilizing greenhouse effect concentrations.

Smog is a synergy, which may be:

- Reducing, also called sulfur dioxide, wet or London type;
- Photochemical, also called an oxidizer or Californian type.

Reducing smog is formed due to the presence, in high concentrations, of SO_2 and soft particles in suspention (smoke) in a calm atmosphere with high humidity, at relatively low temperatures. Component substances that contribute to forming reducing smog are solid particles of soot, cenospheres, sulfur oxides and carbon oxide, derived from incomplete combustion of coals or oils. Cold water vapors condense around the solid particles of smoke, which have a large surface and can act as catalysts to form soft particles of fog. Sometimes reducing smog also contains soft particles of silicon, aluminum as metal oxides.

Photochemical smog is formed due to the presence, in high concentrations, of NO_x and VOCs, in a calm atmosphere, with temperatures over 18°C, in the absence of rain and wind. By the action of solar radiation on primary pollutants, NO_x and VOCs, mainly derived from heavy road traffic, secondary pollutants are formed as NO_2, which decompose under the action of solar radiation in highly toxic compounds, such as peroxiacetil nitrate (PAN).

Acid rain is also a form of pollution induced by pollutants emitted by combustion processes of fuels, SO_x and NO_x. Arrived in the atmosphere, sulphur and nitrogen oxides react with oxygen, water vapors and other substances forming acid compounds, such as dilute solutions of H_2SO_4, HNO_3 and their salts, sulfates and nitrates. These acid compounds reaching the Earth by rain, snow and fog, forming wet acidic depositions, respectively by particles, wind and stormwater forming dry acidic depositions. Acidic depositions have a negative impact on population health status and adverse effects on aquatic ecosystems, on forests and crops. Also, acidic depositions attack buildings and works of art.

Another form of pollution induced by pollutants emitted by fuel combustion processes, is the formation of tropospheric (high altitude) ozone (O_3). Precursors of tropospheric ozone O_3 formation are: nitrogen oxides (NO_x), non-methane volatile organic compounds (NMVOC), carbon monoxide (CO) and methane (CH_4). Because of the ozone in the atmosphere vegetation growth is reduced and aging occurs on the crops and forests. Solid particles in the air are classified by their size, their capacity of sedimentation and the possibility to penetrate the organisms, as following:

- Particles larger than 10 μm are deposited after the law of gravity with an uniform accelerated motion, also known as dust;
- Particles between 10 μm and 0.1 μm that are deposited after Stokes law, consisting mainly of ash and black smoke from burning fuel results;
- Particles smaller than 0.1 μm that are deposited and move after Brownian movement law.

Especially in large steam boilers, through the burning of coals and oils, solid particles driven by exhaust gases may contain non-ferrous metals (Pb, Zn, Ni, Cd, Sn etc.) and metalloids (F and As).

In diesel engines, through incomplete combustion the fumes can appear white, blue respectively, which consist of a suspension in gases of liquid particles of unburned fuel or partially oxidized. Particles with diameter over 1 µm are found in the white smoke and particles with a diameter of about 0.5 µm appear in the blue smoke.

Black smoke from diesel engines contains solid carbonaceous particles, with average diameters of 1 µm.

Some experts consider that solid primary particles ranging in size from 0.01 to 0.05 µm, during the expansion and exhaustion of burning gases, find suited conditions to aggregate in soot particles with dimensions up to 0.6 µm occupying a specific surface of about 200 m^2/g. Experimentally it was found that these particles are formed under certain specific conditions, of: 20-90% amorphous carbon, 2-5% ash and 10-80% carcinogenic hydrocarbons.

In order to diminish air pollution by combustion processes, the following solutions are recommended (Baukal, 2001, Ghia & Gaba,2000):

- reducing the pollutant emission at origin, at the producing sources (primary procedures) by changing-replacing the combustion processes (use of renewable energy, production reducing, increase of production/technological lines efficiency, replace of production technologies), by using pretreatment procedures of fuels and oxidizer substances (changing fuels and oxidizers, improving the fuels characteristics by processing, preparation and use of additives) and by using improvement - changing procedures of combustion processes, both in case of combustion installations - especially the burners and burning chambers with reference to their construction (mixed stages, flue gases recirculation, auto-carburizing, steam or water spraying) and in case of combustion adjustment (control of excess air, burners out of service, adjustment of combustion air temperature);
- collecting and /or reducing the pollutants resulted from flue gasses before their suppression into the air by using the so-called secondary pollution reducing techniques or flue gases treatment (catalytic and non catalytic reactors, active carbon processes, SNOX and DESONOX processes, electrostatic precipitators, bag filters, scrubbers etc).

2. Formation and reduction pollutants in combustion processes

2.1 Carbon monoxide

The emission of carbon monoxide (CO) pollutant in combustion processes is produced when some conditions occur which lead to incomplete burning of fuels that contain carbon. Among heavy users of the most polluting fossil fuels, through emission of CO, are the means of transportation driven by spark ignition engines and compression ignition engines, and the lesser CO polluting are power plants.

If oxygen is missing during the combustion, compared to stoichiometric requirements, in an area of a fuel-air mixture, it will result CO even at a sufficiently high temperature. The possibility of CO subsequent oxidation, when the contribution of necessary oxygen to a lower temperature appears, is slower than forming CO.

In boilers and in furnaces, the presence of CO in the exhausted flue gases highlight essentially a disorder of the combustion equipment: burner, sprayer, combustion air blow installation, etc. Carbon oxide appears usually in the flue gases discharged from old, rudimentary combustion plants and is almost absent in the exhaust gases from modern industrial combustion plants of high-capacity. Also, at these facilities is taken into account that reducing the excess of air, which reduces the emissions of NO_x, can increase CO emission. As a result an optimal value for the air excess coefficient is established (Gaba, 2005).

Ensuring complete combustion is a measure considered to be BAT (best available techniques), which ensure the maintenance of the CO emission levels below 100 mg/Nm^3, for 15% O_2 (for internal combustion stationary engines and gas turbines) and in boilers and furnaces, for 3% O_2, to run on liquid and gaseous fuels and 6% O_2 to run on solid fuels (Directive 2008/1/EC, 2008).

For mobile sources, which are the biggest pollutants with CO, as reduction techniques were developed, among others, multi-point injection engines that work at high pressures and oxidation catalysts (Duca et al., 2002).

The amount of CO emitted by transportation in European Union countries (EU-27) represents 89% of the total CO emissions at their level. Due to equipping vehicles with better catalysts and application in EU-27 of new regulations that produced engines up to Euro-6, CO emissions diminished, in 2008, compared to 1990, with 71% (EEA-TERM 003, 2011).

2.2 Sulfur oxides

Beside some gaseous fuels, the other industrial fossil fuels generally contain sulfur S. For engines, diesel has a low sulfur content, which is null to petrol diesel prompts. Sulfur content in diesel determines SO_2 polluting emissions, which on diesel engines represent about 10% of the polluting emissions of these engines, but are being endangered as a result of production, lately, of sulfur-free diesel fuels.

Maximum sulfur content from main fossil fuels is regulated, e.g. France, $S \leq 4\%$ for industrial liquid fuels and $S \leq 1\%$ for solid fuels.

Primarily coals and oils, which are used in industrial plants to produce electricity and heat, are responsible for environmental pollution by sulfur oxides, strong acidity of air and acid rains. Sulfur dioxide tendency is to reduce the "greenhouse effect" by "albedo" effect.

Crude oils contain variable quantities of sulfur depending on their origin. During refining operations, sulfur tends to gather in waste fractions which can be an important element to the industrial liquid fuels. The difficulty to reduce the sulfur content in crude oils and petroleum residues consist especially in their content of asfaltene and metals which lead to deactivation of catalysts. Researches are made in this respect. At a large scale of solid fuels the sulfur content is close to the values for heavy liquid fuels. Also, desulphurization of solid fuels with high content of sulfur is a major activity, that develops both through practical achievements and through researches.

Discharged in the atmosphere, SO_2 react in proportion of 1-2 ‰/h with oxygen under the action of solar ultraviolet radiation resulting SO_3 which combines with water vapor in the atmosphere to form sulfuric acid. In very wet days, with fog, the degree of conversion to sulfuric acid is up to 15.7%.

Also, starting from SO_2 gas, sulfate aerosols forms but they can be washed by the rain which becomes acid. To ensure a satisfactory dilution and dispersion of the SO_2 gas in the atmosphere, the combustion gases must be evacuated from the funnel with an adequate speed and to a convenient height from the ground. Thus the large thermoelectric oil plants funnels exceed 200m height frequently. Because of this the pollution due to pollutants produced by sulfur in the fuel may occur at great distances from where they were generated.

Generally, the growth of sulfur dioxide quantity in the atmosphere has the effect of increasing the overall morbidity rate of the population. At concentrations of sulfur dioxide in the cities air over 0.046 ppm (annual average) lead to an increasing frequency of respiratory diseases, at concentrations of 0.52 ppm in the presence of other solid particles, overall mortality is increased. Sulfur dioxide and sulfuric acid aggravate the respiratory system of animals, but this is much more toxic than SO_2, its action being dependent on the size of aerosols. In a series of countries, particularly those located in northern area, the specific flora is affected even by the traces of sulfur dioxide, moss and lichens around large cities from Europe and America, disappear. Sulfur dioxide absorbed by plants leads to acute and chronic effects. Acute effects, that follow high concentrations and the relatively short exposures, manifest by changing the color (to yellow-ivory or red-brown) of damaged tissues. Chronic effects, results of prolonged exposure to low concentrations, manifest by perpetual yellowing of the plant foliage following the alteration of the chlorophyll production mechanism. It seems that, to the acute effects, the plant defend itself transforming sulfur dioxide into sulfuric acid and then in sulfates that are deposited in certain portions of tissues, but soon the balance of sulfates is disturbed and sulfuric acid appears in the plant cellular system which attacks its cells. Chronic effects are due to sulfate aggregations. Sulfur dioxide in small concentration (on the order of the size of 100 - 500 $\mu g/m^3$) react synergistically with both ozone and nitrogen dioxide, which affects the plants tissue, especially as these are more sensitive.

Sulfur dioxide and sulfuric acid formed have very corrosive action on metals, constructions, leather, paper and textiles, especially if relative humidity of air exceeds 70%. Attack of construction materials by SO_2 is due to next mechanism: SO_2 in the atmosphere prompts a condensation of water vapors as fog even when their partial pressure is lower than the saturation pressure at that temperature, due to the fact that SO_2 form with water a solution which a lower vapor pressure than pure water, slip by easily as fog state; the phenomenon can be facilitated by existing aerosols as particles in the smoke, which forms condensation nuclei. In soft drops of fog with SO_2 is obtained sulfurous acid and its oxidation to sulfuric acid. Reducing character of SO_2 manifests also by changing the color of some paint pigments. As such it is affected lead white. Calcium carbonate, a constituent of many construction materials, some of which serve to works of arts, is transformed into calcium sulfate. Destructive effect is accelerated by the acid rains.

In 1983 the first symptoms of forests etching have appeared in Western Europe, due to acid rains which gather both acid pollution and photooxidant pollution. It has been established that acidity in the atmosphere, most often originates in a ratio of 2/3 due to emissions of sulfur oxides, less than 1/3 because of NO_x, and the rest due to pollutants as fluorine or chlorine. Photooxidant pollution is due transforming NO_x in the presence of hydrocarbons and solar radiation.

Beside the negative effects outlined above, the presence of sulfur oxides in the air space reduces the visibility when the fog photochemical processes occurring and determines the appearance of an unpleasant smell.

In the atmosphere, oxidation of SO_2 to SO_3 takes place mainly through a photochemical process, although there are some opinions that it could take place catalytic processes also. During these reactions the presence of other pollutants such as nitrogen oxides or metal oxides may play an important role to accelerate the reactions. But there are occuring other factors such as concentration, residence time in the atmosphere, temperature, humidity, intensity and spectral distribution of radiations. In boilers furnaces sulfur trioxide, or sulfuric anhydride SO_3, derives from the oxidation sulfur dioxide. Actually, in the combustion gases is found 3-5% of the initial sulfur contained in the fuel, as SO_3.

Under certain circumstances SO_2 combines with water vapors, giving sulfurous acid gas H_2SO_3 and SO_3 combines still with water vapors, giving gaseous sulfuric acid H_2SO_4. Acid

vapors condensation on cold walls attack metals. Beside corrosion, due to SO_4H_2 vapors condensation, especially in thermal uninsulated funnels, it is favored the agglomeration of solid particles of soot and flying coke, who create acid rain too.
The fight against acid corrosion and pollution is done in three ways:

- Preventing the forming of anhydrides, especially SO_3;
- Neutralizing the formation of acids;
- Preventing the condensation of formed acid vapors.

Avoiding actual forming of anhydrides, means burning fuels with a low sulfur content. Heavy petroleum desulphurization is not generalized to an industrial scale today because of the high prices required. Fuels with low sulfur content can be subjected to industrial operations only when sulfur compounds are extremely harmful. Sometimes it is possible to have them burned out at low excesses of air, thus forming reduced SO_2 and SO_3 quantities. Limiting the excess of air at maximum 3% is difficult to do because:

- the burner must ensure the right mix between combustion air and fuel over the whole tuning;
- industrial boilers being equipped with numerous burners, each burner must have its own system of adjustment and control, which raise the cost of the related equipment. So the solution is applied only to very large boilers from thermoelectric power plants, which have numerous personnel for maintenance and exploitation.

Spraying silica allows to stop the catalytic action of deposits that form on the heating and overheating pipes, which reduces the amount of formed SO_3. Frequent cleaning of depositions of solid particles completes the action mentioned before, but does not reduce the formation of SO_3 from the flame. SO_3 action may be partially neutralized by injection into the combustion products of calcium carbonate, zinc oxide, dolomite and of ammonia, which allows some boilers to reduce significantly the corrosion of air preheaters. Injection technique is sensible and it must be assured a good distribution of the substances mentioned above, avoiding obstruction of the flue gas flow and solid particle entrainment in atmosphere. When using ammonia, neutralization does not stop at the stage of acid sulfate of ammonia, but it reaches the stage of neutral sulfate.
Primary procedures for reduction of SO_2 pollutant emissions make the desulphurization, reducing the content of SO_2 in the flue gas by injecting absorbent substances sprayed into the furnace, such as limestone, lime or dolomite. Desulphurization combustion can be seen as a primary technique for major reduction of SO_2, applied to certain types of boilers. For the first achieved version, when the flame is obtained from a conventional burner, the U.S. has developed an extensive research program LIMB (Limestone Injection in Multistage Burners) with several forms of application.
Desulphurization efficiency depends particularly on the magnitude of the temperatures in the furnace, but meritorious performances are obtained using high consumption of absorbent.
For the second achieved version, the combustion is developing in dense or circulating fluidized bed. Such techniques can achieve efficiencies of combustion and desulphurization with high values having a consumption of absorbent lower than the first version. It should be noted that circulating bed technique is superior as efficiency than the dense bed technique.
Dense fluidized bed is characterized by the flow speed, generally reduced, where the solid particles move relatively slow and require a moderate recirculation. Increasing the speed of flow, the volume occupied by the layer will increase, as the movement speed of solid

particles, which will require an intense recirculation of particles realizing the circulating fluidized bed. These particles are of two types:

- a kind of incompletely burned fuel, because the coal particles introduced into the furnace having a high temperature generate gaseous volatile mater and carbon tailing which require a much longer combustion time than the combustion gases;
- absorbent material, which is mostly limestone, which has three forms: $CaCO_3$, CaO and $CaSO_4$.

Because the system operates in stationary state, the continuously exhausted gases will have to contain very small quantities of combustible substances to obtain the highest possible energy efficiency. Also, the exhaust gases will contain $CaCO_3$ and CaO as well as a very high content of $CaSO_4$ for a good desulphurization. Conventional circulating fluidized bed boilers realize the burning and desulphurization in the same furnace known as reactor that is fed simultaneously with fuel and absorbent material. The differences between these boilers refer to the ways of extracting produced energy and the means of control operation of variation with load. IFP, Lardet-Babcock and Cecar made a boiler called self desulphurization to which combustion and desulphurization are produced in different rooms. This boiler has a circulating fluidized bed furnace which is fed with combustion gases, having the temperature of 500-850°C, instead of air, as conventional boilers are operating. Hot combustion gases, rich in SO_2, from the furnace, are routed through the loop of circulation of solid particles in suspension, where is added limestone. This loop does not have exchange surfaces for useful transfer of heat, so it is protected from the effects of the sulfuric corrosion. The combustion gases thus sweetened are directed to the heat exchange surfaces of a exhaust-heat boiler. Whatever the system, in principle, all types of boilers with circulating layer are operating with relatively soft particles of limestone, of a few hundred microns, rather than a few millimeters bed as the dense bed boilers user. We mention that circulating bed systems allow a better use of calcium than dense systems.

Although there are few studies of the kinetics for desulphurization in circulating bed, there are global and fragmentary results often achieved on industrial boilers where the degree of desulphurization was measured in given operational conditions, knowing the nature of absorbent and the fuel burned and the Ca/C ratio, respectively.

Secondary procedures for reducing pollutant emissions of SO_2 use desulphurization plants of exhaust gases. The most used are dry FGD (flue gas desulphurization) plants or wet FGD plants and to a lesser extent plants combined, for desulphurization and NOx reduction of exhaust gases, as DESONOX, which are regarded as techniques associated with BAT. FGD plants with lime or limestone wet scrubber have a rate of SO_2 reduction by 92-98%, those who dry scrubbers, with spray, by 85-92%.

Depending on the type of installation of boilers, old or (new), as its power, by applying the BAT techniques, the emission levels of SO_2, in mg/Nm^3, are limited, e.g. for liquid fuel (calculated at 3% O_2) to 350 (350) for powers under 100MWt, to 250 (200) for powers under 300MWt and to 200 (150) for powers over 300MWt (Directive 2008/1/EC, 2008).

The amount of SO_2 emitted in producing electricity and heat sector in European Union countries (EU-27) represents 69% of the total SO_2 emissions at their level (energy using industry sector is 10.2% and transportation 3.8%). Due to replacing solid and liquid fuels with high sulfur content with natural gases, the use of technologies of flue gas desulphurization and application in EU-27 of the new regulations which limited the sulfur content of some liquid fuels, the emissions of SO_2 have decreased from approx. 26000 kt, in 1990, by 71%, to approx. 7500 kt in 2007 (EEA -APE 001, 2010).

2.3 Nitrogen oxides

Among heavy users of the most polluting fossil fuels, through the NO_x emissions, are the means of transportation driven by spark ignition engines and compression ignition engines, followed by thermoelectric power plants and then by boilers and furnaces in the sector of energy use in Industry (EEA -APE 002, 2010).

NOx emission depends on the residence time of the molecules in the flame, most of NOx is formed in the second part of flame development, where the temperature is sufficiently high. Usage of the excess air raise the NOx equilibrium value, and the time needed to achieve balance is net superior to the stationary time of the molecules in the furnace, which explains the experimental values of NOx lower than the theoretical front.

There are three possible mechanisms of formation the NOx (De Soete,1982, Kramer, 1986, Skunca, 1981):

- thermal NOx from using the nitrogen from the combustion air;
- fuel NOx, by conversion of chemically bound nitrogen in the fuel;
- prompt NOx.

NOx forming synthesis is detailed schematically as follows:

Nitrogen origin — Reaction medium — Formation mechanism

Nitrogen from the combustion air → Combustion gases → Thermal NO_x

Nitrogen from the combustion air → Flame front

Nitrogen from fuel → Flame front → Fuel NO_x

Flame front → Prompt NO_x

The Zeldovich chain reaction pattern shows the exponential dependence for temperature of the emission of NO_x, until the temperature threshold is reached. The emission is directly proportional with the residence time in the furnace of the molecules and with the oxygen square concentration (Zeldovich, 1980).

The temperature is the strongest factor to influence the NO_x emission, exceeding significantly the influence of the oxygen concentration and the stationary time of the molecules. Stepped combustion is mainly intended to avoid forming high temperatures for reactants before reaching equilibrium, thus nitrogen oxides overall emission to be lower.

The wrong placement of the burners may determine through the jets interaction phenomenon local increases in temperature, which can lead to increased emissions of nitrogen oxides. Fuel nitrogen is present as organic compounds in the amines (NH and NC) and cyanides (CN) family.

These compounds react in two directions by:

- Reaction with substances that contain oxygen, forming NO and
- Reaction with substances containing nitrogen, forming molecular nitrogen.

Therefore not all the fuel nitrogen from combustion passes in NO_X.

Thermal NO formation occurs in the flame in the post-reaction area, according to distinct mechanisms corresponding to mixture of fuel with poor or rich air.

The main factors that influence the development of thermal NO production reactions are: the atomic oxygen concentration, the residence time of molecules and the temperature of the furnace, with higher values at 1300°C, exert the strongest influence.

The reaction of molecular nitrogen with atomic oxygen is the slowest. Dissociation of molecular oxygen at normal combustion temperatures is insignificant. Thus, for example at 2000 K, atomic oxygen is formed having a concentration below 10 ppm.

NO formation from fuel nitrogen occurs in the flame after a complex mechanism, partially unknown. It is known that first CN radicals are formed, whose evolution in the presence of oxygen leads to the formation of NO. The main factors that influence the forming of NO in this case are the nitrogen content of the fuel, oxygen concentration in the fuel, residence time and the flame temperature.

NO formation from the nitrogen contained in the fuel flows a little faster than the formation of thermal NO, but is considerably slower than the prompt NO formation.

NO_2 formation occurs in the exhaust ducts of combustion gases, funnel and free atmosphere at temperatures below 650°C, the main factors being: the temperature value, molecular oxygen O_2 concentration, residence time, air pollution and solar radiation.

Primary procedures for reducing pollutant emissions of NO_x have known three generations in development (ECE NOx Task Force, 1992)

- First generation consisting in reducing the temperature of the combustion air preheating (RAP), flue gas recirculation application in the furnace (FGR), use of low excess of combustion air (LEA) and burners-out-of-service (BOOS);
- The second generation consisting in production of low NO_x burners 1 (LNB) or air staging at burner, flue gas recirculation at burner (FGR) and over fire air (OFA);
- Third generation consisting in production of low NO_x burners 2 (LNB) or air and fuel - staging at burner and in furnace NOx reduction (IFNR) as well as burners improvement for NO_x decreased by increasing the internal flue gas recirculation.

In addition to these techniques there can be mentioned:

- Emulsification of oil with water;
- Injection of aerosols;
- Water-steam injection;
- Addition of fuel, combustion air, or spraying additives in flame (Ghia & Gaba, 2000).

NO_x emission depends on the size of volume of the furnace in which take place the development of flames produced by burners. Large furnaces allow the existence of higher temperatures in and after the area occupied by the flame, so the NO_x emission will increase, and the phenomenon is amplified by increasing the residence time. As a result, thermal charges of the volume and the cross section of the furnace represent important criteria for assessing the emissions of NO_x.

In large furnaces, which burn gaseous fuels, the NO_x emission is reduced by lowering the preheating temperature of the combustion air. Using this technique to oil burning is not recommended because it increases the percentage of unburned, especially soot production, as a result of lowering the burning temperature.

The amount of preheating temperature of the combustion air must have an optimum determined by the combustion to be more completely, NO_x emission below the normalized limit and high thermal efficiency for the heat production plant (Ghia & Gaba, 2000).

For industrial furnaces, where high temperatures in the furnace are required by the technological process, the amount of preheated temperature of the combustion air is dependent on this need taking into account, firstly, the type of the fuel used. Primary reduction of NO_x emission is done especially by removing the temperature peaks that are achieved mainly by the combustion gases recirculation.

Flue gases recirculation in the combustion air represent an efficient method of reduction of NO_x emissions for fuels and furnaces that allow development of high temperatures in the flame. The researches have shown that the most effective method of decreasing the levels of

NO_x is the recirculation gases mixture with the primary air or gaseous fuel, because in this case it acts on the maximum of temperature of the flame core. Thus the burning rate is reduced, which increase the length of the flame and reduces the NO_x content. As result there are necessary constructive changes to the burners and limiting the quantity of recirculation gases, so the complete combustion to be assured.

But it is noticed that by increasing the recirculation degree, the flue gas CO content is increasing, while the NO_x content is decreasing. Reducing the temperature of the flame decrease the useful transfer of the heat by convection, because of the increase of the gases flow from the combustion of fuel with recirculation flue gas flow.

To the Hrenox process, recirculation gases mixture with some cold combustion air is introduced into the burner separately from the rest of the combustion air that is preheated. For a high degree of flue gas recirculation of 15% and a proportion of cold air up to maximum 10%, the NO_x emission was below 200 mg/m^{3}N for the combustion of gaseous fuels and below 250 mg/m^{3}N for the combustion of high quality liquid industrial fuels.

Another way to reduce NO_x emissions is in using tertiary air injected through special slits, located above or below them.

The burners will work with decreased excess of air, the combustion being done by tertiary air which represent 15-30% of total combustion air. Often the application of this combustion technique is related to the existence of large furnaces, equipped with several rows of burners located at different heights, where only to the upper burners it may be introduced by blowing tertiary air.

Also, to the individual burners of natural gas, or oil, the burning in steps is a safe technique reduce the NO_x emissions. Combustion by-steps can be obtained by introducing fuel gradually by FS (air- fuel staged) process. The fuel added outside the primary burner forms a more under-stoichiometric flame after which, in the second stage NO_x is reduced by NH_3, HCN and CO radicals in N_2. Optimal proportion of the secondary fuel from total fuel consumption is 20-30%. The third stage follows when the process completes when the final combustion with reduced NO_x is done.

Insertion by-steps of the combustion air or fuel, reduces the NO_x emissions, but this process must be implemented by optimal constructive solutions, whereas the NO_x reduction processes are contrary to those of decreasing the CO and partially oxidized hydrocarbons.

The time required for combustion and reduction being relatively high, the FS process is recommended to apply to large furnaces. At the oil combustion, the FS process application may produce significant soot quantities in the primary flame, so it must be applied judiciously in case of combined oil-gas burners.

INFR combustion by-steps process implies the fuel injection in the furnace above the main combustion zone thereby causing a secondary under-stoichiometric area, after which it is added downstream the secondary air which completes the combustion. As a result, the combustion is divided into three zones. Hydro-carbonate radicals, formed in the second zone, occur to a temperature of over 1200°C in the reduction atmosphere, reacting with nitrogen oxides produced in the main combustion zone so the N_2 and other components to be formed. In practice, reducing NO_x emissions by spraying water is rarely used, in large furnaces and only with superior fuels that form relatively high combustion temperatures and who have virtually no sulfur in the composition.

In the case of heat engines, primary measurements to reduce NO_x emissions refer to optimizing the combustion chamber, the injection or carburetion systems, engine operation with an appropriate adjustment, using recirculation of exhaust gases, to which it is added the use of fuels treated by different methods (desulphurization, emulsifying with water, ultrasonic treatment).

Most of the researches conducted at the boilers to reduce the emissions of NO_x produced by burners referred to the oil - natural gas mixed burners to which the constructive solutions adopted are independent of fuel that they use. Primarily it is used the combustion by-steps and the combustion gases internal recirculation, as the ASR (Axial Stage Return Flow) burner is done. Part of the combustion air flow passes through the central tubing of the burner as primary air. The secondary air is repressed from the burner box through pipes equidistant distributed on a peripheric circumference, each pipe having two outputs for secondary air I and secondary air II, the supply being performed by a separate cold air fan. To delay the combustion with primary air, due to the ejection caused by axial repression with higher speed of secondary air, flue gases are absorbed from the furnace through a annular transversal section, forming a separating layer between the primary step and the secondary step. In this burner there were integrated essential components of techniques to reduce NO_x emissions:

- Supply with combustion air in three stages;
- Aspiration of the flue gas from the furnace ensuring internal recirculation;
- Separation of primary air from secondary air through a stream of combustion inert gases that reduces the burning rate;
- Liquid and gaseous fuels flows are adjustable allowing the control of the combustion processes to conditions of reduced NO_x.

Very low NO_x emissions were obtained by using the ASR burner alone and by applying additional techniques such as the addition of tertiary air, external flue gas recirculation and special spray nozzles for liquid fuel. To the combustion of oils with high sulfur content, according to the quality of these fuels, for entry into the solid particles and sulfur oxides emissions standards it is required the application of post-combustion control techniques that reside in the use of desulphurization systems, filters and cyclones.

For the furnaces of boilers burning solid fuel with low volatile content and a lot of bituminous ash having the liquid discharge of ash, NOx emissions level is relatively high. Since it is necessary to maintain a minimum value of the combustion gases temperature required for the liquids discharge of ash, NO_x emission control techniques are restricted in operation. Cyclone furnaces with second chamber downstream offer very good conditions for combustion by-steps. When inserted into the cyclone furnace the required air for combustion there are obtained high temperatures for oxygen excess, so the NOx emissions have high values to 1500-1800 mg/Nm^3. To the combustion by-steps the procedure is as follows:

- in the cyclone furnace is inserted about 80% of stoichiometric air;
- it is injected in the secondary chamber, as secondary and tertiary air, the remaining air required for fuel combustion. The optimal conditions for injection with tertiary air are met when the injection nozzles are designed and located so to provide the intimate mixture of the flue gas from the cyclone furnace and air. Respecting the above conditions, it was obtained a reduction in NO_x emissions of 30-40% depending on the operating mode. Further reduction in NO_x emissions was achieved by applying the external flue gas recirculation.

For mixed gas-oil burners the reduction of NO_x emissions is done by the swirling the combustion air to a turbulence degree of $n<3$ to prevent high temperature peaks as a result of excessive shortening of the flame.

Intensifying the internal recirculation of combustion gases is done with repressed secondary air with great speed through nozzles located at the end of the entry into splay, creating an important vacuum that absorb primary combustion products. Combustion air admission is done best in three steps to avoid local temperature increase to high values. It is recommended that the primary air to represent 55-60% of the total combustion air. Oil spraying is done especially with slightly overheated steam using Y-type injectors.

To reduce NO_x emissions and pollutants in general, it can be used the self-carburizing process for both boilers and furnaces (Gaba, 2010).

A low burner of original conception, with solid carbon particles formation in flame, is fitted on three condensation boilers for steam and hot water. The main burner elements provide the formation of annular jet of combustion air mixed with natural gas and of central fuel jet, which, by thermal cracking, forms subsequently particles of solid carbon. These particles lead to increase of flame emissivity. Having in view the NOx emission, the characteristics of this burner are the following:

- use of natural gas as fuel;
- the solid carbon particles formation in flame;
- three stages mixing of the fuel in the combustion air;
- the swirling motion of the ventilated combustion air.

The low NOx burner for boilers demonstrated excellent NOx and CO performance, producing emission levels below 60 ppm for CO(1.05 air ratio at nominal power and 1.09 air ratio at minimum power) and below 50 ppm for NOx (1.03 air ratio at nominal power and 1.09 air ratio at minimum power), operating with natural gas (Gaba, 2010).

Another low NOx burner of original conception, with solid carbon particles formation in flame, is fitted on forge and treatment furnaces (Gaba, 2009). The main constructive elements of this low NOx recuperative burner are: a ceramic quarl, a zone for the natural gas-ventilated air mixture, a zone for fuel atomization by collision with the compressed air, an air preheater and an ejector for exhaust gases. The ceramic quarl burner has a cylindrical combustion chamber and in the peripherical zone a few cylindrical channels for exhaust gases entrances. Between combustion chamber and cylindrical channels are another channels for exhaust gases recirculation. The zone for the natural gas-ventilated combustion air includes three concentric pipes: the central pipe with different orifices and nozzles for natural gas and two pipes for divided combustion air in primary, secondary and tertiary air. The air preheater ensures the heat exchange from exhaust gases to combustion air. Having in view the NO_x emission, the characteristics of this burner are the following:

- use of natural gas or heavy fuel;
- three stages mixing of the fuel with the combustion air;
- the solid carbon particles formation in flame;
- the exhaust gases recirculation.

The low NO_x recuperative burner demonstrated excellent NO_x and CO performance, producing emission levels below 30[50] ppm for CO(1.03 [1.08]air ratio, and 700^{0}C furnace temperature) and below 120 [160] ppm for NO_x (1.30 [1.35] air ratio, and 1200^{0}C furnace temperature), operating with natural gas[heavy fuel] (Gaba, 2009).

In conclusion, the solutions for upgrading the combustion plants for fossil fuels for boilers and furnaces can be grouped into:

- the completion of the measurement, control and automation equipment and establishing optimal operating parameters;

- the development of new types of burners;
- the development of new combustion installations with high complexity.

The following can be a part of the first type of solutions:

- operation with optimal values of the coefficient of combustion excess;
- operation with optimal values of the preheating temperature of the combustion air and the gaseous fuel, like BFG (blast furnace gas) that is used;
- operation with an optimal number of burners with a given unitary capacity.

The second group refers to making of:

- burners with the insertion by-steps of combustion air or fuel;
- burners with internal recirculation of flue gas;
- burners having gradual combustion.

In the last group of solutions there are:

- making of combustion plants with external recirculation;
- making of combustion plants with water-fuel liquid emulsions or additived aerosols for fluid ionization in the furnace;
- replacing of the solid fuels with liquid or gaseous fuels that are less pollutant. By applying primary techniques, the rate of reduction of NO_x emissions is 10-44% for operating with reduced air excess, 10-70% at the insertion by-steps of the fuel or oxidizer, 20-50% at flue gas recirculation, 20-30% at low preheating of the combustion air and 25-60% to use low NO_x burners.

Secondary procedures for reducing NO_x pollutant emissions use flue gases treatment plants. The most commonly used installations are SNCR (selective non-catalytic reduction) and SCR (selective catalytic reduction). In a less extent combined plants are used for desulphurization and NO_x reduction of exhaust gases, DESONOX and active carbon plants (ECE NO_x Task Force, 1992). Selective Catalytic Reduction occurs at temperatures of 300-400ºC in the presence of catalysts, by injecting ammonia (NH_3). Catalysts used: TiO_2, WO_3, V_2O_5.

The rate of reduction of nitrogen oxides is 80-95% for SCR installations. Selective non-catalytic reduction occurs at temperatures of 900-1000ºC in the absence of catalysts, by ammonia injection. The rate of reduction of nitrogen oxides is 30-50% (ECE NO_x Task Force, 1992).

The reactions presented in the literature are selective, indicating oxidation of ammonia and sulfur dioxide (SO_2) may not occur, the presence of oxygen being essential in the development of some reactions.

Treatment of exhaust gases by passing them in the activated carbon reactor is done at temperatures of 100-150ºC.

Depending on the type of installation, old or (new), by applying BAT (best available techniques), the emission levels of NO_x, calculated as NO_2, in mg/Nm^3, are limited to 100 (100) for boilers, operating on gaseous fuels (calculated at 3% O_2), to 90 (50) for gas turbines (calculated at 15% O_2) and to 100 (75) for internal combustion engines, stationary, for gaseous fuel, calculated at 15% O_2. In case of using liquid fuels on boiler installations, depending on the type of installation, old or (new), and its power by applying the BAT techniques, emission levels of SO_2, in mg/Nm^3, calculated at 3% O_2, are limited to 350 (350) for power below 100MWt, 250 (200) for power below 300MWt and 200 (150) for power over 300MWt (Directive 2008/1/EC, 2008).

The amount of NO_x emitted in the means of transport sector in European Union countries (EU-27) represent 46.3% of the total of NO_x emissions at their level (electricity and heat generation sector represent 20.5% and energy using industry represents 13.9%). Due mainly to use low NO_x combustion technology, to the replacement of solid fuels to natural gases, to the use flue gases treatment (selective non-catalytic and selective catalytic) emissions of SO_2 decreased from approx. 16900 kt, in 1990, with 31% to about 10300 kt in 2007 (EEA - APE 002, 2010).

2.4 Solid particles

Particles released into the atmosphere come both from combustion processes and the entrainment by the combustion gases from the raw material, e.g. rotary drum furnaces for clinker, limestone, dolomite, etc.

Result of the combustion processes, the black smoke is a heterogeneous mixture and variable in structure, between soft solid particulates, water vapors and gases that result from the incomplete combustion of fuels. As a result of the reduced size of suspended solid particles, the smoke refract light and color depending on the concentration and color of substances component in it. That is why the smoke comes in large range of dark colors from the weak gray to black, depending especially by the degree of perfection of the combustion and the quality of burned fuels. The determination of the index that characterizes smoke color is used for technical control of combustion quality, in combating environmental pollution, resulted in particular from burning of solid fuels and lower oils (Bacharach no.).

Generally, the formation of unburned solid materials in fossil fuels flames is pointing out especially the wrong mixing between fuel with air that is insufficient to achieve combustion even before carbon oxide is obtained. To liquid fuels without mineral content these solid particles can virtually disappear by burning that if the oxygen is in adequate proportions, in warm enough areas of the furnaces and for a sufficient residence time. In black smoke it is noticed the emergence of two categories of unburned solid: very soft particles of soot and particles much larger than the first, sometimes called cenospheres from intermediate and heavy fuel combustion, or especially flying coke by combustion of solid fuels.

The mechanism of the formation process of soot particles in gaseous fuels flames is explained by theories that belong to three main groups. Thus, the first group includes theories that explain formation by thermal decomposition of hydrocarbons in fuel in carbon and hydrogen, with subsequent polymerization of the carbon. The second group of theories explains soot formation by oxidation of hydrocarbons until peroxides are formed and then by decomposition of peroxides free radicals are divided, which favors higher hydrocarbon formation, which under the high temperatures in the flame, are decomposing to form solid carbon particles. The third group of theories involves the formation of soot particles in flame by polymerization of the C_2 radical.

All these theories have a qualitative character, experiments verifying only certain assumptions that are made. The main factors that prompt the formation, quantity, quality and size of soot particles are as following (De Soete & Feugier, 1974):

- the degree of fuel enrichment (dosage), or lack of required air for stoichiometric combustion;
- physico-chemical properties of fuels, determined by the gravimetrically report C/H of carbon and hydrogen from fuels and by the molecular mass;
- the conditions of mixing fuel with combustion air which may be sufficient but unevenly distributed in relation to fuel gas;
- the temperature level in the combustion chamber, as the temperature rises the polymerization and hydrocarbon cracking processes intensify;

- the pressure in the combustion chamber which, by increasing, it intensifies the formation of soot process in flame.

The burner of industrial liquid fuels create the fuels spraying as drops, firstly, to increase their surface in contact with the oxidizer. Spraying of liquid fuels in the form of drops jets that ignite in a gaseous environment that lacks the oxygen required for complete combustion, can generate soft soot particles and relatively large solid particles with average diameter up to 100 μm.These last ones having a spongy aspect and spherical form, thus called cenospheres, are more numerous as the liquid fuel is more heavy, meaning that it contains molecules with more carbon atoms.

The cenospheres result from cracking hydrocarbons in liquid phase as soft drops, through decomposition reactions that lead to release of gaseous products and building of solid residue rich in carbon, but which also contains minerals from the original fuel. Similarly, the flying coke is made from soft coal particles, sprayed with air blast at coal dust operating burners. The heavy liquid fuels, in particular, build solid tailings from sediments and organ-metallic compounds that associate with carbon cracking residues. Overall mass of solid particles thus formed for heavy fuels may be in the order of a few decigrams per kilogram of fuel.

Cenospheres are in smaller numbers than particles of soot, but their overall mass is generally much bigger. Essentially, the presence of these relatively large solid particles, sometimes make the regulations compliance on pollution from the combustion of heavy liquid fuels difficult (Ghia & Gaba, 2000).

Given the size of cenospheres, their combustion is slower than the combustion of soot particles. This combustion is essentially controlled by the diffusion phenomena. Thus, its influence is decisive on: diffusion rate of oxygen to particle, diffusion rate of gases resulted from combustion by the ambient gaseous environment, the particle temperature and the activation energy of substances at work. It should not be missed the fact that the hydrocarbon skeleton of cenospheres contains mineral elements originally present in the fuel. These elements may be particularly harmful in contact with the heating furnaces charge, with the elaboration furnaces bath, refraction bricks from the furnace's walls, boilers or chemical industry ovens pipes. A negative example is the harmful action of vanadium salts on steel at temperatures above 600°C. In flame, the vanadium is present mainly as V_2O_5 because V_2O_4, V_2O_3 and V_3O_5 oxides can be considered as intermediate steps to V_2O_5. When using a good quality combustion plant and properly exploited, the solid particles mostly disappear by burning, leaving behind ashes of the initial mineral mass of fuel. To the heat generators operating on industrial liquid fuels is imposed that solid suspensions should not exceed 100-200 mg/kg of fuel. This prevents the excessive pollution of the environment and the excessive increase of maintenance costs associated with deposition cleaning on the heat change pipes. Oil use with high sulfur content, requires operation at low excess of air close to the stoichiometric value, to reduce corrosion at low temperature. It is favored this way the increase of losses by unburned particles related to the soot and cenospheres and flying coke respectively. Reducing these polluting emissions is directly influenced by the ratio between carbon C and hydrogen H of the initial fuel and the presence of asfaltenes in oil.

Gaseous fuels with a high ratio C/H produce soot easier than those with the lower C/H ratio. Soot production in the flame makes this to be brighter and to transfer more useable heat by radiation. As a result of this advantage, there was developed a process (called self - carburizing) by which carbon particles produced in the initial zone of the flame are completely oxidized to the top of the flame. This procedure has been used on some burners of boilers and furnaces, proving at the same time with the increasing efficiency very effective to reduce the emissions of CO, NO_x and partially oxidized hydrocarbons (Gaba, 2010).

Primary procedures for reducing particles emissions, are to replace the solid and heavy liquid fuels (oil) with light liquid fuel combustions and to use complete combustion methods.

Primary processes that reduce particle emissions resulting from incomplete combustion of liquid fuel are:

- prior physical treatment of liquid fuel consisting in preheating, filtration, centrifugation to remove the largest part of sediments, water, soft particles of minerals;
- high quality spraying of which are essential choosing of optimal spraying system, well adjusted preheat temperature of the fuel. After spraying by mechanical injection under pressure of heavy liquid fuels, a variation of the viscosity of dozen centistokes modify essentially the spraying features, with great repercussions on the process of combustion;
- fuel combustion with preheated air with a moderated air excess (below 30%), using especially overheated steam spraying;
- Improvement, by addition, of the combustion characteristics of industrial fuels used;
- Use of modern combustion techniques such as aerosols combustion, or water - liquid fuels emulsions combustion (with 3% - 5% water).

Secondary procedures for reducing particles emissions, most used by boilers and furnaces are:

- dry mechanical refinement, by which the particles settling is done under the action of gravity and/or centrifugal force;
- Wet refinement by passing the gases under a liquid mist;
- Refinement by filtration, by passing gases through porous materials;
- Electrical refinement by gas ionization and dispersed particles sedimentation in a high-voltage electrical field. There are considered BAT techniques the use of electrical filters (ESP), textile fabric filters (FF), the wet scrubbers and cyclones, when the latter are used together with ESP or FF. The retention efficiency of ESP for particles larger than 1μm, is 96.5%, the FF is 99.95%, the wet scrubber is 98.5%, and the cyclone is over 85% for particles larger than 5μm (Directive 2008/1/EC, 2008).

Depending on the type of installation of boilers, old or (new), also on its power, by applying the BAT techniques, the levels of particles emission, in mg/Nm^3, are limited to 30 (20) for powers below 100MWt, 25 (20) for powers below 300MWt and 20 (10) for powers over 300MWt, for liquid fuel (calculated at 3% O_2) and for solid fuels (coal and lignite), calculated at 6% O2. For internal combustion engines, stationary, liquid fuel, BAT associated filtering techniques can lead to a maximum value of 30 mg/ Nm^3, gas oil and 50 mg / Nm^3, heavy liquid fuel (Directive 2008/1/EC, 2008).

The amount of PM 10 particles, with diameters less than 10μm, in European Union countries (EU-27), as primary products (directly discharged into the atmosphere) and secondary (derived from the precursors NO_x, SO_2 and NH_3, transformed by photochemical reactions into particles) were reduced by 53% between 1990-2005. The direct amount of PM10 were reduced from 4779 kt in 1990 to 2491 kt in 2005, and the secondary from 33588 kt to 19629 kt. The sector of combustion processes represents 70% of the overall direct emissions of particulates and 90% of the total secondary products. The sector of energy using industry represented 39% of the overall direct emissions of PM10 particulates in 1990 and reached at 13% in 2005, the industry of power generating represented 14% of the overall emissions of PM10 particles in 1990 and reached 10% in 2005, and the means of transport reached from 13% to 21%. The sector of energy using industry represented 39% of the total secondary emission of PM10 particles in 1990 and reached 26% in 2005, the industry of power generation represented 15% of the total secondary emissions of PM 10 particles in 1990 and reached 11% in 2005, and the means of transport reached from 26% to 29%. These substantial reductions were done, due to the dusting techniques of the exhaust gases, the replacement of solid and liquid fuels with high content in sulfur with natural gases, the introduction of the primary techniques for achieving combustion with reduced NO_x emissions, the use of flue gases treatment and the application in EU-27 of new regulations which limited the content of sulfur from some liquid fuels and which used advanced catalysts in internal combustion engines for transportation (EEA– EN 07,2010).

2.5 Volatile organic compounds (VOCs) and persistent organic pollutants (POPs)

Volatile organic compounds are aliphatic and aromatic hydrocarbons, alcohols, ketones, esters, aldehydes, benzene, toluene, acetones, methanol and formaldehydes.

POPs are mainly composed of polycyclic aromatic hydrocarbons (PAHs), resulted in combustion processes of fuels in a smaller extent, but with particularly harmful effects, from dioxins and furans, resulted from the incineration of waste, pesticides and products of chemical industry.

Partially oxidized hydrocarbons are generated by the lack of oxygen (combustion air) in combustion process, or from slow oxidation.

Lack of oxygen in the combustion process can be explained locally or temporarily, because combustion installation from industry are supplied with excess of combustion air. Due to poor mixing of fuel-combustion air, in the flame can occur rich zones in fuels which are formed unsaturated hydrocarbons, aldehydes and acids and, in some cases, the process can continue until the formation of soot. These unburned elements may evolve, behind the flame front, depending on thermodynamic parameters and the interaction of oxidants.

To achieve sufficiently high oxidation rates the following are necessary:

- to set the mixture of fuel gas - combustion air between flammability limits;
- achieving a level of temperature high enough ahead of the flame front;
- creating a bigger turbulence;
- ensuring that the various intermediate elements stay a sufficient period of time in the flame (Gaba, 2005).

In combustion processes, the objective of reducing VOCs is their volatilization and oxidation to CO_2 and H_2O, by thermal oxidation, catalytic or bio-oxidation. In the case of incinerators, especially, is practiced burning in secondary outbreaks, in which the catalysts can be introduced, after the primary focus.

Among the sectors of fuel combustion processes that produce VOCs, transport is the most significant with 18,6% of total EU emissions, the rest being insignificant. Between 1990 and 2007, road transport has reduced VOCs by 60%, as a result of using the three-way catalytic converter and the improvement of diesel quality (EEA – APE 004, 2010).

Among the sectors of fuel combustion processes that produce PAHs, residential heating sector is the most significant with 41% of total EU emissions, and transport represents 15%. Between 1990 and 2007, residential heating sector reduced the production of PAHs by 14% and road transport by 47%, mainly due to the advanced catalyst (EEA – APE 006, 2010).

2.6 Atmospheric greenhouse gas concentrations

Among the gases that produce greenhouse effect, the carbon dioxide has the most important contribution, its percentage being 65%. It appears in an overwhelmingly proportion after fuel combustion processes. In the year 2009 the concentration of CO_2 in the atmosphere was 387ppm. Carbon dioxide recently considered as pollutant, prevents the discharge of heat energy from the surface of the Earth which is acquired by radiation from the sun. As the time passes, even the increase with a few degrees of the temperature could lead to reduction of the land ice cap, the spread of deserts, raising sea and oceans levels, as well as important climate changes. The concentration of CO_2 in the atmosphere has increased by 38% compared to the pre-industrial age, when it was 280 ppm (EEA – CSI 013, 2010).

In 1990, a report of a United Nations committee has confirmed the tendency of heating of the globe and considered that, if the world will not take measures to stabilize the gas flow released in the atmosphere (carbon dioxide, methanol, gases from chemical fertilizers, various hydrocarbons), the temperature will rise by three degrees until 2000. A United Nations Framework Convention on Climate Changes was negotiated between 1991 - 1992 and signed at Rio Conference in 1992. The Convention entered into force in 1995. The main objective of the Convention was to stabilize the greenhouse gas concentrations from atmosphere, especially carbon dioxide, at a level that will not endanger climate system.

Kyoto Protocol, signed in 1997, with the exception of USA, is an extension of the UNFCCC (United Nations Framework Convention on Climate Change) from 1993 and is the first step in an effort to reduce the long-term changes in emissions to prevent climate changes. Both at the global (UNFCCC, 2009) and the EU level (October 2008 Environment Council conclusions) this "dangerous anthropogenic interference" has been recognized by formulating the objective of keeping the long-term global average temperature rise below 2°C compared to pre-industrial times.

There is a consensus in EU-27 countries for collective action to reduce greenhouse effect gas emanations by: improving techniques for burning fossil fuels, the use of renewable sources for energy production (water, sun, wind, geothermal energy), improving industrial technologies, decreasing the consumption of fuels for road transport, increasing the share of public transport, reduce emanation of animal manure, reduce the fertilizers based on nitrogen, stopping the clearance of forest and accelerated action of forestation, reducing waste and gases from municipal waste dumps, etc. The commitment taken by the EU-27 for 2008-2012 compared to 1997, is a reduction of 8% versus 5%, as was stipulated in the Kyoto Protocol (EEA- EN 07, 2010).

3. Determination of pollutant emissions in combustion processes

Quantitative determination of pollutant emissions refers to:

- The determination of pollutant concentration, c_i [mg/Nm³], representing the concentration of pollutant i in the gases released in atmosphere, required to screening verification in the emission limits stipulated by the regulations;
- Determination of pollutant quantity emission, which represents the amount of pollutants released into the atmosphere over a specified period, usually one year, E_i [t], required to establish the activities contribution from combustion processes at general air pollution.

Determination of pollutant concentration, c_i [mg/Nm³] and the amount of pollutants E_i [t] produced by a combustion plants requires knowledge of the effluent gaseous composition

in point of measurement. In addition, at determination of pollutant quantity emission, it is necessary to know the amount of fuel consumed during the period, in m^3N, for gaseous fuels, or kg for liquid or solid fuels [1].
The measurements are based on national norms, which are prescribed the measuring and calculation methodologies, type of measurement devices that are accepted, as well as the execution range for measurements. For large combustion plants requires continuous monitoring. The automatic devices that are used:

- gas analyzer to determine the volumetric concentrations, from the flue gas mixture, O_2 [% volumetric], CO_2 [% volumetric], CO [ppm], NO_x [ppm], NO_2 [ppm], and SO_2 [ppm];
- automatic device to determine the volatile organic compounds;
- automatic instrument, with isokinetic probe for determining particulate matter (Gaba, 2005).

Gaseous effluent concentrations measured values are required, on the one hand to determine the quantity of emissions of pollutants and on the other hand directly to determine the concentration (in mg/Nm^3) of the emission of pollutants. Calculation of concentrations of different pollutants emissions i, c_i [mg/Nm^3], related to reference conditions, it is required to be done to compare it with the limit values set by national regulations, for periods of time, depending on the type, age and power plant and fuel type consumed.
E.g. if the analyzer indicates a concentration of 60 ppm SO_2, 80 ppm NO_x and 10.5% O_2, to measurement of gaseous effluents produced by burning a liquid fuel, it can be calculated the concentration of pollutant emission of SO_2, CSO_2, NO_x, C_{Nox}, respectively, related to reference conditions, as [1.10]:

$$c_{SO_x}\left(mg/Nm^3\right) = c_{SO_x mas}\left(ppm\right)\cdot k1\cdot\frac{21-c_{O_{2ref}}}{21-c_{O_{2mas}}} \tag{1}$$

$$c_{NO_x}\left(mg/Nm^3\right) = c_{NO_x mas}\left(ppm\right)\cdot k2\cdot\frac{21-c_{O_{2ref}}}{21-c_{O_{2mas}}}$$

where:
$c_{SO2\ mas}$ - SO_2 emission concentration measured in ppm;
$c_{Nox\ mas}$ - NO_x emission concentration measured in ppm;
k1 - 2.85 - ppm conversion factor → mg/Nm^3 for SO_2;
k2 - 2.05 - ppm conversion factor →mg/Nm^3 for NO_2;
$C_{O2\ ref}$ – 3% volumetric - reference oxygen concentration for a gaseous fuel according to EU norms;
$C_{O2\ mas}$ – measured oxygen concentration in the gaseous effluent, in volumetric %.
Replacing in (1) with numeric values it is obtained:

$$C_{SO_x} = 60\cdot 2{,}85\cdot\frac{21-3}{21-10{,}5} = 293{,}1\ mg/Nm^3 \tag{2}$$

$$C_{NO_x} = 80\cdot 2{,}05\cdot\frac{21-3}{21-10{,}5} = 281{,}1\ mg/Nm^3$$

The amount thus calculated is compared with the emission limit for SO_2 and NO_x respectively, according to national norms, which are usually set for some period of time, depending on the type, age and power of the plant and fuel type consumed.
In the case of large combustion plants, which consume liquid fuels, in Table 1 are presented associated BAT limit values, for NO_x emissions calculated as NO_2 and SO_2, in mg/Nm^3, for 3% O_2, depending on the type and power of the plant (Directive 2001/80/EC, 2001).

Crt. No.	Plant type	Pollutant	Measure unit	Thermal power, MWt		
				<100	100-300	>300
1	Old	SO_2	mg/Nm^3	350	250	200
2	Old	NO_x	mg/Nm^3	450	200	150
3	New	SO_2	mg/Nm^3	350	200	150
4	New	NO_x	mg/Nm^3	300	150	100

Table 1. Limit values of SO_2 and NO_x emissions for electricity and heating generation.

It is found that the values calculated for the exemplified installation would not exceed the limits unless it would have a thermal power less than 100 MW.
To determine the exact amount of a pollutant i, Ei, in t for a certain period, the general relation can be used:

$$E_i = ki \cdot V_{gu} \cdot B \cdot c_i \cdot 10^9 [t] \tag{3}$$

where:
ki - conversion factor for the i pollutant, ppm → mg/Nm^3;
V_{gu} - volume of dry burned gases produced by burning the fuel,in m^{3}N/m^{3}N or m^{3}N/kg;
B - fuel consumption, in Nm^3 of consumed gas in the respective period, or kg of liquid or solid fuel;
c_i - measured concentration of i pollutant, in ppm
The specialty papers shows calculating relations for the volume of dry flue gases resulting from combustion of various fuels [10]. For example, for combustion of gaseous fuels to find the volume of dry flue gases produced from burning a Nm^3 of fuel, use the following formula:

$$V_{gu} = \frac{CO_2^{(c)} + CO^{(c)} + \Sigma m C_m H_n^{(c)}}{CO_2^{(g.a)} + CO^{(g.a)} + \Sigma m C_m H_n^{(g.a)}} \tag{4}$$

Values in the numerator results from the initial analysis of fuel gas composition and the denominator values are measured at the exhaust of plant flue gases, for example with a flue gas analyzer.
For less accurate calculations, but faster, it is used the relationship:

$$E_i = fi \cdot B \cdot PCI \cdot 10^6 [t] \tag{5}$$

where:
fi – emission factor, in g/GJ;
B – fuel consumption, in Nm^3 of consumed gas in the in the respective period, or kg of liquid or solid fuel;
PCI – low calorific power of the fuel, in GJ/Nm^3, GJ/kg respectively.
The factors *fi* (in g/GJ, respectively $lb/10^6$ Btu), are used in the EU and U.S. respectively, to calculate the amount of pollutants emitted into the atmosphere over a period of time and are tabled according to the many types of fuels, installation, etc. (US EPA- AP 42, 1995). In Table 2 are summarized the approximate amounts of various pollutants to be emitted into the atmosphere worldwide, calculated with the emission factors *fi* (taking the values for coals, oils and natural gas) for fuel consumption in 2030 provided by the World Energy Outlook -2008.

crt. no.	Fuel type		Pollutant				
			SO_2	NO_x	CO	PM10	VOCs
1	Coals	Emission factor [g/GJ]	10746.5	4083.67	1308.49	3094.99	21.49
		Pollutant quantity [kt]	2207331	838786	268766	635713	4414
2	Oils	Emission factor [g/GJ]	7307.62	1590.48	143.57	343.89	38.68
		Pollutant quantity [kt]	1562457	340064	30697	73528	8270
3	Natural gas	Emission factor [g/GJ]	0.0	601.80	150.88	12.89	25.79
		Pollutant quantity [kt]	0.0	92430	23173	1979	3961

Table 2. Approximate amounts of various pollutants to be emitted into the atmosphere worldwide, calculated with emission factors fi, for fuel consumption in 2030.

To these annual amounts of pollutants emission the amount of CO_2 is added, approximate for the year 2030, to approx. 30 million kt, which would lead to a planetary temperature exceeding by more than 2°C.To reduce the quantity of air pollutants emissions it is necessary that reduction measures and techniques to be generalized on a planetary scale Also there should be mentioned the EU-27 achievements in the period 1990-2007, that have significantly reduced emissions of pollutants into the atmosphere: CO by 63%, NO_x by 31%, SO_2 by 69%, VOCs by 41%, PAHs by 63% and PM10 by 53%. For the future, the key to success in reducing the air pollution is increasing energy efficiency and using regenerative resources. In 2009, the EU agrees a Climate and Energy Package to:

- reduce greenhouse gas emission by 20% by 2020;
- increase the share of renewable energy by 20% by 2020;
- improve energy efficiency by 20% by 2020 (EEA – EN 07, 2010).

4. References

Baukal, C. (2001). *The John Zink Combustion Handbook*, CRC Press, New York.

De Soete, G. & Feugier, A. (1974). *Connaissance actualles sur les mecanismes de formation des principaux polluants dus à la combustion*. Technip, Paris.

De Soete, G. (1982). *Physikalisch-chemische Mechanismen bei der Stickstoffoxidbildung in industriellen Flammen*. Gas Wärme International, vol. 31, no.1, (Jan. 1982), pp.15-23,

Directive 2008/1/EC - *Integrated Pollution Prevention and Control, Reference Document on Best Available Techniques for LargeCombustionPlants*,July2006. www.eippcb.jrc.es/reference

Duca, Gh., Crăciun, A., Sajin, T., Gaba, A., Păunescu, L. (2002). *Modern combustion and air pollution reduction techniques*. USM, ISBN 9975-70-077-2, Chisinau.

ECE (Economic Commission for Europe) NOx Task Force. (1992). *Operating Experience with NOx Abatement at Stationary Sources*. IIP University of Karlsruhe.

EEA (European Environment Agency). 2010. *Sulfur dioxide SO2 emissions* (APE 001), Feb. 2010. www.eea.europa.eu/APE.

EEA. 2010. *Emissions of nitrogen oxides NOx* (APE 002), Feb. 2010. www.eea.europa.eu/APE

EEA. 2010. *Non-metane volatile organic compounds (NMVOC) emissions* (APE-004), Feb. 2010. www.eea.europa.eu/APE

EEA. 2010. *Persistent organic pollutant (POP) emissions* (APE-006), Feb. 2010. www.eea.europa.eu/APE

EEA. 2010. *Atmospheric greenhouse gas concentration* (CSI-013), Nov. 2010. www.eea.europa.eu/CSI

EEA. 2010. *Energy related particulate matter emissions* (EN-07), Jan.2010. www.eea.europa.eu/EN

EEA. 2011. *Transport emissions of air pollutants* (TERM 003), Jan. 2011. www.eea.europa.eu/TERM

IEA (International Energy Agency). (2008). *World Energy Outlook-2008*. www.iea.org/text base/nppdf (www. worldenergyoutlook.org.)

IEA. (2010). *World Energy Outlook-2010*. www.iea.org/text (www. worldenergyoutlook.org.)

Gaba, A. (2005). *Combustion of fuels and environment*. Bibliotheca, ISBN 973-712-061-2, Targoviste.

Gaba, A. (2009). *Low NOx recuperative burner for forge and treatment furnaces*. Metalurgia International, vol. 14, no.8, (August, 2009), pp.23-30, ISSN 1582-2214.

Gaba, A. (2010) *Air pollution reduction by using low NOx burners for furnaces and boilers*. Environmental Engineering and Management Journal, vol. 9, no.1, (January, 2010), pp.165-170, ISSN 1582-9596.

Ghia, V. & Gaba, A. (2000). *Primary measures for air pollution reduction in combustion processes*. AGIR, ISBN 973-8130-06-09, Bucharest.

Kramer, H. (1986). *Grundlangen de NOx Entstehung und Minderung*. Gas Wärme International, vol. 35, no. 5, (Mai-June, 1986), pp.239-247, ISSN 0020-9384.

Skunca, I. (1981). *Vereinfachten Verfahren zur Berechnung der Gleichgewichts- konzentrationen von Stickstoffmonoxid und Kohlenmonoxid in Flammen*. Gas Wärme International, vol. 30, no. 1, (Jan., 1981), pp.48-52, ISSN 0020-9384.

U.S. EPA, AP-42. (1995).*Compilation of Air Pollutant Emission Factors*, Jan. 1995. http://www.epa.gov/ttn/chief/ap42/index.html
Zeldovici, Ia. B. (1980). *Matematicescaia teoria gorenia i vrzva*, Nauka, Moskva.

7

Monitoring of Gaseous Air Pollution

Ryszard J. Katulski[1], Jacek Namieśnik[2], Jarosław Sadowski[1], Jacek Stefański[1] and Waldemar Wardencki[2]
[1]Gdansk University of Technology,
[1]Department of Radiocommunication Systems and Networks
[2]Gdansk University of Technology, Department of Analytical Chemistry
Poland

1. Introduction

Air pollution can be defined as the presence in the atmosphere of one or more contaminants in such quality and for such duration as is injurious, or tends to be injurious, to human health or welfare, animal or plant life. It is the contamination of air by the discharge of harmful substances. Industrialization have led to air getting more and more polluted over the years. Industries, vehicles, increase in the population, and urbanization are some of the major factors responsible for air pollution. The following industries are among those that emit a great deal of pollutants into the air: thermal power plants, cement, steel, refineries, petro chemicals, and mines.

Air pollution can cause health problems and it can also damage the environment and property. It has caused thinning of the protective ozone layer of the atmosphere, which is leading to climate change. The source of pollution may be in one country but the impact of pollution may be felt elsewhere. The discovery of pesticides in Antarctica, where they have never been used, suggests the extent to which aerial transport can carry pollutants from one place to another.

The most important tools in environmental protection is monitoring and modeling. Both the monitoring and modeling of air pollution is essential to provide a picture of the damage humans are doing to the environment, and to enable pollution problems to be discovered and dealt with. An environmental monitoring could be defined as a system of detection, measurements, evaluations and forecasts of environmental states, and the collecting, processing and spreading of information on the environment.

Air pollution and its control is a global issue demanding international cooperation. Monitoring of air pollution is a very important source of data. However, measurement of the air pollutant concentrations, in comparison to monitoring of other elements in the environment, is the most difficult. The difficulties arise from the large dynamics of the atmosphere, causing that it constitutes the main route of pollution spreading and their transport between remaining environmental compartments and universal exposure for a large population without a chance for isolation, which is possible in the case of polluted waters and soil. Another problem is low concentration of air pollutants and their interaction with other gases.

This chapter is divided into two parts. In the first, a general information on gaseous air pollution, parameters used to characterize the pollution level and equipment applied for air monitoring and analysis is presented. Finally, the mobile monitoring system (MMS), as an example of a new approach for air monitoring along communication lanes, proposed by the authors is described.

2. History of air pollution legislation

Concern about urban air quality is not new. First complaints were recorded in the 13th century when coal was first used in London. Today the emphasis has shifted from the pollution problems caused by industry to the pollution in urban areas. A growing concern over the influence of different air pollutants on human health were the main driving force to develop and implement of air quality criteria and standards.

In 1967, the US Congress enacted the Air Quality Acts, the first modern environmental law.

The Clean Air Act (Environmental Protection Agency, 2008), which was last amended in 1990, requires the United States Environmental Protection (US EPA) to set National Ambient Air Quality Standards (NAAQS) for pollutants considered harmful to public health and the environment. The Clean Air Act established two types of national air quality standards. *Primary standards* set limits to protect public health, including the health of "sensitive" populations such as asthmatics, children, and the elderly. *Secondary standards* set limits to protect public welfare, including protection against decreased visibility, damage to animals, crops, vegetation, and buildings. The EPA has established NAAQS for six principal pollutants, which are called criteria pollutants: sulphur dioxide, particulate matter, nitrogen oxide, carbon monoxide, ozone and lead. These standards are threshold concentrations based on a detailed review of scientific information related to effects.

In Europe, the first international air quality standards were introduced by the European Commission in 1980 for SO_2 and suspended particulates, mainly aimed at protecting human health. A few years later, the World Health Organization, recognizing ecological damage as being relevant to human health, introduced air quality guidelines for Europe which included the former as well as the latter, later revised in 2000 (World Health Organisation, 2000). The newest directive on ambient air quality and cleaner air of the European Union entered into force in June 2008 (European Union, 2008). It merges four earlier directives and one Council decision into a single directive on air quality. The new directive of the European Union on air quality takes into account concerns from the latest WHO air quality guidelines (WHO, 2005) on fine particles. Reflecting the latest WHO air quality guidelines that identify fine particles (PM2.5) as one of the most dangerous pollutants for human health, the new EU directive sets objectives and target dates for reducing population exposure to PM2.5. It also maintains limits for concentrations of coarser particles known as PM10 and other main pollutants already subject to legislation.

Pollutant levels, especially gaseous, which negatively affect life on earth are nowadays well defined. Table 1 presents examples of air quality standards issued by the Environmental Protection Agency (EPA), the Word Health Organization (WHO) and some states (ARMAAG).

3. Regulated air pollutants

The contaminants in ambient air that are of concern are basically categorized as criteria and non criteria pollutants (Grifin, 2007).

Criteria air pollutants are those air contaminants for which numerical concentration limits

Pollutant	Averaging time	WHO	EPA	UE	UK	France	Germany	Poland
SO_2	10-15 min	500	-	-	266 (not more than 35 times)	-	-	-
	30 min	-	-	-	-		-	-
	1h	-	-	350 (not more than 24 times)	350 (not more than 24 times)	350 (not more than 24 times)	350	350 (not more than 24 times)
	3h	-	1300	-	-	-	-	-
	24h	125	365	125 (not more than 3 times)	125 (not more than 3 times)	125 (not more than 3 times)	125	125 (not more than 3 times)
	year	50	80	20	20	20	20	20
NO_2	30 min	-	-	-	-	-	200	-
	1h	200	-	200 (not more than 18 times)	200 (not more than 18 times)	230 (not longer than 0.2% of time)	-	200 (not more than 18 times)
	24h	-	-	-	-	-	100	-
	year	40	100	40	40	46	-	40
PM10	30 min	-	-	-	-	-	-	-
	24h	20	150	50 (not more than 35 times)	50 (not more than 35 times)	50 (not more than 35 times)	50	50 (not more than 35 times)
	year	50	-	40	40	40	40	40
CO	10-15 min	100000	-	-	-	-	-	-
	30 min	60000	-	-	-	-	-	-
	1h	30000	4000	-	-	-	-	-
	8h	10000	1000	1000	1000	1000	1000	1000
	24h	-	-	-	-	-	-	-
	year	-	-	-	-	-	-	-
O_3	30 min	-	-	-	-	-	-	-
	1h	-	235	-	-	-	-	-
	8h	100	157	120	100 (not more than 10 times)	120	120	120 (not more than 25 days)
Pb	24h	-	-	-	-	-	5	-
	3 months	-	1.5	-	0.5	-	-	-
	year	0.5	-	0.5	-	0.5	0.5	0.5
Benzene	year	-	-	5	16.25	8	-	5

Table 1. Comparison of limit values [μm/m^3] for a given averaging time, number of exceedances per year issued by different countries and organizations (ARMAAG)

have been set as the dividing line between acceptable air quality and poor or unhealthy air quality. Criteria pollutants include five gases/vapours and two solids: nitrogen oxides (NOx), sulphur dioxide (SO_2), carbon monoxide (CO), ozone (O_3), benzene, particulate matter (PM10) and lead (Pb). Non criteria pollutants are those contaminants designated as toxic or hazardous by legislation or regulation. They fall in two further subcategories, depending on the legislation that defines them. In general, hazardous air pollutants may pose a variety of health effects, whereas the toxic focus on one physiological response.

4. Characteristics of criteria air pollutants

Air pollutants arise from a wide variety of sources although they are mainly a result of the combustion process (Friedrich & Reis, 2004). The largest sources include power generation, motor vehicle and industry. The emissions of pollutants into the atmosphere badly influence vegetation, human and animal life, agriculture and climate. Emissions of carbon monoxide (CO), nitrogen oxides (NOx) and hydrocarbons (HC) are controlled by catalytic converters on new gasoline driven cars. Emissions of sulphur oxides are being reduced through lower sulphur content in gasoline. However, emissions of particulate matter are not decreasing. Any successful strategies for controlling or countering these problems must be based on reliable air quality monitoring data for management, to make informed decisions on air pollution control.

Volatile organic compounds (VOCs) is a collective name for a very large number of different chemical species, including hydrocarbons, halocarbons and oxygenates that have different physico-chemical properties and are directly emitted from both anthropogenic and natural sources, and which can contribute to the formation of secondary pollutants with different efficiencies. For vehicular emissions, the list of compounds is long and variable depending on fuel, type of engine, and operating conditions. Hydrocarbons such as ethane, ethyne, higher aliphatic hydrocarbons, benzene, toluene, and xylenes (BTX) are typical emissions in most cases. Each of these compounds can be released unreacted or can undergo oxidation reactions. One of them, benzene, is found in the highest concentrations. Ambient concentrations are typically between 1-50 ppb but close to major emissions can be as high as several hundred ppb. In the unreacted state, it has undesirable ecotoxicological properties. Besides causing annoying physiological reactions such as dizziness and membrane irritation, it is known to be a human carcinogen.

The two nitrogen oxides, NO and NO_2 (together called NO_x) from anthropogenic sources are present as a consequence of various combustion processes from both stationary sources, i.e. power generation (21%) and mobile sources, i.e. transport (44%). These species have very short atmospheric lifetimes, around five days, being ultimately converted to nitric acid and removed in rainfalls. However, nitrogen oxide is important because it is a precursor to tropospheric ozone. Whereas NO does not affect climate, ozone does. A typical sea-level mixing ratio of NO is 5 ppt (parts per trillion, $1:10^{12}$) but in urban regions, NO mixing ratios reach 0.1 ppm in the early morning, but decrease to zero by midmorning due to their reaction with ozone. A major source of NO_2 is oxidation of NO, being intermediary between NO emission and O_3 formation. Nitrogen dioxide is one of the six criteria air pollutants for which ambient standards are set by the US EPA under CAAA70 (Clean Air Act Amendments of 1970). In urban regions, the mixing ratio of NO_2 ranges from 0.1 to 0.25 ppm. It is more prevalent during midmorning than during midday or afternoon because sunlight breaks down most NO_2 past midmorning. Exposure to high concentrations of NO_2

harms the lungs and increases respiratory infections. It may trigger asthma by damaging or irritating and sensitizing the lungs, making people more susceptible to allergens. At higher concentrations, it can result in acute bronchitis or death.

Sulfur dioxide (SO2) is a strong smelling, colorless gas that is formed by the combustion of fossil fuels, smelting, manufacture of sulfuric acid, conversion of wood pulp to paper, incineration of refuse and production of elemental sulphur. Power plants, which may use coal or oil high in sulfur content, can be major sources of SO_2 accounting for about 50 % of annual global emissions. SO_2 and other sulfur oxides contribute to the problem of acid deposition and can be major contributors to smog. Natural background levels of SO_2 are about 2 ppb. Hourly peak values can reach 750 ppb on infrequent occasions. Sulphur dioxide can lead to lung diseases. SO_2 is a criteria air pollutant.

Ozone (O_3) is not directly emitted from both anthropogenic and natural sources. Its only source into air is chemical reaction. In urban air, ozone mixing ratios range from less than 0.01 ppm at night to 0.5 ppm (during afternoons in the most polluted cities world wide), with typical values of 0.15 ppm during moderately polluted afternoons. Ozone causes headaches at concentrations greater than 0.15 ppmv, chest pains at mixing ratios greater than 0.25 ppm, and a sore throat and coughing at mixing ratios greater than 0.30 ppm. Exceeding a level of 0.30 ppm decreases lung functions. Respiratory symptoms include coughing and breathing discomfort. Ozone can also accelerate the aging of lung tissue. It also interferes with the growth of plants and deteriorates organic materials, such as rubber, textiles and some paints and coatings. Furthermore, ozone increases plant and tree stress and their susceptibility to disease, infestation and death.

Particulate matter (PM) in the atmosphere, frequently described simply as particle pollution, arises from natural sources, such as windborne dust, sea spray, and volcanoes and from anthropogenic activities, such as combustion of fuels. Particle pollution in the air includes a mixture of solids and liquid droplets and come in a wide range of sizes. Those less than 10 micrometers in diameter (PM10) are so small that they can enter the lungs, potentially causing serious health problems. Particles less than 2.5 micrometers in diameter are called fine particles. These particles are so small they can only be detected with an electron microscope. Sources of fine particles include all types of combustion, including motor vehicles, power plants, residential wood burning, forest fires, agricultural burning, and some industrial processes. Particles between 2.5 and 10 micrometers in diameter are referred to as coarse. Sources of coarse particles include crushing or grinding operations, and dust stirred up by vehicles traveling on roads. After being released into the air, particles can change their size and composition by condensation of vapour species or by evaporation, by coagulating with other particles or by chemical reaction. Particles smaller than 1 μm in diameter generally have atmospheric concentrations in a range from around ten to several thousand per cm^3; those exceeding 1 μm diameter are usually found at concentrations less than 1 cm^{-3}.

Carbon monoxide (CO) is a colourless, odourless gas that is produced by the incomplete burning of carbon-based fuels including petrol, diesel and wood. It is also produced from the combustion of natural and synthetic products such as cigarettes. Natural background levels of CO fall in a range of 10-200 ppb. Levels in urban areas are highly variable, depending upon weather conditions and traffic density. 8-hour mean values are generally less than 10 ppm but sometimes can be as high as 500 ppm. Carbon monoxide lowers the amount of oxygen that enters the blood. It can slow human reflexes and may make people confused and sleepy.

5. Air pollution monitoring networks

For the purpose of monitoring and reporting of air pollution, most industrialised countries have been divided into regions or zones and urban areas or agglomerations, e.g. in Europe in accordance with EC Directive 96/62/EC (European Union, 1996). This Directive sets a framework for ways to monitor and report ambient levels of air pollutants. Other directives set ambient air limit values for particular pollutants:

- Directive 99/30/EC for nitrogen dioxide and oxides of nitrogen, sulphur dioxide, lead and particulate matter,
- Directive 2000/69/EC for benzene and carbon monoxide,
- Directive 2002/3/EC for ozone.

The monitoring sites are organized into automatic and non-automatic networks (national, regional and local) that gather a specific kind of information using a particular method. For example, across the UK there are over 1500 monitoring sites which monitor air quality and these are organized into networks (automatic and non-automatic) that gather a specific kind of information, using a particular method. The pollutants measured and the method used by each network depend on the reason for setting up the network, and what the data is to be used for. In Poland, the monitoring of air quality has been performed systematically since 1992, mainly by using automatic air monitoring stations. National network comprises 81 specialized monitoring stations, where basic pollutants are measured everyday (CO, CO_2, SO_x, NO_x, HCl and particulate matter PM10 and PM2,5). Regional and local networks comprises 390 stations localized in cities over 20 000 citizens and additionally 60 stations in spas.

The obtained information concerns different types of concentrations of investigated pollutants depending on applied sampling techniques and a measuring period. The results of measurements may be referred to in real time (instantaneous concentrations) or to a selected period of time (e.g., 1-hr, 8 hrs, 24-hrs, month, year). Final measurements represent averaged concentrations. The obtained data from air monitoring is used in air quality inventories and bulletins.

Considering the frequency of discrete sampling, periodic and instantaneous measurements are distinguished. Taking into account space, parameter measurements are divided to a point, averaged along a defined part of space and averaged on the selected area. Point monitoring is inadequate to measure poorly mixed gases such as fugitive emissions over large areas. If the point instrument is wrongly placed, measurement results are not representative. Final measurements enable the determination of weighted average concentrations over the sampling period.

6. General characteristic of instruments for air monitoring

The earliest air quality measurements used relatively simple, manual techniques. Often a sample was collected at a site and then returned to a laboratory for analysis. This is, in fact, still a common practice today, due to the relative low cost and high reliability of these types of measurements. Since seventieth of the XX century the automatic analyzers were introduced to the monitoring networks. These had the benefit of low labour costs and could provide highly resolved, continuous measurements. These continuous measurements became a requirement for regulatory purposes and so, in many cities urban monitoring network was established to monitor compliance with the emerging EC Directive limit values on air quality. The networks were subsequently expanded, following commitments by different governments to expand urban monitoring in particular countries and improve public availability of air quality information.

The basic requirement of the analyzers for air monitoring is high measurement sensitivity, i.e. a low limit of detection (LOD) and low limit of quantitation (LOQ). It gives a chance for the detection of pollutants at required levels. Instruments which should acquire analytical data in real time or only with a small time delay, have to be equipped with the following additional capabilities:

- provide high resolution data (characterized by a low response time),
- provide automatic calibration and zeroing,
- extended operation without service or maintenance.

The last demand means that they should be equipped with an independent power supply and be able to automatically regenerate or exchange worn out filters and, depending on the type of detector (sensor) used, fulfill special demands, e.g., for electrochemical sensors, exchange or supplement the working solution and reagents, and in devices with FID or FPD detection, protect against flame extinguishing.

Analytical instruments currently used for the detection and determination of atmospheric pollutants can be classified according to various criteria (Wardencki et al., 2008).

Recently, measuring techniques based on a physical (or physicochemical) principle, have been used more frequently in the assessment of air quality. Such methods involve the direct determination of a physical property of a pollutant, sometimes after its interaction with another compound. In this approach, better stability, sensitivity and reliability may be easily achieved. Furthermore, the practical application requires less maintenance. Instruments based on this principle can be easily automated, which enables their use in providing continuous measurements needed for up-to-date assessments of air quality. It is especially relevant to environmental monitoring because many existing standards refer to a specified period of time, i.e. 1 hour, 24 hours or a year.

According to the location where measurements are taken, instruments can be stationary or *on-site*. In the first case, analysis is performed in the laboratory and sophisticated instruments are used, such as mass, electron mobility or X-ray fluorescence spectrometers. *On-site* systems enable measuring pollution levels in the field. Since access to a sophisticated laboratory is not required, the devices (usually uncomplicated, relatively cheap and portable) hold great promise for use in remote locations. The main advantage of on-site analysis is the potential for rapid assessment and response to a particular problem.

All monitoring systems can be classified as mobile or stationary. Most existing systems which monitor gaseous pollutants of atmospheric air and ambient aerosols, both automatic and manual, usually perform stationary measurements, i.e. they are directly linked to a specific point or space in the vicinity of that point. Based on the data obtained from a single monitoring site, it is not possible to assess spatial and temporal variations of air pollutants.

Mobile refers to a continuous-monitoring instrument that is portable or transportable. They are usually designed to perform analytical measurements without preliminary operations. Portable refers to a self-contained, battery-operated device, which may be worn or carried by the person using it, or may require the use of special vehicles for placement in a specific area to be monitored. Transportable gas monitors can be mounted on a vehicle such as a car, plane, balloon, ship or space shuttle, but not to a mining machine or an industrial truck.

For mobile systems, registered values of pollutants have to be correlated with information about the geographical site and actual meteorological conditions (temperature and humidity).
Several contributions published during the last decade have proven the advantages of mobile systems in getting information concerning the spatial and temporal distribution of atmospheric trace gases, without the need of a dense network of stationary stations. Most of the proposed systems are based on the application of mobile laboratories (Bukowiecki et al., 2002), (Gouriou et al., 2004), (Pirjola et al., 2004), equipped with appropriate monitors. There are also systems which allow the measurement of pollutants in a stream of vehicles, but a measuring unit is installed on any vehicle (Katulski et al., 2006, 2007) rather than attached to a dedicated van (Seakins et al., 2002), (Bogo et al., 1999).
The general trend in the field of creating instruments for air quality assessment is combining several instruments into one system and forming so called hybrid multisensor systems, controlled by a microprocessor capable of transferring the obtained data to a central station, frequently using a wireless system. In the central station, the data is collected both from single objects (houses, plants) or from larger areas. Many systems are equipped with devices for testing the sensors and for providing a diagnosis of the whole instrument. They frequently have alarms which warn the user of any dangerous situation due to the breeching of some value limit. Such systems are battery-powered and able to work continuously for several days or months.
The environments in which analyzers are used differ from the relative calm of the laboratory. Analyzers have to withstand wide ambient temperatures, fluctuations and vibrations. Due to this, many systems are completely sealed so as to operate independent of outside conditions and to be able to withstand the onslaught of monkey-wrench mechanics.

7. Description of mobile system for on-road measurements of air pollutants

The developed prototype of monitoring system consists of an original equipment measuring in motion the gaseous pollution of atmospheric air, further called ARPOL (the acronym stems from ARMAAG - Agency of Regional Monitoring of Atmosphere of Gdańsk Agglomeration and Gdańsk POLytechnic), which is designed to test the pollution of atmospheric air with the following chemical compounds: C_6H_6, NO_2, NO_x, CO and CO_2. The equipment can be installed on any mobile vehicle like a bus or taxicab, relay the results of concentration measurements in air of the substances outlined above to a server which plays the role of a database, along with information on vehicle location and speed as well as the temperature and humidity of air at the measurement spot. Data transmission between the equipment and the server is effected with the use of a radio modem operating in GSM/GPRS systems.
An ARPOL monitoring station has been developed and built in the laboratory of the Department of Radiocommunication Systems and Networks at Gdańsk University of Technology. The set of commercially available gas concentration sensors in air has been selected by researchers from the Chemical Faculty of Gdańsk University of Technology and the ARMAAG Foundation. The construction of the system falls into line with requirements of the sensors and the anticipated application. In the result of decisions taken it has been established to use semiconductor (thick-film) sensors of the newest generation, made of nanostructures with grain-size distribution from 30 to 50 nm. After preliminary investigations the following sensor (all from Figaro Engineering Inc from Osaka, Japan)

were chosen: TGS2442 for CO, TGS4161 for CO_2, TGS823 for benzene, TGS2106 for NO_2 and TGS2201 for NO_x. Some of these sensors were designed for measurement of gases in ppm (parts per million, 10^{-6}) concentration, but the concentration of NO_x, CO or C_6H_6 in typical air condition is of the order of ppb (parts per billion, 10^{-9}). However, comparison of results from several months of measurements proved that after proper calibration and compensation of air temperature and humidity, they were enough stable and precise for determining the sources and regions of highest pollution caused by road traffic. All of the gas concentration sensors are heated using internal heaters as a substrate for sensing structure. The temperature of substrate cannot be controlled because no temperature sensor is built in the substrate of gas-sensing element. Correction of substrate temperature change caused by change in temperature of air flowing around the sensor is made by measuring the temperature of air using additional semiconductor sensor and proper calibration equations. Proper heating of gas concentration sensors takes up to several tens of minutes. During this period results of measurements cannot be considered as a reliable results. Also any break in powering the sensors makes results unstable for some unpredictable time (up to several minutes) so monitoring device with such sensors need to provide uninterrupted heating of sensors during whole measurement campaign. Special electronic circuits ensure high sensitivity and high accuracy of measurements.

In the design of ARPOL a microcontroller of the AtMega series (supplier: Atmel Corporation from San Jose, USA) with built-in analog-to-digital converters has been used; the microcontroller circuit has been connected to a ready-made GSM/GPRS modem board (supplier: Enfora from Richardson, USA) and a GPS signal receiver (supplier: Trimble Navigation Ltd. from Sunnyvale, USA), as well as a series RS232 interface (RS232 signal converter based on Maxim chip, supplier: Maxim Integrated Products Inc. from Sunnyvale, USA) and the required additional supply circuits. The main board has been also provided with a connector for the attachment of a board with possible additional sensors not foreseen in the original design, and transoptor circuits and actuators which allow to apply control signals.

To ensure highest immunity of the equipment against shocks it was decided that the whole electronic circuitry of the main part of ARPOL will be assembled on one printed board. As some devices of the equipment (power supply circuits in particular) may heat up during normal operation, it has been decided to place all chemical sensors and the external temperature and air humidity sensors at one edge of the main board, so that the microprocessor part with power supplies could be separated in the housing from the sensors by means of an appropriate dividing partition.

The equipment has been enclosed in a housing which provides protection of the system from the influence of precipitation, while simultaneously allowing access of atmospheric air to the gas-concentration, temperature and humidity sensors. The housing permits the unit to be mounted both in a car and stationary at selected points in town. Part of the electronic board with sensors has been enclosed in a plastic channel terminated with an air in- and outlet, provided with additional "roofs" for protection from rain. In order to reduce the effect of exhaust gas of the engine the unit has forced sampling system opening towards the driving direction. To ensure an uniform air exchange, in the channel with sensors a blower has been placed, provided with revolution control – a blower failure or reduction of its efficiency resulting in a decrease of air flow in the measurement channel is registered by a microcontroller as an alarm state.

A schematic description of the operation of the mobile monitoring system (MMS) for testing and analysis of atmospheric air pollution along road traffic arteries is presented in Fig.1. This system consists of an ARPOL unit mounted on a vehicle and a WWW server in which a database for collection and processing of measurement data has been established. The values of concentration of chemical compounds polluting the atmosphere are transmitted through radio, through the mobile GSM/GPRS network and the Internet network to a WWW server located anywhere, e.g. at Gdańsk University of Technology (Gdańsk Polytechnic). There, these data are processed and verified as to their credibility. Subsequently, they are sorted according to the measurement date and, if needed, displayed on the monitor of an authorized user by a common WWW page browser.

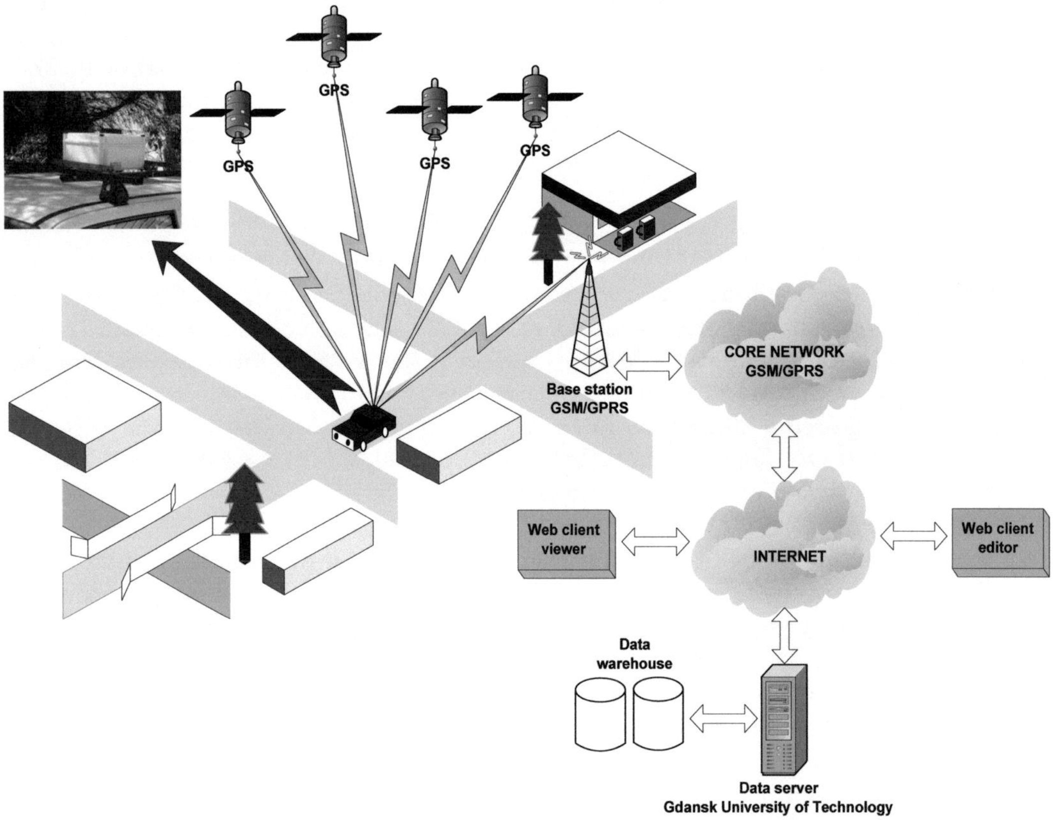

Fig. 1. The concept of the mobile monitoring system for chemical agents control in the air.

To achieve simplicity of operation of the monitoring system, the configuration process of ARPOL unit and the WWW server is kept to a minimum. This means that this system can be implemented on a wide range of servers.

The proposed monitoring unit can be described as maintenance-free. Properly configured and set-up monitoring station can operate for several months without intervention. Information about the state of a device and possible errors in the functioning of electronic components or problems with data collection are transmitted to respective person via short messages (SMS).

8. Sensors description

Road traffic is responsible for emission of several air pollutants, the most important being nitrogen oxides (NO and NO_2, together called NO_x) which may harm the lungs and trigger asthma, sulphur dioxide (SO_2), particulate matter (PM), carbon monoxide (CO), carbon dioxide (CO_2) and volatile organic compounds (VOCs) which can contribute to the formation of secondary pollutants with different efficiencies. The measurement of particulate matter in air requires some air filtering and mass measuring during specified time which cannot be made using semiconductor sensors and is not applicable to mobile monitoring units. Other air pollutants can be measured using thick film semiconductor sensors. Prototypes of monitoring units for proposed systems use selected sensors made by Figaro:

- TGS2442 – CO sensor, response: resistance of SnO_2 layer on electrical insulation layer. Response time: tens of seconds;
- TGS4161 – CO_2 sensor, response: electromotive force which is the result of electrochemical reaction of CO_2 and electrodes made of gold and lithum carbonate with cation (Na+) solid electrolyte. Response time: up to 1 minute, recovery time: up to 2.5minutes;
- TGS823 – VOC sensor. This sensor reacts to the presence of several volatile organic compounds, such as: Benzene (C_6H_6), Ethanol, n-Hexane, Acetone, Isobutane and others, but is used mostly for detection of Benzene which is found in highest concentrations. Response: resistance of SnO_2 layer, response time: tens of seconds;
- TGS2106 – NO_2 sensor, response: resistance of metal-oxide semiconductor layer. Response time: several seconds;
- TGS2201 – NO_x and CO sensor (double) used only for NO_x measurement, response: resistance of metal-oxide semiconductor layer. Response time: tens of seconds.

Some of these sensors were designed for measurement of gases in ppmv (parts per million,10^{-6}) concentration, but the concentration of NO_x, CO or C_6H_6 in typical air condition is of the order of ppbv (parts per billion, 10^{-9}). However, comparison of results from several months of measurements proved that after proper calibration and compensation of air temperature and humidity, they are enough stable and precise for determining the sources and regions of highest pollution caused by road traffic.

9. Air monitoring unit

Block diagram of mobile monitoring unit is presented in Fig.2. This version of measurement device is made of sensor unit, A/D converters and microcontroller to convert sensor's electric response to gas concentrations. Microcontroller is also responsible for linking measurement results with geographic coordinates from GPS and sending these data to server using GSM/GPRS link. The interior of one of monitoring units built in Gdansk University of Technology laboratory is shown in Fig.3.

The gas concentration sensors together with temperature and humidity sensors are put inside manifold (at the top of Fig.3). Air flow inside manifold is forced by fan, which rotate speed is controlled by microprocessor because change in flow rate will change gas sensor's substrate temperatures as air is cooling the casings of sensors.

Electrical response of gas sensors is converted into voltage signals using very high impedance operational amplifier (for EMF output) or by resistive voltage divider (for

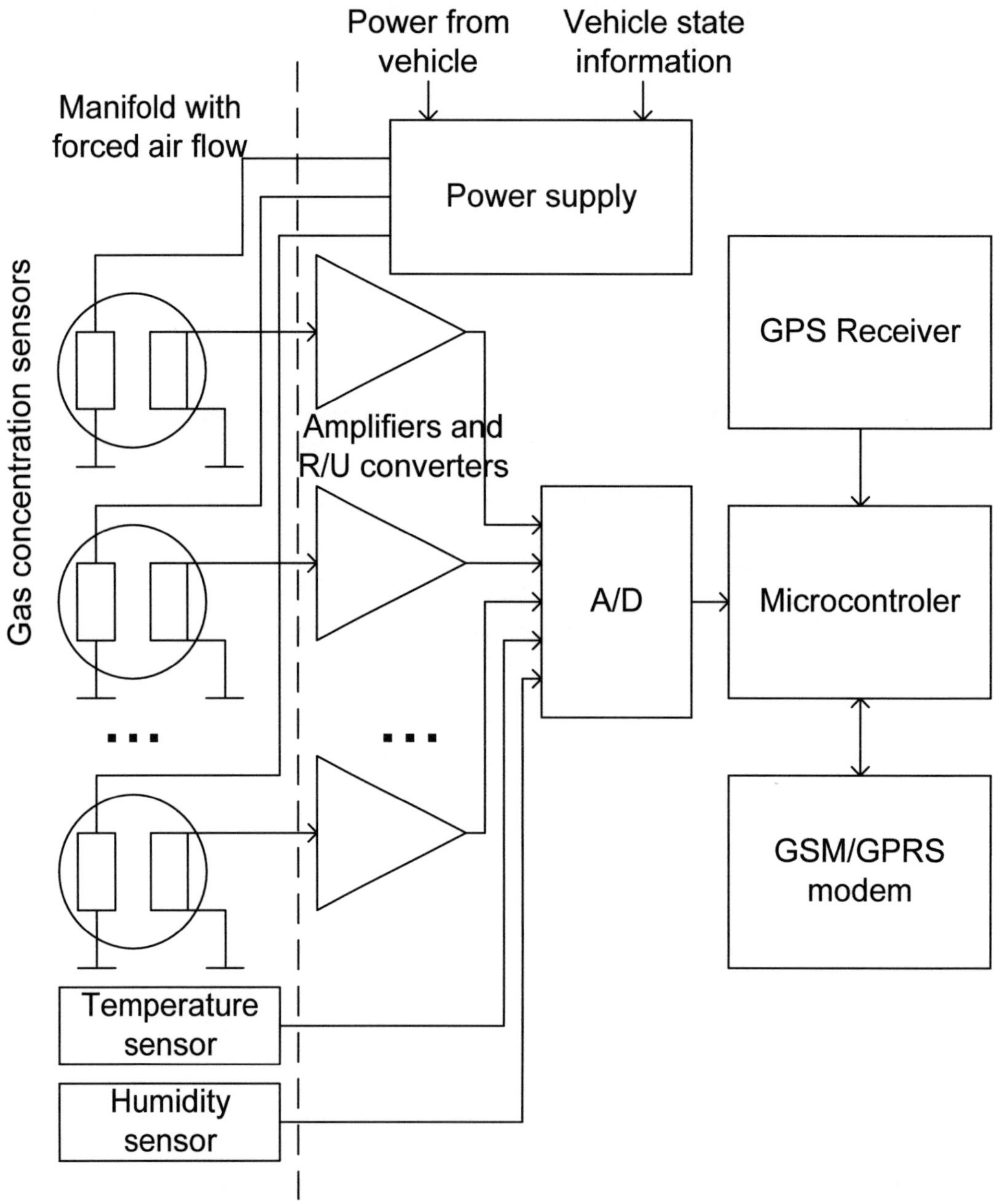

Fig. 2. Block diagram of mobile monitoring unit.

Fig. 3. Mobile monitoring unit prototype.

resistive output). Then the signals are passing low-pass filters with cutoff frequency about 10Hz for all sensors except TGS2442 which requires pulsed resistance measurement and therefore filter cutoff frequency has to be greater than 100Hz. Additional noise-filtering is made by microprocessor which averages digital values after A/D conversion during 10 seconds period. Averaged digital values are being transmitted to server without any additional conversion. Calculation of gas concentration is made by server using nonlinear equations derived during calibration procedure. Other data sent to server in every packet are:

- monitoring unit identification number;
- geographical coordinates of measurement point (from GPS receiver);
- speed and direction of movement (from GPS receiver);
- date/time of measurement (from GPS receiver);
- temperature and humidity of air in manifold;
- temperature of electronic circuits outside manifold;
- external or internal power supply voltage;
- state of vehicle engine in case of mounting the device on car/bus;
- error flags which indicate condition of monitoring unit.

Data transmission uses GPRS/EDGE radio link and TCP/IP protocol stack. To avoid writing special software at server side, data records are encoded as special HTTP request. Such requests are easily handled by PHP script and after decoding raw data records are written to MySQL database.

10. Conclusions

Polluted air negatively influences health and in some cases may even lead to death. Therefore, the issue of air quality is now a major concern for many countries which have been working to improve air quality by controlling emissions of harmful substances into the atmosphere, improving fuel quality, and by integrating environmental protection requirements into the transport and energy sectors. Despite these improvements in air quality over recent years, the problem of air pollution still remains. Therefore, more needs to be done at local, national, European and international levels.

Monitoring of air pollution is a prerequisite of air quality control and is carried out by a wide variety of analytical methods employing different measurement instruments which have different sensitivities and specificities. Monitoring plays a critical role in protecting the environment and is a key element of all actions related with the management and protection of ambient air.

According to existing law, member countries of the European Union are obliged to carry out measurements of atmospheric air cleanness and to introduce preventive measures against excessive pollution exceeding allowable threshold limits. To meet these obligations, the member countries are required to prepare inventories of zones and agglomerates where transgression of limiting norms occurs, and to prepare and introduce plans and programs how to adhere to the parameters required by law. Thus the mobile monitoring system described here, called ARPOL, unique on the national scale, can upgrade the existing system of measuring atmospheric air cleanness and facilitate taking appropriate decisions through analysis of various emission scenarios.

The developed device forms part of an expert system which undoubtedly contributes to the enhancement of ecological safety in a given area. It may be a cheap and useful tool (considering low investment and exploitation costs) for assessing ecological effects of new efforts for limiting pollutants emissions rather than for assessment of air quality. It contributes to more careful environmental planning by extending the measurement basis with information from non-stationary sources.

11. References

Environmental Protection Agency (2008). Clean Air Act, in: *U.S. Code*, 03.2011, Available from http://www.epa.gov/air/caa/

World Health Organisation (2000). Air Quality Guidelines for Europe, 2nd edition, WHO Regional Publications European Series No.91, Copenhagen, 2000.

European Union (2008). Directive 2008/50/EC of the European Parliament and of the Council of 21 May 2008 on ambient air quality and cleaner air for Europe, Official Journal of the European Union, 11.6.2008, L. 152/1-152/44

WHO (2005). WHO Air quality guidelines for particulate matter, ozone, nitrogen dioxide and sulfur dioxide - Global update 2005 - Summary of risk assessment

ARMAAG. Emission limits values (in Polish), 03.2011, Available from http://armaag.gda.pl/normy.htm

Griffin, R.D. (2007). *Principles of air quality management*, Taylor & Francis Group, Boca Raton, 2007

Friedrich, R. & Reis, S. (2004). *Emission of Air Pollutants*. Springer_Verlag Berlin Heidelberg, 2004

European Union (1996). Council Directive 96/62/EC of 27 September 1996 on ambient air quality assessment and management . Official Journal L 296 , 21/11/1996 P. 0055 - 0063

Wardencki, W., Katulski, R. Stefański J. & Namieśnik, J. (2008). *The state of the art in the field of non-stationary instruments for the determination and monitoring of atmospheric pollutants*. Critical Reviews in Analytical Chemistry, 2008, 38, 1-10.

Bukowiecki, N., Dommen, J., Prevot, A.S.H, Richter, R., Weingartner, E. & Baltensperger U. (2002). *A mobile pollutant measurement laboratory-measuring gas phase and aerosol ambient concentrations with high spatial and temporal resolution*. Atmospheric Environment, 2002, 36, 5569-5579.

Gouriou, F., Morin, J-P. & Weill, M-E. (2004). *On-road measurements of particle number concentrations and size distributions in urban and tunnel environments Atmospheric Environment*, 2004, 38 (18), 2831-2840.

Pirjola, L., Parviainen, H., Hussein,T., Valli, A., Hameri, K., Aaalto, P., Virtanen, A., Keskinen, J., Pakkanen,T.A., Makela,T. & Hillamo, R.E. (2004). *''Sniffer''-a novel tool for chasing vehicles and measuring traffic pollutants*. Atmospheric Environment, 2004, 38, 3625-3635.

Katulski, R., Stefański, J., Wardencki, W. & Żurek J. (2006). Concept of the mobile monitoring system for chemical agents control in the air, Proceedings of the *IEEE Conference on Technologies for Homeland Security – Enhancing Transportation Security and Efficiency*, June 7-8, 2006, Boston, USA, 181-184.

Katulski, R., Stefański, J., Wardencki, W. & Żurek J. (2007). The Mobile Monitoring System (MMS) – a Useful Tool for Assessing Air Pollution in Cities, Proceedings of the *Pittsburgh Conference on Analytical Chemistry and Applied Spectroscopy PITTCON'2007*, February 25 – March 2, 2007, Chicago, USA.

Seakins, P.W, Lansley, D.L, Hodgson, A., Huntley A. & Pope, F. (2002), *New Directions: Mobile laboratory reveals new issues in urban air quality*, Atmospheric Environment 2002, 36, 1247-1248.

Bogo H., Negri R.M. & San Roman, E. (1999). *Continuous measurement of gaseous pollutants in Buenos Aires city*. Atmospheric Environment, 1999, 33, 2587 – 2598.

8

Influence of the Air Engine on Global Warming Issues - 21st Century Fuel Technology

Bharat Raj Singh[1] and Onkar Singh[2]
[1]*SMS Institute of Technology, Lucknow*
[2]*Harcourt Butler Technological Institute, Kanpur*
India

1. Introduction

Global warming or climate change is undoubtedly one of the most important challenges for our future generation, and quite possibly any generation in history. The worldwide scientific community is unanimous in its agreement that global warming is happening, that is our fault, and that the opportunity to stop it is slipping away. If we let it get out of our control, the consequences - which are already evident in most of our lifetimes - will be catastrophic. For example some of the consequences that can be reasonably expected are rising sea levels, frequent and severe natural disasters, large-scale food and water shortages, plagues, massive species extinctions, unprecedented numbers of refugees, intensified ethnic and political tensions, and a global economic depression the likes of which no one has ever seen.

1.1 Current situation

The situation is still within our grip, but we must act now, we must act strongly, and we must act together. Individuals, companies, and governments across the globe must each do what they can to reverse climate change. We will never get a second chance.

1.2 What can we do?

There are plenty of things that one can reduce the alarming carbon footprint shown in **Fig.1**. The key word is *reduce*. Anyone can greatly lessen the impact on climate change by using the planet's resources more responsibly. There are many things which can be reduced, and many ways that can reduce them, but three of the major ones are: reduce the amount of animal products being consumed (e.g., meat, dairy, eggs, leather, etc.), reduce the amount of fuel being used in cars, air travels, etc., and reduce the amount of electricity being used for domestic as well as industrial purposes utilizing some alternatives. There are plenty of good resources as detailed below encourages to-do the research.

Since the use of transport has become a part of life in the current civilization of developed and developing countries, the population of transport vehicles are increasing 2 to 3 times in every 5-7 years. Approximately 80% transport population in the developing countries such as; India, China, Taiwan, Romania, Bulgaria etc., are two wheelers, that is also adding about 77.8% of total emission in the atmosphere.Hence the transport sector alone, is responsible to the faster consumption of hydrocarbon fuel and releases heavy tail pipe emissions, thereby adding billion tonnes of excessive carbon dioxide, carbon monoxide and unburnt hydrocarbon etc. (see **Table-1**), in the atmospheric air every day. This is causing a serious threat to the global warming. Also the fuel reserves are depleting very fast.

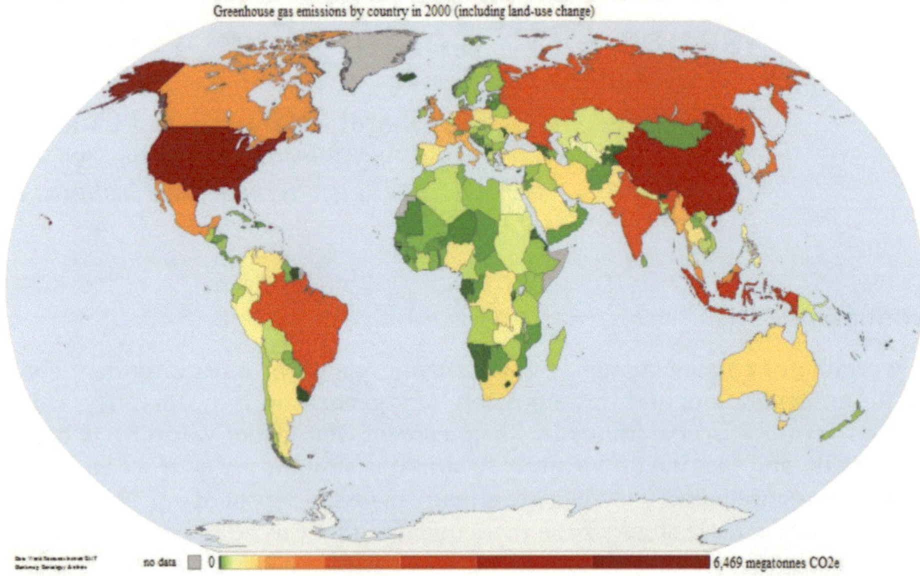

Fig. 1. Contrywise Green House Emission (Carbon Foot Print)

S.No.	Name of Major Countries	Population In Millions	Total CO2 Yearly Release (In Million Tonnes)	Remarks
1.	China	1334.8	6,126.7	
2.	USA	311.3	5,983.0	
3.	**Russia**	**138.9**	**1,572.4**	
4.	**India**	**1183.3**	**1,526.5**	
5.	**Japan**	**126.6**	**1,316.6**	
6.	**Germany**	**82.1**	**878.5**	
7.	Canada	33.8	561.0	
8.	Iran	77.5	519.3	
9.	Italy	57.9	486.4	
10.	South Korea	48.7	475.8	
11.	Mexico	113.5	440.0	
12.	South Africa	49.2	415.3	
13.	France	64.6	409.6	
14.	Australia	21.6	392.0	
15.	Saudi Arabia	26.0	385.6	

(Source: www.Breathingearth.net Feb'2011)

Table 1. List of 15-Countries Contributing Highest CO_2

In 1956, the Marion King Hubbert a noted geophysicist predicted that US Fuel reserves may peak by 1975 and fuel crisis will be noticed by 1995. He illustrated the projection with a bell shaped *Hubert Curve* based on the availability and consumptions of the fossil fuel. Large fields are discovered first, small ones later. After exploration and initial growth in output, production plateaus and eventually declines to zero (Hubbert, 1995). Thereafter in 2003, Aleklett and Campbell expressed their views that most of the countries will pass through peak oil days by 2010-12 and fuel consumption will reach to 80% by 2020-30 with the current rate of consumption (Alekett & Campbell, 2003).
In India, vehicular pollution is estimated to have increased eight times over the last two decades. This source alone is estimated to contribute about 70 per cent to the total air pollution. With 243.3 million tons of carbon released from the consumption and combustion of fossil fuels in 1999, India is ranked fifth in the world behind the U.S., China, Russia and Japan. India's contribution to world carbon emissions is expected to increase in the coming years due to the rapid pace of urbanization, shift from non-commercial to commercial fuels, increased vehicular usage and continued use of older and more inefficient coal-fired and fuel power-plants. Thus, peak oil year may be the turning point for mankind which in turn lead to the end of 100 year of easy growth, if self-sufficiency and sustainability of energy is not maintained on priority. It may end up a better world as per proceedings of conference held in Architectural Institute, 2004, Concept and Technology- First International Workshop on Sustainable Habitat Systems, at Japan. Presently major thrust is being given to explore wind energy, hydro-power, tidal and nuclear power generation etc. Efforts are also given to energy storage system for the clean energy by conversion system and its better utilization to run prime-moves for light vehicles.
This paper describes about the energy conversion and energy storage system. One of the dominant option is storage of compressed air from easily available atmospheric air which may be compressed by electricity or by alternative energy sources like wind, solar energy etc. or disaster energy sources and stored in the air receiver or storage tanks of suitable size. Such energy could be reutilized as clean energy source for running prime-movers of domestic appliances and light vehicles as a nearly zero pollution fuel sources. The design of air turbine / engine and its different parametric and performance aspect are also elaborated in this article. To maintain sustainability in 21st Century and check the global warming issues to the extent of 50-60%, the storage energy systems and its utilization to run the lighter vehicles or motorbikes by novel and efficient air engine / turbine could be one of the dominant technology. This may also leads to environmentally and ecologically better future.

2. Types of energy storage system

The Power Conversion System (PCS) is a vital part of all energy storage systems. It interfaces the energy storage, the energy storage device and the load (the end-user). PCS cost is significant and it can be greater than 25% of the overall energy storage system. PCS cost range from Rs. 4,700/kW for UPS markets to Rs.56, 000/kW for Standalone markets have been seen. Some of the major PCS markets include:

- Motor drives
- Power supplies
- UPS (uninterruptible power supply)
- Electric vehicles
- Inverters/Converters for solar-hybrid systems

- Micro-turbines
- Fuel cells
- Wind turbines

However, power conversion system technology has been evolving slowly due to the limited distributed energy resource (DER) market. As a result, Energy Storage System cost has been high with low profit margins and the manufacturing volume has been low impacting reliability and quality of the Power Conversion System designs. What is needed is a significant reduction in overall cost with improved reliability, development of state-of-the-art Power Conversion System with multiple uses, which increases production volumes for DER applications, improve controls and adaptability, and improve manufacturing.

2.1 Battery energy storage system

Sandia National Laboratory has a broad technical base of battery expertise focusing on integrated storage systems. These storage systems operate in varying environments and electrical conditions. In these storage systems there are many different types of battery technologies. With the different designs are having advantages under specific operational conditions. It is important to understand the capabilities and limitations of each storage technology.

- Lead-Acid Battery- short cycle life
- Li-Ion -Lithium Ion Battery-High energy and 100% efficiency
- NaS -Sodium Sulfur Battery- can run at high temperature of 300 deg centigrade.
- PSB - Polysulfide Bromide Flow Battery- 75% efficiency
- VRB -Vanadium Redox Flow Battery- 85% efficiency.

2.2 Super capacitor energy storage

Electrochemical capacitors (EC) store electrical energy in the two series capacitors of the electric double layer (EDL), which is formed between each of the electrodes and the electrolyte ions. The distance over which the charge separation occurs is just a few angstroms. The capacitance and energy density of these devices is thousands of times larger than electrolytic capacitors. The electrodes are often made with porous carbon material. The electrolyte is either aqueous or organic. The aqueous capacitors have a lower energy density due to a lower cell voltage but are less expensive and work in a wider temperature range. The asymmetrical capacitors that use metal for one of the electrodes have a significantly larger energy density than the symmetric ones and have lower leakage current.

2.3 Fly wheel energy storage

Most modern flywheel energy storage systems consist of a massive rotating cylinder (comprised of a rim attached to a shaft) supported on a stator by magnetically levitated bearings that eliminate bearing wear and increase system life. To maintain efficiency, the flywheel system is operated in a low vacuum environment to reduce drag. The flywheel is connected to a motor/generator mounted onto the stator that interact with the utility grid. Some of the key features of flywheels are little maintenance, long life (20 years or 10s of thousands of deep cycles) and environmentally inert material. Flywheels can bridge the gap between short term ride-through and long term storage with excellent cyclic and load following characteristics. The choice of using solid steel versus composite rims is based on the system cost, weight, size, and performance trades of using dense steel (200 to 375 m/s

tip speed) vs. a much lighter but stronger composite that can achieve much higher rim velocities (600 to 1000 m/s tip speed). Actual delivered energy depends on the speed range of the flywheel as it cannot deliver its rated power at very low speeds. For example, over 3:1 speed range, a flywheel will deliver ~90% of its stored energy to the electric load.
While high-power flywheels are developed and deployed for aerospace and UPS applications, there is an effort, pioneered by Beacon Power, to optimize low cost commercial flywheel designs for long duration operation (up to several hours). 2kW / 6kWh systems are in telecom service today. Megawatts for minutes or hours can be stored using a flywheel farm approach. Forty 25kW / 25 kWh wheels can store 1MW for 1 hour efficiently in a small footprint.
The stored energy can be approximated by:

$$E = (Iw^2)/2 = (mr^2w^2)/2 = (mv^2)/2$$

where w is the rotational velocity (rad/sec), I the moment of inertia for the thin rim cylinder, m is the cylinder mass and v is linear rim velocity.

2.4 Compressed air energy storage

The Technology of air engine is not new.The Sterling air engine was developed in 1790-1810, but due to its limitation no much work was carried out. In view of fire problems in Coalmines and other volatile places, where high flammable fuel like fossil fuel vehicles are not adviseable, compressed air operated vehicles are normally being put in use. Now from last two decades major thrust is being given by the researchers for development of compressed air engine. Some technical developments, which may be considered to work on 21st Century Energy Storage system and can work on compressed air or hybrid system as an alternative to fossil fuel for running light vehicles, are listed below:-

2.4.1 Reciprocating compressed air engine

Guy Negre, a French Scientist, in 1998 developed compressed air- 4- cylinders engine run on air and gasoline, claims zero pollution cars and got 52- patents registered since 1998 to 2004.The car was demonstrated in Oct.'2004 publically(Negre, 2004).

2.4.2 Rotary hot air motor (Quasi-turbine)

Saint Hilaire G., an inventor developed zero pollution cars using Quasiturbine with a set of 14- engines parameters and disclosed on Sept'2005 using gasoline (Saint-Hilaire, 2005). In the basic single rotor Quasiturbine engine, an oval housing surrounds a four-sided articulated rotor, which turns and moves within the housing. The sides of the rotor seal against the sides of the housing, and the corners of rotor seal against the inner periphery, dividing it into four chambers.

3. Compressed air as alternative and zero emission source to fossil fuel

3.1 History of air engine

The compressed air as an energy and locomotion vector is precisely not a recent technology. In fact at the end of the 19th century the first approximations to what could one day become a compressed air driven vehicle already existed, through the arrival of the first pneumatic locomotives. Yet even two centuries before that Dennis Papin apparently came up with the idea of using compressed air (Royal Society London, 1687).

The first recorded compressed-air vehicle in France was built by the Frenchmen Andraud and Tessie of Motay in 1838. A car ran on a test track at Chaillot on the 9th July 1840, and worked well, but the idea was not pursued further. In 1872 the Mekarski air engine was used for street transit, consisting of a single-stage engine. It represented an extremely important advance in terms of pneumatic engines, due to its forward thinking use of thermodynamics, ensuring that the air was heated, by passing it through tanks of boiling water. Numerous locomotives were manufactured and the first in Nantes in 1879. The H. K. Porter Company in Pittsburgh sold hundreds of these locomotives to coal-mining companies in the eastern U.S. With the hopeful days of air powered street transit over, the compressed air locomotive became a standard fixture in coal mines around the world because it created no heat or spark and was therefore invaluable in gassy mines where explosions were always a danger with electric or gas engines.

Also in 1896, Porter supplied ten compressed air motor cars for the Eckington System in Washington, D.C. There was a tank on the front of the engine and it was recharged at the station. Between 1890 and 1902 ten compressed air trams circulated in Bern, Switzerland. In 1892, Robert Hardie introduced a new method of heating that at the same time served to increase the range of the engine. However, the first urban transport locomotive was not introduced until 1898, by Hoadley and Knight, and they introduced a two stage engine. Later on, in 1912 the American's method was improved by Europeans, adding a further expansion stage to the engine (3 stages).

In 1926, Lee Barton Williams of Pittsburg USA presented his invention: an automobile which he claimed run on air. The motor starts on gasoline, but after it has reached a speed of ten miles an hour the gasoline supply is shut off and the air starts to work. At the first test his invention attained a speed of 62 miles an hour.

In January 1932 what appears to be the first journalistic article ever written about a car driven by compressed air was published. In 1934, a 21-year-old researcher Johannes Wardenier announced the development of world's first fuel-less automobile. For weeks Dutch newspapers reported of an incredible invention that would change the world for ever.

After the Second World War the term "air engine" was never used again in textbooks in reference to compressed air or pneumatic locomotives and, whenever they mentioned it the article would go on to state that these engines are of little use or efficiency.

In 1970's, Willard Truitt presented his invention in McKees Rocks, USA. But because he did not have the financial means to develop his compressed air car further, he gave the rights of his invention to NASA and the US Army in 1982.

In 1979, Terry Miller decided that compressed air was the perfect medium for storing energy. He developed Air Car One, which he built for $ 1,500. Terry's engines showed that it was feasible to manufacture a car that could run on compressed air. He patented his method in 1983 (US4370857).

In the 1980's Carl Leissler developed a motor that was able to function on air. The retired horticulturalist had been working from his garage in Hollywood for over 15 years. He says that to use his motor in a car you might have to use a small electric or gas energy source to help drive the air compressor. 'We might be able to get 2000 miles per gallon; air is a power in itself' Leissler comments.

Until 1987 the German company Arnold Jung Lokomotivenfabrik GmbH produced locomotives functioning on compressed air to be used in mines. In the 1980's they were still selling and renovating locomotives.

Currently the tram association in Bern Switzerland (BTG) is developing a locomotive according to the original plans.
At present (2008) various researchers and industries are developing compressed air engines/motors applicable to transportation, apart from the many industries that produce and commercialize compressed air motors for industrial purposes.

3.2 Advantages / disadvantages of air engine

There are advantages and disadvantages of air engines in comparison to the electric motors.

3.2.1 Advantages

- Air turbines do not need any electric power to run and hence, they can easily be used in volatile atmospheres also.
- Air turbines of the same size as that of electric motor can deliver more power as power density of the air engine is higher and thus, air engine the same size as that of electric motor can deliver more power.
- The air turbines can operate without any type of auxiliary speed reducers unlike electric motors.
- Overloads cause no harm to the air turbines whereas in the case of electric motor, it can easily trip the circuits, which have to be restored.
- Air engines can easily be controlled by simple flow control by means of valves whereas electric motor need electronic speed controls. By regulating the pressure, the torque produced by the air turbines can easily be varied.
- Air turbines do not need any magnetic starters like the electric motors.
- Air turbines generate very less heat as compared to the electric motors.

3.2.2 Disadvantages

- There are some drawbacks of air engines as the flow of air or the pressure is disturbed; the torque being produced by the air engines suffers in a great way.

3.3 Working principles of air engine / air turbine

3.3.1 Thermodynamic principle

The high pressure jet of air when injected into the rotor of vaned air turbine causes its rotation. Rotor movement is caused due to impact of air jet and the expansion of high pressure air admitted in turbine. Thus the high pressure air entering through the inlet passage pushes the vane for producing rotational movement through this vane and thereafter air so collected between two consecutive vanes of the rotor is gradually expanded till the rotation brings vanes in front of exit passage. This can be considered like isobaric admission and adiabatic expansion of high pressure air both contributing in producing the shaft work from air turbine. Compressed air leaving the air turbine after expansion is sent out from the exit passage.
The proposed air turbine is considered to work on the reverse working principle of vane type compressor and transforms the energy of the compressed air into shaft work. The multi-vane type air turbine has casing diameter 100 mm and rotor diameter 75 mm as shown in **Fig. 2**. In this arrangement total shaft work is cumulative effect of isobaric admission of compressed air jet on vanes and the adiabatic expansion of high pressure air.

In an earlier study conducted by authors a prototype of air turbine was developed (Singh & Singh, 2008c).

A cylinder for the storage of compressed air with a capacity of storing air for the requirement of 30 minutes running at initial stage and maximum pressure of 20 bar is used as a source of compressed air. The compressed air storage cylinder is designed to produce constant pressure for the minimum variation of torque at low volumes of compressed air and attached with filter, regulator and lubricator. The clean air then admits into air turbine through inlet nozzle. Vanes of novel air turbine are placed under spring loading to maintain their regular contact with the casing wall to minimize leakage which is proposed as improvement over the currently available vane turbine. A study on high efficiency energy conversion system for liquid nitrogen (Honton,2004) design and verification of airfoil and its tests, influence of tip speed ratios for small wind turbine and parabolic heat transfer and structural analysis were also carried out for conceptualizing the energy conversion system and design of the air turbine (Faglsang, 2004; Gorla, 2004; Knowlen, 1998; Schreck, 2004; Selig, 2004). Studies have shown feasibility of vane type novel air turbine (Singh & Singh 2006; 2007; 2008 a, b; 2009 a-f).

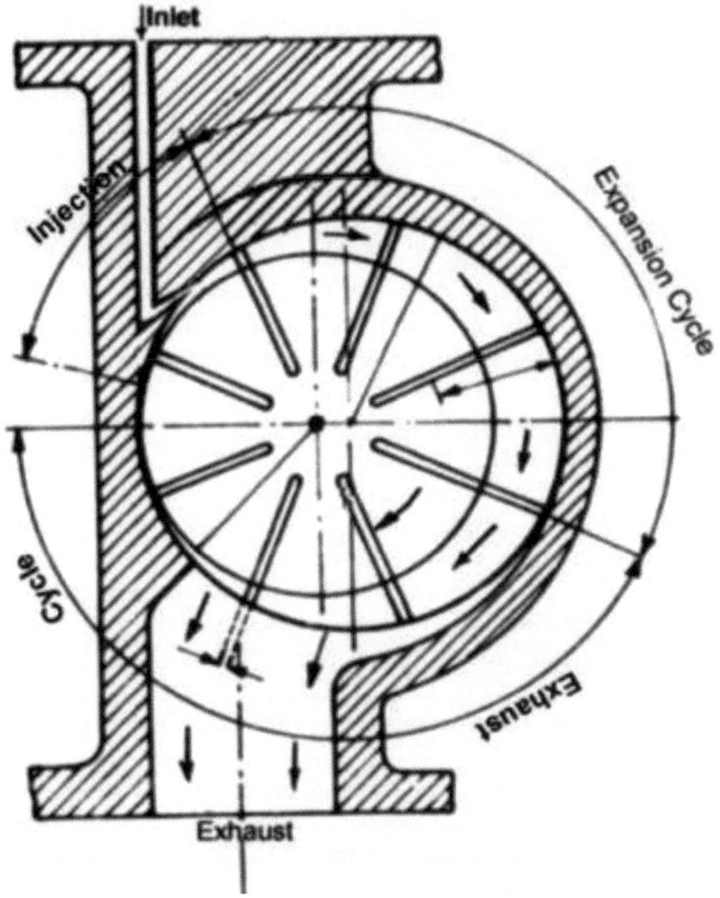

Fig. 2. Air Turbine- Model

The present objective is to investigate the performance of an air turbine with the variation of injection angle, i.e. angle at which air should be admitted into the turbine between first two consecutive vanes. The air turbine considered here has capability to yield output of 5.25 to 6.50 HP at 4-6 bar air pressure for speed of 2000–2500 rpm, which is suitable for a motorcycle.
It is assumed that the scavenging of the rotor is perfect and the work involved in recompression of the residual air is absent. The assumptions made for thermodynamic analysis are as under.

- Admission of air takes place isobarically.
- High pressure air admitted in turbine rotor expands adiabatically.
- There is no residual air in rotor after the expanded air is discharged from rotor. Thus no recompression of air takes place.
- Loss of work due to rubbing friction between vanes and casing walls is neglected.
- Leakage of air across the vane tip and rotor casing contact interface is negligible.
- Mechanical losses due to shaft friction are negligible.
- Trace of the lubricant added to the air being admitted is negligible compared to total mass flow of air.
- Loss due to throttling of air at admission and exit is negligible.
- Exit of expanded air occurs isochorically
- Flow across the turbine is of steady flow type.

3.3.2 Mathematical modeling

The mathematical model shown here is already presented in the author's publications (Singh & Singh, 2010a, b, c, d, e, f, g and h). But for the benefits of readers it is again reproduced here in brief. The high pressure jet of air at ambient temperature drives the rotor in novel air turbine due to both isobaric admission and adiabatic expansion. Such high pressure air when enters through the inlet passage, pushes the vane for producing rotational movement through this vane and thereafter air so collected between two consecutive vanes of the rotor is gradually expanded up to exit passage. This isobaric admission and adiabatic expansion of high pressure air both contribute in producing the shaft work from air turbine. Compressed air leaving the air turbine after expansion is sent out from the exit passage. It is assumed that the scavenging of the rotor is perfect and the work involved in recompression of the residual air is absent.
From **Fig. 3**, it is seen that work output is due to isobaric admission (E to 1), adiabatic expansion (1 to 4) and steady exit flow work (4 to 5). Thus, total work done due to thermodynamic process may be written as:
[Area under (E145CE)] = [Area under (E1BOE) +Area under (14AB1) – Area under (4AOD4) + Steady Flow (45CD4)]
Total Work output = [Thermodynamic expansion work (w_1)] + [Exit steady flow work (w_2)]

$$w = [(w_1) + (w_2)\} \tag{1}$$

Now thermodynamic expansion work (w_1), can be written as:

$$w_1 = p_1.v_1 + \left(\frac{p_1.v_1 - p_4.v_4}{\gamma - 1}\right) - p_4.v_4 \text{ or } w_1 = \left(\frac{\gamma}{\gamma - 1}\right)(p_1.v_1 - p_4.v_4)$$

For adiabatic process, $p.v^{\gamma} = p_1.v_1^{\gamma} = p_4.v_4^{\gamma}$ = constant

$$\text{or } v_4 = \left(\frac{p_1}{p_4}\right)^{\frac{1}{\gamma}}.v_1$$

Thus thermodynamic expansion work output would be:

$$w_1 = \left(\frac{\gamma}{\gamma - 1}\right).p_1.v_1.\left\{1 - \left(\frac{p_4}{p_1}\right)^{\frac{\gamma-1}{\gamma}}\right\} \tag{2}$$

The exhaust of expanded air to atmosphere (4-5) takes place after the expansion process (E-4) as shown in **Fig.3**. In this process; till no over expansion takes place pressure p_4 can't fall below atmospheric pressure p_5. Thus at constant volume when pressure p_4 drops to exit pressure p_5, no physical work is seen. Since turbine is functioning as positive displacement machine, hence under steady fluid flow at the exit of the turbine, the potential work is absorbed by the rotor and flow work (w_2), can be written as:

$$w_2 = \int_4^5 v.dp = v_4(p_4 - p_5) \tag{3}$$

Substituting equations (2), (3) into equation (1), the net work output will be:

$$w = (w_1 + w_2) = \left(\frac{\gamma}{\gamma - 1}\right).p_1.v_1.\left\{1 - \left(\frac{p_4}{p_1}\right)^{\frac{\gamma-1}{\gamma}}\right\} + (p_4 - p_5).v_4 \tag{4}$$

When air turbine is having ***n*** number of vanes, then shaft output can be written as:

$$w_n = n.\left(\frac{\gamma}{\gamma - 1}\right).p_1.v_1\left\{1 - \left(\frac{p_4}{p_1}\right)^{\frac{\gamma-1}{\gamma}}\right\} + n.(p_4 - p_5).v_4 \tag{5}$$

where w_n is work output (in Nm), for complete one cycle.

Therefore, the total power output (work done per unit time) (W_{total}), for speed of rotation N rpm will be mentioned as (Book; Singh O., 2009):

$$W_{total} = n.(N/60).\left(\frac{\gamma}{\gamma - 1}\right).p_1.v_1.\left\{1 - \left(\frac{p_4}{p_1}\right)^{\frac{\gamma-1}{\gamma}}\right\} + n.(N/60).(p_4 - p_5).v_4 \tag{6}$$

where $$W_{\exp} = n.(N/60).\left(\frac{\gamma}{\gamma - 1}\right).p_1.v_1.\left\{1 - \left(\frac{p_4}{p_1}\right)^{\frac{\gamma-1}{\gamma}}\right\}$$

and $$W_{flow} = n.(N/60).(p_4 - p_5)v_4$$

Fig. 2. shows that if vanes are at angular spacing of θ degree, then total number of vanes will be n = $(360/\theta)$. The variation in volume during expansion from inlet to exit (i.e. v_1 to v_4) can be derived by the variable extended length of vane as shown in Figure 3 at every point of movement between two consecutive vanes.

From **Fig. 4**, it is seen that when two consecutive vanes at OK and OL moves to position OH and OB, the extended vane lengths varies from SK to IH and LM to BG, thus the variable length BG at variable α_i is assumed as $X_{at'variable'\alpha}$ can be written from the geometry:

$$BG = X_{at,variable"\alpha"} = (1/2).D.cos\left[\sin^{-1}\left\{\left(\frac{D-d}{D}\right).\sin\alpha\right\}\right] + (1/2).(D-d).\cos\alpha - d/2 \quad (7)$$

where D is diameter of casing and d is diameter of rotor, α is angle $\angle$BOF, β is angle $\angle$BAF and θ is angle $\angle$HOB or $\angle$KOL, between two consecutive vanes and ϕ is angle $\angle$KOJ at which injection pressure enters the air turbine.

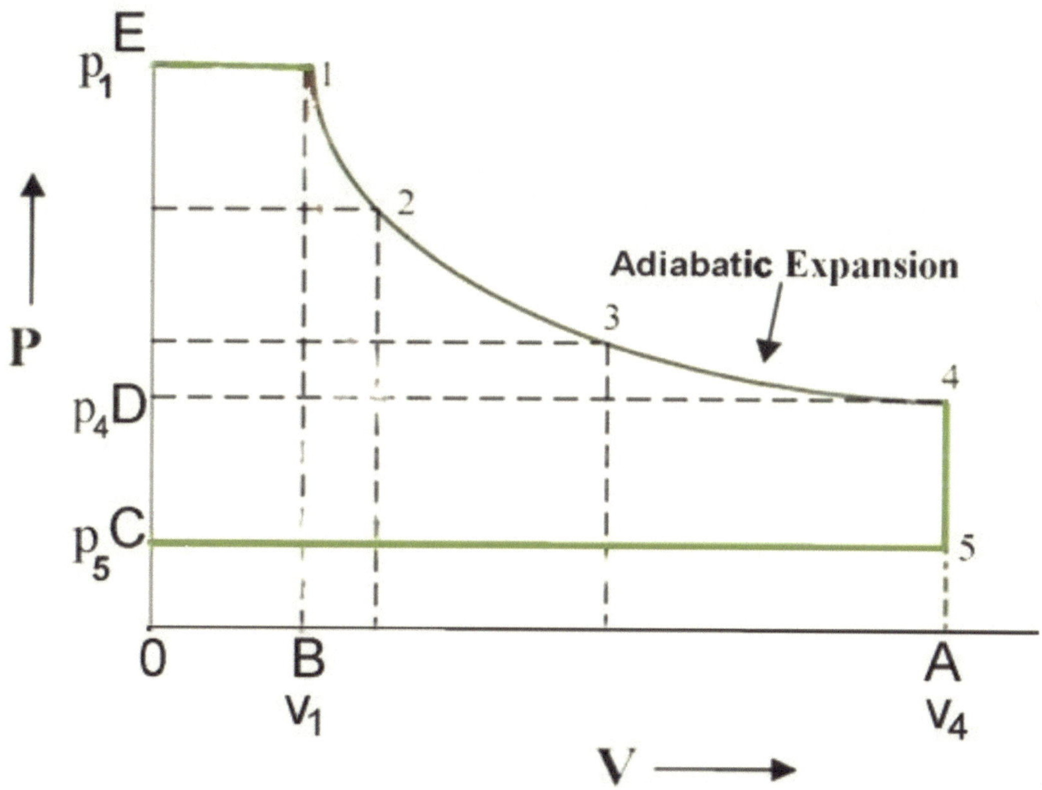

Fig. 3. Thermodynamic Processes (Isobaric, adiabatic and Isochoric Expansion)

Variable volume of cuboids B-G-I-H-B can be written as:

$$v_{cuboids} = L.\left\{\frac{(X_{1i} + X_{2i})(d + X_{1i})}{4}\right\}.\sin\theta \quad (8)$$

Where BG= X_{1i} and IH= X_{2i} are variable projected lengths of vanes when rotor rotates in the turbine as shown in **Fig. 4**.
The lengths (IG, HB and LK, SM.), are considered linear whereas all are chords of circles. This approximation is done in mathematical model which has very least impact on the overall values.
The volume at inlet v_1 or v_{min} will fall between angle $\angle$ LOF= $\alpha_{1min} = (180 - \theta - \phi)$ and angle $\angle$ KOF= $\alpha_{2min} = (\alpha_{1min} + \theta) = (180 - \phi)$ as seen in **Fig. 3**, when air is admits into turbine at angle ϕ .

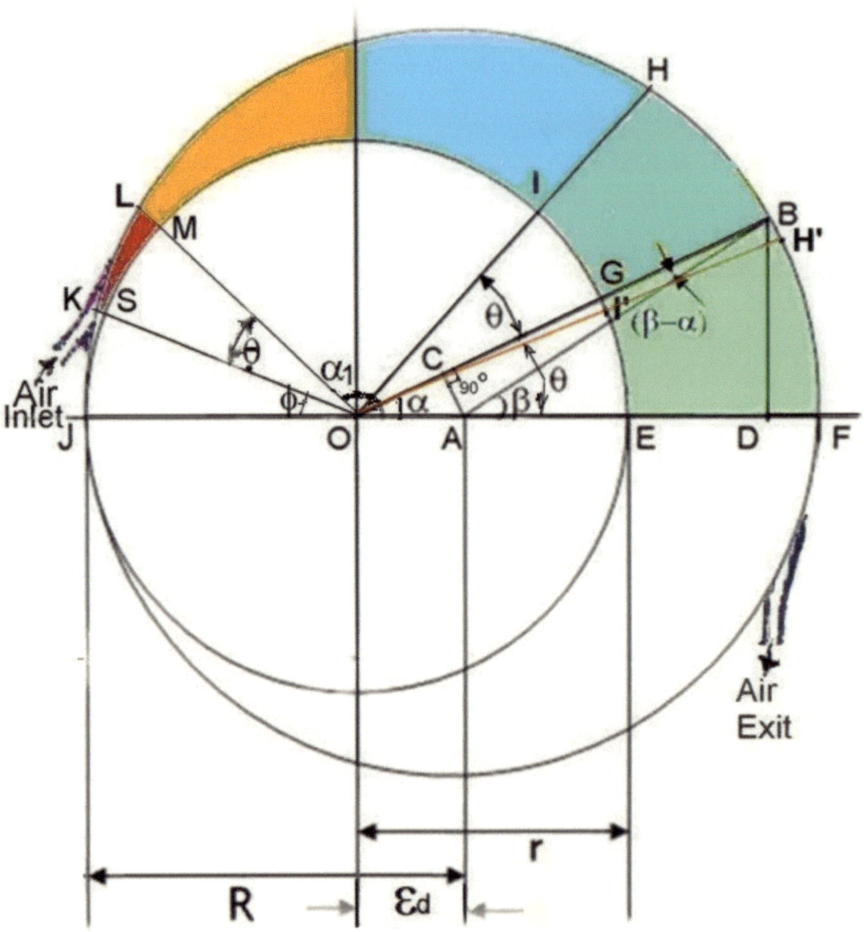

Fig. 4. Variable length BG and IH and injection angle ϕ

Applying above conditions into equations (7), then LM=X_{1min} and SK=X_{2min} can be written as:

$$X_{1\min} = R.cos\left[\sin^{-1}\left\{\left(\frac{R-r}{R}\right).\sin(180-\theta-\phi)\right\}\right]+\left[(R-r).\cos(180-\theta-\phi)-r\right] \quad (9)$$

$$X_{2\min} = R.cos\left[\sin^{-1}\left\{\left(\frac{R-r}{R}\right).\sin(180-\phi)\right\}\right]+\left[(R-r).\cos(180-\phi)-r\right] \quad (10)$$

Applying values of X_{1min} and X_{2min} to equation (8),

$$v_1 = v_{\min} = L.\left\{\frac{(X_{1\min}+X_{2\min})(2r+X_{1\min})}{4}\right\}.\sin\theta \quad (11)$$

The Volume at exit v_4 or $v_{\max}$ will fall between angle $\angle$ BOF $\alpha_{1\max}=\alpha=0$ and angle $\angle$ HOF $\alpha_{2\max}=(\alpha_{1\max}+\theta)=\theta$

Applying above conditions into equations (7), then FE=X_{1max}=Corresponding to BG at α =0 degree and I'H'=X_{2max}=Corresponding IH at $(\alpha+\theta)=\theta$ degree can be written as:

$$X_{1\max} = (D-d) = 2(R-r) \quad (12)$$

$$X_{2\max} = R.cos\left[\sin^{-1}\left\{\left(\frac{R-r}{R}\right).\sin\theta\right\}\right]+\{(R-r).\cos\theta\}-r \quad (13)$$

Applying values of X_{1max} and $X_{2\ max}$ to equation (8),

$$v_4 = v_{\max} = L.\left\{\frac{(X_{1\max}+X_{2\max})(2r+X_{1\max})}{4}\right\}.\sin\theta \quad (14)$$

Substituting values of v_1 and v_4 from equations (11) and (14) to equation (6), the total power output available W_{total} can be written as:

$$W_{total} = n.(N/60).\left(\frac{\gamma}{\gamma-1}\right).\left\{1-\left(\frac{p_4}{p_1}\right)^{\frac{\gamma-1}{\gamma}}\right\}p_1.\left[L.\left\{\frac{(X_{1\min}+X_{2\min}).(2r+X_{1\min})}{4}\right\}.\sin\theta\right]$$
$$+n.(N/60).(p_4-p_5).\left[L.\left\{\frac{(X_{1\max}+X_{2\max}).(2r+X_{1\max})}{4}\right\}.\sin\theta\right] \quad (15)$$

4. Input parameters and assumptions

In this study the vane angle (θ) of air turbine is considered 30°, 45°, 51.4°, 60°, 72° and 90° (i.e. 12, 8, 7, 6, 5 and 4 vanes respectively) and injection angle (ϕ) is kept 30°, 45° and 60°. The outer diameter of Casing is considered 200 mm, 150 mm and 100 mm and correspondingly Rotor diameter is chosen 150 mm, 100 mm and 75 mm correspondingly.

Other various input parameters are listed in **Table 2, 3, 4** for investigation of optimum shaft power output at different vane angles, and injection angles.

4.1 To Optimize the power by selecting various options of vane angles (θ)

4.1.1 Input parameters for investigation of effect of vane angle

Symbols	Parameters
$D=2R$	150 mm (outer)
$d=2r$	100 mm (inner) corresponding
p_1	2bar (=30psi), 3bar (=45 psi), 4bar (=60 psi), 5bar (=75 psi), 6bar (=90 psi)
p_4	$=(v_1 / v_4)^{\gamma} .p_1$ $> p_5$ assuming adiabatic expansion
p_5	1 atm = 1.0132 bar
θ	30°, 36°, 45°, 51.4°, 60°, 72°, 90° angles between 2- consecutive vanes (i.e. rotor contains correspondingly 12, 10, 8, 7, 6, 5, 4 vanes)
N	2500 rpm
L	35 mm length of rotor
γ	1.4 for air
n	Number of vanes (360 / θ)
$\varnothing$	Injection angles at which compressed air enters through nozzle into rotor *Case-1:* 30°, *Case-2:* 45° and *Case-3:* 60°

Table 2. Input Parameters for Investigation of Vane angles

4.1.2 Results and discussion

From **Fig. 5** to **Fig. 7**, it is noted that there is an optimum power output value for each set of injection pressure for different vane angles. Thus the optimum total power output is obtained as:

Case-1. For injection angle 30°

Total work power from **Fig. 5,** is seen to increase with increasing vane angles from 36° to 60° or with the decrease in number of vanes from 12 to 6 on rotor and thereafter it declines from vane angles (θ) = 72° to 90°. With increase in injection pressure the work output increases gradually as shown in graphical patterns. Thus the optimal total shaft output is found at vane angle (θ) = 60° (i.e. 6- vanes).

Case-2. For Injection angle 45°

Similarly, total power output from **Fig. 6,** is seen to increase with increasing vane angle up to 36° to 51.4° or with the decrease in number of vanes from 12 to 7 on rotor and thereafter it declines from vane angles (θ) = 60° to 90°. With increase in injection pressure the work output increases gradually as shown in graphical patterns. Thus the optimal total shaft output is found at vane angle (θ) = 45° and 51.4° (i.e. 8-7 vanes).

Case-3. For Injection angle 60°
Here also the total power output from **Fig. 7**, is seen to increase with increasing vane angle up to 36° or with the decrease in number of vanes from 12 to 10 on rotor and thereafter it declines for vane angles (θ) = 45° to 90°. With increase in injection pressure the work output increases gradually as shown in graphical patterns. Thus the optimal total shaft output is found at vane angle (θ) = 36° (i.e. 10 vanes).

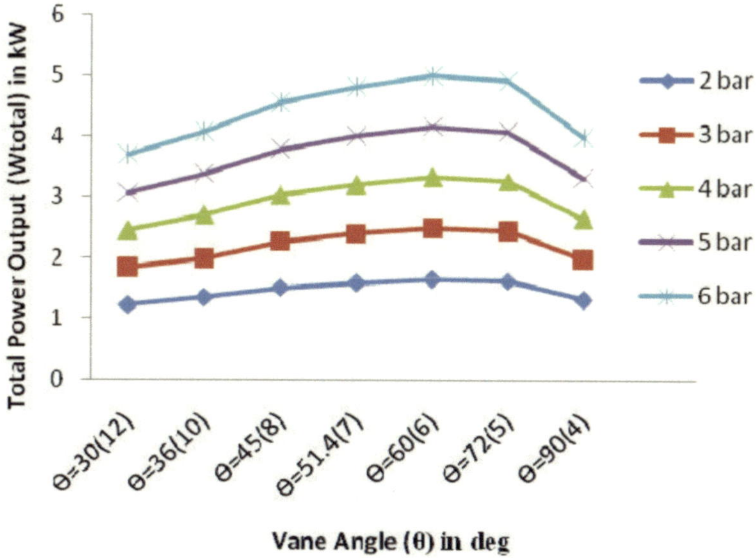

Fig. 5. Total power output versus different vane angles, at injection air pressure 2-6 bar, injection angle of 30° and speed of rotation 2500 rpm

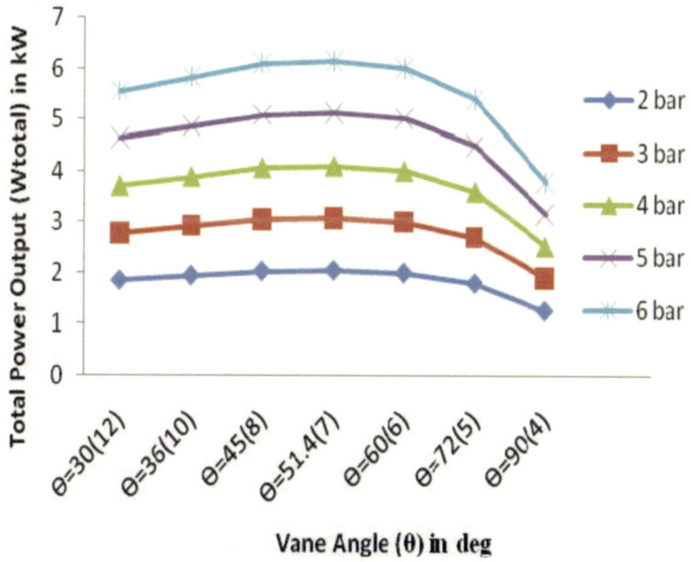

Fig. 6. Total power output versus different vane angles, at injection air pressure 2-6 bar, injection angle of 45° and speed of rotation 2500 rpm

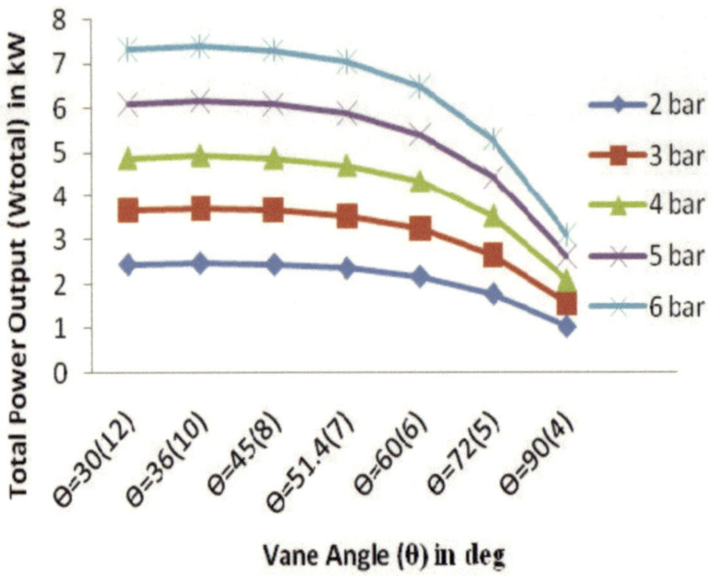

Fig. 7. Total power output versus different vane angles, at injection air pressure 2-6 bar, injection angle of 60° and speed of rotation 2500 rpm

Thus, total optimum power is observed as:
Case-1: 5.00 kW at vane angle (θ) = 60° (i.e. 6 vanes) when injection angle is kept 30°,
Case-2: 6.1-6.15 kW at vane angle (θ) = 45°, 51.4° (i.e. 8-7 vanes) when injection angle is 45° and
Case-3: 7.39 kW at vane angle (θ) = 36° (i.e. 10 vanes) when injection angle is 60°.

4.2 To optimize the power by selection of injection angle (Ø)

4.2.1 Input parameters for investigation of injection angle

Symbols	Parameters
D=2R	200 mm (outer)
d=2r	160 mm (inner) corresponding
p_1	2 bar (=30 psi), 3 bar (=45 psi), 4 bar (=60 psi), 5 bar (=75 psi), 6 bar (=90 psi)
p_5	1.0132 bar (atmospheric pressure)
p_4	$(v_1 / v_4)^{\gamma} .p_1 > p_5$ assuming adiabatic expansion
θ	45° (i.e. rotor contains correspondingly 8 number vanes)
N	2500 rpm (as total power is directly proportion to rpm)
L	35 mm length of rotor
γ	1.4 for air
n	Number of vanes = (360 / θ)
ϕ	10°, 15°, 20°, 25°, 30°, 45°, 60°, 75°, 90° angles at which compressed air enters through nozzle into rotor

Table 3. Input Parameters for Investigation of Injection angle

4.2.2 Results and discussion

From the **Fig. 8** it is observed that total power output is seen to increase with small pace at increasing injection angles from 10° to 30° at 2 bar admission pressure, it gradually increases with increase in injection angles from 15° to 60° and thereafter it suddenly decreases with increase in injection angles from 60° to 90°. Also total power output is seen to be large for higher injection pressure 3 - 6 bar and follow the same trend as shown in graphical patterns, which attributes to the large power capacity.

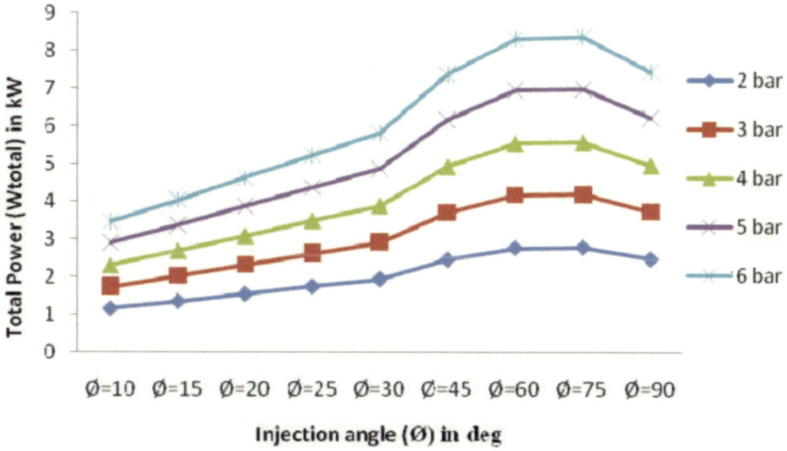

Fig. 8. Total turbine power output versus different injection angles, at injection air pressure 2-6 bar and speed of rotation 2500 rpm

It is also observed that the multi-vane turbine develops maximum shaft power output when injection angle is kept 60° or 75° for all injection pressures in the range of 2-6 bar.

4.3 To optimize the power by critically selecting vane and injection angles

4.3.1 Input parameters for investigation of vane and injection angle

Symbols	**Parameters**
D=2R	150 mm (outer)
d =2r	100 mm (inner) corresponding
p_1	2bar (=30psi), 3bar (=45 psi), 4bar (=60 psi), 5bar (=75 psi), 6bar (=90 psi)
p_4	$=(v_1 / v_4)^{\gamma} .p_1$ assuming adiabatic expansion
p_5	1 atm = 1.0132 bar
θ	30 °, 36 °, 45°, 51.4°, 60 °, 72 ° and 90 ° angles between 2- consecutive vanes (i.e. rotor contains correspondingly 12, 10, 8, 7, 6, 5 and 4 vanes)
N	3000 rpm
L	45 mm length of rotor
γ	1.4 for air
n	(360 / θ) Number of vanes
Ø	30 °, 45 ° and 60 ° Injection angles at which compressed air enters through nozzle into rotor

Table 4. Input Parameters for Investigation of Vane and Injection Angles Both

4.3.2 Results and discussion

From **Fig, 9**, it is noted that optimum total power output is obtained as:

i. 5.0 kW when vane angle (θ) = 60° (i.e. 6 vanes) and injection angle (Ø) = 30°,

ii. 6.1-6.15 kW when vane angle (θ) = 45°-51.4° (i.e. 8-7 vanes) and injection angle (Ø) = 45 ° and

iii. 7.39 kW at vane angle (θ) is 30°-45° (i.e. 12-8 vanes) and injection angle (Ø) = 60°.

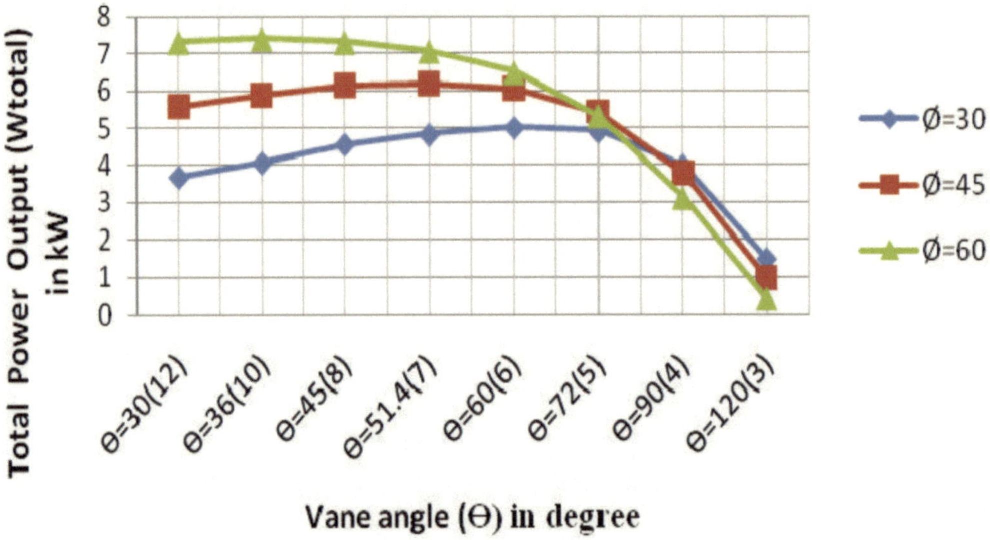

Fig. 9. Total power output versus different vane angles at air injection pressure 6 bar, at injection angle of 30°, 45°, 60°, and speed of rotation 3000 rpm

Thus the maximum shaft power output 6.1 kW to 7.39 kW is achievable at 60° injection angle when number of vanes of rotor is kept between 12-8 (vane angle is 30°-45°). Such type turbine can be used as replacement to internal combustion engines and it can easily be used to run light vehicle or motorbikes with zero pollution.

5. Conclusions

From the above study, it is noticed that the heavy consumptions of fossil fuel in the transport sector is causing serious threat to mankind causing global warming and damage to the environment and ecology. Following conclusions are observed from the study:

- Fossil fuel reserves are depleting due to growing energy requirements.
- Vehicular pollution is estimated to increase many fold in near future.
- Energy storage system could be utilized for the clean energy future.
- The shaft power output of small air engine is achievable from 6.1 kW to 7.39 kW, which is sufficient to run motorbikes or any light vehicle.
- Compressed air technology is going to be the most economical and dominating energy source for running light vehicles / motorbike as tail pipe emission is zero.

The air engine technology, if implimented widely in developing countries to run the light vehicles / motobikes, it might reduce 50-60% global warming issues which is a big challenge of 21st century.

6. Acknowledgement

The authors would like to express their sincere thanks to the SMS Institute of Technology, Lucknow and Harcourt Butler Technological Institute, Kanpur for extending the support of laboratory setup and other facilities.

7. Nomenclature

d	: diameter of rotor (2r) in meter
D	: diameter of outer (2R) cylinder in meter
L	: length of rotor having vanes in meter
n	: no. of vanes = (360/θ)
N	: no. of revolution per minute
p	: pressure in bar
$p_1 v_1$	: pressure and volume respectively at which air strike the Turbine
p_4, v_4	: pressure and volume respectively at which maximum expansion of air takes place
p_5	: pressure at which turbine releases the air to atmosphere.
v	: volume in cum
w	: theoretical work output in (J) Joules
W	: theoretical power output (W) Watts
X_{1i}	: variable extended lengths of vane at point 1
X_{2i}	: variable extended lengths of vane at point 2

Subscripts

1, 2...4, 5	: subscripts - indicates the positions of vanes in casing
exp	: expansion
min	: minimum
max	: maximum

Greek symbols

α	: angle BOF (see Fig.3)
α_1	: angle LOF (=180 - ϕ) (see Fig.3)
α_2	: angle KOF (=180 - θ - ϕ) (see Fig.3)
β	: angle BAF (see Fig.3)
γ	: 1.4 for air
θ	: angle between 2-vanes (BOH) (see Fig.3)
ϕ	: angle at which compressed air enters into rotor through nozzle
ξ_d	: eccentricity *(R-r)*

8. References

Aleklett K. and Campbell C.J., 2003, The Peak and Decline of World Oil and Gas Production-Minerals and Energy, Raw Materials Report, Volume 18, Number 1, 2003, pp. 5-20(16).

Architectural Institute, 2004, Concept and Technology- First International Workshop on Sustainable Habitat Systems - Kyushu Branch, Japan, December, 10, 2004 (10:00-18:00), Venue: ACROS Fukuoka.

Fuglsang P., Bak C. and Gunna M., 2004, Design and verification of the Ris0-B1 Airfoil-family for Wind Turbines, *Journal of Solar Energy Engg.*, ASME- Nov'2004, Vol.126 pp 1002-1008.

Gorla R., and Reddy, S., 2005, Probabilistic Heat Transfer and Structural Analysis of Turbine Blade, *IJTJE*, Vol. 22, pp 1- 11.

Honton E. J., 2004, Hydrogen Fuel Cell Car, presented at *15th Annual US Conference and Hydrogen Expo*, April'2004, USA.

Hubbert, M. K., 1956, Nuclear energy and the fossil fuels, Amer., Petrol Inst. Drilling and Production Practice, *Proc. Spring Meeting*, San Antonio, Texas, pp 7-25.

Knowlen C., Bruckner A. P., Mattick A.T. and Hertzberg A., 1998, High Efficiency Energy Conversion Systems for Liquid Nitrogen Automobiles, *Society of Automotive Engineers, Inc., AIAA* 98-1898.

Negre Guy and Negre Cyril, 2004, Compressed Air-The Most Sustainable Energy Carrier for Community Vehicles, Speech in front of assembly at Kultur gathered for *Fuel Cells World*, Tuesday 29th June '2004.

Royal Society London, (1687), The History of the Compressed Air Vehicles, *Website: http://www.aircarfactories.com/air-cars/compressed-air-history.html.*

Saint Hilaire G., Saint Hilaire R. and Saint Hilaire, Y., 2005, Quasiturbine zero pollution car using gasoline. *Festival at Le Lundi, Montreal Gazette*, 26 September 2005.

Schreck S. and Robinson M., 2004, Tip Speed Ratio Influences on Rationally Augmented Boundary Layer Topology and Aerodynamic Force Generation, *Journal of Solar Energy Engg.*, ASME-Nov' 2004-Vol.126 pp1025-1033.

Selig Michel S. and Bryan D. McGranahan, 2004, Wind Tunnel Aerodynamics Tests of Six Airfoils for use on Small Wind Turbines, *Journal of Solar Energy Engg.*, ASME-Nov'2004, Vol.126, pp986-1000.

Singh B.R. and Singh Onkar, 2006, Necessity and Potential for Bio-Diesel Use in India, *International Conference on Bio-Fuel Vision-2015*, October'13th -15th, 2006 at Bikaner, India- Proceedings pp 71-89.

Singh B.R. and Singh Onkar, 2007, Uses of Wind Power as a Non-Conventional / Renewable Energy for Sustainability, *National Conference on State of Art Technology in Mechanical Engineering, STEM-2007*, held on October 29-31, 2007 at College of Technology, G.B. Pant University, Pant Nagar, UP-Proceedings pp 503-515.

Singh B.R. and Singh Onkar, 2008a, A concept for Development of a Vaned Type Novel Air Turbine, *12th International Symposium on Transport Phenomena and Dynamics of Rotating Machinery* - held on February 17-22, 2008 at Pacific Center of Thermal-Fluids Engineering, Sheraton Mohana Surfrider Hotel Honolulu, Hawaii - Paper No. ISROMAC-12-20046.

Singh B.R. and Singh Onkar, 2008b, A Study to Optimize the Output of Vaned Type Novel Air Turbine, *4th International Conference on Energy Research and Development*, held on 17-19 November, 2008 at State of Kuwait, Kuwait- Paper No. ICERD - 4 -1353.

Singh B.R. and Singh Onkar, 2008c, Development of a vaned type novel Air Turbine, International Journal of Mechanical Engineering Science (The manuscript was received on 21st December 2007 and was accepted after revision for publication on 03rd June 2008), International Journal of IMechE, Part C, Vol.222, 2008, pp 2419-2426.

Singh B.R. and Singh Onkar, 2008d, Energy Storage System to meet Challenges of 21st Century- an Overview, *All India Seminar on Energy Management in Perceptive of Indian Scenario*-held on October 17-19, 2008 at Institution of Engineer (India), State Centre, Engineer's Bhawan, Lucknow-Proceedings Chapter15, pp 157-167.

Singh B.R. and Singh Onkar, 2008e, Parametric Evaluation of Vane Angle on performance of Novel Air Turbine, *Journal of Science, Engineering and Management*, SITM , December, 2008,Vol. 2, pp 7-18.

Singh B.R. and Singh Onkar, 2009a, Analytical Study on a Vaned Type Novel Air Turbine for Different Conditions of Casing and Rotor Diameters, *2009 ASME International Conference on Energy Sustainability* - held on July 17-23, 2009, at San Francisco, California, USA, Paper No. ES2009 -90207.

Singh B.R. and Singh Onkar, 2009b, Applications of Compressed Air as an Alternative Energy to Meet Challenges of 21st Century- Global Warming, *International Conference on Engineering Congress on Alternatives Energy Applications: Option or Necessity?,* held on 3-5 November, 2009 at State of Kuwait, Kuwait- Paper No. EC2009 Kuwait-1082.

Singh B.R. and Singh Onkar, 2009c, Numerical Analysis of Pressure Admission Angle to Vane Angle Ratios on Performance of a Vaned Type Novel Air Turbine, *International Journal of Natural Science and Engineering, IJNSE,* France, Vol. 1-1-2009, pp 20-27.

Singh B.R. and Singh Onkar, 2009d, Optimization of Power Output of a Vaned Type Novel Air Turbine With Respect to Different Injection Angles-Under Ideal Adiabatic Expansion, *International Journal of Mechanical Engineering, Serials Publications, New Delhi, India,* Vol. 2 Number 2, July-December'2009, pp 205-211.

Singh B.R. and Singh Onkar, 2009e, Parametric Evaluations of Injection Angles and Vane Angles on Performance of a Vaned Type Novel Air Turbine, *International Journal of Mathematical, Physical and Engineering Science, IJMPES,* France, Vol. 3-4-2009, pp 226-233.

Singh B.R. and Singh Onkar, 2009f, Theoretical Investigations on Different Casing and Rotor Diameters Ratio to Optimize Shaft Output of a Vaned Type Air Turbine , *International Journal of Natural Science and Engineering, IJNSE,* France, Vol. 1-1-2009, pp 28-35.

Singh B.R. and Singh Onkar, 2010a, Effect of Different Vane Angle on Rotor - Casing Diameter Ratios to Optimize the Shaft Output of a Vaned Type Novel Air Turbine, *International Journal of Engineering Science and Technology, Chennai, India, IJEST*-ISSN-0975-5472,Vol. 2, Number 3 (2), pp 114-121.

Singh B.R. and Singh Onkar, 2010b, Study of Effect of Injection Angle to Rotor-Casing Diameter Ratios on Performance of a Vaned Type Novel Air Turbine, *International Journal of Engineering Science and Technology, Chennai, India, IJEST*-ISSN-0975-5472,Vol. 2, Number 4 (10), pp 409-417.

Singh B.R. and Singh Onkar, 2010c, Critical Effect of Rotor Vanes with Different Injection Angles on Performance of a Vaned Type Novel Air Turbine"- *International Journal of Engineering and Technology, Chennai, India,* IJET-ISSN: 0975-4024, Vol. 2 Number 2(28), 2010, pp. 118-123.

Singh B.R. and Singh Onkar, 2010d, Study of Effect of Rotor Vanes to Rotor-Casing Dimensions on Performance of a Zero Pollution Vane Type Novel Air Turbine,

International Journal of the Physical Sciences, 5170-00200, Nairobi-73023 Victoria Island, Lagos, Vol.5(5), 2010, pp. 547-556..

Singh B.R. and Singh Onkar, 2010e, Analytical Investigations on Different Air Injection Angles to Optimize Power Output of a Vaned Type Air Turbine, *International Journal of Power and Energy* (The manuscript was received on 11th June' 2009 and was accepted after revision for publication on 07th October' 2009), *Proceedings of* IMechE, Part A, Vol.224, 2010, pp 305-311.

Singh B.R. and Singh Onkar, 2010f, Compressed Air Energy Storage Power Plant and its Feasibility in India- An Overview, *Proceedings of National Conference on Advances in Management of Energy Efficiency and Clean Environment,* April 3-4, 2010, pp 55-61.

Singh B.R. and Singh Onkar, 2010g, Study of Influence of Vane Angle on Shaft Output of a Multi Vane Air Turbine, *International Journal of Renewable and Sustainable Energy,* AIP, New York, USA, Vol.2 Number 3, pp. 033101-16.

Singh B.R. and Singh Onkar, 2010h, Performance Investigations for Power Output of a Vaned Type Novel Air Turbine, *MIT- International Journal of Mechanical Engineering,* Moradabad, India, ISSN : 2230-7699; (Accepted: Dec. 17, 2010; Published: Jan 11. 2011), Vol.1(1),2011, pp 9-16 .

Singh Onkar, 2009, Reciprocating and Rotary Compressor, Applied Thermodynamics, *New Age International (P) Ltd., Publishers, New Delhi, India,* ISBN: 978-81-224-2583-3, Feb2009, pp 727-728.

9

Air Pollution in the Niger Delta Area: Scope, Challenges and Remedies

Godson Rowland Ana
Department of Environmental Health Sciences, Faculty of Public Health, University of Ibadan, Nigeria

1. Introduction

Air pollution is one of the major environmental problems confronting the Niger Delta Area (NDA) yet information regarding this is very scanty. Aside from data collected by a few individuals and corporate organizations at scattered locations, there is no comprehensive and empirical database on the magnitude of the hazard and its deleterious effects on the ecosystems and people in the region. Also the existing network of meteorological stations is too coarse to provide data covering the whole of the region. The NDA is Nigeria's most endowed region in terms of oil mineral reserves and one of the most industrialized after Lagos, Nigeria's former administrative capital. The operations of these industries especially the upstream and downstream petroleum sectors as well as a variety of other anthropogenically related activities including biomass combustion, refuse burning and traffic emissions releases a barrage of substances like volatile organics, oxides of carbon, nitrogen, sulphur, particulate matter, heavy metals and other toxics at levels that most times exceed both the national and international guidelines. Apart from compromising the quality of the atmosphere most of the air pollution in the region is observed to have local and regional effects such as the formation of acid rain, water pollution, soil pollution, impacts on plants and wild life, effects on materials and artifacts and recently contributing to the global warming effects. While most of these claims may be true a great majority has not been substantiated scientifically and is rather based on assumptions or being speculative. There are numerous challenges facing air quality studies in NDA. Like in most developing nations the issues are multifaceted and the most acute have been lack of equipment, inadequate expertise, lack of infrastructure and weak policy frame work. This and many other factors has been the bane to achieving the much desired goals towards preserving the quality of the ecosystems and safeguarding the health of the public in the region. In this chapter a conscious effort has been made to give a brief description of the Niger delta area, the possible sources of air pollution in the region with highlights of a few case scenarios of air pollution from community automobile traffic and industries as well as common related health problems. Again as a way forward to most of the problems and drawbacks hitherto mentioned several models and schemes have been advocated.

2. The Niger Delta area

The Niger Delta area in Nigeria (Fig 1) is situated in the Gulf of Guinea between longitude 50E to 80E and latitudes 40N to 60N.It is the largest wetland in Africa and the third largest in the world consisting of flat low lying swampy terrain that is criss-crossed by meandering and anastomosing streams, rivers and creeks. It covers 20,000 km^2 within wetlands of 70,000 km^2 formed primarily by sediment deposition. It has an equatorial monsoon climate influenced by the south west monsoonal winds (maritime tropical) MT airmass coming from the South Atlantic Ocean. It is home to 20 million people drawn from nine states namely Abia, Akwa Ibom, Bayelsa, Cross River, Delta, Edo, Imo, Ondo and Rivers states with 40 different ethnic groups. This floodplain makes up 7.5% of Nigeria's total land mass. The Delta's environment can be broken down into four ecological zones: coastal barrier islands, mangrove swamp forests, freshwater swamps, and lowland rainforests. This incredibly well-endowed ecosystem, contains one of the highest concentrations of biodiversity on the planet, in addition to supporting abundant flora and fauna, arable terrain that can sustain a wide variety of crops, lumber or agricultural trees, and more species of freshwater fish than any ecosystem in West Africa. The vegetation of the Niger Delta consists mainly of forest swamps. The forests are of two types, nearest the sea is a belt of saline/brackish Mangrove swamp separated from the sea by sand beach ridges. Numerous sandy islands occur with fresh water vegetation. Fresh water swamps gradually supersede the mangrove on the landward side. More than 70% of Nigeria's crude oil and gas production is from the area. The region produces over 90% of Nigeria’s foreign earnings through oil exploration activities. It plays host to most of the upstream and downstream oil related industries and non oil related industries that release tons of pollutants into the ecosystems. The pollution from the Niger Delta on a scale could be regarded as one of the worst among similar delta areas in the world.

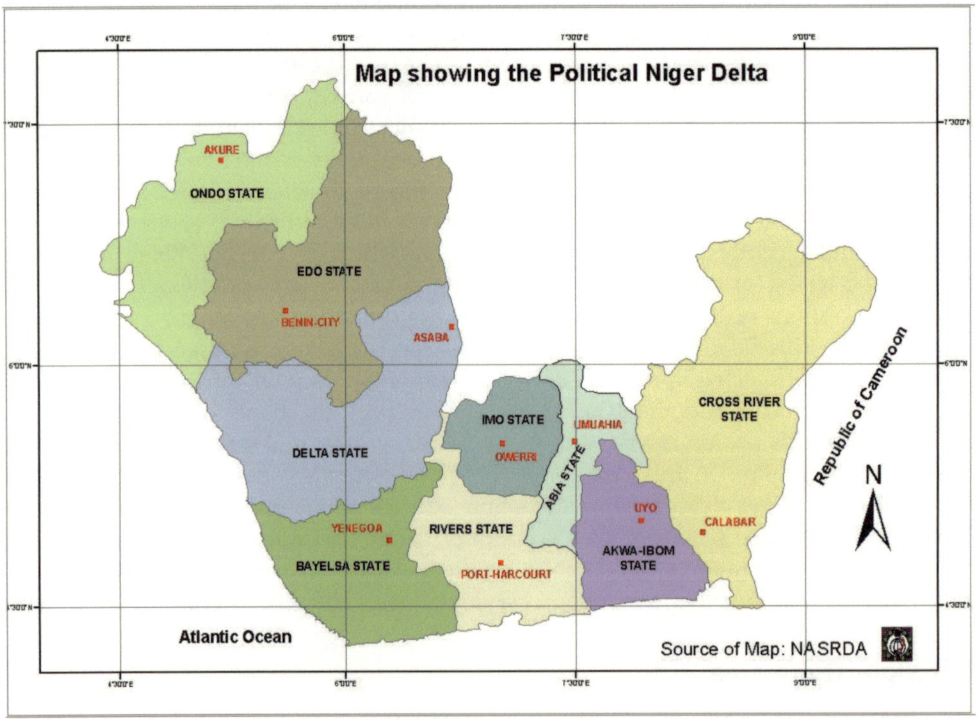

Fig. 1. Political Map of the Niger Delta Area

3. Air pollution sources

3.1 Biomass combustion

Biomass in form of firewood, coal, bamboo trunks and dead leaves are commonly used sources of cooking fuel in the region. Of all these the most frequently used is firewood. This fuel woods are usually logged from nearby bushes and forests or are collected as dead branches within the residential vicinities by the women and sometimes children and are used to generate energy for cooking(Pate 1). The combustion of firewood releases gaseous pollutants and particulate matter. From literature the gaseous pollutants from cooking emissions are carbon monoxide CO, carbon dioxide CO_2, sulphur dioxide SO_2, nitrogen dioxide NO_2, volatile organic compounds VOCs and particulate matter. The particulate matter generated is in the form of carbon black, sooth and fly ash which are major components of smoke and are most often within the 10μm size range.

Plate 1. Cooking Fuel emissions from biomass combustion

3.2 Bush burning

Bush burning is a common phenomenon in the Niger delta communities. Most times the burning of bush is done prior to land cultivation as one of the initial steps of land preparation. This constitutes part of the pre planting preparation. The process of bush burning leads to the release of various types of gaseous pollutants and particulate matter. Very often the gas stream is inundated with volatile organics and oxides of carbon (COx), sulphur(SOx) and nitrogen(NOx) depending on the fuel composition and intensity of the flame. Particulate matter usually within the 10μm size range is also produced in the course of the combustion process.

3.3 Refuse burning

Refuse disposal is a major environmental problem in the Niger delta communities particularly in the urban areas (Plate 2). The refuse is usually from multiple sources including domestic, municipal, agricultural and industrial sources. One of the environmentally unfriendly methods of managing the waste is by open burning either on nearby lands or open dumps within the residential vicinities. The composition of the refuse, age of the dump and intensity of the flame usually determines the nature of the air pollutants. Often times the air within refuse burning sites is inundated with VOCs, COx, SOx, NOx, total hydrocarbons(THCs), as well as various classes of toxic and hazardous compounds viz polycyclic aromatic hydrocarbons(PAHs), dioxins, PCBs(Polychloro Biphenyls) and heavy metals such as lead, nickel and mercury.

Plate 2. Emissions from open burning of solid waste

3.4 Traffic emissions

Over 600 million people globally are exposed to hazardous level of traffic - generated pollutants UN, (1998). Human exposure to these air pollutants is believed to have posed severe health problems especially in urban areas where pollution levels are on the increase. Pollution due to traffic constitute up to 90 - 95% of the ambient CO levels, 80 - 90% of NOx, hydrocarbon and particulate matter in the world, posing a serious threat to human health Savile, (1993). Research has shown that transportation sources in the USA were responsible for 77% of CO levels, 80 -90% of NOx, 36% of volatile organic compounds and 22% of particulate matter USEPA, (1993). Similarly, in UK the average concentration of NO_2 was found to increase by 35% from 1986 to 1991 due to increase in vehicular emission CEC, (1992). On the global scene, Seneca and Tausig, (1994) concluded that transportation is the major culprit of air pollution accounting for over 80% of total air pollutants.

In Nigeria much attention is focused on general industrial pollution and pollution from the oil industries, with little attention on the effects of air pollution from mobile transportation sources (Faboye, 1997; Iyoha, 2000 and Magbabeola, 2001).Increased pollution from mobile

Plate 3. Traffic emissions from busy urban roads

sources is on the increase with per capita increase in vehicle ownership(Plate 3). The consequence of this is the congestion of most Nigeria city roads and a corresponding increase in the burden of air pollutants and their associated effects. Studies conducted by Akpan and Ndoke, (1999) in Northern Nigeria show higher values of CO_2 concentration (1780ppm- 1840ppm) in heavily congested areas in Kaduna and (1160ppm-1530ppm) in Abuja. A study of the impacts of urban road transportation on the ambient air was conducted by Koku and Osuntogun, (1999) in three cities in south western Nigeria. Air quality indicators namely CO, SO_2, NO_2, and total suspended particulates (TSP) were determined. The highest levels obtained for the air pollution indicators were CO- 233ppm , SO_2-2.9ppm , NO_2-1.5ppm and total particulates 852ppm in Lagos . In Ibadan the highest levels obtained were CO-271ppm, SO_2 -1.44 ppm and NO_2- 1.0ppm. In Ado-Ekiti the highest levels obtained were CO-317ppm, NO_2 -0.6ppm and SO_2-0.8ppm. These results were found to be higher than FEPA limits for CO-10ppm, SO_2-0.01ppm, NO_2-0.04-0.06ppm.

A comparative study of emission levels in Lagos and the Niger Delta area was reported by Jerome, (2000). Two major cities, Port-Harcourt and Warri were considered for the NDA. The results obtained showed that the concentrations of TSP, NOx, SO_2, and CO in Lagos and the Niger Delta communities were above the FEPA recommended limits. The CO levels for Lagos 10 - 250ppm were higher than the levels 5.0 - 61.0ppm and 1.0 - 52ppm recorded in the two selected communities in the Niger Delta. The TSP concentrations were also high for both locations when compared to WHO standard. A similar traffic related emission study was carried out in Calabar, a major city in NDA and the results presented in the box indicate that the overall average value(OAV) of each pollutant for the 72 hours experimental duration was found to be in the following range: CO :5.0 – 6.1ppm, NO_2 :0 -0.05ppm, PM10: 202 - 230μg/m^3.

Lagos Area Pollutant	Non-Traffic Urban zone	Traffic Zone	Niger Delta Area Oil Communities	Cities	FEPA Standards
TSP μ/m^3	31.4-746.5	72 – 950	92.2 – 348.5	396.8 – 583.3	250
NOX(ppm)	81 – 81.5	34 – 131.6	22.0 – 295.0	35 – 370	40 – 60
SO_2(ppm)	0.5 – 43	20 – 250	7.0 – 97.0	16 – 300	100
CO(ppm)	0.5 – 3.9	10 – 250	5.0 – 61.0	1.0 – 52	10
CO/NO_X(ppm)	0.0 – 6.0	50 – 200	20	15 – 130	-

Source: Jerome 2000

Table 1. Ambient air pollutants in Lagos and Niger Delta Area

At the National level, available data on the total number of vehicles registered in Nigeria shows an increase from 38,000 to 1.6 million between 1950 and 1992(Enemari, 2001). Data from the Federal Road Safety Commission (FRSC) of Nigeria however, indicates that between 1999 and 2004 about six million vehicles (6,000,000) were registered in Nigeria (Fig 2) of which 70% of the registered vehicles were cars and 30% busses and trucks. The reason for the large increase of registered vehicles in 1999 was due to the deadline given for the registration of all vehicles in the country by FRSC. The subsequent years indicate vehicles coming in for the first time into the country. The increase recorded in 2002 was the climax due to favourable government policies on importation of fairly used vehicles at the inception of democratic rule in Nigeria. Although there was a decline in 2004, it is obvious that the number of vehicles plying Nigerian roads had continued to increase. In Lagos alone about 223,764 vehicles were registered in 2008 compared to 160,134 vehicles registered in 2007. Also in the first quarter of 2009 more than 58,000 vehicles were registered and a total of 27,587 motorcycles were also registered between July 2006 and July 2009 (http//alafrica.com/stories). From the above data it is apparent that the average emission concentration from motor vehicles and motorcycles in the country is in the increase.

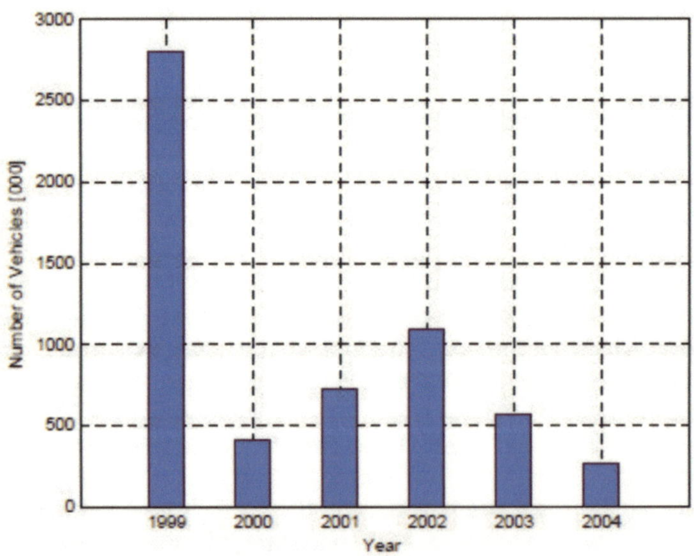

Fig. 2. Total number of vehicles Registered in Nigeria (1999-2004)

Box 1: Traffic Emission Scenario in Calabar

The overall average values (OAV) of each pollutant for the 72 hours experimental duration was found to be in the following range: CO: 5.0-6.1ppm, NO2 - 0 -0.05ppm, PM10 - 202 - 230 $\mu g/m^3$ and noise level was 60.5 - 63dB. Comparing these values with AQI CO, range from moderate to very poor; PM10 and SO_2 were very poor. The OAV value of NO_2 is constant for the three experimental days comparing with AQI rating the value is still poor. The reason for this constant OAV value of NO_2 could be the recent government policy of replacing the intercity transport system with new efficient cars with catalytic reduction system. The overall AQI rating for ambient air quality data indicates that the air quality standard in Calabar is below the USEPA standard for ambient air quality rating. .

3.5 Industrial emissions

Apart from Lagos which was the former political capital but still regarded as Nigeria's economic capital, the next most industrialized part of the country is the Niger Delta Area. The region boasts of oil and non oil related industries including refinery, petrochemical, liquefied natural gas, chemical fertilizer, aluminum smelter, paper, cement, flour, wood, battery and textile industries etc which emit various kinds of air pollutants (Box 2). The pollution from these industries adds to the burden of gaseous and particulate pollutants in the air. In this chapter information regarding air pollution from three of these industries namely the chemical fertilizer plant, petroleum refinery plant and the petrochemical complex would be reported.

Box 2: Major Air pollutants in NDA

Gases: CO_2, CH_4, SO_2, N_2O, NO_2, NH_3
VOCs

Particles: PM10, PM2.5, PAHs, Dioxins
PAHs, Heavy metals

3.5.1 Chemical fertilizer industry

The chemical fertilizer industry in Nigeria is still in its infant stage. The National Fertilizer Company of Nigeria (NAFCON) described here is located at Onne (4.49^0 and 4.50 N and 6.59^0 and 7.00 East of Greenwich Meridian), 30 km away from Port Harcourt in the Niger delta area. It utilizes a variety of raw materials such as natural or synthetic gas (methane), atmospheric nitrogen, steam and sand as filler material in the production of various formulations. The major products (per day) are ammonia (1,000 tonnes), urea (1,500 tonnes), and NPK formulations (1,000 tonnes).The main sections where gaseous and particulate emissions are reported are the ammonia plant, urea plant, NPK plant, bulk blending plant, bulk storage plant, and the bagging area. Potential sources for leaks in the ammonia plant are the refrigeration loop, storage areas, flanges, valve packing, and the pump and compressor seals. The gaseous emissions from the ammonia plant include reformer and boiler flue gases, excess carbon dioxide, condenser stripper vapour and ammonia discharge. The gases coming from the reformer are vented into the atmosphere with a mixture of CO_2 (20%) and O_2 (3-4%). Most of the CO_2 is recovered and recycled in the process. The condensate of the process water contained some CO_2, methanol and other dissolved gases, which are usually stripped. In the urea plant there are two possible sources of emissions, viz. the high-pressure scrubber and the granular stack. Some amounts of methane, hydrogen, ammonia and CO_2 are released. The wet scrubbers also release some particulate urea through the vent. In the NPK plant the tail gas scrubber is the only source of emission of ammonia, particulates and small quantity of fluoride. In various operations, e.g. granulation, drying, mixing and cooling, some amounts of dust and fumes are generated. In a study carried out by Ana et al,2005 the highest mean levels of ammonia (459.1ppm), was recorded at the Urea synthesis plant and the lowest levels (7.29ppm) recorded at a control point about 5km away from the point. For particulate matter, the highest concentration of 260875.7 $\mu g/m^3$ and the least concentration of 7.29ug/m^3 were recorded.

3.5.2 Refinery and petrochemical complex

The quality of air in Nigeria's largest oil refinery (PHRC) and petrochemical complex (EPCL) was assessed based on key priority pollutants in a study carried out by Ana et al(2009)(Plate 4).Three points each were observed per industrial location At the refinery, the highest PM10 level (130.3$\mu g/m^3$) was recorded while at the petrochemical complex the highest PM10 level (81.3 $\mu g/m^3$) was recorded. The study indicated that the average PM10 levels at PHRC were higher than that recorded at EPCL .In terms of the level of heavy metals, the highest Pb and Ni levels of 0.20 mg/m^3 and 0.86 mg/m^3 respectively was recorded at the PHRC. At the petrochemical complex, the highest Pb and Ni levels of 0.16 mg/m^3 and 0.05 mg/m^3 respectively was recorded at EPCL. Overall, the average heavy metal concentrations were higher at PHRC.The highest concentration of benzo (a) pyrene (1.63x102ng/m^3) was recorded at PHRC compared to (1.61x102 ng/m^3) that was recorded at EPCL. However, the total PAH concentration taken as the sum of benzo(a)pyrene and indeno(123)-cd pyrene was found to be higher at EPCL when compared with PHRC(Table 2).

3.6 Gas flaring

According to Cedigaz (2000), Nigeria holds the highest record (19.79%) of natural gas flaring globally and is responsible for about 46 % of Africa's total gas flared per tonne of oil produced(Plate 5). Until present there are not less than 123 flaring sites in the region making Nigeria one of the highest emitter of green house gases in Africa (Uyigue and Agho, 2007). Similarly, analytical assessment of the statistical bulletin of the Central Bank of Nigeria, (2004) showed that the average rate of gas flaring in Nigeria during the period 1970-1979 stood at 97%, while 97% and 95% were flared between 1980 and 1989, and 1990 to 1999 respectively. Between 2000 and 2004 51% was flared. In the same vein, between 1970 and 2004 Nigeria has flared an average of 76% of the total gas produced. Much of the natural gas extracted in oil wells in the Niger Delta is immediately flared into the environment at a rate that approximates 70 million /m^3 per day. This is equivalent to 40% of African natural gas consumption and forms the single largest source of greenhouse gas emissions on the planet (Wikipedia 2007, Moffat and Linden, 1995). Orubu, (2002b), who undertook a comparison of concentrations of ambient air pollutants in the region and Lagos State concludes that

Plate 4. Emissions form a Petrochemical plant

Parameter	PHRC1 (Mean ± SD)	PHRC2 (Mean ± SD)	PHRC3 (Mean ± SD)	Average/ Location	Guideline Limits
PM_{10} (µg/m³)	130 ± 3.32	36.5 ± 1.27	27.9 ± 2.33	64.8	100(FMENV)
Pb (mg/m³)	0.16 ± 0.12	0.20 ± 0.03	0.11 ± 0.10	0.16	-
Cd (mg/m³)	0.008 ± 0.002	0.003 ± 0.001	0.009 ± 0.006	0.007	-
Ni (mg/m³)	0.004 ± 0.002	0.009 ± 0.004	0.86 ± 0.34	0.291	-
Benzo (a) pyrene ng/m³	-	-	1.63×10^2	54.3	<0.1–100ng/m³ (WHO, 1997)
Indeno (1,2,3-cd) pyrene ng/m³	1.53×10^3	2.53×10^3	-	0.0014	<0.1–100ng/m³ (WHO, 1997)
ᵀPAH*(ng/m³)	1.53×10^3	2.53×10^3	1.63×10^2	54.3	<0.1–100ng/m³ (WHO, 1997)

* Total PAH {sum of benzo(a)pyrene + indeno(123-cd)pyrene}

Table 2. Levels of Priority Air pollutants in Nigeria's Refinery and Petrochemicals pollutant concentrations are highest in the Niger Delta and argues that some of the green house gases (such as methane and carbon dioxide) emitted at flare sites contribute to global warming. The largest proportion of these flare sites are located in the Niger Delta.

Plate 5. Gas flaring from an oil exploration process

3.7 Pipeline explosion

The explosion of pipelines occurs either accidentally or by sabotage(Plate 6). In the Niger Delta area much of the pipeline explosions are a product of the later and they are usually accompanied most times with fire outbreak. The burning flame and smoke from the oil pipelines releases large concentrations of gaseous substances and particulate matter. The substances in most cases include COx, NOx VOCs, THCs, carbon black, soot and some heavy metal residues.

3.8 Multiple air pollution sources

Within the Niger delta communities there are diverse sources of air pollution. The common air pollution sources are biomass combustion, bush burning, automobile emissions, generator emissions, pipeline explosions, industrial emissions and gas flaring. The measurement of air quality for some specific priority pollutants viz ammonia, particulate matter and Polycyclic aromatic hydrocarbons at specific locations within the communities has been carried out using standard procedures (Ana et al, 2005, 2009). The mean values of the parameters are presented in Table 3.

3.9 Meteorological conditions in the Niger Delta

The Niger Delta has a humid, semi-hot equatorial climate. Temperature ranges are small and constant throughout the year. It's hottest month records 28 °C (82.4 °F) and its coolest month 26 °C (78.8 °F) with the temperature range of not more than 2 °C (5 °F). Precipitation

Plate 6. Emissions from fire outbreak following pipeline explosion

Air quality at Eleme and Ahoada East

Sample Parameter	Eleme Mean ± SD	Range	Ahoada East Mean ± SD	Range	National Guideline Limits	*t*-statistic	*p*-value
TSP(mg/m^3)	123.4 ± 14.9	37.2–260.6	53.5 ± 26.3	8.84–161.4	100 (mg/m3)	−6.118	0.0001**
Fe (mg/m^3)	0.09 ± 0.06	0.09–0.10	0.027 ± 0.006	0.01–0.05	None	−2.764	0.0171*
Zn (mg/m^3)	0.05 ± 0.03	0.02–0.10	0.05 ± 0.02	0.03–0.42	-do-	0.000	1.000
Pb (mg/m^3)	0.06 ± 0.03	0.03–0.10	0.14 ± 0.03	0.03–0.13	-do-	4.939	0.0003**
Cd (mg/m^3)	0.009 ± 0.01	0.002–0.02	0.006 ± 0.003	0.004–0.01	-do-	−0.760	0.4618
Ni (mg/m^3)	0.03 ± 0.02	0.02–0.05	0.03 ± 0.02	0.01–0.05	-do-	0.000	1.000
†PAH(ng/m^3)	7.99×10^3	3.48×10^{-1}–	1.21×10^{-3}	2.50×10^{-5}–	<0.1 –	7.62×10^6	0.000**
	$\pm 1.31 \times 10^4$	1.88×10^4	$\pm 2.77 \times 10^{-3}$	7.49×10^{-2}	100 ng/m^3		
	(1.91×10^2)		(6.43×10^{-4})				

* Significant ** Highly Significant

† Values in brackets represent median values for PAH in the various locations

Table 3. Levels of Air Quality Parameters in Two Niger Delta Communities is heavy, between 1824 millimeters to over 4000 millimeters along the coast. Rain falls throughout the year with a short break in August and a longer break from December to January. The heavy rainfall experienced in the area leads to floods in urban areas and communities along the rivers. Also due to the heavy and frequent rainfalls, soils consisting mainly of silt and clay become saturated, reducing infiltration to the barest minimum and encouraging run-off. Rainfall induced run-off is directed mainly to the back swamps in the upper and middle parts of the delta. In the coastal zone, there is an even smaller topographic gradient that makes run-off difficult.

4. Environmental and health problems associated with air pollution in the delta area

4.1 Environmental problems

4.1.1 Formation of acid rain

From the release of large concentration of oxides of nitrogen and sulphur in the Niger delta environment it is glaring that some cases of acidified rain occur in the region. Although literature is deplete in this area it is obvious that such phenomenon is experienced. Olobaniyi and Efe (2007) also show elevated levels of lead (0.56 mg/l) and low pH values ranging from 5.10 - 6.35 in rain water collected in Warri and environs. Akpoborie, *et al*, (2000) also report low pH values from water obtained from shallow hand dug wells in Ughelli, Warri, and Okurekpo all in Delta State.

4.1.2 Visibility Impairment

Most combustion processes lead to the release of oxides of nitrogen and lower fractions of hydrocarbons and in the presence of uv radiation the resultant smog could lead to decrease in visibility. Although not readily reported this is a common phenomenon in a region like the Niger delta that has large hydrocarbon deposits and various episodes of combustion processes.

4.1.3 Ambient thermal conditions

Some 45.8 billion kilowatts of heat are discharged into the atmosphere of the Niger-Delta from 1.8 billion cubic feet of gas everyday (Aaron, 2006). Ogbuigwe (1998), reports that temperatures produced at flare sites could be as high as 1,600°C. Temperature as high as 400 °C at an average distance of 43.8metres from flare sites in Isoko, Delta State have been recorded.

4.1.4 Effects on vegetation and animal life

Studies by Alakpodia (1989, 1995) showed that flares have negative effects on vegetation growth, animal life and ecological equilibrium in the Niger Delta area. Heat Production kills vegetation around the heat area. It suppresses the growth of flowering plants and reduces agricultural productivity and wild life biodiversity.

4.1.5 Corrosion effects on materials and other artifacts

The presence of acidified rain water in the environment increases the corrosion rate of roofing sheets, monuments and other economic structures. In the Niger delta area there is glaring evidence of the impacts of corrosion on several building structures and arts work and these cases have been observed to deteriorate at rather alarming rates.

4.1.6 Coastal erosion and flooding

Awosika and Foluronsho, (2006) documented projected impacts of sea level rise SLR on the Nigerian coastal area and resources to include: large scale inundation, increased coastal erosion, salt water intrusion into coastal aquifers, habitat modification with direct effects on wild life distribution, increased frequency of high intensity rainfall events and associated increased run off, elevated erosion rates, flash floods and increased frequency of ocean storm surges. Other models projected a loss of up to 15,000 km^2 by the year 2100 as a result of a 1m (SLR) while Magbagbeola, (2002) projects flooding of over 18,000km^2 land area within the delta in an SLR of 1m in the next 100 years.

4.1.7 Air quality impairment

Oluwole et al.(1996) in a typical air quality assessment of the Niger Delta showed that the levels of volatile oxides of carbon, nitrogen and sulphur and total particulates exceed existing Federal Environmental Protection Agency's (FEPA,1991) standards. Also Olobaniyi and Efe (2007) showed elevated levels of lead at concentrations of 0.56 mg/l in the atmosphere.

4.2 Health effects

4.2.1 Common health problems in the general population

A survey was carried out on the health conditions of populations' resident and working in the Niger Delta communities (Ana et al, 2009).A summary of the prevalent air pollution-related morbidities is shown in Box 3. Two communities, one with high industrial presence, Eleme and the other, Ahoada East, with low industrial presence were selected. Questionnaire and hospital records were employed for this survey and morbidities that have direct bearing with air pollution were focused on. The questionnaire survey indicated that at Eleme (39, 60.9%) as compared to Ahoada East (10, 4.5%) reported contaminated air as the major reason for ill health among the residents. Also there were more reported cases of skin outgrowths among Eleme residents as compared to Ahoada East. The morbidity conditions from hospital records for respiratory disorders showed 3.85% in males and 4.39% in females (Fig 3a). At Ahoada East respiratory disorder was(3.68% male; 4.18% female) (Fig 3b). In terms of respiratory disease, there was a high significant difference ($p < 0.0001$) between the male and female population at Ahoada East only. Also between Eleme and Ahoada East there was a significant difference among the male population ($p = 0.04$) and the female population ($p = 0.05$) (Figures 4 and 5).In addition the results indicated that for the combined male and female population, there was a significant difference between the two communities for skin disorders ($p = 0.023$) and disease of the respiratory tract ($p = 0.045$).

Box 3: Prevalent Air Pollution related Morbidities

Breathing Difficulties
Bronchitis
Aggravation of Asthma
Cardio-respiratory disorders
Pulmonary edema
Eye disorders
Skin disorders

4.2.2 Common health problems in the industrial community

In a study carried out by Ana et al (2005) in which 384 plant workers were interviewed in a chemical fertilizer industry at Onne in the Niger Delta, 70.5% spend 8 hours per day at work. 66.1% reported respiratory disorders, 24.4% reported skin disorders and 22.6% reported eye disorders. There were strong associations ($p < 0.05$) between eye and respiratory disorders and the industrial activities carried out by the workers. Also Ana et al (2009) in another survey carried out among refinery (PHRC) and petrochemicals (EPCL) plant workers reported that 70.8% workers at PHRC as compared to 67.2% at EPCL reported various health problems related to poor air quality conditions. In both cases, the major cause was attributed to gas flaring. Symptoms associated with exposure to fumes and sprays such as respiratory disorder were widely reported among 40.8% PHRC and 27.6% EPCL workers. There were reports of respiratory symptoms associated with exposure to dust and smoke, leading to irritation among 65.7% PHRC and 57.1% EPCL workers. Both the eyes and the skin were among other parts of the body also affected by the irritants. Further statistical analysis indicated that the duration of stay of PHRC workers in their residential communities was significantly associated with respiratory health problem (p=0.000), with Cancers (p=0.000). At EPCL, the duration of residence in the community was significantly associated with miscarriages (p=0.000), with deformed children (p=0.000) and with symptoms related to health effects from air contaminants (p= 0.000).

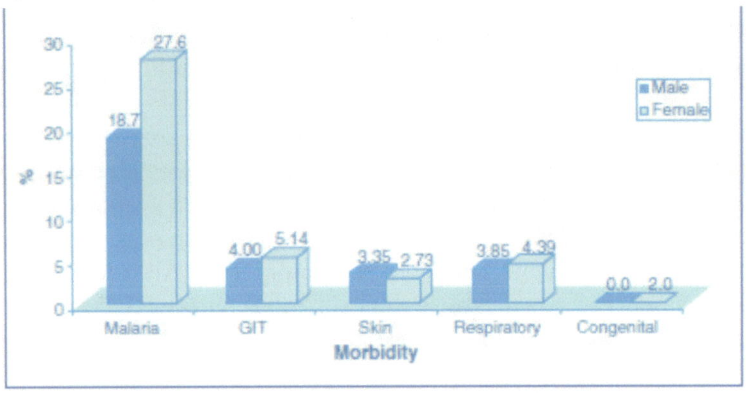

Fig. 3a. Typical morbidity pattern in a more industrialized NDA community

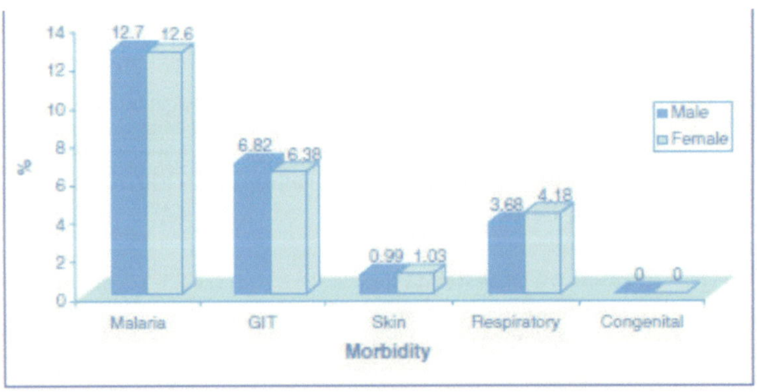

Fig. 3b. Typical morbidity pattern in a less industrialized NDA community

4.2.3 Emerging health problems

The exposure of populations to a barrage of pollutants and pollution episodes in the Niger delta area is believed to cause several health problems, of which etiologies of most cannot be ascertained. Cancer prevalence is believed to be on the increase. Studies by Ana et al(2010) revealed that of the two cancer reference centres Ibadan in the South west and Port Harcourt in the Niger delta the ratio of reporting was 1:4 for UPTH (904) and UCH (3521) respectively. The results indicate that apart from prostate and breast cancers that were higher in Ibadan (79.1%) than in Port Harcourt (75.4%) both the lung and skin cancers were more prevalent in Port Harcourt than in Ibadan(Table 4). This observation is consistent with studies by Ana et al.(2009) which indicated increased lung cancers in the Port Harcourt environment due likely to exposure to atmospheric insults. Similarly, the higher percentage of skin cancer in Port Harcourt above that recorded in Ibadan could be explained though with some degree of uncertainty by the increased environmental risk factors in the more industrialized Port Harcourt area.

Cancer types	Port Harcourt (N=904)	Ibadan(N=3521)	X^2	p- value
Lung	33(3.7)	113(3.2)	0.44	0.508
Skin	172(19.0)	365(10.4)	50.59	<0.001
Eye	15(1.7)	258(7.3)	39.92	<0.001
Others	684(75.4)	2787(79.1)	5.18	0.023

Table 4. Comparison of Cancer cases between two prominent cities in Nigeria

5. Challenges of air quality studies in the Delta area

Air quality studies in Nigeria and particularly the Niger Delta area is still in its infant stage and encumbered with several challenges. According to Taiwo (2005) air pollution studies in Nigeria are few and independently carried out. The same author also indicated that government is not involved in systematic and consistent air quality assessment programmes as is being done in other parts of the world such as that carried out by the Environmental Protection Agency (EPA) in the United States. There is the problem of insecurity and difficulty in terrain that militates against most community based air sampling initiatives and then the lack of requisite and adequate technical manpower to carry out the multifaceted and complex air quality studies in the region. Other obvious and prominent drawbacks are as follows

- Lack of emissions inventory/database due to lack of consistent and systematic measurements
- Unavailability of air pollution and GHG monitoring stations in the Niger Delta. This is based on information from World Data Centre for Greenhouse Gases
- Few independent and research-based measurement data are not readily available for general public use
- Lack of collaboration between key regulatory authorities
- Laxity in the enforcement of emission regulations
- Air quality assessment and air pollution studies have focused mainly on urban centres (Ajao& Anurigwo, 2002)

6. Conclusions

Air pollution has been identified as one of the most critical environmental problems confronting the Niger delta Area. Traffic, industry and gas flaring are the major air pollution sources in the region. Impact on socioeconomic, ecosystems health, properties and climate have been linked with the pollution episodes in the area but which requires substantial scientific and empirical evidence. Multiple factors especially lack of equipment, inadequate skilled personnel and poor policy frame work has militated against effective and qualitative air quality studies in the area. To achieve long term goals for the region in this regard that would lead to the overall benefit of the people in the area, there is need therefore to employ a holistic and integrated approach to air pollution management that will involve all the stakeholders.

7. Future directions

In order to improve on the current air quality monitoring and assessment programmes in the Niger Delta area there is need to embark on the following:

- Develop monitoring mechanisms, regulations and enforcement measures
- Institute planning policies to minimize pollution that may be caused by future development.
- Government agencies such as the Niger Delta Development Commission(NDDC) should collaborate with other multinationals and stakeholders in air pollution management to come up with a comprehensive AQM scheme for the region.
- There should be a focus on the reduction of pollution levels from vehicles, industry, gas flaring and domestic burning of timber, to permissible levels as defined in national and international standards.
- The impact of air pollution from industrial and vehicular sources on the health of the communities in the region and its biodiversity needs to be researched in-depth.
- Existing air quality monitoring programmes should be re-examined and new ones introduced to determine the most effective means of mainstreaming national programmes with regional projects to improve air quality.
- Research on air quality should focus on source apportionment of the pollutants in the region
- Indepth epidemiological and toxicological studies using risk and exposure assessment tools need to be carried out to establish causalities between the air pollution exposure factors and the associated health problems.
- Motor vehicles annual testing and other regulations must be created or re-introduced and strictly enforced.
- Focus should be on air pollution models, real-time monitoring of pollutants and speciation of pollutants from the particulate matter.
- There is a need to engage in renewable energy, clean energy and cleaner air initiatives.
- Usage of emissions abatement control mechanisms by polluters should be enforced.

8. References

Aaron, K. K. (2006). Human Rights Violation and Environmental Degradation in the Niger-Delta in Elizabeth Porter and Baden Offord(eds), Activating Human Rights, Oxford, Barne, New York.

Ajao EA. and Anurigwo S(2002). Land – based sources of pollution in the Niger Delta, Nigeria. Ambio Journal of the human environment. 31 (5):442 - 445.

Akpan, U.G., Ndoke, P. N. (1999). Contribution of vehicular traffic emission to CO2 emission in Kaduna and Abuja.Federal University of Technology Minna Nigeria.

Akpoborie, I.A.; Ekakite, A. O.; Adaikpoh, E.O.(2000), The Quality of Groundwater from Dug wells in Parts of the Western Niger Delta, Knowledge Review 2: 72-79.

Alakpodia, I J (1989), The Effects of Gas Flaring on the Micro-climate and Adjacent Vegetation in Isoko Area of Bendel State" (Unpublished M.Sc Thesis, University of Ibadan).

Alakpodia, I. J. (1995), "The Oil Industry and the Economic Environment of the Niger Delta"Paper presented at the 30th Annual Conference.Nigeria Geographical Association, at the University of Benin. Benin City. Nigeria.

Ana, Godson R.E.E, Mynepalli K.C.Sridhar and Elijah A. Bamgboye (2009) Environmental risk factors and health outcomes in selected communities of the Niger Delta Area, Nigeria. Perspectives in Public Health. Vol 129(4); 183-191

Ana, G.R.E.E and Sridhar, M.K.C(2009) Industrial Emissions and Health Hazards among selected factory workers at Eleme, Nigeria. Journal of Environmental Health Research, Vol9(1):43-51

Ana, Godson R. E. E, Sridhar, Mynepalli K.C and Asuzu, Michael C(2010) Environmental risk factors and hospital based cancers in two Nigerian cities Journal of Public Health and Epidemiology Vol. 2(8), pp. 216-223,

Awosika, L.; Foluronsho, R. (2006), Climate Change and Impact on the coastal environment of Nigeria, In, Sustainable Environmental Management in Nigeria, Eds. Ivbijaro *et al.*Mattivi Productions, Ibadan, pp 85-103.

CEC, 1992. The state of the environment in the European Community. Overview, Vol. 3. Commission of the Communities, Brussels and Belgium.

Cedigaz, (2000) www.cedigaz.org.Central Bank of Nigeria. (2004), Statistical Bulletin 15

Enemari, E., 2001. Vehicular emissions: Environmental and health implications. National conference on the phase-out leaded gasoline in Nigeria.

Faboya, O.O., 1997. "Industrial pollution and waste management" pp 26-35 in Akinjide Osuntokun(ed),Dimensions of Environmental problems in Nigeria, Ibadan Davidson press.

Federal Environmental Protection Agency, (1991), National Policy on the Environment, The Presidency, Abuja, Nigeria

Gill, L. S.; Nyawuame, H. G. K.; Ehikhametalor, A. O. (1992), Effect of Crude oil on growth and anatomical features of *Chromolaena odorata* L. Newsletter 5:46-50 http//alafrica.com/stories.

Iyoha, M. A., 2009. The Environmental effects of oil industry activities on the Nigerian Economy:A theoretical Analysis: Paper presented at National Conference on the management of Nigeria's petroleum Resources, organised by the Department of Economics, Delta State University.

Jerome, A. 2000. ''Use of Economic instruments for Environmental Management in Nigeria'' Paper presented at workshop on Environmental Management in Nigeria and Administration (NCEMA).

Koku, C.A., Osuntogun, B.A. 2007. Enviromental-Impacts of Road Transportation in Southwestern States of Nigeria. Journal of Applied Sciences 7 (16): 2536-2360, 2007.

Magbagbeola, N. O. 2001. The use of Economic Instruments for Industrial pollution Abatement in Nigeria: Application to the Lagos Lagoon. Selected papers, Annual Conferences of the Nigerian Economic Society held in Port-Harcourt.

Magbagbeola, I (2002), Environmental Underdevelopment of the Niger Delta: An Eclectic view, In: The Petroleum Industry, Economy and the Niger-Delta Environment. (Eds), Orubu. C, Ogisi, D.O. and Okoh R.N. 32-40.

Moffat, D; Linden O (1995), "Perception and Reality Assessing Priorities for Sustainable Development in the Niger Delta, Journal of the Human Environment, 24 No 7-8 327-538.

Ogbuigwe, A (1998), "Judicial Activism in the Enforcement of Environmental Regulations in the Petroleum Industry: Past, Present and the Future." Proceedings of 1998 International Conference on the Petroleum and the Nigerian Environment 83-123.

Olobaniyi, S.B.; Efe, S.I. (2007), Comparative assessment of rainwater and groundwater quality in an oil producing area of Nigeria: environmental and health implications, Journal of Environmental Health and Res. (6)2: 111-118.

Oluwole, A.F.; Olaniyi, H.B.; Akeredolu, F.A.; Ogunsola, O.J.; Obioh, I.B. (1996), Impact of the Petroleum Industry on Air Quality in Nigeria, Presented at the 8th Biennial International Seminar on the Petroleum Industry and the Nigerian Environment, Port Harcourt, 17-21 November.

Orubu, C O (2002b), Oil Industry activities, Environmental Quality, and the Paradox of Poverty in Niger Delta. In: The Petroleum Industry, the Economy and the Niger-Delta Environment. (Eds), Orubu, C.O., Ogisi, D.O. and Okoh, R.N. 17-31.

Saville, S. B., 1993. Automotive options and quality Management in developing Countries Industrial Environment. 16(1-2); 20, 32.

Seneca, J.J. and M.K. Tausig. 1994. *Environmental Economics*, Engle Wood Cliffs, Prentice Hall.

Taiwo, Olukayode(2005). "The case of Lagos –Air quality improvement project." LAMATA.

USEPA. 1993. Guide to Environmental Issues, Doc. No 520/B-94-01. United States Environmental Protection Agency, Washington, DC, USA.

United Nation. Prospect of World Urbanization 1989 (Population Study No. 112) New York.

Uyigue, E. and Agho, M. (2007). Coping with Climate Change and Environmental Degradation in the Niger-Delta of South Nigeria. Benin. Community Research and Development Centre (CREDC).

Wikipedia (2007), http://en.Wikipedia.org/Wiki/Environmental Issues in Nigeria/Niger Delta.

World Bank (1995), Defining an Environmental Strategy for the Niger Delta. Washington D.C.Industry and Energy Operations Division (West/Central Africa Department

10

Emerging Air Pollution Issues in Changing Pearl River Delta of South China

Mang Lin[1], Iat-Neng Chan[2], Chuen-Yu Chan[*1],
Xue-Mei Wang[1] and Han-Ying Dong[1]
[1]*School of Environmental Science and Engineering, Sun Yat-sen University*
[2]*Faculty of Science and Technology, University of Macao*
P.R. China

1. Introduction

1.1 Background

The Pearl River Delta (PRD) in Southern China is one of the most developed regions in China. Administratively, the delta region is constituted by nine cities in Guangdong Province, and the Special Administrative Regions of Hong Kong and Macao. The PRD has developed into a manufacturing center of the world since the onset of China's economic reform in the late 1970's. The important role of this region in China is highlighted by the high percentage contribution of gross domestic product (GDP) of Guangdong Province, which is mostly contributed by the PRD cities in the mainland, to that of the nation (>10%, National Bureau of Statistics China, 2009).

As an important manufacturing center and city cluster, the PRD region is inevitably affected by severe air pollution problems, which attract much public attention. In response to this situation, a regional air quality monitoring network compositing of 16 automatic stations has been established by the joint effort from Guangdong and Hong Kong governments since late 2005 with an aim to examine the air pollution problem of the region (GDEPMC & HKEPD, 2005). A series of large scale research programs including comprehensive aircraft and ground-base measurements and modeling exercises have been conducted by the scientific community to better characterize and analyse the problem and to find scientific evidence for formulating possible control strategies (e.g. the Program of Regional Integrated Experiments on Air Quality over the PRD of China-PRIDE-PRD in 2004 and 2006) (Zhang et al., 2008; Hua et al., 2008).

1.2 Changing characteristics of the PRD region

Fig. 1 depicts the land use change of the PRD from 1992 to 2004. The major urban areas were mainly found in the capital Guangzhou and the former British colony Hong Kong in the early 1990's. However, in the 21st century, almost the whole PRD region was characterized by urban areas. This dramatic urban expansion was found to be one of the most key factors affecting the air quality and climate of the region (Wang et al., 2007; Kaufmann et al., 2007). Furthermore, the steady increasing trends of GDP and electricity consumption also

*corresponding author

highlight the tremendous progress of urbanization and industrialization of this region (Fig. 2). Concurrent atmospheric composition changes and air quality degradation have been observed by space-borne satellite (e.g., Richter et al., 2005; Van der A et al., 2006; Ghude et al., 2009) and limited long-term ground-based measurements. Chan et al. (2004) first deduced that the regional background ozone (O_3) concentration in south China including the PRD region had been increased at a rate of 1.5% per year from 1984 to 1999 based on the urban measurements in Hong Kong. This finding was recently proven by Wang et al. (2009) using the data recorded from a background station in Hong Kong during the period 1994 to 2007. Besides, the metropolitan Guangzhou was also inevitably suffered from the air quality and visibility degradation during recent half century (Deng et al., 2008). The China's "12th five-year's plan" intends the PRD to transform its labor intensive and resource consuming light industry to heavy industry, telecom, equipment manufacture, auto and petrochemical industries for examples. With such a strategy, the traditional industry and enterprises are forced to move out from this region. Moreover, "cleaner" fuels such as liquefied petroleum gas (LPG) and low sulfur gasoline have been promoted by the government. Such massive changes of industry structure and prevalent fuel are anticipated to have strong effects on air quality. For instance, Tang et al. (2008) reported that there have been around 1.5 times increases in propane concentration in Guangzhou due to the introduction of LPG to the city in 2003. This pilot study highlights the scientific importance of clear understanding of the emerging air pollution issues along with these economic and industrial changes.

1.3 Aim and objectives

The objective of this chapter is to unveil the long-term air pollution changes in three typical commercial and industrial cities of the PRD region, Hong Kong, Macao and Foshan. The possible impacts of these changes on the atmospheric environment and their implications on the regional and urban air quality control strategy will be discussed. These cities are selected simply because they are the cities that we have accessible data. Hong Kong and Macao are located at the southern part of the PRD (Fig. 1). With the second-largest stock market in Asia, Hong Kong is well-known as an international financial center, while Macao is famous as a gambling city with flourishing tourism. On the other hand, Foshan is located at the central-western of the region, where urbanization process is most rapid (Fig. 1). This city is highly industrialized as reflected by its high industrial production value compared with those of Hong Kong and Macao (Table 1). The city is especially famous for ceramic industry, which produced approximately one fourth of the world's ceramic products in 2007 (Peng 2007; Yin 2008). A statistical summary on the general information of these three cities and data available periods are shown in Table 1.

2. Air pollution patterns and occurrence of episodes

Fig. 3 and Fig. 4 summarize the mean concentrations of major air pollutants and the occurrence of episode days in Hong Kong, Macao and Foshan. As expected, the highest SO_2 and PM_{10} concentrations were found in the industrial city Foshan due to the high SO_2 and PM_{10} emissions from industrial activities. Note that the average SO_2 concentration (60 μg m^{-3}) reached the National Ambient Air Quality Standard (Level II) (GB3095-1996) (NAAQS II) of 60 μg m^{-3} (monthly mean). In fact, more than 50% of the days during the period 2001-2007 in Foshan had daily SO_2 concentrations exceeding the NAAQS I of 50 μg m^{-3} (Table 2 and Fig. 4). Besides, the high percentages of PM_{10} episode days in Foshan highlight the severe

	Hong Kong	Macao	Foshan
Population[a]	7.0 million	0.5 million	5.7 million
Area[a]	1100 km^2	21 km^2	3800 km^2
Vehicle number[a]	0.70 million	0.18 million	0.60 million
Value of industrial production[b]	160 billion (HKD)	16 billion (MOP)	940 billion (CNY)
Visibility data	1968.01-2008.12	1990.01-2007.12	1998.01-2008.12
SO_2 data	1984.01-2008.12	1998.04-2007.12	2001.01-2008.12
NO_2 data	1984.01-2008.12	1998.04-2007.12	2001.01-2008.12
O_3 data	1984.01-2008.12	1998.04-2007.12	n.a.[c]
PM_{10} data	1993.01-2008.12	1998.04-2007.12	2001.01-2008.12

a. statistical data at 2008; b. statistical data at 2007; c. n.a. stands for data not applicable

Table 1. Summary of the general statistic information and the periods with data available for analysis

particle pollution at this industrial region. Hong Kong and Macao are both commercial cities with limited manufacturing industries. The SO_2, PM_{10} and O_3 concentrations of Macao were slightly higher, which could be attributed to the fact that Macao is located at the southeast of Hong Kong (Fig. 1). The prevailing northeasterly wind from September to April will carry aged and polluted air from Hong Kong and other PRD cities like Shenzhen and Dongguan to Macao to deteriorate air quality.

The average NO_2 concentration was highest in Hong Kong and lowest in Foshan. This result is reasonable as Hong Kong is a commercial city with dense traffic (Table 1), which results in the elevated NO_x emission and ambient NO_2 concentration. However, it is noted particularly that the average NO_2 concentration of Foshan during the period of 2006-2008 (60 μg m^{-3}, not shown) was higher than that in Hong Kong and Macao, which suggested that the remarkable increases of vehicle number and vehicular emission in Foshan was becoming more important (Wan et al., 2011). Moreover, although the number of episode day with daily NO_2 concentration larger than 80 μg m^{-3} (NAAQS I & II) in Foshan was close to that in Macao (~7.3%), the frequency of daily NO_2 concentration exceeding 120 μg m^{-3} (NAAQS III) in Foshan was significantly higher than that in Macao (Fig. 3). This fact emphasizes the needs of implementation of stricter NO_2 control measures in Foshan.

	Level I	Level II	Level III
SO_2 (Daily Mean)	50	150	250
NO_2 (Daily Mean)	80	80	120
O_3 (Hourly Mean)	120	160	200
PM_{10} (Daily Mean)	50	150	250

Table 2. Summary of the National Ambient Air Quality Standard (GB3095-1996) in China (unit: μg m^{-3})

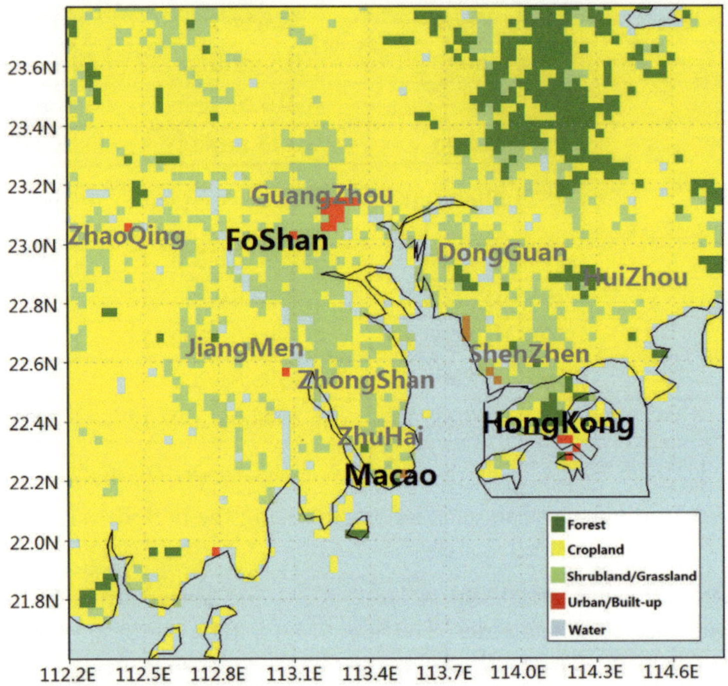

(a)

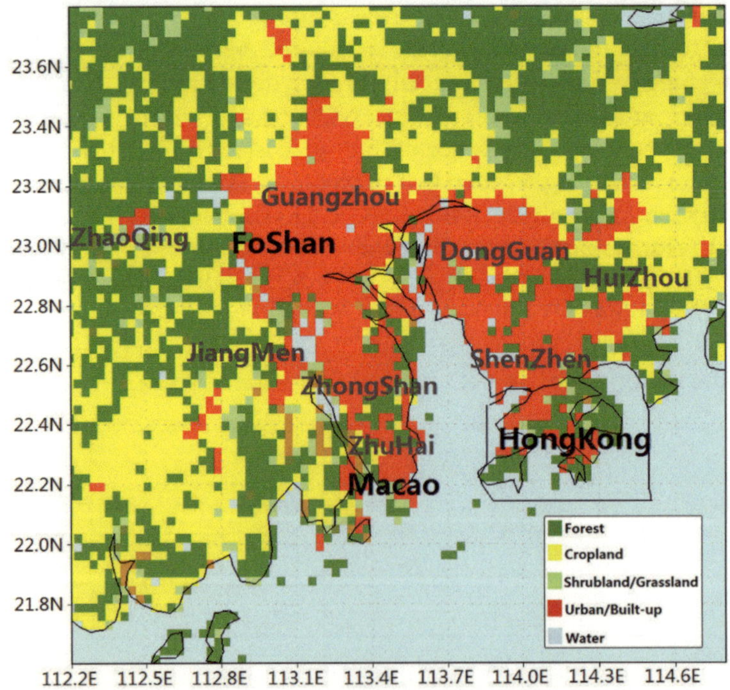

(b)

Fig. 1. Urban expansion in the PRD region from (a) 1992 to (b) 2004

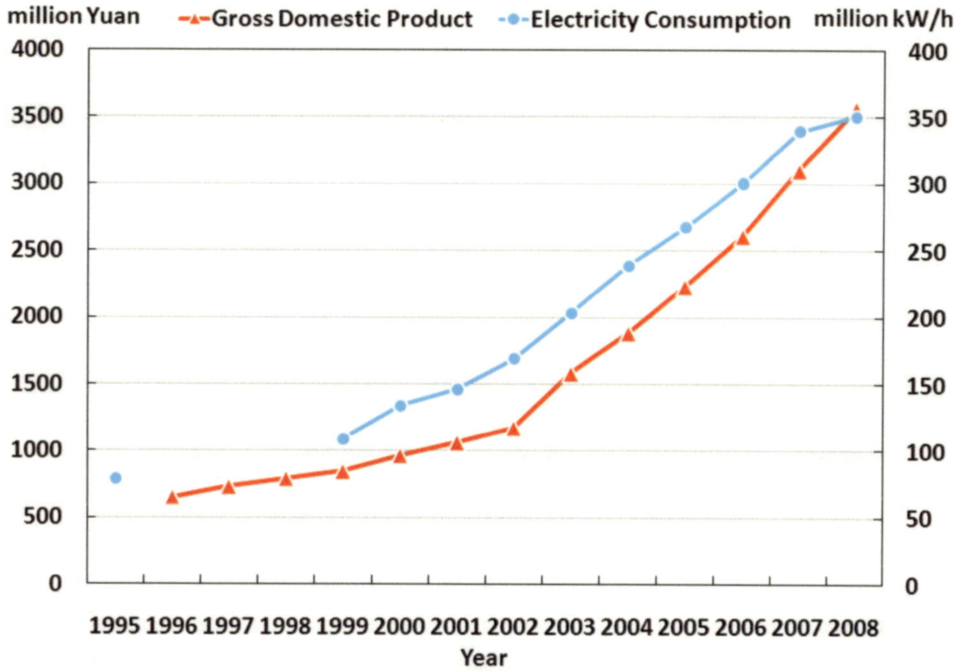

Fig. 2. Trends of annual gross domestic product and electricity consumption in Guangdong Province from 1995 to 2008.

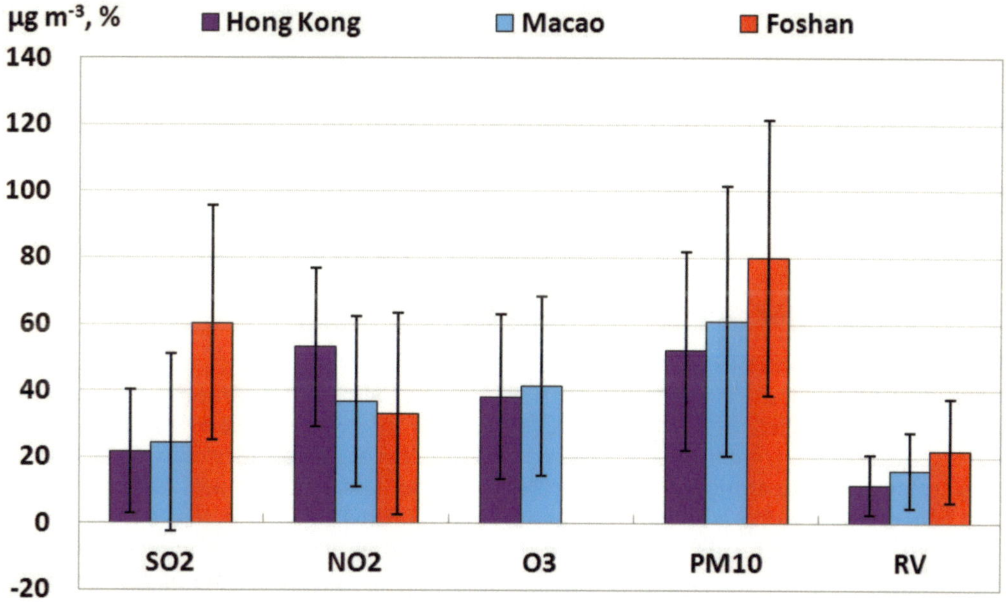

Fig. 3. Comparison of average air pollutants concentrations (2001-2007) and reduced visibility percentage (1998-2007) in Hong Kong, Macao and Foshan (error bars stand for one standard deviation)

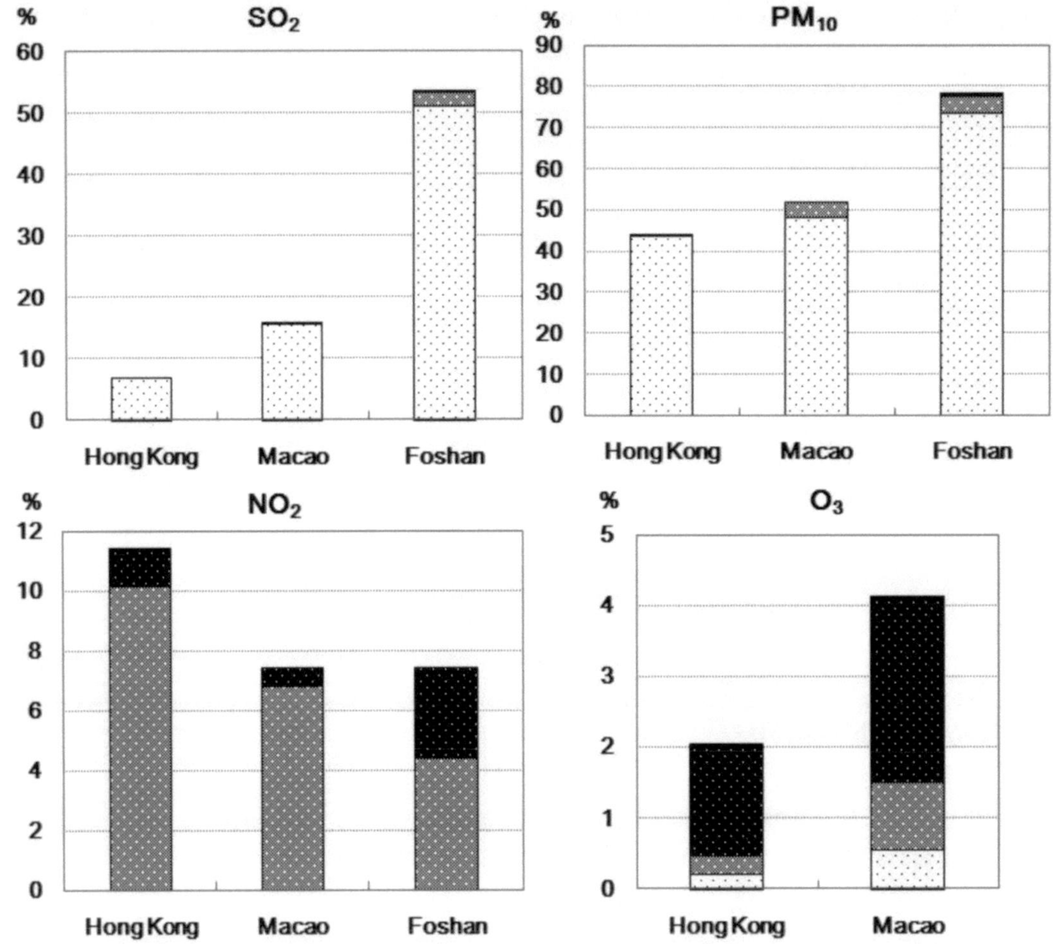

Fig. 4. Comparison of percentages of SO_2, PM_{10} and NO_2 exceeding days as well as O_3 exceeding hours in Hong Kong, Macao and Foshan during the period of 2001-2007 (White, gray and black bars stand for the percentages exceeding NAAQS I, II and III respectively)

3. Changes of major atmospheric compositions

3.1 Surface ozone and nitrogen dioxide

Ozone concentrations in Hong Kong and Macao both showed substantial increases with rates of 1.00 (R^2=0.752) and 1.06 (R^2=0.324) µg m^{-3} per year respectively (Fig. 5). The observed increasing trends are mostly caused by the regional increases in surface O_3 concentrations due to transport of its precursors from south China and East Asia as the urban area of Hong Kong is shown to be a net sink of inflowing O_3 due to the titration effect of nitrogen monoxide freshly emitted from traffic (Chan et al. 2004; Wang et al. 2009). NO_2 concentration in Macao showed a smoothly increasing trend with a rate of 1.26 µg m^{-3} per year (R^2=0.712). We believe that this change is probably accounted by the simultaneous increasing trend of vehicle number in Macao, which increased from 0.11 million in 2000 to 0.18 million in 2007 (Macao Statistics and Census Service, www.dsec.gov.mo). In contrast, NO_2 in Hong Kong kept a rather stable level after the 1990s.

Although long-term O_3 data were not available in Foshan, NO_2 concentration showed sharp increases from 18 μg m^{-3} in mid 2005 to 80 μg m^{-3} in late 2006 contradicting to the general decreases before 2005 (Fig. 5). This increasing rate reached 440% although NO_2 decreased to lower levels from 2007 to 2008. The overall increasing rate during 2001-2008 was 6.34 μg m^{-3} per year (R^2=0.496). The continuous increases in coal burning and road traffic, as well as lack of an efficient NO_2 control strategy, led to the remarkable increases of NO_2 level. More detailed discussions can be found in Wan et al (2011).

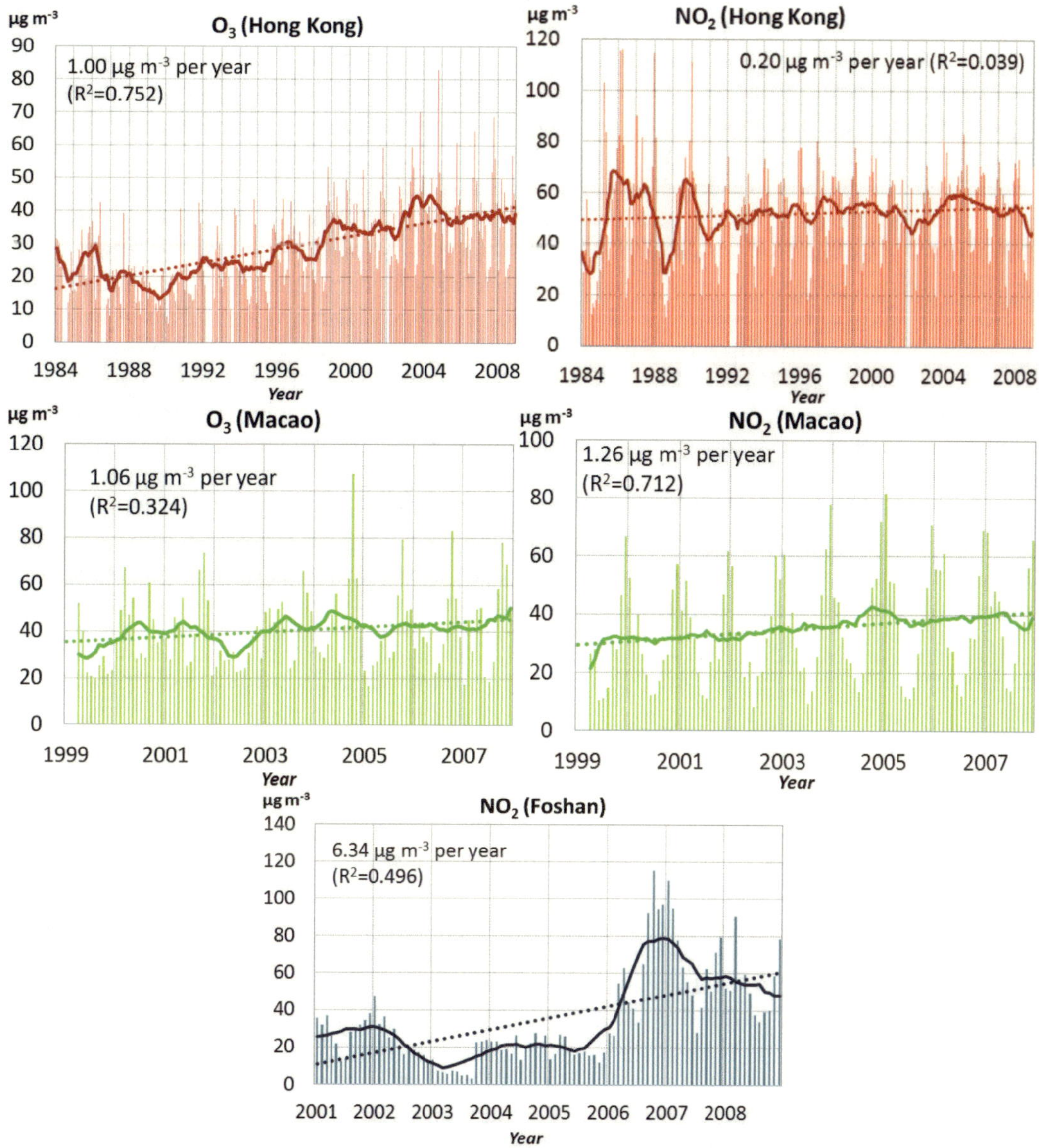

Fig. 5. Trends of O_3 and NO_2 concentrations in Hong Kong, Macao and Foshan (bars indicate the monthly means; solid lines indicate 12 months running averages; dot lines indicate linear regression lines)

3.2 Particulate matter and sulfur dioxide

In Macau, PM_{10} concentrations showed a steady increasing trend with a rate of 2.39 μg m^{-3} per year (R^2=0.716, Fig. 6). As high portion of organic carbon, which is one of the important species in particle primarily emitted by automobile in urban area, was found in PM_{10} in Macao (Wu et al., 2003), the increases of PM_{10} concentration were probably attributed to the increases of vehicle emission as mentioned above. Apart from the local emission, the airborne particulate matter in Macao was also influenced by transported secondary aerosols of regional scale (Wu et al., 2002, 2003). On the contrary, PM_{10} concentrations in Foshan displayed a steady decreasing trend with an overall rate of 3.47 μg m^{-3} per year in 2001-2008, which was believed to be due to continuous implementation of stricter emission control measures on various industrial sources. For instance, many high emitting factories including cement plants and small-scale ceramic plants were forced to close down by the city government in the past decades (Wan et al., 2011).

The SO_2 concentrations in Hong Kong showed a growth rate of 0.28 μg m^{-3} per year in the period of 1984-2008 (Fig. 6). Specifically, the decreases from 1993 were due to the effective banns of high-sulfur fuel usage and reduction of power plant emissions in Hong Kong (HKEPD, www.epd.gov.hk), while the increases in recent years were owing to the increases in SO_2 emissions in both Shenzhen city and Hong Kong (Chan and Yao 2008). Apart from the local SO_2 emission, the increasing trend of regional SO_2 emission also needed to be considered (Lu et al., 2010). Although the SO_2 concentration in Foshan was high when compared to other two commercial cities, a slightly decreasing trend (0.28 μg m^{-3} per year, R^2=0.364) was observed in the 21st century, which was due to the wide application of flue-gas desulfurization (FGD) devices in power plants in response to a new policy of China's government (Lu et al., 2010). In fact, the overall SO_2 removal efficiency of exhaust stream passing the treatment system in various major industrial sectors in Foshan was up to 64% (Fig. 7). However, the continuous growth in coal consumption triggered the substantial increases in 2006 (Wan et al., 2011).

4. Enhancement of atmospheric oxidizing power and its possible impacts

Given that O_3 is a vital atmospheric oxidant and NO_2 is one of the most important O_3 precursors, the remarkable increases of O_3 in Hong Kong and Macao as well as the tremendous augment of NO_2 in Foshan highlight the enhancement of atmospheric oxidizing power in the PRD region. It is reasonable to hypothesize that such changes would exert impacts on atmospheric environmental quality and chemistry. In this section, we used visibility as a proxy of atmospheric environmental quality changes to investigate the impacts of air pollution changes on the atmospheric chemical processes. Following the definition of the Hong Kong Observatory, the observed visibility data were transformed to Reduced Visibility (RV) which is defined as the percentage of hours with visibility below 8 km, excluding the data of rainy days or those affected by fog or high relative humidity (≥95%) (Leung et al., 2004). With this definition, the resulted RV would better indicate the frequency of visibility degradation caused by elevated pollutant level in the atmosphere. As expected, the highest RV was found in Foshan while the lowest RV occurred in Hong Kong (Fig. 3).

Three multiple regression models and the partial correlations between RV and each air pollutant were calculated. The model results were subjected to the F-test and student t-test with a significant level $p<0.05$. After that, the air pollutants with significant partial correlation were kept at the model equations while others were removed from the equations

(Table 3). Generally, visibility in Hong Kong is primarily influenced by concentration levels of O_3, SO_2 and NO_2 in recent 16 years. Particularly, RV would increase 0.28%, 0.27% and 0.15% if concentrations of SO_2, NO_2 and O_3 increase 1 μg m^{-3} respectively. In Macao, the RV was significantly affected by PM_{10} concentration with a value of increasing 0.33% of RV per 1 μg m^{-3} of PM_{10}. Similar relationship exists in Foshan with a value of increasing 0.50% of RV per 1 μg m^{-3} of PM_{10}. Besides, the RV in Foshan correlated with NO_2 concentration with a value of increasing 0.14% of RV per 1 μg m^{-3} of NO_2.

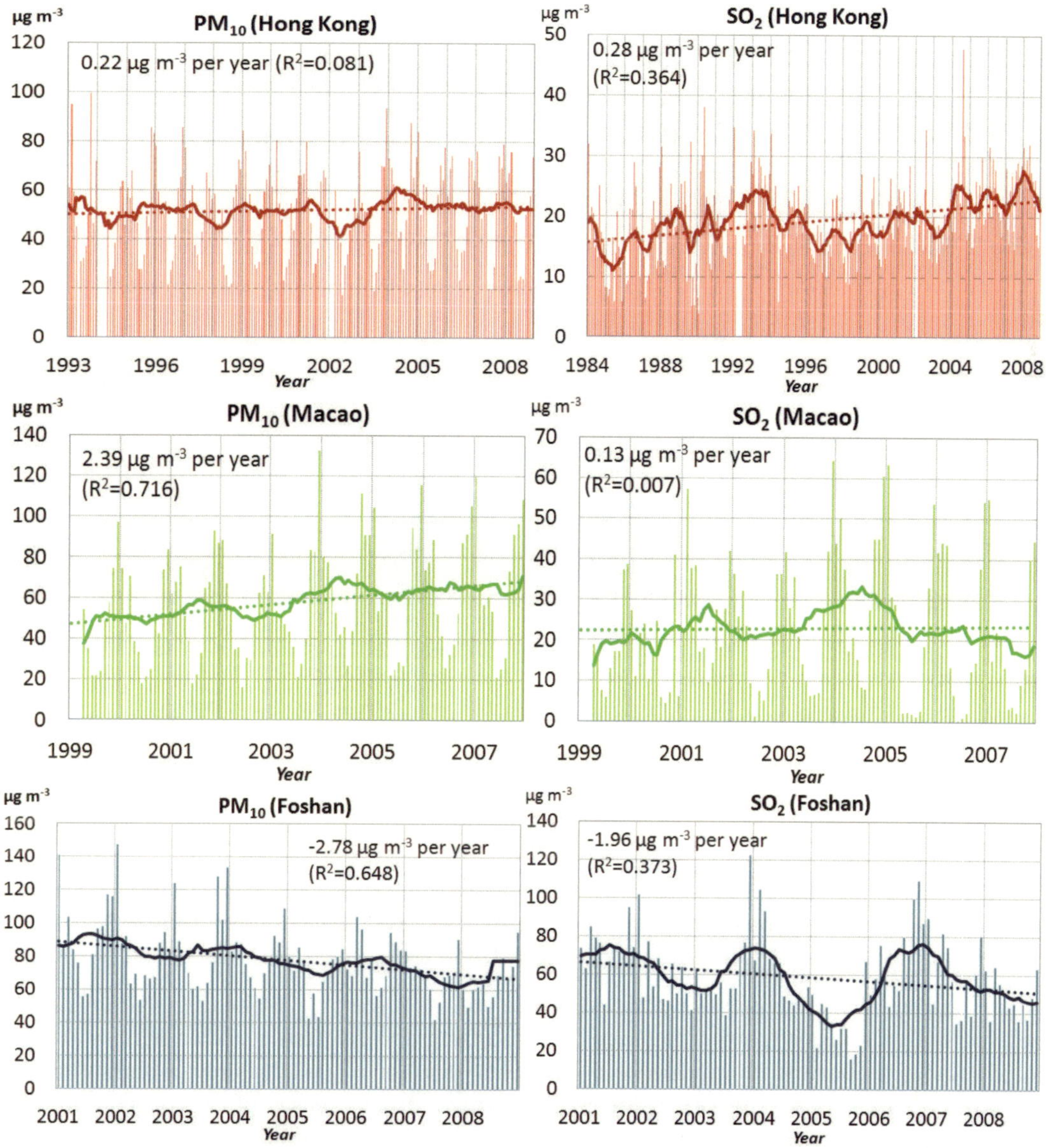

Fig. 6. Trends of PM_{10} and SO_2 concentrations in Hong Kong, Macao and Foshan (bars indicate the monthly means; solid lines indicate 12 months running averages; dot lines indicate linear regression lines)

It is well-known that visibility impairment is always proportion to the loading of fine particles owing to the scattering capacity of aerosol. Besides, NO_2 is another important species in the atmosphere which can degrade visibility due to its strong blue-light-absorbing capacity. Therefore, it is easy to explain the correlation between RV and PM_{10} as well as NO_2.

Regression Model	Independent Variables	Equation[a]
1 (Hong Kong 1993 – 2008)	O_3, SO_2, NO_2, PM_{10}	$[RV] = 0.28\,[SO_2] + 0.27\,[NO_2] + 0.15\,[O_3] - 15.80$
2 (Macao 1999 - 2007)	O_3, SO_2, NO_2, PM_{10}	$[RV] = 0.33\,[PM_{10}] - 5.70$
3 (Foshan 2001 - 2008)	SO_2, NO_2, PM_{10}	$[RV] = 0.14\,[NO_2] + 0.50\,[PM_{10}] - 21.24$

a. Units of RV and air pollutant are % and $\mu g\ m^{-3}$, respectively.

Table 3. Summary of equations calculated by multiple linear regression method

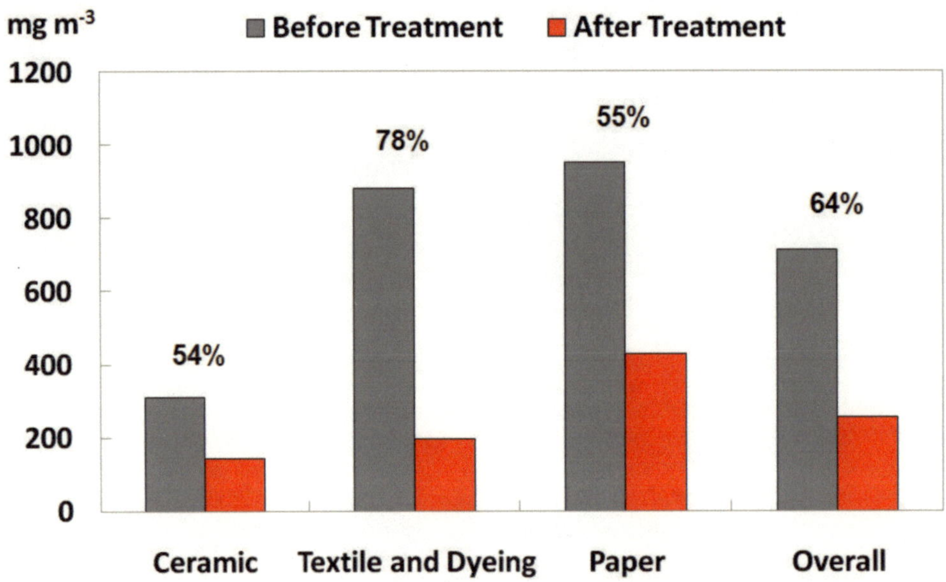

Fig. 7. Emission data of major industrial sectors in 2007 and 2008: comparison of SO_2 concentrations before and after emission control treatment (Source: Wan et al., 2011)

However, the correlation between RV and SO_2 as well as O_3 can hardly be directly interpreted because both O_3 and SO_2 are rather weak in absorbing and scattering visible light, and they thus will not produce strong influence on visibility directly. In fact, this result highlights the changes of atmospheric chemistry due to the tremendous increases of O_3 concentration in Hong Kong and Macao. In the enhanced O_3 atmosphere, the increasing oxidation rate would transform SO_2 to secondary fine sulfate particles more rapidly, which could effectively degrade visibility (Fig. 8, Meng et al. 1997, Shao et al. 2006). The chemical mechanism of S(IV) to S(VI) transformation has been extensively studied in literatures (e.g. Penkett et al, 1979;

Stockwell and Calvert, 1983; Seinfeld and Pandis, 1998; Liang and Jacobson, 1999). The enhanced formation of secondary sulfate in the O_3-rich atmosphere mentioned above actually agrees with the findings from a recent study, in which SO_4^{2-} concentrations in East Asia were found to increase at a greater rate than SO_2 emissions and SO_2 concentrations because "East Asia is a less oxidant-limited area than other parts of the world" (Lu et al. 2010).

In addition, although long-term O_3 data in Foshan were not available, the elevated level of NO_2 concentration more or less highlights the enhancement of oxidizing power of atmosphere in this region as well. It is well known that O_3 is generated from the photolysis of NO_2 and reacts mainly with NO to regenerate NO_2 quickly. If CO or VOCs are present, the simply photostationary state mentioned above is modified in a way that NO_2 can be regenerated through the reaction between NO and HO_2 (or RO_2). Net formation of O_3 occurs because the conversion of NO to NO_2 is accomplished by the HO_2 (or RO_2) radical rather than by O_3 itself. Therefore, the enhanced NO_2 concentration in the atmosphere also reflects the presences of other secondary pollutants like fine particle which significantly contributes to visibility impairment. The significant correlations between RV and NO_2 concentration in Hong Kong and Foshan thus underline the important role of NO_2, or even secondary pollutants, play in visibility reduction. In Macao, RV did not correlate well with O_3 and NO_2 concentrations. However, the high portion of secondary aerosols in PM_{10} and steady increasing trends of O_3 and NO_2 also emphasize the vital role of high O_3 levels play in photochemical formation of secondary pollutants such as sulfate, which would exert notable impacts on visibility, public health and climate.

5. Summary and discussion

Changes of major atmospheric compositions and air pollution levels due to the rapid economic and industrial development have been observed in the PRD region. The data such as those presented in this paper suggest that these long-term air pollution trends have vital implications on the formulation of regional air quality control strategies. Generally, the elevated concentration levels of NO_2 and O_3 in the three commercial and industrial cities highlight the enhancement of atmospheric oxidizing power in the PRD region, which may exert profound effects on atmospheric chemistry, public health as well as regional climate.

Specifically, both of SO_2 and O_3 concentrations in Hong Kong show significant increasing trends. Given that SO_2 would be oxidized to secondary sulfate more easily in the O_3 enhanced atmosphere, the simultaneously increasing O_3 and SO_2 are reasonably expected to trigger the formation of secondary sulfate and resulted in the pronounced visibility degradation. Wang et al. (2009) have suggested that short-term O_3 control strategies should be aimed at O_3 precursor sources in Hong Kong and the adjacent PRD while the long-term ones need to consider distant sources. Given that the increase of SO_2 concentration was due to the enhanced local SO_2 emissions of Hong Kong and Shenzhen (Chan and Yao, 2008) as well as elevated regional SO_2 emissions in East Asia (Lu et al., 2010), similar strategy should be adopted for SO_2 control.

In Macao, the increasing trends of secondary pollutants like O_3 and NO_2 highlight the influences of increasing O_3 precursor emission in its upwind regions such as eastern PRD and the east Chinese coast. Therefore, the air pollution controls of Hong Kong, as well as other upwind regions are vital. Furthermore, the augment of NO_2 concentration also implies that the local vehicular emission control of Macao should not be neglected. As PM_{10} in Macao was mainly affected by local vehicular emission and regional transport, its increasing trend also emphasized the importance of both local and regional controls.

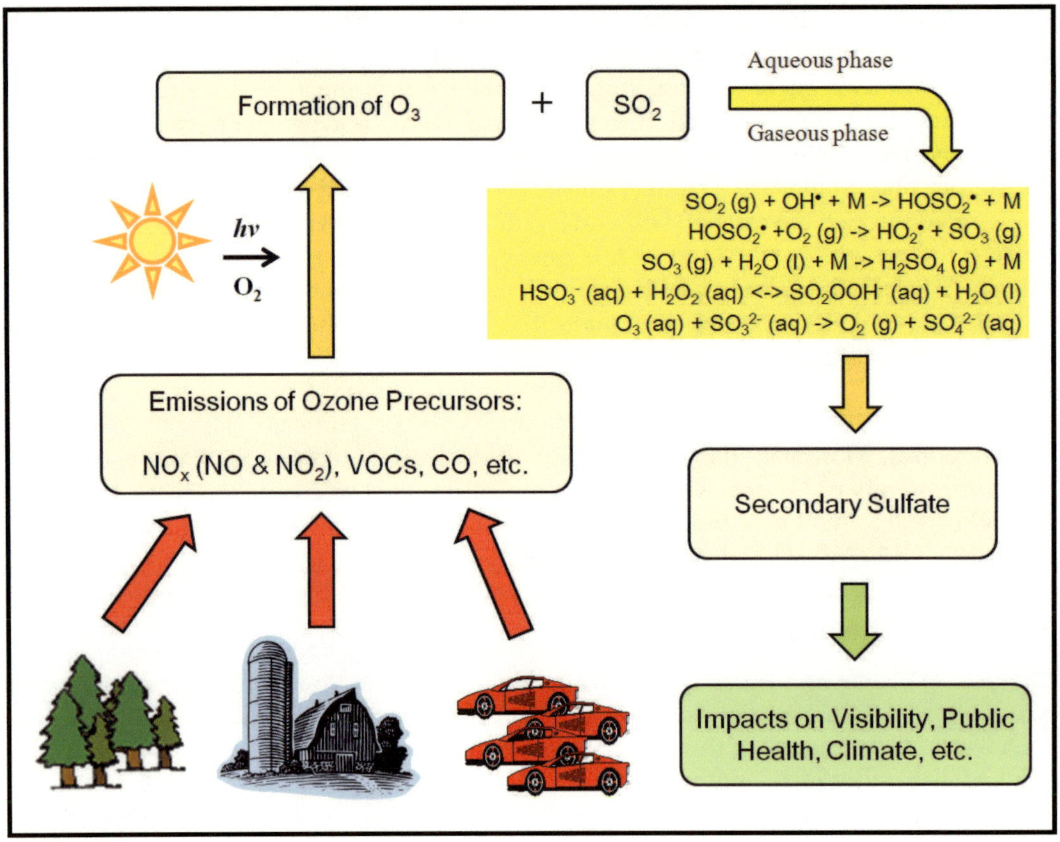

Fig. 8. Schematic diagram showing photochemical ozone formation from its precursors emitted from anthropogenic and natural sources and its impact on secondary sulfate formation and visibility degradation through aqueous and gaseous phase chemical processes in the ozone-enhanced atmosphere.

On the other hand, decreasing trends of PM_{10} and SO_2 were observed in Foshan. These are due to the efforts of more efficient control in industrial sectors. In a recent paper, Wan et al. (2011) emphasized that the implementation of stricter PM_{10} and SO_2 control measures in industrial exhaust had resulted in reduced ambient air pollution level. However, this industrial city is still severely polluted compared to the commercial cities like Hong Kong and Macao due to its high coal burning and industrial emissions. Such fact suggested that the SO_2 as well as particles emission control measures must be strengthened. In contrast, the surprising increases of NO_2 concentration due to the continuous increasing in road traffic and lack of efficient NO_2 control in industrial sectors should be noticed. Such increasing trend is an evident of atmospheric oxidizing power enhancement in this region. If the air quality control strategies are not changed, the NO_2 concentration will keep increasing especially with the fact that the vehicle number keeps on increasing sharply in recent years.

Overall, due to the rapid economical and industrial development, the nature of air pollution in the PRD region has been changing sharply. Essentially, the findings reported in this chapter indicated that the atmospheric oxidizing power in the PRD region has been continuously increasing. Even in the industrial city, which was more affected by primary pollutants like SO_2, the vehicle emission and resulted photochemical pollution should be noticed. The enhancing atmospheric oxidizing power could change the atmospheric chemistry by triggering the secondary pollutant formation via increasing oxidation rate. The increasing secondary pollutants may exert a number of significant impacts on public health and regional climate. As most of secondary pollutants were related to transport of aged air mass, such findings emphasize the critical impacts of trans-boundary transportation and regional air pollutions. With an aim to control the secondary pollutants, cooperation with other provinces and even countries become more and more important. Besides, due to the enhanced atmospheric oxidation capacities, primary pollutants like SO_2 and some of volatile organic compounds will be transformed to secondary fine particles more efficiently, and therefore, stricter implementation of local primary pollutant control was also necessary. In summary, in order to efficiently improve the regional air quality in the PRD region, local and regional efforts on air pollution control are both required.

6. Acknowledgment

We thank Hong Kong Observatory, Hong Kong Environmental Protection Department, Meteorological and Geophysical Bureau of Macao, Environmental Protection Bureau of Foshan and Meteorological Bureau of Foshan for providing the data. This study was supported by a key project of the Natural Science Foundation of Guangdong Province, China (No. 825102501000002), the National Natural Science Foundation of China (No. 40875075), a joint fund of the National Natural Science Foundation of China and Natural Science Foundation of Guangdong Province, China (No. U0833001), and Rosa Luxemburg Foundation.

7. References

Chan, C.Y. & Chan, L.Y. (2000). Effect of meteorology and air pollutant transport on ozone episodes at a subtropical coastal Asian city, Hong Kong. *Journal of Geophysical Research*, 105(D16), 20707-20724.

Chan, C.Y., Chan, L.Y. & Harris, J.M. (2004). Urban and background ozone trend in 1984-1999 at subtropical Hong Kong, South China. *Journal of Ozone: Science and Engineering* 25, 513-522.

Chan, C.K. & Yao, X.H. (2008). Air pollution in mega cities in China. *Atmospheric Environment* 42, 1-42.

Chan, L.Y., Chan, C.Y. & Qin, Y. (1998). Surface ozone pattern in Hong Kong. *Journal of Applied Meteorology*, 37, 1153-1165.

Che, H.Z., Zhang, X.Y., Li, Y., Zou, Z.J. & Qu, J.J. (2007). Horizontal visibility trends in China 1981-2005. *Geophysical Research Letters* 34, doi:10.1029/2007GL031450.

Che, H.Z., Zhang, X.Y., Li, Y., Zou, Z.J., Qu, J.J. & Hao, X.J. (2009). Haze trends over the capital cities of 31 provinces in China, 1981-2005. *Theoretical and Applied Climatology* 97(3-4), 235-242.

Deng, X.J., Tie, X.X., Wu, D., Zhou, X.J., Bi, X.Y., Tan, H.B., Li, F. & Jiang, C.L. (2008). Long-term trend of visibility and its characterizations in the Pearl River Delta (PRD) region, China. *Atmospheric Environment* 42(7), 1424-1435.

Guangdong Provincial Environmental Protection Monitoring Center & Hong Kong Environmental Protection Department (GDEPMC & HKEPD) (2005). Pearl River Delta Regional Air Quality Monitoring Network Operational Report for the Second Half of 2005. Report number : PRDAIR-2005-1.

GDEPMC & HKEPD (2008). Pearl River Delta Regional Air Quality Monitoring Network Report. Report number : PRDAIR-2008-2.

Ghude, S.D., Van der A, R.J., Beig, G., Fadnavis, S. & Polade, S.D. (2009). Satellite derived trends in NO_2 over the major global hotspot regions during the past decade and their inter-comparison. *Environmental Pollution* 157, (1873-1878).

Hua, W., Chen, Z.M., Jie, C.Y., Kondo, Y., Hofzumahaus, A., Takegawa, N., Chang, C.C., Lu, K.D., Miyazaki, Y., Kita, K.,Wang, H.L., Zhang, Y.H., Hu, M., 2008. Atmospheric hydrogen peroxide and organic hydroperoxides during PRIDE-PRD'06, China: their concentration, formation mechanism and contribution to secondary aerosols. Atmospheric Chemistry and Physics 8, 6755e6773.

Kaufmann, R.K., Seto, K.C., Schneider, A., Liu, Z., Zhou, L. & Wang, W. (2007). Climate response to rapid urban growth: evidence of a human-induced precipitation deficit. *Journal of Climate*, 20: (2299-2306).

Leung, Y.K., Wu, M.C. & Yeung, K.K., 2007. A preliminary study on the relationship between visibility and atmospheric suspended particulate concentration in Hong Kong. Hong Kong Observatory Reprint No. 692. (In Chinese with English abstract)

Liang, J.Y. & Jacobson, M.Z. (2000). Effects of subgrid segregation on ozone production efficieny in a chemical model. *Atmospheric Environment* 34(18), (2975-2982).

Lu, Z., Streets, D.G., Zhang, Q., Wang, S., Carmichael, G.R., Cheng, Y.F., Wei, C., Chin, M., Diehl, T., & Tan, Q. (2010). Sulfur dioxide emissions in China and sulfur trends in East Asia since 2000. *Atmospheric Chemistry and Physics*, 10, 6311-6331.

Meng, Z., Dabdub, D. & Seinfeld, J.H. (1997).Chemical coupling between atmospheric ozone and particulate matter. *Science*, 277, 5322, 116-119.

National Bureau of Statistics China (2009). China Statistical Yearbook 2008. Beijing: China Statistics Press.

Peng, J.N. (2007). The ceramic production transfer in Foshan. Available at URL: http://finance.sina.com.cn/roll/20071206/15211840962.shtml. (In Chinese)

Penkett, S.A., Jones, B.M.R. & Eggleton, A.E.J. (1979). A study of SO_2 oxidation in stored rainwater samples. *Atmospheric Environment* 13(1), 139-147.

Qiu, J.H. & Yang, L.Q. (2000). Variation characteristics of atmospheric aerosol optical depths and visibility in North China during 1980–1994. *Atmospheric Environment* 34(4), 603-609.

Richter, A., Burrows, J.P., Nüß, H., Granier & C., Niemeier, U. (2005). Increase in tropospheric nitrogen dioxide over China observed from space. *Nature* 437, 129-132.

Seinfeld, J.H. & Pandis, S.N. (1997). Atmospheric chemistry and physics: From air pollution to climate change. New York: Wiley-Interscience.

Shao, M., Tang, X.Y., Zhang, Y.H. & Li, W.J. (2006). City clusters in China: air and surface water pollution. *Front Ecol Environ* 4(7): 353-361.

Sisler, J.F. & Malm, W.C. (2000). Interpretation of trends of $PM_{2.5}$ and reconstructed visibility from the IMPROVE network. *Journal of the Air & Waste Management Association* 50, 775-789.

Stockwell, W.R. & Calvert, J.G. (1983). The mechanism of the $HO\text{-}SO_2$ reaction. *Atmospheric Environment* 17(11), 2231-2235.

Tang, J.H., Chan, L.Y., Chan, C.Y., Li, Y.S., Chang, C.C., Wang, X.M., Zou S.C., Barletta, B., Blake, D.R. & Wu, D. (2008). Implications of changing urban and rural emissions on non-methane hydrocarbons in the Pearl River Delta region of China. *Atmospheric Environment* 42, 3780-3794.

Van der A, R.J., Peters, D.H.M.U., Eskes, H., Boersma, K.F., Van Roozendael, M., De Smedt, I. & Kelder, H.M. (2006). Detection of the trend and seasonal variation in tropospheric NO2 over China. *Journal of Geophysical Research* 111, doi:10.1029/2005JD006594.

Wan, J.M., Lin, M., Chan, C.Y., Zhang, Z.S., Engling, G., Wang, X.M., Chan, I.N. & Li, S.Y., (2011). Change of air quality and its impact on atmospheric visibility in central-western Pearl River Delta. *Environmental Monitoring and Assessment*, 172, 339-351.

Wang, X.M., Lin, W.S., Yang, L.M., Deng, R.R. & Lin, H. (2007). A numerical study of influences of urban land-use change on ozone distribution over the Pearl River Delta region, China. *Tellus B*, 59,3 633-641.

Wang, T., Wei, X.L., Ding, A.J., Poon, C.N., Lam, K.S., Li, Y.S., Chan, L.Y. & Anson, M. (2009). Increasing surface ozone concentrations in the background atmosphere of southern China, 1994-2007. *Atmospheric Chemistry and Physics Discussions* 9, 10429-10455.

Wu, Y., Hao, J.M., Fu, L.X., Wang, Z.S., Tang, U.W. (2002). Vertical and horizontal profiles of airborne particulate matter near major roads in Macao, China. *Atmospheric Environment* 36, 4907-4918.

Wu, Y., Hao, J.M., Fu, L.X., Hu, J.N., Wang, Z.S., Tang, U.W. (2003). Chemical characteristics of airborne particulate matter near major roads and at background locations in Macao, China. *The Science of the Total Environment* 317, 159-172.

Yin, H. (2008). Analysis and prospect of the tile industry in 2008. *China Ceramics*, 44(5), 3-4. (In Chinese with English abstract)

Zhang, Y.H., Hu, M., Zhong, L.J., Wiedensohler, A., Liu, S.C., Andreae, M.O., Wang, W. & Fan, S.J. (2008). Regional Integrated Experiments on Air Quality over Pearl River Delta 2004 (PRIDE-PRD2004): Overview. *Atmospheric Environment* 42, 6157-6173.

Zheng, J., Zhong, L., Wang, T., Louie, P.K.K. & Li, Z. (2010). Ground-level ozone in the Pearl River Delta region : Analysis of data from a recently established regional air quality monitoring network. *Atmospheric Environment* 44, 814-823.

11

Causes and Consequences of Air Pollution and Environmental Injustice as Critical Issues for Science and Environmental Education

Anastasia Dimitriou[1] and Vasilia Christidou[2]
[1]*Democritus University of Thrace*
[2]*University of Thessaly*
Greece

1. Introduction

1.1 Humanity facing complex environmental issues

Modern societies suffer from a diversity of problems involving crucial environmental aspects. These problems concern biodiversity conservation, clean water and clean unpolluted air availability, healthy food, personal and public health and well-being. The causes of these problems lay in the way individuals -as well as entire societies- interact with the natural and form the man-made environment by adopting specific practices and developmental choices. They are also connected with the way people perceive the relationship between human beings, nature and society, in other words the way in which we perceive nature, ourselves and others. For decades -even for centuries- the kind of this relationship has been determined by the conviction of man's dominance over nature and of the more powerful over the weak parts of the population. Thus, in order to face environmental problems and protect the environment it is essential that the relationship between human beings, nature and society is revised. As Einstein aptly put it *"we can't solve problems by using the same kind of thinking we used when we created them"*.

Confronting environmental problems is not an unequivocal process, since they involve multiple dimensions and are determined by a variety of factors. They entail physical, chemical, biological, ecological, social, economic, and political dimensions, intrinsically interrelated and interacting, and often conflicting. Also, environmental problems are usually ideologically 'charged', incorporate different social groups' views and interests, and are connected with decisions taken at economic, political and social level.

Therefore, facing environmental problems is a complex procedure requiring –among others- determining their causes and consequences, their management, the development of alternative solutions and decision-making. It also requires disentangling their components and characteristics and critically determining their inherent interrelations, mechanisms, and conflicts. For this procedure to be effective, an understanding of the environment, its functions and constituent systems –both natural and social- as a whole is imperative. Hence, addressing environmental issues calls for interdisciplinary, systemic and holistic approaches. It also calls for promoting citizens' critical and systemic thinking, in order for them to identify the causes –and not simply the symptoms- of these issues, to understand the conflicts embedded in their consideration and treatment by different social, political, or interest groups (Dimitriou, 2009; Flogaitis, 2006; Orr, 1992; Scott & Oulton 1998; Sterling, 2004; Tilbury & Ross, 2006).

In this context, the role of science and environmental education is important. These fields of study concentrate on mechanisms and phenomena of the natural and man-made environment and provide students –as future citizens- the knowledge, skills and methodological tools necessary for studying the world around them. Both science and environmental education can contribute to the development of citizens capable of analytical and synthetic thinking, of investigating their position in the world they live in, of understanding the ways in which people change their environment and are affected by it and its changes, and conscious of the relationships between the individual, society, and the environment. Citizens who will be equipped with knowledge and skills essential for appreciating contemporary environmental problems, making relevant decisions and taking actions for their solution in the perspective of a sustainable future. Moreover, students, as future citizens could be supported in perceiving themselves as equivalent in their relationships with other people and natural entities, with a view to counterbalancing the dominating view of man's dominance over nature.

1.2 Sustainability: A key idea to cope with environmental degradation and social injustice

It is widely acknowledged that modern lifestyles and developmental models exceed earth's ability to support human beings and other organisms in terms of providing clean and adequate natural resources. An important concept related to modern lifestyles and the resulting environmental degradation is that of the 'ecological footprint'. Ecological footprint is a measure of people's demand on natural ecosystems. It is a measure of the area of biologically productive[1] land and water an individual, population or activity needs to produce the resources it consumes and absorb the waste it generates using current technology and resource management practices (Global Footprint Network, 2011). In other words, ecological footprint represents the land area an individual, a city or a nation take to support their lifestyles. It is tightly connected with the amounts of waste and pollutants produced by human activities as well as their energy demands. Probably the personal ecological footprint differs from one person to another as well as from one city (or country) to another. More particularly, the ecological footprint of developed countries is much larger than that of developing ones (Global Footprint Network, 2005). This indicates that ecological footprint is related with the notion of environmental injustice.

This has lead societies to recognize the necessity of changing the way of thinking about human activities, focusing on practices that sustain and protect the environment. Adopting sustainable everyday practices and developmental interventions is a crucial challenge for modern societies. Sustainability means the capacity to endure and refers to both physical and man-made systems, living and non-living things. Thus the term "sustainability" is used as determinative for that kind of development that ensures healthy ecosystems and the potential of long-term preservation of humanity.

---

[1] The land and water area that supports significant photysynthetic activity and biomass accumulation used by humans.

The terms of sustainability and sustainable development have been defined in various ways. The most common definition is given in the report of the World Commission on Environment and Development entitled "Our Common Future", also known as the Brundtland Report, according to which *"Sustainable development is development that meets the needs of the present without compromising the ability of future generations to meet their own needs"* (WCED, 1987, paragraph 27). In particular, as it is quoted (WCED, 1987, paragraph 27):
"The concept of sustainable development does imply limits -not absolute limits but limitations imposed by the present state of technology and social organization on environmental resources and by the ability of the biosphere to absorb the effects of human activities. But technology and social organization can be both managed and improved to make way for a new era of economic growth. The Commission believes that widespread poverty is no longer inevitable. Poverty is not only an evil in itself, but sustainable development requires meeting the basic needs of all and extending to all the opportunity to fulfil their aspirations for a better life. A world in which poverty is endemic will always be prone to ecological and other catastrophes".
Sustainability contains within it, at least, three crucial concepts:

- the fulfilment of human needs -and in particular the essential needs of the world's poor- that support improvement of people's lives;
- the fulfilment of the needs of present and future generations that support the value of intergeneration equity;
- the ability of the biosphere to absorb the effects of human activities.

So, sustainability is an evolving concept that contains the key ideas of improving everyone's quality of life (including present and future generations), in the frame of the limits that are posed by the carrying capacity of the ecosystems that support lives. The carrying capacity of an ecosystem is the population size of the species that the ecosystem can support indefinitely, providing the necessities (food, habitat, water, air, and ground) without being damaged (Hui, 2006).
A crucial factor for accomplishing sustainability is the appreciation of social injustice resulting from environmental degradation. Both researchers and international organisations agree that the degree to which the quality of life of individuals and societies is influenced by environmental degradation is different between members of a society as well as between societies (Bard et al 2007; Cifuentes, & Frumkin, 2007; UNICEF, 2008; WHO, 2003a, 2006, 2008a).
These differences create or amplify existing inequalities both between and within countries, related to the quality of basic sanitary infrastructures, food sufficiency, or people's access to clean air and potable water. These commodities constitute fundamental rights for life and health. Nowadays -despite technological and cultural achievements- people suffer from lack of clean, potable water, food insufficiency, polluted air, and deficient sanitary infrastructure. Conditions intensifying environmental injustice threaten fundamental human rights. Klaus Toepfer, Executive Director of the United Nations Environment Programme appositely described this reality at the 57th Human Rights Commission meeting in Geneva in 2001 (Shelton, 2002):
"Environmental conditions clearly help to determine the extent to which people enjoy their basic rights to life, health, adequate food and housing, and traditional livelihood and culture. It is time to recognize that those who pollute or destroy the natural environment are not just committing a crime against nature, but are violating human rights as well. Human rights cannot be secured in a degraded or polluted environment. The fundamental right to life is threatened by soil degradation and deforestation and by exposures to toxic chemicals, hazardous wastes, and contaminated drinking water".

The study of environmental quality in relation to social inequalities produced by environmental degradation is supported and promoted by various researchers. Some underline the role of environmental science in this direction and focus on the development of scientific methods to estimate pollution of environmental recipients, i.e. air, ground, and water (Stephens, 2007). Others emphasise the relation of environmental quality with quality of life and particularly with human health (Donohoe, 2003; Heynen, 2003; Mitchell & Dorling, 2003).

In this chapter we aim at elaborating on the idea of environmental justice as central in organising and implementing relevant educational interventions in the context of science and environmental education at school, as a means to cope with environmental and social injustice and promote sustainability. In the following sections of this chapter we discuss the problem of air pollution and further develop the issues of environmental justice and injustice. More particularly, air pollution, its causes and consequences are discussed, with an emphasis on human health and quality of life. Data demonstrating environmental injustice caused by air pollution are presented and the idea of environmental justice is introduced as a means to counterbalance it. The discussion then focuses on the contribution of science and environmental education to facilitate students' and (future) citizens' understanding of the causes and consequences of air pollution as critical dimensions of environmental injustice. Suggestions and implications for developing relevant educational programs at school are also provided, aiming at enhancing students' scientific knowledge and skills necessary to understand air pollution and determine actions for resolving it in view of sustainability and environmental justice.

2. Air pollution and environmental injustice

Air pollution is one of the most pressing environmental problems and is among the critical challenges facing modern societies. Air pollution is responsible for major harmful effects on human health, animal lives, natural ecosystems and the man-made environment. It is also responsible for climate change due to the enhanced greenhouse effect, acid rain, and the depletion of the ozone layer that constitute important global environmental problems. Air pollution occurs both outdoors and indoors and is caused by human activities and natural mechanisms. Although air quality degradation because of air pollution is a global problem that threatens humanity's wellbeing overall, it is largely associated with environmental injustice.

2.1 Outdoor air pollution

Outdoor air pollution is defined as the release of several substances -called air pollutants- into the atmosphere, in concentrations that threaten the wellbeing of living organisms or disrupt the function of the environment as a system leading to human health damages in various ways. Air pollutants can be gaseous, solid particles, or liquid droplets. They can be products of either natural processes or human activities. Natural sources of air pollutants include volcanic activity, forest fires, organic decay or soil dispersion into the air by the wind. The main anthropogenic sources of pollutants released in the atmosphere are human activities such as transportation (motor vehicles, aircrafts), burning coal or other fossil fuels for energy demands, industrial processes, or use of chemicals in agriculture, and facilities like power plants, incinerators, landfills for waste deposition (EEA, 2006a; 2006b; EPA, 2007; Valent et al., 2004).

The major outdoor air pollutants produced by human activities include, among others (EEA, 2006a; 2006b; EPA, 2007):

- Carbon oxides, especially carbon monoxide (CO) and carbon dioxide (CO_2) produced by the transportation sector (motor vehicle exhaust) and combustion of fossil fuels;
- Nitrogen oxides (NO_X), especially nitrogen dioxide (NO_2) emitted from high temperature fossil fuel combustion and electricity production;
- Sulphur oxides (SO_X), produced in various industrial processes such the smelting of sulphur-bearing ores for extracting metals and electricity production;
- Volatile organic compounds (hydrocarbons, VOCs) that include a variety of substances released from power plants and from industries producing numerous products such as painting colours, cleaning products, pesticides, building materials, and furniture;
- Particulate matter (PM), that is solid or liquid air pollutants mainly emitted by power plants and the transportation sector (aircrafts, motor vehicles), mining, and incinerators;
- Ground level ozone (O_3), an air pollutant that results from photochemical reactions between nitrogen oxides (NOx) and volatile organic compounds (VOCs) in the presence of sunlight;
- Toxic metals particularly lead (Pb), cadmium (Cd) and cooper (Cu) that are emitted from the transportation sector (motor vehicle exhaust), as well as from industrial procedures (production of painting colours, mining processes).

Outdoor air pollution is a major problem mainly in urban areas because of overpopulation. Increasing population[2] in urban areas results in increasing demands in transportation, industrial production and energy, which constitute the main sources of outdoor air pollution (UN, 2004). Moreover, this problem is intensified due to inadequate green open spaces in towns and cities and their restricted possibility to improve air quality and reduce air pollution (Givoni, 1991).

2.2 Indoor air pollution

Indoor air pollution refers to the amount of chemical, biological and physical contaminants in the air inside a building. Indoor air pollution can often be worse than outdoor air pollution. There is a wide range of indoor air pollutant sources in houses including building materials, pressed wood products and furniture, central heating and cooling systems, several personal care or household cleaning products, painting colours, solvents, heating or cooking appliances (stoves, wood and gas burning fireplaces, gas heaters), tobacco smoke, office machines and a variety of other products used in daily activities (EPA, 2007, 2008, 2009). In addition, outdoor pollutants are also traced in indoor spaces, such as radon (a natural trace component of soil and ground). The major source of indoor air pollution is household combustion of coal or biomass for cooking and heating. It is estimated that more than half of the world's population relies on animal dung, wood, crop waste, or coal to meet their most basic energy needs (WHO, 2005).

The major indoor pollutants that can practically be traced in any home are -among others- formaldehyde, asbestos, carbon monoxide, nitrogen oxides, sulphur oxides, benzene, polyaromatic compounds, particulate matter, toxic metals (lead, mercury, cadmium, and chromium), and volatile organic compounds (EPA, 2007, 2008, 2009; WHO, 2005).

---

[2] It is estimated that in 2010, 50.5 per cent or 3.5 billion of the people on Earth were living in cities (UN, 2004).

Several factors affect indoor air quality and are related to the type and maintenance of their sources, the level of hazard of the emitted pollutants, the quality of house ventilation and the quality of the building as well. For example, an old air pollutant source (a stove, a gas heating system, etc.) that is not properly maintained can emit significantly larger amounts of air pollutants than a source that is properly maintained (EPA, 2009). Indoors air temperature and humidity, biological contaminants (bacteria, mold, pollen, and viruses) as well as overcrowding are also important factors determining the quality of indoors air. In particular, high air temperature and humidity can affect the rate of indoors pollutant emissions. Moreover, these factors combined with inadequate building ventilation can further degrade air quality since indoors pollutants cannot be removed. Indoor air degradation in buildings may result in several health problems such as headaches, tiredness, hypersensitivity, coughing, eye, nose, and throat irritation, skin and mucosal dryness, dizziness, nausea, lethargy, inability to concentrate and mental fatigue. The occurrence of these symptoms is defined as the "sick building syndrome"(Burge, 2004; Stolwijk, 1991).

The "sick building syndrome", is also related to workplaces. In order to ensure healthy conditions at workplaces international organizations have defined upper limits of both environmental and chemical factors that could cause serious problems or disturbances to human health. Among these factors thermal conditions (air temperature and humidity) as well as the levels of dust and chemicals concentration in the air are recognized as important factors affecting health and safety conditions at workplace (ILO, 1993; OSHA, 2009).

2.3 Air pollution, health and environmental injustice

Numerous outdoor and indoor air pollutants affect human health and pose significant threats to individuals worldwide, such as cardiovascular or respiratory disorders, asthma and lung cancer, which can be fatal (WHO, 2002, 2003a, 2006, 2009). Given that an individual inhales more than 14.000 litres of air per day, it becomes obvious that air pollutants pose significant dangers for human health. According to the World Health Organization (WHO, 2006, 2009) more than two million premature deaths each year can be attributed to the effects of urban outdoor and indoor air pollution. In particular, every year indoor air pollution is responsible for the death of 1.6 million people while 800.000 deaths from lung cancer, cardiovascular and respiratory diseases worldwide are attributed to outdoor air pollution (Valent et al., 2004; WHO, 2005). It is estimated that currently air pollution reduces average life expectancy of Europeans by 9 to 24 months (CEC, 2006).

Do urban populations equally enjoy environmental quality? Intense social injustice is often located within cities and is related to the way in which the man-made environment is formulated, functions and develops in time and space. This evolution is a result of various social, cultural, economic, and political processes, along with environmental components and their interrelations, which combine to shape space consumption in an urban environment determining its residential quality from an environmental perspective. These processes are typically regulated by individuals belonging in higher socio-economic strata, having access to decision-making and economic activity centres, which allows them to act in favour of financial profit and against lower socio-economic status populations (Bard et al., 2007; FOE, 2001; Harner, Warner, Pierce & Huber, 2002).

Some dimensions of environmental quality of an urban area entailing social injustice include the spatial distribution of functions and activities, the distribution of outdoor and green spaces and the quality of residential and natural environments. Distribution of functions and activities in urban spaces frequently involves environmental injustice, creating environmentally degraded areas. Noisy facilities (industries, landfills, biological treatment facilities, highways) serving large urban populations are usually located in degraded residential areas. These facilities are responsible for pollutant emissions, noise, and also entail other dangers (e.g. explosions, fires) threatening nearby residents' physical and mental health. Air quality in these regions is typically low due to toxic gaseous substances emitted (Bard et al., 2007). According to international epidemiological studies, the populations living in the vicinity of large-scale urban facilities like industrialized regions exhibit higher frequencies of respiratory and cardiovascular incidents compared to inhabitants of other regions of the same cities. These incidents are related with pollutants traced in the air of these regions (Bard et al., 2007; Harner, Warner, Pierce & Huber, 2002; WHO, 2003a, 2003b).

As far as green and free open spaces are concerned, apart from being generally inadequate, they are not evenly distributed in all city districts; lack of public open spaces is more intense in densely inhabited and degraded quarters. Green space development requires free open spaces that are not always available, mainly because of the density of an urban area. Nevertheless, density is differentiated from one district to another within the same city, which results in further degradation of unprivileged neighbourhoods compared to privileged ones. As Heynen (2003, p. 980) points out: *"Social production of urban environments explicitly leads to uneven urban environments and environmental injustice. Environmental inequalities clearly exist within cities".*

Furthermore, air pollution occurs in both rural and urban areas because of transportation, industrial emissions and energy use. What is common between highly polluted rural and urban areas is that they are both correlated with low income residential quarters, which indicates another facet of social and environmental injustice (FOE, 2001; Mitchell & Dorling, 2003). In other words, higher socio-economic status social groups' actions influence life quality of lower socio-economic status groups within countries. In a study implemented in London respiratory problems have been found to concentrate in the poorest areas and to correlate with high traffic levels even though the responsibility for the causes of these problems is unequal -car ownership was lower in areas with worse traffic levels (Stevenson et al., 1999).

Environmental differences and inequalities related to air pollution also appear between countries. In developing countries the issue of air quality is mainly related with indoor air pollution. In most of these countries organic fuel, such as wood and coal are the most common energy resources available for domestic use –e.g. for heating and cooking. Their combustion releases harmful pollutants inside houses, especially solid particles (WHO, 2005). Chronic exposure to solid particles is considered to contribute to the development of cardiovascular and respiratory diseases, as well as of lung cancer, while it is related with mortality mostly among children (WHO, 2006). Actually, 20%[3] of child mortality for children under 5 years is attributed to indoor pollution (WHO, 2008a). Also, it is estimated that indoor air pollution associated with using biomass fuels (such as wood and grasses) causes an average of 871,500 deaths annually from childhood pneumonia (Valent et al., 2004).

As already indicated, dangers of air pollution are particularly increased in the case of children, who –in proportion to their body weight and surface- inhale larger quantities of air than adults, thus taking in larger amounts of the dangerous substances they are being exposed to. Moreover, their body systems are not mature, but are still evolving during

---

[3] This percentage corresponds to two million children under five years of age annually.

childhood. This implies that children are more vulnerable to incoming toxic substances than adults. Their tissues directly absorb these substances, and -due to their immaturity- they are not able to restore the damages caused. So, if their brain, immune, or endocrine system are affected by toxic substances, the damage caused can be permanent and irreversible (Landrigan et al., 1998).

Data relate the development of leukemia in children living in areas with high concentrations of suspended particles. Also, children suffering from asthma living in areas with increased air pollution-especially pollution from nitrogen oxides, acid vapours and particulate matter- exhibit a higher probability of deterioration (e.g. development of bronchitis symptoms). These data are further supported by research indicating that when children suffering from respiratory disorders leave areas with high particulate air pollution, their lung function improves, while this is not the case for those who remain in polluted areas (WHO, 2002, 2003a).

In addition, epidemiological studies in European countries indicate that morbidity and mortality are higher in low socio-economic status populations[4] suffering unhealthy environmental conditions. This especially concerns populations living in industrialised regions and nearby power plants (Bard et al., 2007; Cifuentes & Frumkin, 2007; UNICEF, 2008; WHO, 2003a, 2006, 2008a, 2009). Usually these regions are inhabited by low socio-economic status populations intensively experiencing the consequences of environmental degradation without being responsible for its emergence. On the contrary, they suffer from choices made by individuals of higher socio-economic status, pursuing economic growth. These choices concern investments entailing excessive use or degradation of natural resources. As a result, those who produce environmental degradation enjoy profits, while others suffer from its consequences. In this frame air pollution poses a threat to the fundamental right of life in a healthy environment and creates environmental injustice.

Another issue related with air quality degradation which entails social injustice involves air quality in working environments. These constitute local environments frequently containing harmful indoor air components, which may directly influence employees' health. They include various physical, chemical and biological components[5] released during the production process and degrade the quality of the air inhaled causing severe problems (ILO, 1993). It is estimated that in the US approximately 135 employees die every day from diseases caused by long-term exposure to toxic[6] substances at work (Faber & McCarthy, 2002). According to data available by European and World organisations high rates of morbidity and mortality are recorded among employees due to their exposure to environmental dangers in their workplaces. Obviously, low income employees –typically in lower ranking occupations- are those who suffer indoor air pollution in their workplaces the

[4] In Western Europe danger of disease caused by degraded environmental quality is 1.5-2.5 times higher in low socio-economic status populations.

[5] Physical environmental components include light, noise, radiation (ionizing or non-ionizing), dust, temperature, and humidity of indoor air. Chemical components are substances belonging in the production process and may be dangerous gaseous, toxic, erosive, or explosive substances. Biological components involve pathogens, bacteria, or viruses that may be traced in the air of a working environment.

[6] Toxicity is the degree to which a substance is able to externally or internally damage an exposed organism. Toxicity danger is the probability of harmful effects caused by the use of a chemical compound.

most compared to those in highly ranked occupations –i.e. close to management (FACTS 29, 2002; WHO, 2003b, 2004).

2.4 Environmental justice

The issues raised in the previous paragraphs suggest that air pollution is directly related with social injustice. People are exposed to polluted air and their health and wellbeing are affected in socially unjust ways. Mitigating and overcoming this injustice does not mean equal exposure to environmental danger; it means elimination of every kind of pollution endangering the environment and living organisms. The challenge here lies in ensuring air quality, health and wellbeing for everyone. Health is not simplistically meant as absence of illness, but as a broader concept involving prevention as well as promoting physical, mental, and social wellbeing. It is a basic human value and a fundamental right of every individual regardless of race, religion, political beliefs, social, or economic status. It is the foundation of modern citizens', and accordingly our societies' and civilization's development and progress (WHO, 1992).

Environmental justice can thus bridge the gap between environmental quality and social injustice. Environmental justice is the fair treatment and meaningful involvement of all people with respect to the development, implementation, and enforcement of environmental laws, regulations, and policies. It will be achieved when everyone enjoys the same degree of protection from environmental and health hazards and equal access to decision-making processes to have a healthy environment in which to live, learn, and work (EPA, 2001).

The demand for environmental justice is expressed by organised social groups in the frame of the 'environmental justice movement'[7] which has been increasingly established and reinforced in people's consciousness. It originates from hunan rights movements and is organised on the basis of three central principles: equality in environmental danger distribution, recognition of otherness and diversity between individuals, and active participation in political processes designing and applying environmental policies (Faber, & McCarthy, 2002; Harner, Warner, Pierce & Tuber, 2002; Scholosberg, 2004).

Environmental justice thus encompasses a variety of environmental, social and political dimensions and components. It involves issues such as natural resources management and distribution, social justice, development for all citizens, the right to participation in co-constructing an environment of quality. It is therefore apparent that environmental justice and the relevant issues are considered as directly connected to human rights. The UN Declaration of Human Rights and the Environment recognized the strong relationship between human rights and environmental quality that is reflected in the following principles (UN, 1994, pp. 3-4):

1. Human rights, an ecologically sound environment, sustainable development and peace are interdependent and indivisible.
2. All persons have the right to a secure, healthy and ecologically sound environment. This right and other human rights, including civil, cultural, economic, political, and social rights, are universal, interdependent and indivisible.
3. All persons shall be free from any form of discrimination in regard to actions and decisions that affect the environment.

---

[7] The Environmental Justice Movement started in the United States in the early '80s and has since spread in European and other countries.

4. All persons have the right to an environment adequate to meet equitably the needs of present generations and that does not impair the rights of future generations to equitably meet their needs.
5. All persons have the right to freedom from pollution, environmental degradation and activities that adversely affect the environment, threaten life, health, livelihood, well-being or sustainable development within, across or outside national boundaries.

3. Environmental quality in the context of science and environmental education for the promotion of environmental justice

3.1 Environmental education and environmental quality

Environmental education has long been established as a premise for the development and promotion of education about the environment. In the beginning -during the '70s- it focused on improving environmental quality in view of the urgent and unprecedented environmental problems. During that period the environment was identified with 'nature' and its protection was a moral imperative. In this context, environmental education concentrated on the development of knowledge about nature and its functions; on the development of the skills necessary to face environmental problems on the basis of identifying their causes; and on encouraging appropriate behaviours for resolving them. In other words, environmental education focused on causal relationships and suggested that a change in behaviour (the cause) could amend the effect for the benefit of the environment (UN, 1973; UNESCO, 1976, 1978).

During the '80's and '90's environmental education was associated with developmental issues, since environmental problems were considered as instigated by developmental choices made by modern societies. Environmental education was enriched with concepts related with development, economy, and society, and aimed at supporting learners' capacity to investigate further from the physicochemical components of environmental problems, the socio-economic dimensions inherent in their creation in the framework of the relationships between individuals, society and the environment. The role of human activities and choices in environmental quality was acknowledged as critical in the formation and resolution of environmental problems (UNEP, 1992; UNESCO, 1988; UNESCO-EPD, 1997). At the same time the notion of sustainability –involving social processes, individual and collective values- was dynamically adopted as central in organising and achieving environmental education objectives. As indicated in the Declaration of Thessaloniki (UNESCO-EPD, 1997): *"The concept of sustainability encompasses not only environment but also poverty, population, health, food security, democracy, human rights and peace. Sustainability is, in the final analysis, a moral and ethical imperative in which cultural diversity and traditional knowledge need to be respected"* (article 10).

Since the early '00's the idea of sustainability has been further promoted as a crucial organising principle redefining the scope of environmental education. In the context of sustainability, development is not merely conceptualised as economic growth, or economic profit, but also takes into account environmental quality and social cohesion. Therefore, sustainable development encompasses populations, animal and plant species, ecosystems, natural resources (water, air, energy, agriculture, biodiversity), climate change, rural development, urbanization, and integrates concerns such as poverty reduction, human rights, health, human security, culture diversity and intercultural understanding (UNESCO, 2005, 2007).

In order to accomplish sustainable development environmental education aims at preparing individuals for being capable of co-constructing their future in terms of environmental and social sustainability by integrating values, activities and principles inherently linked to sustainable development and encourage a change in attitudes, behaviours and values to ensure a more sustainable future in social, environmental and economic terms (UNESCO, 2007). For this general aim to be accomplished particular objectives are considered as crucial for each of the following domains (UNESCO, 2005, p. 6):

- Society: an understanding of social institutions and their role in change and development, as well as the democratic and participatory systems which give opportunity for the expression of opinion, the selection of governments, the forging of consensus and the resolution of differences.
- Environment: awareness of the resources and fragility of the physical environment and the effects on it of human activity and decisions, with a commitment to factoring environmental concerns into social and economic policy development.
- Economy: a sensitivity to the limits and potential of economic growth and their impact on society and on the environment, with a commitment to assess personal and societal levels of consumption out of concern for the environment and for social justice.

Finally, sustainability is organised with reference to the notions of ecological limits, social justice, environmental justice, and individual participation in environmental actions. The formation of developmental processes based on environmental quality, balance of ecological processes, and quality of socio-political structures is imperative for ensuring a sustainable future aiming at the improvement of life quality of citizens (Huckle, 2006; Jacobs, 2004; Orr, 1992; Scott, & Gough, 2004). In this framework, both individuals and society should respect the boundaries set by ecological processes within and between natural systems. This requires an appropriate understanding of the concepts of carrying capacity and ecological footprint (see subsection 1.2) that determine the sustainable function of ecosystems and the degree to which human societies exert pressure on them in order to fulfill their diverse needs.

Therefore, environmental education can play an important role in supporting individuals to critically consider environmental decisions of contemporary societies and enabling them to investigate and challenge the social structures and political decisions that preserve or reinforce social and environmental injustice. At the same time, it can strengthen students' roles as future citizens by encouraging participation in the design and implementation of actions to resolve environmental problems, training them in democratic processes, and involving them in issues concerning their local communities. Such orientations build up citizens' confidence, mutual help and solidarity for the improvement of life quality. These values are the foundations of a democratic society with social and environmental justice.

3.2 Science education and environmental quality

Science has developed tools and knowledge for understanding the complex physicochemical mechanisms underlying environmental problems, the ways in which human interventions and everyday habits affect natural ecosystems and the ways in which ecosystems react to these interventions. Ecological footprint, ecosystems' carrying capacity, atmospheric and hydrosphere mechanisms are among these important tools for studying environmental problems, and estimating the degree to which human activities, practices and developmental interventions create or enhance them (AAAS, 1989; Hodson, 2003). Science teaching can address these issues to support students in constructing the environmental knowledge necessary for understanding environmental problems and undertaking appropriate actions for their resolution. Scientific knowledge and methodological tools can therefore enhance students' ecological literacy, which involves understanding natural phenomena and their interrelations, the interdependence between society, economy, the natural environment and the ways in which human activities affect the environment and are affected by it (Orr, 1992). In this framework natural systems are considered as constituents of the environment (also including the social, man-made, and economic environment). Therefore, science teaching can enhance students' ability to connect scientific knowledge with everyday issues, such as environmental problems which nowadays –more than ever before- are part of individuals' and society's daily life.

Besides, such an integration of scientific knowledge with the multiple components of environmental problems is expected to stimulate students' involvement in science, since these problems require an interdisciplinary connection of science teaching with education for the environment; on the one hand integrated scientific knowledge provides the means for understanding environmental issues by means of a holistic and systemic approach; on the other it provides insight to the human-environment interaction that enables environmental sustainability. From this perspective science teaching may become more meaningful for students, since it also incorporates a strong affective dimension and provides possibilities to develop evaluative judgments and take up environmentally appropriate actions in real life situations (AAAS, 1989; Brody, 1991; Bybee, 1993; Gough, 2002; Hodson, 2003; Iozzi, 1989; Schneider et al., 2002).

In this context science teaching enhances students' critical thinking, essential for understanding the complexity of environmental problems. Critical thinking requires that an individual is both mentally and emotionally engaged when facing a problem in order to end up with rational conclusions, judgments, and choices. It is a process of data collection, processing, analysis and synthesis with existing knowledge to explain concepts and phenomena, to understand personal conceptions and evaluate them against the ideas of others, and to question prejudices (Jones, Merritt & Palmer, 1999; Tilbury & Ross, 2006). Furthermore, data analysis and interpretation to extract conclusions are basic components of critical thinking, and are also considered as central in science education (Facione, 2007).

As already mentioned, systemic thinking is crucial for understanding environmental problems. Understanding the environment –as a 'system' interacting with other systems – involves conceptualising its organisation in time, its influxes, the decisions and measures taken by individuals and organisations about its management, and the way in which it reacts to the actions it is imposed to (Ragou & Souchon, 2000). Thus, a systemic approach of environmental problems requires systemic thinking, which is substantiated by means of the complementary processes of analytical and synthetic thinking (Bartlett, 2001; Espejo, 1994; Sterman 1994; Richardson, 1994; Richmond, 1993, 1994). The first process is essential in understanding the parts of a problem; the second enables comprehension of the way in which these parts interact within a system, as well as with surrounding systems. According to Richmond (1994) systemic thinking is a dynamic way of reasoning and learning, the 'art' and science of creating reliable conclusions about a system's behaviour by deeply understanding the structures regulating it. It is therefore evident that science education can provide essential knowledge and tools of thinking for studying natural systems and their interrelations with other –natural or man-made- systems.

Moreover, the study of environmental problems through the prism of environmental justice can emphasise the social role of science education in improving life quality and ensuring individual and public health. More particularly, an environmental perspective in science education allows for

- Application of scientific tools and knowledge in estimating environmental degradation and its consequences on health, and in revealing pertinent social injustice;
- Development of instruments to consider alternative ways of environmental management that ensure everyone's right to security and health;
- Application of scientific knowledge in considering the environment as a system comprising other natural and social systems and in understanding their interrelations as well as the relations producing social injustice;
- Development of scientific method skills to support critical and systemic thinking and promote the idea of the social role of science and technology in dealing with environmental injustice.

3.3 Connecting environmental and science education to promote environmental justice

The preceding analysis indicates that the connection of science with environmental education can elucidate environmental and social injustice inherent in environmental problems. This connection can serve basic objectives regarded as preconditions for students' engagement in understanding environmental problems and dealing with environmental injustice. These objectives include (Brody, 1991; Bybee, 1993; Orr, 1992; Peloso, 2007; Scott, 2007):

1. Supporting students in developing thinking skills (e.g. systematic observation and investigation, problem solving, critical synthesis of different parameters in decision making) to allow them satisfy basic personal needs and use natural resources rationally aiming at achieving environmental justice;
2. Supporting students in developing communication and collaboration skills so that they are capable of formulating and comparing different views, of constructing new, synthetic and consensual ideas ensuring environmental quality for all citizens;
3. Enhancing students' interest in science by revealing its relevance to real and important issues concerning life quality;
4. Developing learning processes that support students in constructing more adequate explanatory models about the natural world and the environment;
5. Revealing interrelations between science, technology, the environment, and society, and promoting positive attitudes towards the environment, aiming at improving its quality to the benefit of all people;
6. Presenting science and technology as useful means for the adaptation of humanity to the environment, since they provide rationally structured explanations of reality -as interpreted in every period and developmental phase- to ensure environmental quality for all;
7. Making clear that science and technology entail -apart from the necessary information and knowledge- value and ethical issues.
8. Supporting students' understanding of the importance of concepts like interdependence, sustainable development, conservation of natural resources, population control. Awareness of the sensitive environmental equilibrium could support the development of a 'universal ethos' to counteract social injustice and adopt sustainable actions for protecting, conserving and improving environmental quality.

4. Air pollution in the framework of science and environmental education to promote environmental justice

As already mentioned air pollution is related with environmental injustice inherent in its causes as well as its consequences. On the one hand, the causes of air pollution lie in people's lifestyles, which are differentiated according to their socio-economic possibilities. On the other, they are related to developmental choices and financial investments made by higher socio-economic strata, and to political decisions forming their institutional context. At the same time air pollution is associated with the way in which cities are organised, with their infrastructures and green spaces available determining the quality of urban environment and producing inequalities between residents. Thus, air quality is recognised as one of the priorities for healthy and sustainable cities.

Apart from knowledge about scientific concepts and the complexity of environmental phenomena, the development of educational programs related to environmental issues - such as air pollution- also requires that students' ideas and ways of thinking are taken into account. An analysis of learners' thinking and conceptions can provide valuable insight into the way in which individuals or social groups come to organise their reasoning and concern about the environment and the relevant problems. For instance, if one attributes environmental problems to the development of industry and believes that they can be faced through technological applications, s/he adopts a technocratic approach. On the other hand, those who attribute environmental problems to human activities embrace a more politicised stance (Wals, 1992). In the first case an individual may not develop further interest or an intention to act to protect the environment, given that technology can solve any problem. Therefore, students' thinking about environmental problems can guide the development of teaching material and activities that depart from their conceptions.

Students' ideas about air pollution have been recorded by a number of studies. Their main findings can be summarised as follows:

- Pupils conceptualise air pollution as presence of harmful substances into the atmosphere (Ali, 1991). From 10-11 years on they appreciate that air is polluted with 'extra gas(es)', 'fumes', or 'smoke', some of them identifying specific pollutants such as CFCs, carbon monoxide and dioxide, sulphur dioxide, and nitrogen oxides (Batterham et al., 1996; Boyes and Stanisstreet, 1997; Thornber et al., 1999).
- Students describe cars and transportation, industry, waste, fires and sprays as the primary anthropogenic pollution sources (Boyes and Stanisstreet, 1997; Thornber et al., 1999). They regard air pollution as exclusively man-made (Ali, 1991; Boyes and Stanisstreet, 1994; Brody, 1991; Brody, 1994; Dove, 1996; Thornber et al., 1999), adopt a restricted and local view of the issue, being unable to appreciate its global dimension (Brody, 1994), and perceive air pollutants as inert, i.e. not participating in physical or chemical processes in the atmosphere (Marinopoulos and Stavridou, 2002).
- Air pollution is considered to cause plant and animal diseases and/or death, as well as respiratory problems to humans (Ali, 1991; Batterham et al., 1996; Brody, 1994; Thornber et al., 1999). Some students attribute other global environmental problems to air pollution (Batterham et al., 1996). In this case, they often hold a generalised idea of global environmental issues, confusing them at the levels of causes, mechanisms and consequences (Boyes and Stanisstreet, 1997; Christidou & Koulaidis, 1996; Francis et al., 1993; Hillman et al., 1996; Koulaidis & Christidou, 1999; Plunkett and Skamp, 1994; Rye

et al., 1994). They often consider air pollutants in general as enhancing the greenhouse effect and ozone depletion.

- Research outcomes from a Greek study (Dimitriou & Crhistidou, 2007) exploring students' understanding of air pollution indicate that they have a generalised understanding of the problem. They are familiar with a variety of its components - such as sources of pollution, pollutant types, consequences on living things and on the environment- but they do not seem to comprehend the relationships between these components. At the same time they do not have an adequate understanding of the degree to which individual and social decisions and practices influence the intensity of air pollution. On the other hand, they seem to have a more appropriate understanding of the effects of air pollution on human health and explicitly express their concern about them.

The preceding analysis reveals that students are not aware of all the dimensions of air pollution and therefore cannot fully understand the interrelations and interactions between its components. Understanding the framework of relations between natural and human dimensions composing an environmental phenomenon such as air pollution is -in our view- a prerequisite for developing critical reasoning about it and should constitute a key educational objective for supporting future citizens in thinking about the world. Therefore, students should become competent in recognising and interpreting the framework of physical, ecological, social, economic and political relations underlying environmental issues, the contradictions inherent in those components, as well as possible alternative explanations and solutions to these issues. They should also become capable of sincerely examining their personal views, practices, prejudices and stereotypes, make appropriate decisions and implement changes at individual and collective level. In other words, students should be equipped with knowledge, skills and thinking tools necessary for interpreting the world around them in order to change it, since the world cannot change unless it is adequately explained (Huckle, 2004).

In this context, educational programs on air pollution should ultimately aim at determining all relevant components, indicating their interrelations and the ways in which these interrelations enhance or reduce the problem and its consequences. In other words, educational programs need to be designed in the framework of a systemic approach for air pollution, considering all the systems with which it interacts. Such a systemic approach would also contribute to understanding and resolving environmental injustice inherent in air pollution and related with people's rights, lifestyles, financial activities, along with state and governmental decisions. These systems reflect the social, political, and financial dimensions of air pollution and are tightly connected with the way individuals and organised societies (institutions, legislation, etc.) decide and act. The functions of these systems correspond to influxes to the air pollution 'system' and in turn influence the way it functions (at the level of physicochemical and biological mechanisms), the way it reacts to these influxes and its effects on the natural and man-made environment.

This approach provides the possibility of a holistic study of air pollution in order to understand natural and human components causing it and their interrelations. At the same time it provides an opportunity for understanding the temporal evolution of this issue due to human interventions -be it in terms of managerial practices or protection practices for its resolution.

The previous discussion allows for developing educational programs about air quality degradation and the relevant social injustice by studying indoor and outdoor local environments. Such programs would enhance students' critical understanding of environmental quality in the context of individual and social activities, practises and political actions. Within this context, environmental justice in the perspective of sustainability is a crucial component for maintaining a healthy environment for all, and for creating sustainable places to live and work, now and in the future. By integrating scientific knowledge with life experiences students would be encouraged to make personal decisions and become active members of society to ensure better air quality and environmental justice. More particularly, educational programs about air quality degradation and the relevant social injustice could aim at:

- **Investigating different lifestyles and the ways these affect air quality and enhance – or reduce-air pollution**
- Lifestyle describes the way an individual lives. It includes, among others, individual consumption behaviours and everyday life practises. Almost all of our daily activities and habits eventually produce indoors and outdoors air pollutants. Investigating their different lifestyles would give students the opportunity to understand how themselves and others affect air quality in their daily lives. Moreover, students could be encouraged to investigate alternative habits that would contribute to air pollution reduction, such as using alternative energy sources (e.g. solar, or wind energy), saving energy, or using eco-friendly products.
- **Assessing the personal ecological footprint on air and comparing ecological footprints of different socio-economic status individuals**
- This approach would provide students the opportunity to explore the way individuals of different economic status consume natural resources consequently affecting air quality. Since socio-economic status is plays a crucial role in determining consumption behaviours and everyday practices, such an assessment would aid students develop a deeper understanding of social and environmental injustice related to air pollution.
- **Estimating the factors influencing urban space quality (urban design, transportation networks, green spaces) and evaluating its quality in privileged and underprivileged areas within the same city to reveal ensuing injustice**
- Involving students in investigating urban space gives them the opportunity to realise that the way a city is organised affects the quality of citizens' lives. They are also assisted in recognising the relationship between environmental quality and human wellbeing and to locate environmental injustices in their local environments to concretise that urban quality is not the same within the same city. Students living in underprivileged areas, in areas of injustice, should be empowered to develop and participate in communities of resistance and planning (Peloso, 2007). As Bullard, one of the major researchers and organizers in the environmental justice movement, suggested: *"We have to educate young people that it is their right to have access to open space, green space, parks, outdoors, as opposed to people thinking that their supposed to be living in an area where the only park is a basketball court with no net. We have to give people this idea that it's their right to have access to open space and green space and we have to provide funds to make sure that we get them early on and take them on field trips, take them to a wilderness area, a refuge, a reserve, to a park-a real park and to integrate this information into our curriculum"* (Schweizer, 1999).

- **Estimating the effects of air pollution on human health and the distribution of relevant diseases in respect to environmental quality in privileged and underprivileged areas within the same city**
  Estimating the distribution of diseases that are attributed to urban air pollution gives students the opportunity to understand that environmental justice is the most significant tool for a healthy environment for all people regardless of socio-economic status.
- **Exploring pollution sources and their distribution in the local environment in relation to the socio-economic status of residents**
  This approach gives students the opportunity to realize that people of low income usually live in degraded residential areas where the main air pollution sources, such as industries, landfills, biological treatment facilities, highways, are usually located.
- **Determining the factors influencing air quality in different workplaces**
  Students could be encouraged to investigate air quality at different workplaces to locate related environmental injustice experienced by specific groups of employees. In this framework, the "sick building syndrome" could also be explored based on real case studies.
- **Measuring air pollutant concentrations in the school, local and broader environment**
  This activity would give students the opportunity to locate air pollutant sources in their immediate environment and suggest alternative interventions to their school in order to render it sustainable. Such interventions could include -among others- using building material, heating sources, furniture, and equipment with a low impact on air quality. Such actions would also enable all students participate in school life and in decision-making processes, to cooperate with local citizens and broaden their activities to improve the quality of the local environment and thus improve wellbeing of the local community.
- **Studying national and international legislation referring to air quality and engaging students in proposing legislative directions for the formation of urban environments, improvement of urban air quality and elimination of environmental injustice**
  Student's involvement in studying environmental legislation provides the opportunity of deeper understanding of the political dimension of air pollution and environmental injustice.
- **Familiarising students with national and local institutions in order to understand their role in the formation and conservation of the local environment**
  Student's familiarisation with institutions enables greater interaction with the community. It is also expected to increase their interest in local issues and their willingness to work together on reducing air pollution and promoting sustainability.

To determine environmental injustice at an international level, educational programs to assist students in understanding injustice between developed and developing countries could be developed. Indicative objectives of such programs would include:

- **Recording differences in energy sources used for satisfying basic daily necessities of developed and developing countries' inhabitants and identifying indoor air pollution in each case;**
- **Studying indoor air pollution effects on human health and especially on children;**
- **Studying geographical distribution of diseases related with air pollution in developed and developing countries;**
- **Comparing personal ecological footprints on air between inhabitants of developed and developing countries;**

- **Understanding poverty experienced by people in developing or unprivileged countries as determinative factor for the environmental injustices they experienced.**

Probably, one could object that studying local environmental problems and pertinent injustices solutions is much more meaningful to students than grasping -for instance- environmental injustices between development and developing countries. Are students able to cope with these injustices and promote air quality for the people in developing countries? We strongly believe that education has to promote global citizenship and allow students to understand that environmental problems are linked to the social, economic and political situations in countries all over the world. Therefore, science and environmental programs about air pollution could start from the local environment and deal with real, everyday and concrete problems facing students and their community and subsequently be enriched with activities that enable students to develop a deeper and global understanding of poverty as a critical dimension of environmental injustice (UNCSD, 2003). After all, "education is one of the most effective forces to bring about the changes in knowledge, values, behaviour and lifestyles required to achieve sustainability and stability within and among countries, and to guarantee democracy, human security and peace" (UNESCO, 2005b, p. 10).

5. Conclusions

In this chapter we attempted to present environmental quality in the context of environmental injustice. The discussion was based on data provided by international institutions mapping environmental degradation, its consequences on human health, wellbeing and life quality. These data reveal that environmental injustice within countries is particularly experienced by low socio-economic level groups of the population living in underprivileged areas. Moreover, environmental injustice between countries is more intense and degrades health quality of their population, indeed contributing to increase of mortality.

The chapter particularly focused on the environmental problem of air pollution aiming at highlighting environmental injustice caused by the types, sources, and distribution of air pollutants along with their effects on human health and wellbeing. These issues were discussed in the scope of sustainability, which in our view is the only means to ensure environmental quality and justice for all. This is a requirement and a fundamental challenge for modern societies.

Sustainability and environmental justice can only be accomplished by means of active involvement of citizens in environmental actions. In other words engaging people in making crucial decisions about the quality of their environment is a prerequisite for environmental sustainability and justice to occur. Therefore, individuals should be supported in demanding life quality based on respect and justice for all human beings; in ensuring equal access to natural resources and health for all; in deciding upon a development of their cities in harmony with nature and with respect to the cultures of different communities.

In this direction we propose the design and implementation of relevant educational programs deriving knowledge, methods, and modes of thinking from science and environmental education. These two fields provide the necessary tools for understanding environmental degradation and pursuing sustainability and environmental justice. More particularly, science education supplies tools and methods for estimating environmental quality and protection, and for creating a healthy environment ensuring life quality and health for human beings. Environmental education can complement this endeavour by critically supporting individuals in understanding environmental challenges and their causes, developing actions, and adopting practices to resolve them. Injustice, inherent in environmental degradation, can constitute an organising principle in designing and implementing educational programs for the environment, and more particularly in relation to air pollution.

The conjunction of science and environmental education organised along the concepts of sustainability and environmental justice also opens new grounds for research. Application of the general principles and stances outlined in this chapter in real school environments in the context of specific educational activities would allow a systematic and reliable estimation of their value. More specifically, evaluation of educational applications of the ideas presented in this chapter could investigate students' understanding of air pollution; their relevant speculations; the alternative solutions they propose for reducing air pollution along with pertinent environmental injustice; their views of the role of science in this enterprise. Systematic research on these topics is expected to yield documented, valuable, and productive outcomes to further promote education for sustainable development.

6. References

Ali, I. (1991). How do English pupils understand pollution? *Environmental Education and Information*, Vol. 10, No. 4, pp. 203-220, ISSN 0144-9281.

American Association for the Advancement of Science (AAAS) (1989). *Project 2061: Science for all Americans*, 03.01.2010, Available from http://www.project2061.org

Bard, D., Laurent, O., Filleul, L., Havard, S., Deguen, S., Segala, C. et al. (2007). Exploring the joint Effect of atmospheric pollution and socioeconomic status on selected health outcomes: An overview of the PAISARC project. *Environmental Research Letters*, Vol. 2, No. 4, 28.09.2008, Available from http://iopscience.iop.org/1748-9326/2/4/045003/pdf/1748-9326_2_4_045003.pdf

Bartlett, G. (2001). *Systemic Thinking, a Simple Thinking Technique for Gaining Systemic Focus*, 10.03.2007, Available from http://probsolv.com/systemic-thinking.doc

Batterham, D., Stanisstreet, M., & Boyes, E., (1996). Kids, cars and conservation: children's ideas about the environmental impact of motor vehicles, *International Journal of Science Education*, Vol. 18, No. 3, pp. 347-354, ISSN 0950-0693

Boyes, E., & Stanisstreet, M. (1994). The ideas of secondary school students concerning ozone layer damage. *Global Environmental Change*, Vol. 4, No. 4, pp. 311-324, ISSN 0959-3780

Brody, M. (1991). Understanding of pollution among 4th, 8th and 11th grade students. *International Journal of Science Education*, Vol. 22, No. 2, pp. 24-33, ISSN 0950-0693

Brody, M. (1994). Student science knowledge related to ecological crisis. *International Journal of Science Education*, Vol. 16, No. 4, pp. 421-435, ISSN 0950-0693

Brody, M., Chipman, E., & Marion, S. (1989). Student knowledge of scientific and natural resource concepts concerning acidic deposition. *Journal of Environmental Education*, Vol. 20, No. 2, pp. 32-42, ISSN 0095-8964

Bybee, R. W. (1993). *Reforming Science Education - Social Perspectives and Personal Reflections.* Teachers College Press, New York

Christidou, V., & Koulaidis, V. (1996). Children's Models of the Ozone Layer and Ozone Depletion. *Research in Science Education*, Vol. 26, No. 4, pp. 421-436, ISSN 157-244X

Cifuentes, E., & Frumkin, H. (2007). Environmental injustice: case studies from the South. *Environmental Research Letters*, Vol. 2, 26.09.2008, Available from http://iopscience.iop.org/1748-9326/2/4/045034/pdf/1748-9326_2_4_045034.pdf

Commission of the European Communities (CEC) (2006). *Cohesion Policy and cities: the urban contribution to growth and jobs in the regions.* COM (2005)718

Dimitriou, A. (2009). *Environmental education: Environment, Sustainability. Theoretical and Pedagogical Approaches*. Epikentro, ISBN 978-960-458-214-3, Thessaloniki (in Greek)

Dimitriou, A., & Christidou, V., (2007). Pupils' understanding of air pollution. *Journal of Biological Education*, Vol. 42, No. 1, pp. 24-29, ISSN 0021-9266

Donohoe, M. (2003). Causes and health consequences of environmental degradation and social injustice. *Social Science & Medicine,* Vol. 56, pp. 573–587, ISSN 0277-9536

Dove, J. (1996). Student teacher understanding of the greenhouse effect, ozone layer depletion and acid rain. *Environmental Education Research*, Vol. 2, No. 1, pp. 89-100, ISSN 1350-4622

EEA (European Environment Agency) (2006a). *Air pollution at street level in European cities* Copenhagen, EEA

EEA (European Environment Agency) (2006b). *Application of the emmisions trading directive by EU member states. Technical report No 6.*, Copenhagen, EEA

Environmental Protection Agency (EPA) (2001). *Environmental Justice*, 20.03.2007, Available from http://epa.gov/swerosps/ej/index.html

Environmental Protection Agency (EPA) (2007). *Terms of environment*. Office of Communications, Education, and Public Affairs, Washington DC

Environmental Protection Agency (EPA) (2008). Care for Your Air: A Guide to Indoor Air Quality, 20.02.2011, Available from http://www.epa.gov/iaq/pdfs/careforyourair.pdf

Environmental Protection Agency (EPA) (2009). Residential Air Cleaners. Indoor Air Quality, 10.01.2010, Available from http://www.epa.gov/iaq/pdfs/residential_air_cleaners.pdf

Espejo, R. (1994). What is systemic thinking? *System Dynamics Review*, Vol. 10, No. 2-3, pp. 199-212, ISSN 1099-1727

Faber, D., & McCarthy, D. (2002). The evolving structure of the environmental justice movement in the United States: New models for democratic decision-making. *Social Justice Research*, Vol. 14, No. 4, pp. 405-421, ISSN 0885-7466

Facione, P. A. (2007). *Critical thinking: what it is and why it counts*, 15.07.2008, Available from http://www.insightassessment.com/t.html

FACTS 29, (2002). *Health Care Sectors.* European Agency for Safety and Health of Work, Geneva

Flogaitis, E. (2006). *Education for the environmenta and sustainability.* Ellinika Grammata, ISBN 960-442-238-3, Athens (in Greek)

Francis, C., Boyes, E., Qualter, A., & Stanisstreet, M. (1993). Ideas of elementary pupils about reducing the 'Greenhouse Effect'. *Science Education*, Vol. 77, No. 4, pp. 375-392, ISSN 1098-237X

Friends of the Earth (FOE) (2001). *Pollution and Poverty - Breaking the Link.* Friends of the Earth, London

Givoni, B. (1991). Impact of planted areas on urban environmental quality: A review. *Atmospheric Environment, Part B, Urban Atmosphere*, Vol. 25, Issue 3, pp. 289-299, ISSN 1352-2310

Global Footprint Network (2005). EUROPE 2005. *The Ecological Footprint*. Belgium, WWF

Global Footprint Network (2011). *The Ecological Footprint*, 04.03.2011, Available from http://www.footprintnetwork.org/en/index.php

Gough, A. (2002). Mutualism: a different agenda for environmental and science education. *International Journal of Science Education*, Vol. 24, No. 11, pp. 1201-1215, ISSN 0950-0693

Harner, J., Warner, K., Pierce, J., & Huber, T. (2002). Urban environmental justice indices. *The Professional Geographer*, Vol. 54, No. 3, pp. 318-331, ISSN 0033-0124

Heynen, N. (2003). The scalar production of injustice within the urban forest. *Antipode*. Vol. 35, No. 5, pp. 980-998, ISSN 0066-4812

Hodson, D. (2003). Time for action: Science education for an alternative future. *International Journal of Science Education*, Vol. 25, No. 6, pp. 645-670, ISSN 0950-0693

Huckle, J. 2006. *Education for Sustainable Development*. A briefing paper for the Training and Development Agency for Schools, 15.06.2007, Available from http://john.huckle.org.uk

Hui, C. (2006). Carrying capacity, population equilibrium, and environment's maximal load. *Ecological Modelling*, Vol. 192, No. 1-2, pp. 317-320, ISSN 0304-3800

International Labour Organization (ILO), (1993). *Encyclopedia of occupational health and Safety*. ILO, Geneva

Jacobs, M. (2004). Sustainable development: A contested concept. In: *Fairness and futurity. Essays on Environmental sustainability and social justice* A. Dobson (Ed.), 21-54, Oxford University Press, London

Jones, P. C., Merritt, Q., & Palmer, C. (1999). Critical thinking and interdisciplinarity in envrionmental higher education: The case for epistemological and values awareness, *Journal of Geography in Higher Education*, Vol. 23, No. 3, pp. 349-357, ISSN 0309-8265

Landrigan, P. J., Carlson, J. E., Bearer, C. F., Spyker Cranmer, J., Bullard, R. D., Etzel, R. A., et al. (1998). Children's health and the environment: A new agenda for preventive research, *Environmental Health Perspectives*, Vol. 106, No. S3, ISSN 0091-6765

Iozzi, A. L. (1989). What research says to the educator. Part One: Environmental education and the affective domain, *Journal of Environmental Education*, Vol. 20, No. 3, pp. 3-9, ISSN 0095-8964

Koulaidis V., & Christidou V. (1999). Models of Students' Thinking Concerning the Greenhouse Effect and Teaching Implications. *Science Education*, Vol. 83, pp. 559-576, ISSN 0036-8326

Marinopoulos, D., & Stavridou, H., (2002). The influence of a collaborative learning environment on primary students' conceptions about acid rain. *Journal of Biological Education*, vol. 37, No. 1, pp. 18-25, ISSN 0021-9266

Mitchell, G., & Dorling, D. (2003). An environmental justice analysis of British air quality. *Environment and Planning A*, Vol. 35, No. 5, pp. 909 -929, ISSN 0308-518X

Orr, D. (1992). *Ecological Literacy: Education and the transition to a postmodern world*. SUNY Press, ISBN 0-7914-874-4

Occupational Safety and Health Administration (OSHA) (2009). *Assigned Protection Factors for the Revised Respiratory Protection Standard,* 15.01.2010, Available from http://www.osha.gov/Publications/3352-APF-respirators.pdf

Peloso, J. (2007). Environmental justice education: Empowering students to become environmental citizens. *Perspectives on Urban Education,* Vol. 5, No. 1, 02.02.2010, Available from http://www.urbanedjournal.org

Ragou P. et Souchon C. (2000). L'approche systémique: Une méthode d' apprentissage et un outil d' évaluation en education relative à l' environnement. *Education Relative à l'Environnement: Regards, Recherches, Reflexions,* Vol. 2, pp. 195-200, ISSN 1373-9689

Richardson, G. (1994). Introduction: System thinkers, system thinking. *System Dynamics Review,* Vol. 10, No. 2-3, pp. 95-99, ISSN 1099-1727

Richmond, B. (1993). Systems Thinking: Critical thinking skills for the 1990s and beyond. *System Dynamics Review,* Vol. 9, No. 2, pp. 113-133, ISSN 1099-1727

Richmond, B. (1994). Systems thinking/System dynamics: Let's just get on it. *System Dynamics Review,* Vol. 10, No. 2-3, pp. 135-157, ISSN 1099-1727

Scholosberg, D. (2004). Preconceiving environmental justice: Global movements and political theories. *Environmental Politics,* Vol. 13, No. 3, pp. 517-540, ISSN 0964-4016

Scott, W. (2007). *Raising Standards: Making Sense Of The Sustainable Schools Agenda,* Specialist Schools and Academies Trust, UK,

Scott, W., & Gough, S. (2004) (Eds.), *Key Issues in Sustainable Development and Learning. A Critical Review.* Routledge Falmer, ISBN 0-415-27649-7, London and N. York

Schneider, R. M., Krajcik, J., Marx, R. W., & Soloway, E. (2002). Performance of students in project-based science classrooms on a national measure of science achievement, *Journal of Research in Science Teaching,* Vol. 39, No. 5, pp. 410-422, ISSN 0022-4308

Schweizer, E. (1999). *Environmental justice: An interview with Robert Bullard.* 04.03.2007, Available from http://www.ejnet.org/ej/bullard.html

Shelton, D. (2002). *Human Rights, Health & Environmental Protection: Linkages in Law and Practice.* Notre Dame London Law Centre, London

Stephens, C., Bullock, S., & Scott, A. (2001). Environmental justice: Rights and means to a healthy environment for all, *Special Briefing Paper Economic and Social Research Council (ESRC) Global Environmental Change Programme* (Brighton: ESRC Global Environmental Change Programme, University of Sussex), 04.06.2008, Available from www.foe.co.uk/resource/reports/environmental_justice.pdf

Stephens, C. (2007). Environmental justice: A critical issue for all environmental scientists everywhere. *Environmental Research Letters,* Vol. 2, No. 1-2, pp. 1-3, ISSN 1748-9326

Sterman, J. (1994). Learning in and about complex systems. *System Dynamics Review,* Vol. 10, No. 2-3, pp. 291-330, ISSN 1099-1727

Stevenson, S., Stephens, C., Landon, M., Fletcher, T., Wilkinson, P., & Grundy, C. (1999). Examining the inequality of car ownership and the effects of pollution and health outcomes, presented at the "Healthy Planet Forum", June, Environmental Epidemiology Unit, School of Hygiene and Tropical Medicine, London

Stolwijk, J. A.(1991). Sick-Building Syndrome. *Environmental Health Perspectives,* Vol. 95, pp. 99-100, ISSN 0091-6765

Thornber J., Stanisstreet, M. & Boyes, E. (1999). School students' ideas about air pollution: Hindrance or help for learning? *Journal of Science Education and Technology,* Vol. 8, No. 1, pp. 67-73, ISSN 1059-0145

Tilbury, D., & Ross, K. (2006). *Living Change: Documenting Good Practice in Education for Sustainability in NSW*. Macquarie University, Sydney, and Nature Conservation Council, NSW

United Nations UN, (Ed.) (1973). *Report of the United Nations Conference on the Human Environment* (Stockholm 5-16 June 1972). UN, New York

United Nations (UN) (1994). Draft Principles on Human Rights and the Environment *E/CN.4/Sub.2/1994/9, Annex I*

United Nations. (2004). World population to 2300 United Nations Department of Economic and Social Affairs, Population Division

United Nations Education Scientific and Cultural Organisation (UNESCO), (1976). *The international workshop on environmental education*. Belgrade, Yougoslavia, 13-22 Oct 1975. Final report. UNESCO, Paris

United Nations Education Scientific and Cultural Organisation (UNESCO), (1978). *Intergovernmental Conference on Environmental Education (Tiblisi USSR, 14-26 October 1977)*, UNESCO, Paris

United Nations Education Scientific and Cultural Organisation (UNESCO), (1982). *Environmental Education: Selected Activities of Unesco-UNEP International Environmental Education Programme 1975-1982*, UNESCO, Paris

United Nations Education Scientific and Cultural Organisation (UNESCO), (1988). *International strategy for action in the field of environmental education and training for the 1990s*. UNESCO, Paris

United Nations Education Scientific and Cultural Organisation (UNESCO), (1990). *Environmental Education: Selected Activities of Unesco-UNEP International Environmental Education Programme 1975-1990*. UNESCO, Paris

United Nations Education Scientific and Cultural Organisation (UNESCO), (1992). *United Nations Conference on Environmental and Development: Agenda 21*. UNESCO, Switzerland

United Nations Education Scientific and Cultural Organisation (UNESCO- EPD), (1997). *Declaration of Thessaloniki. UNESCO-EPD97/CONF.401/CLD.2*, UNESCO, Paris

United Nations Education Scientific and Cultural Organisation (UNESCO), (2002). *Teaching and learning for a sustainable future: A multimedia teacher education programme*, 23.04.2006, Available from http://www.unesco.org/education/tlsf/.

United Nations Education Scientific and Cultural Organisation (UNESCO), (2005). UN Decade of Education for Sustainable Development 2005-2014 International Implementation Scheme. DRAFT. UNESCO, Paris

United Nations Education Scientific and Cultural Organisation (UNESCO), (2005b). *UNESCO and sustainable development*, UNESCO, Paris

United Nations Commission on Sustainable Development (UNCSD), (2003). *Plan of Implementation of the World Summit on Sustainable Development*, New York: United Nations.

United Nations Education Scientific and Cultural Organisation (UNESCO), (2007). The UN Decade of Education for Sustainable Development (DESD 2005-2014). The First Two Years. UNESCO, Paris

UNICEF, (2008). *Childhood poverty in industrial world*, 10.10.2008, Available from http://www.unicef.gr.

Valent, F., Little, D., Tamburlini, G., & Barbone, F. (2004). *Burden of disease attributable to selected environmental factors and injuries among Europe's children and adolescents.*

World Health Organization, (WHO Environmental Burden of Disease Series, No. 8), Geneva

World Commission on Environment and Development (WCED) (1987). *Our Common Future.* Oxford, Oxford University Press.

Wiedmann, T., & Barrett, J. (2010). A review of the ecological footprint indicator—perceptions and methods, *Sustainability,* Vol. 2, 1645-1693, ISSN 2071-1050

World Commission on Environment and Development (WCED), (1987). *Our Common Future,* Oxford University Press, Oxford

World Health Organization, (WHO), (1992). *Ottawa Charter for health promotion.* World Health Organization, Geneva,

World Health Organization (WHO), (2002). *World health report 2002. Reducing risks, promoting healthy life.* World Health Organization, Geneva,

World Health Organization (WHO), (2003a). *Health Aspects of Air Pollution with Particulate Matter, Ozone and Nitrogen Dioxide.* World Health Organization, Geneva

World Health Organization, (WHO), (2003b). *Technical meeting on exposure - response relationships on health.* World Health Organization, Regional Office for Europe

World Health Organization, (WHO), (2004). *Development of Environment and Health Indicators for European Union Countries: Results of a Pilot Study.* Report on a WHO Working Group Meeting. Bonn, Germany

World Health Organization, (WHO), (2005*). Indoor air pollution and health. Fact sheet N°292, 10.03.2011,* 12.11.07, Available from http://www.who.int/mediacentre/factsheets/fs29

World Health Organization (WHO), (2006). Air quality guidelines for particulate matter, ozone, nitrogen dioxide and sulfur dioxide. Global update 2005. Summary of risk assessment. World Health Organization, Geneva

World Health Organization (WHO), (2008a). *The world health report 2008: primary health care now more than ever.* World Health Organization, Geneva

World Health Organization (WHO), (2008b). Water, sanitation and hygiene links to health, 10.10.2008, Available from http://www.who.org

World Health Organization (WHO), (2009). Global health risks: mortality and burden of disease attributable to selected major risks. World Health Organization, Geneva

Part 3

The Impact of Air Pollution on Plants, Agricultural Sources and Methods of Resistance

12

The Effects of Air Pollutants on Vegetation and the Role of Vegetation in Reducing Atmospheric Pollution

Iuliana Florentina Gheorghe[1] and Barbu Ion[2]
[1]*Ecological University of Bucharest, Faculty of Ecology and Environmental Protection*
[2]*Forest Research and Management Institute*
Romania

1. Introduction

The main air pollutants are represented by gases forms, particles in suspension, different ionizing radiation and noise.
The gases forms are: oxidized and reduced forms of carbon (CO_2, CO, CH_4), of nitrogen (NO_2, NO, N_2O_4, NH_3, NH^{4+}), SO_2, O_3, C_6H_6 vapours, Hg, volatile phenols, Cl_2, etc.
The particulate forms are: PM10 and PM2.5 particulate matter, heavy metals with toxic effect (Pb, Ni, Cd, As), polycyclic aromatic hydrocarbons PAHs, etc.
Atmospheric pollutants have a negative effect on the plants; they can have direct toxic effects, or indirectly by changing soil pH followed by solubilization of toxic salts of metals like aluminum. The particulate matters have a negative mechanical effect. They cover the leaf blade reducing light penetration and blocking the opening of stomata. These impediments influence strongly the process of photosynthesis which rate declines sharply.
Also the leaves of the trees have an important role in retention of the particulate matters; they are mostly affected when the wet and dry atmospheric deposition increase.
The vegetation plays an important positive role in atmospheric purification and air pollutants reduction.
The primary producers represented by plants are an important component in biogeochemical cycles. The vegetation made exchanges with a part of the atmospheric gases by photosynthesis, respiration processes, and the final stage of litter decomposition which mineralization.
The plants play an important role in reducing atmospheric CO_2 content, by photosynthesis. This reduction of atmospheric CO_2 content has an important role in reducing of greenhouse gases, participating in reducing greenhouse effect and its consequences on climatic changes. The carbon stored in plants is the result of balance between carbon fixed by photosynthesis and carbon released in the atmosphere by respiration.
As the structure of vegetation is more complex, the carbon stock in plants biomass is higher and the period of storage is longer. The most efficient type of vegetation in storing carbon in terms of carbon stored in plants alive is the temperate-continental forest; and in terms of carbon stored in dead organic matter are peat lands.

Trees have also been planted to reduce the intensity of ionizing radiation and noise in different urban and industrial areas. The existence of vegetation in an area creates a microclimate where the temperature differentials between day and night are buffered.
This prevents the occurrence of warmer temperatures which stimulate the production of volatile pollutants into the atmosphere.

2. General information about air pollution

Environmental pollution is any discharge of material or energy into water, land, or air that causes or may cause acute (short-term) or chronic (long-term) detriment to the Earth's ecological balance or that lowers the quality of life. Pollutants may cause primary damage, with direct identifiable impact on the environment, or secondary damage in the form of minor perturbations in the delicate balance of the biological food web that are detectable only over long time periods.
Air pollution is the process which the substances and the energy forms are not present in normal atmospheric composition reach the atmosphere, or are present but in much lower concentrations.Air pollution is the introduction of chemicals, particulate matter, or biological materials that cause harm or discomfort to humans or other living organisms, or cause damage to the natural environment or built environment, into the atmosphere.
More than 3,000 substances that are not part of the atmospheric composition, falling in the atmosphere can be considered air pollutants.
Some substances that are normally present in the atmosphere in a certain concentration can be considerate pollutants because their concentration is much higher than usual concentration.
Also certain substances that are normally present in certain layers of the atmosphere (e.g. ozone in the stratosphere), once arrived in the troposphere is pollutant.
Some gases, such as oxides of nitrogen may have beneficial effect on vegetation, after hydration may affect the leaf fertilizer.
The air pollutants factors can be chemical (chemicals), mechanics (particles in suspension) physical (ionizing radiation) and acoustic (noise).

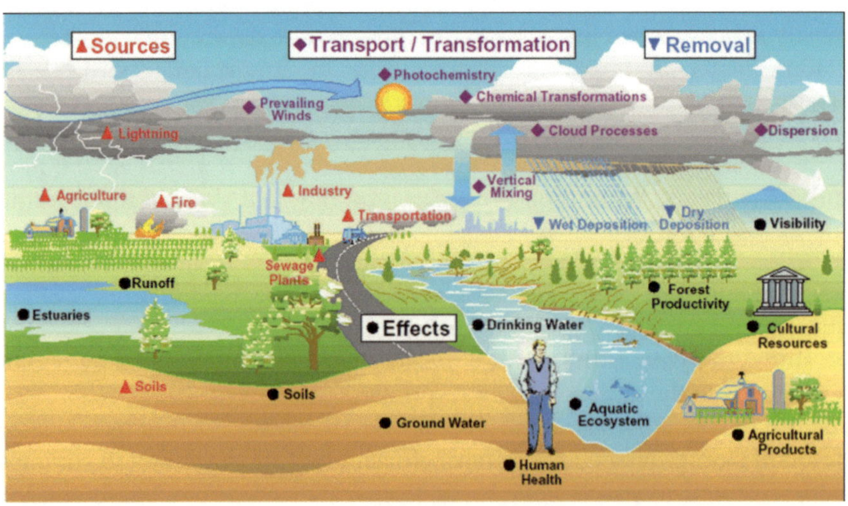

Fig. 1. Global polluants circuit (www.cleartheair.nsw.gov.au, 2011)

Pollutants describe a global circuit; they are produced by different sources, are transported and transformed into atmosphere, some of them being removed, another part is reaching the earth having different effects on different biocoenosis of ecosystems (fig. 1).

An analysis done at the global level revealed a diversification of pollutants agents and sources of air pollution. This diversification and increasing concentrations are in strict correlation with industrialization and the increasing of amounts use as fossil energy (non-renewable sources).

At the beginning, the pollution has been felt in urban areas and the forms of relief that favored the accumulation of pollutants and long stay (depressions, closed valleys, etc.). Currently, air pollution has become a larger area, sometimes to disperse across multiple continents.

Air pollution can be analyzed on three spatial scales: global pollution, regional pollution and local pollutants.

The global pollution is the result of cumulative effects of various sources, located on the entire surface of the globe, manifested by global effects: the stratospheric ozone depletion; greenhouse effect - emission of greenhouse gases (CO_2, methane, CFCs, etc.); formation of aerosols (pollutant clouds which suspended particles and chemical compounds).

The regional pollution is in part the result of local air pollution--including that produced by individual sources, such as automobiles - that has spread out to encompass areas of many thousands of square kilometers. Meteorological conditions and landforms can greatly influence air-pollution concentrations at any given place, especially locally and regionally. For example, cities located in bowls or valleys over which atmospheric inversions form and act as imperfect lids are especially likely to suffer from incidences of severe smog. Oxides of sulfur and nitrogen carried long distances by the atmosphere and then precipitated in solution as acid rain, can cause serious damage to vegetation, waterways, and buildings.

The local pollutants (smog) can be loosely defined as a multi-source, widespread air pollution that occurs in the air of cities. Smog, a contraction of the words smoke and fog, has been caused throughout recorded history by water condensing on smoke particles, usually from burning coal.

In terms of the effects of pollutants can be acidifying agents - sulphur dioxide (SO_2), nitrogen oxides (NOx) ammonia (NH_3) fluoride and Cl_2, hydrogen chloride (HCl) - and oxidizing agents - carbon monoxide (CO), PAN (peroxyacetylnitrate-$CH_3CO.O_2.NO_2$), ozone (O_3).

3. Sources of pollutants

Air pollution comes from natural and anthropic sources; these sources generate pollutants with different effects at global level or on individuals of plants and animals (tab. 1).

Natural processes that affect air quality include volcanoes, which produce sulfur, chlorine, and ash particulates. Wildfires produce smoke and carbon monoxide. Cattle and other animals emit methane as part of their digestive process. Even pine trees emit volatile organic compounds (VOCs).

Many forms of air pollution are **human-made**. Industrial plants, power plants and vehicles with internal combustion engines produce nitrogen oxides, VOCs, carbon monoxide, carbon dioxide, sulfur dioxide and particulates. In most mega-cities, cars are the main source of these pollutants. Stoves, incinerators, and farmers burning their crop waste produce carbon monoxide, carbon dioxide, as well as particulates. Other human-made sources include aerosol sprays and leaky refrigerators, as well as fumes from paint, varnish, and other solvents. One important thing to remember about air pollution is that it doesn't say in one place. Winds and weather play an important part in transport of pollution locally,

Name of pollutants	Origin	Effects
Natural sources		
sulfur, chlorine, and ash particulates, smoke and carbon monoxide methane volatile organic compounds (VOCs) Aerosol from deforestation and burning: CO, CO_2, NO, NO_2, N_2O, NH_4	Volcanoes, wildfires, cattle and other animals, pine trees	- acid rain, - smog, - respiratory irritant - increased respiratory - diseases - damage cell membranes of plants The effects are high only for volcanoes.
Anthropic sources		
Carbon monoxide, carbon dioxide, sulphur dioxide, nitrogen oxides, fluorides and substances with fluorine, chlorine (Cl_2), bromine (Br_2) and iodine (I_2), small dust particles, VOC, methane, ammonia and radioactive radiation.	Industry: the mining industry, oil and natural gas extraction, the energy industry based on fossil fuels - coal, oil, natural gas, the production of brick, tile, enamel frit, ceramics, and glass; the manufacture of aluminium and steel; and the production of hydrofluoric acid, phosphate chemicals and fertilizers. central heating, chemical and metallurgical industry, engineering internal combustion machinery industry, industrial waste, noises	- respiratory irritant, - acid rain, - smog, - increased respiratory - formation of secondary pollutants (PAN, O_3) - effect on soil fertilizer - Respiratory diseases - toxic effects on living cells - greenhouse gas effect - toxic effects - carcinogenic proprieties - accumulation in tissues - blocking of different processes - stratospheric ozone depletion
CO, CO_2, NO, NO_2, NH_3, CH_4, SO_2, oxides of heavy metals, H_2SO_4, SPM, HC, VOC, background aerosols: sea salt oxidation of sulphur containing gases, same organics, nitrous oxide (N_2O) pesticides	Agriculture: the vegetation fire, the denitrification process, in soils excessively fertilized and excessive use the pesticides , paddy field, intensive husbandry, deforestation	- formation of secondary pollutants (PAN, O_3) - effect on soil fertilizer - respiratory diseases - greenhouse gas effect - toxic effects - acid rain, - stratospheric ozone depletion
Aerosols from transport and constructions NOx, CO, HCl, Lead and other heavy metals , SPM	The motor vehicle pollution, noises	- smog - increased respiratory diseases - damage cell membranes of plants - carcinogenic proprieties - accumulation in tissues - blocking of different processes - stratospheric ozone depletion
Domestic aerosols CFC, HC, FC, H_2S, CH_4 CO_2	sewage plans, authorized landfill site	- carcinogenic proprieties - accumulation in tissues - blocking of different processes - stratospheric ozone depletion

Table 1. Type of pollutants, origin and effect at global level or on plants end animals individuals

regionally, and even around the world, where it affects everything it comes in contact with. The major anthropic sources of air pollution are:

- industry and conventional energies (the mining industry, the energy industry based on fossil fuels - coal, oil, natural gas, central heating, chemical and metallurgical industry, engineering internal combustion machinery industry, industrial waste, noises, etc);

- agriculture (the vegetation fire, denitrification in soils excessively fertilized, paddy field, intensive husbandry, deforestation, etc)
- transportation (motor vehicle pollution, noises, etc)
- and urbanization (sewage plans, authorized landfill site, etc) (fig. 2)

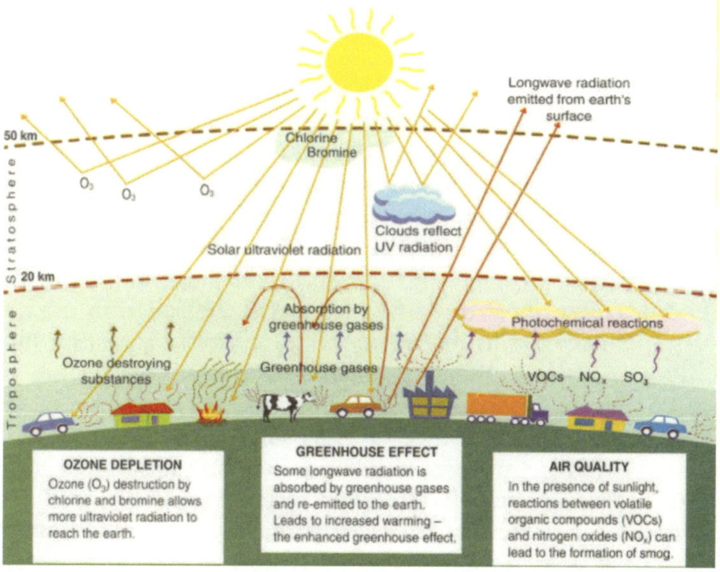

Fig. 2. The major anthropic sources of air pollution (www.cleartheair.nsw.gov.au, 2011)

4. The most important atmospheric pollutants

The air pollutants are represented by gases forms, particles in suspension, different ionizing radiation and noise.

5. Major gaseous pollutants

Sulphur dioxide (SO_2) is the most important and common air pollutant produced in huge amounts in combustion of coal and other fuels in industrial and domestic use. It is also produced during smelting of sulphide ores. Sulphur dioxide concentrations in air have decreased in the past two decades, mainly because we use more non-sulphur-containing fuels for the generation of energy. Sulphur dioxide is a stinging gas and as a result it can cause breathing problems with humans. In moist environments, sulphur dioxide may be transferred to sulphuric acid. This acid causes acidification and winter smog.
Nitrogen oxides (NO_x) and nitrous oxide (N_2O) Using catalysers in car exhausts can prevent emissions of nitrogen oxides. Nitrogen oxides are gasses that react with other air pollutants when they are present in air. For example, nitrogen oxides play an important rolein the formation of ozone in the lower atmosphere, and in acidification and eutrophication processes. They can deeply penetrate the lungs and damage human lung functions.
Fluorides Common gaseous fluoride pollutants are HF, SiF_6, CF_4 and F_2. Particulate fluoride pollutants include Ca_3AlF_6 (Cryolite), CaF_2, NH_3F, AlF_6, CaSiF, NaF and Na_2SiF_6. Aerosols are often formed from NaF, $NaAlF_6$ and AlF_6. Chief sources of fluoride pollutants are brickworks, aluminium factories, glassworks, steelworks, ceramic factories, phosphate

fertilizer plants and uranium smelters. Some fluorine pollution also occurs during combustion of coal. Most injurious fluoride pollutant is gaseous hydrogen fluoride (HF).
Fluoride is released into the air in large quantities by aluminum reduction plants, phosphate processors, steel mills, coal burning operations, brick and tile manufacturers, and various less significant sources[1]. It can cause adverse effects when ingested by domestic animals or absorbed by plants. There are also reports that fluoride air pollution can adversely affect human health, though these are less well documented than those concerning sensitive animals and plants. Fluorides are released into the air in both a gaseous state (as hydrogen fluoride and silicon tetra-fluoride) and in solid particles. The particles fall on, and the gases are absorbed by, vegetation near the polluting industry. If this vegetation includes forage crops which are fed to cattle, sheep, horses, or pigs, serious problems may ensue, since these animals, particularly the cattle are vulnerable to fluoride [2]. In fact, according to the U.S. Ninety-six percent of the ingested fluoride that accumulates in the bodies of animals is incorporated into the crystal structure of bone and tooth mineral [3], [4]. When fluoride is ingested with food or water, most of that which is not deposited in the bones, teeth, and other calcified tissue is excreted in the urine within hours of ingestion [5]. Thus it is not surprising that fluoride mainly affects the bones and teeth. Teeth are more markedly affected by ingested fluoride than are bones, but their high sensitivity is limited to the period of their formation. Thus a cow that has not been exposed to excessive fluoride before the age of two and one-half to three years will not develop the severe dental lesions which would occur in the same animal exposed at a younger age [6]. The developing tooth exposed to small amounts of fluoride may experience color variations ("mottling") that have little or no effect on the animal's ability to eat. Higher levels of fluoride result in more serious dental abnormalities, ranging from small, brittle, chalky areas on the tooth surface to pitting of enamel and easily eroded teeth [2]. Even more serious effects, including severe pain and the wearing down of the tooth right to the gum, can prevent the cattle from drinking cold water or eating. Localized or generalized enlargement of certain bones in the legs (metacarpals and metatarsals) and the lower jaw (mandible) of cattle are common symptoms of excessive fluoride ingestion [7]. As highly abnormal bone tissue replaces normal bone, [8] overall enlargement occurs, and the normally smooth bone surfaces take on a chalky, white, irregular appearance [2]. Hard ground can cause fluorotic hoof (pedal) bones to fracture, resulting in severe lameness [6]. Cattle with advanced fluorosis may also be crippled by mineralization of ligaments, tendons, and the structures surrounding the joints [9]. Enlargement of the joints themselves may also contribute to lameness. Fluoride-induced tooth destruction, lameness, and stiff joints affect the animal's ability to stand, eat, and graze, and all tend to lower the milk yield of dairy cattle or the weight of beef cattle.
Chlorine (Cl_2) Although chlorine concentrations change very rapidly in the atmosphere due to atmospheric chemistry and light rain can remove all the chlorine from the air in a very short time, chlorine injury can occur to plants near the source of pollution.
The impact of chlorine pollution increases in bright sunlight and decreases in drought and low temperature.
Many particulate and gaseous fluorides are produced when ores containing fluorine are processed and used in industries. Common gaseous fluoride pollutants are HF, SiF_6, CF_4 and F_2. Particulate fluoride pollutants include Ca_3AlF_6 (Cryolite), CaF_2, NH_3F, AlF_6, CaSiF, NaF and Na_2SiF_6. Aerosols are often formed from NaF, $NaAlF_6$ and AlF_6. Chief sources of fluoride pollutants are brickworks, aluminium factories, glassworks, steelworks, ceramic factories, phosphate fertilizer plants and uranium smelters. Some fluorine pollution also occurs during combustion of coal. Most injurious fluoride pollutant is gaseous hydrogen fluoride (HF).

Hydrogen chloride (HCl) HCl gas is released in large quantities in combustion of PVC and all chlorinated hydrocarbon material in large fires or incinerators. The HCl gas is very hygroscopic and quickly changes to hydrochloric acid by reacting with atmospheric moisture and forms aerosol droplets.

Ammonia (NH_3) Continuous releases of ammonia from the sources are rarely high enough to cause acute injury but occasional high release or spillage may cause ammonia pollution. High concentrations of ammonia are sometimes found around intensive farm units e.g. chicken batteries. Extent of injury reduces rapidly with increase in distance form the source. Under certain conditions the ammonia may remain as a cloud above ground level causing more injury to trees than to the ground flora. Injury symptoms may take up to 9 days to develop. In most plant species, recovery may occur in about 2 weeks after exposure is stopped.

Ammonia forms during agricultural activities. EEA-32 emissions of NH_3 have declined by 24% between the years 1990 and 2008. Agriculture was responsible for 94% of NH_3 emissions in 2008. The reduction in emissions within the agricultural sector is primarily due to a reduction in livestock numbers (especially cattle) since 1990, changes in the handling and management of organic manures and from the decreased use of nitrogenous fertilizers. The reductions achieved in the agricultural sector have been marginally offset by the increased emissions which have occurred during this period in sectors such as transport and to a lesser extent the energy industry and other (non-energy) sectors.

Environmental context: NH_3 contributes to acid deposition (plays an important role in acidification) and eutrophication. The subsequent impacts of acid deposition can be significant, including adverse effects on aquatic ecosystems in rivers and lakes and damage to forests, crops and other vegetation. Eutrophication can lead to severe reductions in water quality with subsequent impacts including decreased biodiversity, changes in species composition and dominance, and toxicity effects. NH_3 also contributes to the formation of secondary particulate aerosols, an important air pollutant due to its adverse impacts on human health.

VOC (Volatile Organic Compounds) VOC can be a range of different contaminants, such as carbohydrates, organic compounds and solvents. These compounds usually derive from petrol and gasoline reservoirs, industrial processes and fuel combustion, paint and cleanser use, or agricultural activities. VOC play an important role in ozone shaping in the lower atmospheric layer, the main cause of smog. VOC can cause various health effects, depending on the kind of compounds that are present and their concentrations. Effects can vary from smell nuisance to decreases in lung capacity, and even cancer.

Organic gases (Ethylene) and Methane (CH_4) Among organic gaseous pollutants, ethylene is most common. Other organic gases are propylene, butylenes and acetylene. Ethylene is continuously emitted from many sources involving combustion or processing of petroleum or its products or burning of organic materials e.g. straw burning. Other organic gases are also produced in various chemical industrial processes.

Ethylene is a natural plant growth substance so the injury effects produced by it on plants are very similar to growth abnormality symptoms. Other organic gases also produce symptoms similar to those of ethylene pollution. However, the sensitivities of species to different gases are variable.

Ethylene is a byproduct of automobile exhaust and can be a noticeable problem in urban environments.

Chlorofluorocarbons (or CFCs) is an organic compound that contains carbon, chlorine, and fluorine, produced as a volatile derivative of methane and ethane. A common subclass is the **hydro-chlorofluorocarbons** (HCFCs), which contain hydrogen, as well. They are also commonly known by the DuPont trade name **Freon**. The most common representative is dichlorodifluoromethane (R-12 or Freon-12). Many CFCs have been widely used as refrigerants, propellants (in aerosol applications), and solvents. The manufacture of such compounds is being phased out by the Montreal Protocol because they contribute to ozone depletion.

Minor gaseous pollutants Many other air pollutants which are highly injurious to animals and human beings also cause damage to plants. However, plants are affected by these gases at quite higher concentrations than the animals. Common such gaseous pollutants are CO, H_2S, Br_2, I_2 and Hg - vapors.

Hydrogen sulphide (H_2S) It is a colorless, very poisonous, flammable gas with the characteristic foul odor of rotten eggs at concentrations up to 100 parts per million. It often results from the bacterial breakdown of organic matter in the absence of oxygen, such as in swamps and sewers (anaerobic digestion). It also occurs in volcanic gases, natural gas, and some well waters. The human body produces small amounts of H_2S and uses it as a signaling molecule.

Carbon monoxide (CO) **carbon dioxide** in excess (CO_2), this gas consists during incomplete combustion of fuels. When we let a car engine run in a closed room, carbon monoxide concentrations in the air will rise extensively. Carbon monoxide contributes to the greenhouse effect, smog and acidification. The gas can bind to hemoglobin in blood, preventing oxygen transport through the body. This results in oxygen depletion of the heart, brains and blood vessels, eventually causing death.

It is highly toxic to humans and animals in higher quantities, although it is also produced in normal animal metabolism in low quantities, and is thought to have some normal biological functions.

Carbon monoxide consists of one carbon atom and one oxygen atom, connected by a triple bond which consists of two covalent bonds as well as one dative covalent bond. It is the simplest ox-carbon. In coordination complexes the carbon monoxide ligand is called carbonyl.

Carbon monoxide is produced from the partial oxidation of carbon-containing compounds; it forms when there is not enough oxygen to produce carbon dioxide (CO_2), such as when operating a stove or an internal combustion engine in an enclosed space. In the presence of oxygen, carbon monoxide burns with a blue flame, producing carbon dioxide [10]. Coal gas, which was widely used before the 1960s for domestic lighting, cooking and heating, had carbon monoxide as a significant constituent. Some processes in modern technology, such as iron smelting, still produce carbon monoxide as a byproduct [11]. Worldwide, the largest source of carbon monoxide is natural in origin; due to photochemical reactions in the troposphere which generate about 5×10^{12} kilograms per year [12], other natural sources of CO include volcanoes, forest fires, and other forms of combustion.

In biology, carbon monoxide is naturally produced by the action of heme oxygenase 1 and 2 on the heme from hemoglobin breakdown. This process produces a certain amount of carboxyhemoglobin in normal persons, even if they do not breathe any carbon monoxide. Following the first report that carbon monoxide is a normal neurotransmitter in 1993, as well as one of three gases that naturally modulate inflammatory responses in the body (the other two being nitric oxide and hydrogen sulfide), carbon monoxide has received a great deal of clinical attention as a biological regulator . In many tissues, all three gases are known to act as anti- inflammatory, vasodilators and promoters of neo-vascular growth [13]. Clinical trials of small amounts of carbon monoxide as a drug are on-going.

Bromine (Br_2) and Iodine (I_2)

At high temperatures, organo-bromine compounds are easily converted to free bromine atoms, a process which acts to terminate free radical chemical chain reactions. This makes such compounds useful fire retardants and this is bromine's primary industrial use, consuming more than half of world production of the element. The same property allows volatile organo-bromine compounds, under the action of sunlight, to form free bromine atoms in the atmosphere which are highly effective in ozone depletion. This unwanted side-effect has caused many common volatile brominated organics like methyl bromide, a pesticide that was formerly a large industrial bromine consumer, to be abandoned. Remaining uses of bromine compounds are in well-drilling fluids, as an intermediate in manufacture of organic chemicals, and in film photography.

Iodine and its compounds are primarily used in nutrition, the production of acetic acid and polymers. Iodine's relatively high atomic number, low toxicity, and ease of attachment to organic compounds have made it a part of many X-ray contrast materials in modern medicine. Like the other halogens, iodine occurs mainly as a diatomic molecule I_2, not the atom. In nature, iodine is a relatively rare element, ranking 47th in abundance. It is the heaviest essential element utilized in biological functions. Its rarity in many soils has led to many deficiency problems in land animals and inland human populations, with iodine deficiency affecting about two billion people and being the leading preventable cause of mental retardation [14]. As a component of thyroid hormones, iodine is required by higher animals. Radioisotopes of iodine are concentrated in the thyroid gland. This property of thyroid-concentration, along with its mode of beta decay, makes iodine-131 one of the most carcinogenic nuclear fission products.

Hg and Mercury vapors Pre-industrial deposition rates of mercury from the atmosphere may be about 4 ng/ (1 l of ice deposit). Although that can be considered a natural level of exposure, regional or global sources have significant effects. Volcanic eruptions can increase the atmospheric source by 4–6 times [15]. Natural sources, such as volcanoes, are responsible for approximately half of atmospheric mercury emissions. The human-generated half can be divided into the following estimated percentages:

- 65% from stationary combustion, of which coal-fired power plants are the largest aggregate source (40% of U.S. mercury emissions in 1999). This includes power plants fueled with gas where the mercury has not been removed. Emissions from coal combustion are between one and two orders of magnitude higher than emissions from oil combustion, depending on the country [15].
- 11% from gold production. The three largest point sources for mercury emissions in the U.S. are the three largest gold mines. Hydro-geochemical release of mercury from gold-mine tailings has been accounted as a significant source of atmospheric mercury in eastern Canada [16]
- 6.8% from non-ferrous metal production, typically smelters.
- 6.4% from cement production.
- 3.0% from waste disposal, including municipal and hazardous waste, crematoria, and sewage sludge incineration. This is a significant underestimate due to limited information, and is likely to be off by a factor of two to five.

- 3.0% from caustic soda production.
- 1.4% from pig iron and steel production.
- 1.1% from mercury production, mainly for batteries.
- 2.0% from other sources [15], (EPA report, 2007).

The above percentages are estimates of the global human-caused mercury emissions in 2000, excluding biomass burning, an important source in some regions [15]. Current atmospheric mercury contamination in outdoor urban air is (0.01–0.02 μg/m^3) indoor concentrations are significantly elevated over outdoor concentrations, in the range 0.0065–0.523 μg/m^3 (average 0.069 μg/m^3) [17]. Mercury also enters into the environment through the improper disposal (e.g., land filling, incineration) of certain products.

6. Particulate pollutants

Particles in suspension

Dust particles. Dust particles form a complex of organic compounds and minerals. These can derive from natural sources, such as volcanoes, or human activities, such as industrial combustion processes or traffic. Particles are categorized according to particle size. The smallest particles have the ability to transport toxic compounds into the respiratory tract. Some of these compounds are carcinogenic. The upper respiratory tract stops the larger dust particles. When they are released into the environment, dust particles can cause acidification and winter smog.

Cement-kiln dust Cement factories are the chief source of cement dust pollution. The composition of such dust varies with the source. Main component of cement dust is CaO and varying amounts of K_2O, Na_2O and KCl and traces of Al, Fe, Mn, Mg, S and silica. Dust with more than 24% CaO is more injurious to plants. Fine particles cause more damage than larger particles. Cement-kiln dust is alkaline in nature and dissolves in atmospheric moisture forming a solution of pH 10-12.

Lime and gypsum Lime and gypsum processing industries and mining deposits are chief sources from where fine particles of these substances are blown away to great distances.

Soot Burning of fossil fuels, organic matter or natural forest fires produce huge quantities of fine carbon particles which form the soot pollution. Soot can be dispersed over a quite wide area and transported to great distances by blowing winds.

Magnesium oxide Magnesium roasters are the chief sources of such pollution. The magnesium oxide dust may be carried by winds and deposited even at a distance of 5 km from the source. In the atmosphere, magnesium sulphate ($MgSO_4$) combines with carbon dioxide and water to form $Mg(CO_3)_2$. Both these compounds are alkaline and slightly soluble in water.

Boron Boric acid and borax are common raw materials in many industries. Oven and refrigerator manufacturing industries are chief sources of boron pollution.

Chlorides of sodium, potassium and calcium Sodium and calcium chlorides are commonly used in cold countries on the roads during winters to melt ice and snow. Potash industry produces aerial emission of KCl and NaCl in ratio of 3:1. All such chlorides are carried away by winds and deposited on the soil and plants.

Sodium sulphate, with an annual production of 6 million tones, it is a major commodity chemical and one of the most damaging salts in structure conservation: when it grows in the pores of stones it can achieve high levels of pressure, causing structures to crack.

Sodium sulfate is mainly used for the manufacture of detergents and in the Kraft process of paper pulping. About two-thirds of the world's production is from mirabilite, the natural mineral form of the decahydrate, and the remainder from by-products of chemical processes such as hydrochloric acid production.

Pesticides, insecticides and herbicides Pesticide use in the agricultural industry began in earnest in the early 1940s. Although pesticide use had been quite popular for more than twenty years, government officials first became aware of the potential danger of pesticide runoff to humans in the early 1960s when Rachel Carson's famous and influential *Silent Spring* was published. Though this book warned mainly of the detrimental effects of DDT (a popular insecticide developed in the early 1940s) for birds and other non-human victims, Carson's work inspired health officials to speculate about the effects of pesticide runoff on humans. Recently, exposure to DDT was linked to Parkinson's disease. Because of concern over DDT's adverse effects on the environment and on people, this pesticide was banned in 1972. Despite the ban of DDT, pesticide use continues, and the effects of some modern insecticides and herbicides can be just as debilitating. Even through careful use, runoff from pesticides continues to makes its way into drinking water sources.

7. Secondary pollutants

Photo-oxidants In presence of strong sunlight and in hot weather a series of complex chemical reactions involving nitrogen oxides and hydrocarbons may produce certain photo-oxidant chemicals. These chemicals do not have any specific anthropogenic source but are formed over wide areas in which suitable environmental conditions are prevailing. Two such photo-oxidants that can reach ambient concentrations toxic to plants are PAN (Peroxyacetylnitrate) and ozone.

PAN (Peroxyacetylnitrate-$CH_3CO.O_2.NO_2$) Impact of this secondary pollutant is not affected by humidity. However, the impact decreases with lowering of temperature and increasing drought conditions. The impact also increases in the morning and in bright sunlight. Young plants and young rapidly expanding leaves are more sensitive to this pollutant. PAN interacts with SO_2 and O_3 in complex manner producing variable impact conditions.

Ozone (O_3) is the main pollutant in the oxidant smog complex.

Ozone is formed in the troposphere when sunlight causes complex photochemical reactions involving oxides of nitrogen (NOx), volatile organic hydrocarbons (VOC) and carbon monoxide that originate chiefly from gasoline engines and burning of other fossil fuels. Woody vegetation is another major source of VOCs. NOx and VOCs can be transported long distances by regional weather patterns before they react to create ozone in the atmosphere, where it can persist for several weeks. Seasonal exposures at low elevations consist of days when ozone concentrations are relatively low or average, punctuated by days when concentrations are high. Concentrations of ozone are highest during calm, sunny, spring and summer days when primary pollutants from urban areas are present. Ozone concentrations in rural areas can be higher than in urban areas while ozone levels at high elevations can be relatively constant throughout the day and night.

Middle aged leaves and young plants are more sensitive to ozone. This pollutant interacts with SO_2, NO_2, PAN and heavy metals in complex manner.

Ozone is created through photochemical transfer of oxygen. This process takes place under the influence of ultra violet sunlight (UV), aided by pollutants in the outside air (fig. 2). Ozone causes smog and contributes to acidification and climate change. Ozone is an aggressive gas. This can easily penetrate the respiratory tract, deeply. When humans are exposed to ozone, the consequences may be irritation of the eyes and the respiratory tract.

Acid deposition Various acid gases, aerosols and other acidic substances released into the atmosphere from the industrial or domestic sources of combustion of fossil fuels eventually come down to the ground. These substances are deposited directly on the water bodies. In addition, these substances also reach the water bodies along with run-off rainwater from the polluted soil. Deposition of acidic substances causes acidification of water by lowering its pH below 6.0. The sulphates, nitrates and chlorides have been reported to make water bodies like lakes, rivers and ponds acidic in many countries.

Acid deposition is not merely characterized as acid rain; it can also be snow and fog or gas and dust. Acid deposition mainly forms during fossil fuel combustion. When emissions of sulphur dioxide and nitrogen oxides come in contact with water, they will become sulphuric acid and nitric acid.

When acidifying agents, such as sulphur dioxide, nitrogen oxides and ammonia, end up in plants, surface water and soils, this has a number of consequences:

- availability of nutrients and metal spores is likely to decrease
- when acidity is high more metals will dissolve in water. This can cause surface water to become polluted, which has serious health effects on aquatic plants and animals. For example, high aluminum (Al) concentrations can complicate nutrients uptake by plants. This makes aluminum one of the prior causes of forest decay. Mercury can be dispersed by transport through surface water, causing it to accumulate in fish. Mercury can bio magnify up the food chain, to be taken up by humans eventually
- Buildings and monuments may be damaged through erosion. Sulphur dioxide breaks down limestone by reacting with calcium carbonate, causing limestone to absorb water during rainfall. Limestone will than fragment

Noise pollution has a relatively recent origin. It is a composite of sounds generated by human activities ranging from blasting stereo systems to the roar of supersonic transport jets. Although the frequency (pitch) of noise may be of major importance, most noise sources are measured in terms of intensity, or strength of the sound field. The standard unit, one decibel (dB), is the amount of sound that is just audible to the average human. The decibel scale is somewhat misleading because it is logarithmic rather than linear; for example, a noise source measuring 70 dB is 10 times as loud as a source measuring 60 dB and 100 times as loud as a source reading 50 dB. Noise may be generally associated with industrial society, where heavy machinery, motor vehicles, and aircraft have become everyday items. Noise pollution is more intense in the work environment than in the general environment, although ambient noise increased an average of one dB per year during the 1980s. The average background noise in a typical home today is between 40 and 50 decibels. Some examples of high-level sources in the environment are heavy trucks (90 dB at 15 m/50 ft), freight trains (75 dB at 15 m/50 ft), and air conditioning (60 dB at 6 m/20 ft).

Radiation pollution is any form of ionizing or no ionizing radiation that results from human activities. The most well-known radiation results from the detonation of nuclear devices and the controlled release of energy by nuclear-power generating plants (see nuclear energy). Other sources of radiation include spent-fuel reprocessing plants, by-products of mining operations, and experimental research laboratories. Increased exposure to medical x rays and to radiation emissions from microwave ovens and other household appliances, although of considerably less magnitude, all constitute sources of environmental radiation.

Radioactive radiation. Radioactive radiation and radioactive particles are naturally present in the environment. During power plant incidents or treatments of nuclear waste from a war where nuclear weapons are used, radioactive radiation can enter the air on account of humans. When humans are exposed to high levels of radioactive radiation, the chances of serious health effects are very high. Radioactive radiation can cause DNA alteration and cancer.

8. Flow of atmospheric pollutants at global level

The air pollutants are produced by different sectors of the economy like: industry, agriculture, transports and urbanization. The burning of hydrocarbons in motor vehicle engines gives rise to CO_2, CO, SO_2 (sulfur dioxide), NO_x (NO [nitrogen monoxide]) and NO_2^-- in varying proportions-and C_2H_4 (ethylene), as well as a variety of other hydrocarbons. Additional SO_2 originates from domestic and industrial burning of fossil fuels. Industrial plants, such as chemical works and metal-smelting plants, release SO_2, H_2S, NO_2, and HF (hydrogen fluoride) into the atmosphere. Tall chimney stacks may be used to carry gases and particles to a high altitude and thus avoid local pollution, but the pollutants return to Earth, sometimes hundreds of kilometers from the original source.

Photochemical smog is the product of chemical reactions driven by sunlight and involving NO_x of urban and industrial origin and volatile organic compounds from either vegetation (*biogenic* hydrocarbons) or human activities (*anthropogenic* hydrocarbons). Ozone (O_3) and peroxyacetylnitrate (PAN) produced in these complex reactions can become injurious to plants and other life forms, depending on concentration and duration of exposure. Hydrogen peroxide, another potentially injurious molecule, can form by the reaction between O_3 and naturally released volatiles (terpenes) from forest trees.

The concentrations of polluting gases, or their solutions, to which plants are exposed are thus highly variable, depending on location, wind direction, rainfall, and sunlight. Experiments aimed at determining the impact of chronic exposure to low concentrations of gases should allow plants to grow under near-natural conditions. One method is to grow the plants in open-top chambers into which gases are carefully metered, or where plants receiving ambient, polluted air are compared with controls receiving air that has been scrubbed of pollutants.

These pollutants emitted into the atmosphere can react with components of the atmosphere and transform into more or less aggressive or toxic compounds. The air pollutants can accumulate and manifest directly effects in the atmosphere (greenhouse effect, ozone layer depletion, etc) or can to transform in other pollutants and manifest indirectly effects on ecosystem biocoenosis, plants, animals and human health (fig. 3)

9. Effect of pollutants on vegetation, direct effects, and indirect effect, gas toxicity, wet and dry deposition, and deposition mixtures

Dust pollution is of localized importance near roads, quarries, cement works, and other industrial areas. Apart from screening out sunlight, dust on leaves blocks stomata and lowers their conductance to CO_2, simultaneously interfering with photosystem II. Polluting gases such as SO_2 and NO_x enter leaves through stomata, following the same diffusion pathway as CO_2. NO_x dissolves in cells and gives rise to nitrite ions (NO_2^-, which are toxic at high concentrations) and nitrate ions (NO_3^-) that enter into nitrogen metabolism as if they had been absorbed through the roots. In some cases, exposure to pollutant gases,

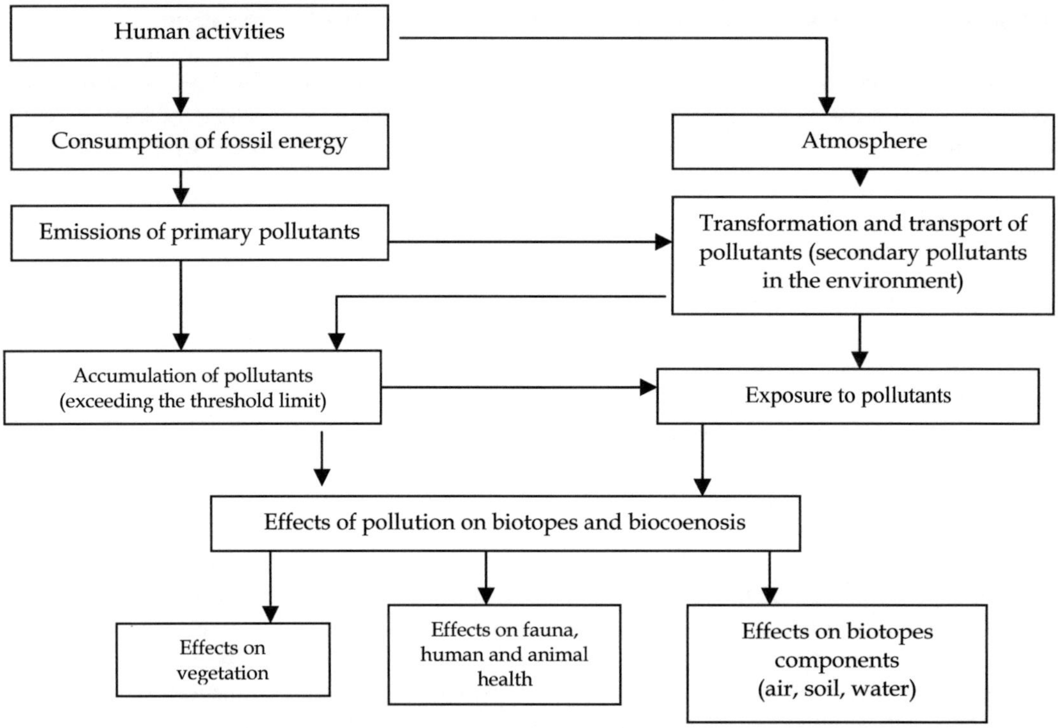

Fig. 3. The air pollutants flow diagram

particularly SO_2, causes stomatal closure, which protects the leaf against further entry of the pollutant but also curtails photosynthesis. In the cells, SO_2 dissolves to give bisulfite and sulfite ions; sulfite is toxic, but at low concentrations it is metabolized by chloroplasts to sulfate, which is not toxic. At sufficiently low concentrations, bisulfite and sulfite are effectively detoxified by plants, and SO_2 air pollution then provides a sulfur source for the plant. In urban areas these polluting gases may be present in such high concentrations that they cannot be detoxified rapidly enough to avoid injury. Ozone is presently considered to be the most damaging phytotoxic air pollutant in North America [18], [19]. It has been estimated that wherever the mean daily O_3 concentration reaches 40, 50, or 60 ppb (parts per billion or per 10^9), the combined yields of soybean, maize, winter wheat, and cotton would be decreased by 5, 10, and 16%, respectively. Ozone is highly reactive: It binds to plasma membranes and it alters metabolism. As a result, stomatal apertures are poorly regulated, chloroplast thylakoid membranes are damaged, rubisco is degraded, and photosynthesis is inhibited. Ozone reacts with O_2 and produces reactive oxygen species, including hydrogen peroxide (H_2O_2), superoxide (O_2^-), singlet oxygen ($^1O_2^*$), and the hydroxyl radical (^-OH). These denature proteins and damage nucleic acids (thereby giving rise to mutations), and cause lipid peroxidation, which breaks down lipids in membranes. Reactive oxygen species form also in the absence of O_3, particularly in electron transport in the mitochondria and chloroplasts, when electrons can be donated to O_2. Cells are protected, at least in part, from reactive oxygen species by enzymatic and nonenzymatic defense mechanisms [20], [21]. Defense against reactive oxygen species is provided by the scavenging properties of molecules, such as ascorbic acid, α-tocopherol, phenolic compounds, and glutathione. Superoxide dismutases (SODs) catalyze the reduction of superoxide to hydrogen peroxide.

Hydrogen peroxide is then converted to H_2O by the action of catalases and peroxidases. Of particular importance is the ascorbate-specific peroxidase localized in the chloroplast. Acting in concert, ascorbate peroxidase, dehydroascorbate reductase, and glutathione reductase remove H_2O_2 in a series of reactions called the *Halliwell–Asada pathway*, named after its discoverers. Glutathione is a sulfur-containing tripeptide that, in its reduced form, reacts rapidly with dehydroascorbate and becomes oxidized in the process. Glutathione reductase catalyzes the regeneration of reduced glutathione (GSH) from its oxidized form (GSSG) in the following reaction:

$$GSSG + NADPH + H^+ \rightarrow 2\,GSH + NADP^+$$

Exposure of plants to reactive oxygen species stimulates the transcription and translation of genes that encode enzymes involved in protection mechanisms. In Arabidopsis, exposure for 6 hours per day to low levels of O_3 induces the expression of several genes that encode enzymes associated with protection from reactive oxygen species, including SOD, glutathione S-transferase (which catalyzes detoxification reactions involving glutathione), and phenylalanine ammonia lyase (an important enzyme at the start of the phenylpropanoid pathway that leads to the synthesis of flavonoids and other phenolics) [22].

In transgenic tobacco transformed with a gene from *Escherichia coli* to give additional glutathione reductase activity in the chloroplast, short-term exposure to high levels of SO_2 is much less damaging than for wild-type tobacco [23]. Environmental extremes may either accelerate the production of reactive oxygen species or impair the normal defense mechanisms that protect cells from reactive oxygen species. In water-deficient leaves, for example, greater oxygen photoreduction by photosystems I and II increases superoxide production, and the pool of glutathione, as well as the activity of glutathione reductase, increase-presumably as part of the cell defense mechanism. In contrast, levels of ascorbate, another antioxidant, generally decline with mild water stress. Transgenic plants overexpressing mitochondrion superoxide dismutase (Mn-SOD), the isozyme localized in the mitochondrial matrix, show less water-deficit damage and, significantly, improved survival and yield under field conditions [24]. In other experiments, transgenic alfalfa overexpressing Mn-SOD was found to be more tolerant of freezing. Conversely, winter rye, wheat, and barley acclimated at 2 °C for several weeks, were found to have developed resistance to the herbicides, paraquat and acifluorfen, which generate reactive oxygen species. Such investigations support the hypothesis that tolerance of oxidative stress is an important factor in tolerance to a wide range of environmental extremes. Many deleterious changes in metabolism caused by air pollution precede external symptoms of injury, which appear only at much higher concentrations. For example, when plants are exposed to air containing NO_x, lesions on leaves appear at a NO_x concentration of 5 ml/l, but photosynthesis starts to be inhibited at a concentration of only 0.1 ml/l. These low, threshold concentrations refer to the effects of a single pollutant. However, two or more pollutants acting together can have a synergistic effect, producing damage at lower concentrations than if they were acting separately. In addition, vegetation weakened by air pollution can become more susceptible to invasion by pathogens and pests. Unpolluted rain is slightly acidic, with a pH close to 5.6, because the CO_2 dissolved in it produces the weak acid, H_2CO_3. Dissolution of NO_x and SO_2 in water droplets in the atmosphere causes the pH of rain to decrease to 3 to 4, and in southern California polluted droplets in fog can be as acidic as pH 1.7. Dilute acidic solution can remove mineral nutrients from leaves, depending on the age of the leaf and the integrity of the cuticle and surface waxes. The total annual contributions to the soil of acid from acid rain (*wet deposition*) and from particulate matter falling on the soil plus direct absorption from the atmosphere (*dry deposition*) may reach 1.0 to 3.0 kg H^+ per hectare in parts of Europe and the northeastern United States [25]. In soils that lack free calcium carbonate, and therefore are not strongly buffered, such additions of acid can be harmful to plants. Furthermore, the added acid can result in the release of aluminum ions from soil minerals, causing aluminum toxicity. Air pollution is considered to be a major factor in the decline of forests in heavily polluted areas of Europe and North America. There are indications that fast-growing pioneer species are better able to tolerate an acidifying atmosphere than are climax forest trees, possibly because they have a greater potential for assimilation of dissolved NO_x, and more effective acid buffering of the leaf tissue cell sap. Air pollution injury to plants can be evident in several ways. Injury to foliage may be visible in a short time and appear as necrotic lesions (dead tissue), or it can develop slowly as a yellowing or chlorosis of the leaf. There may be a reduction in growth of various portions of a plant. Plants may be killed outright, but they usually do not succumb until they have suffered recurrent injury.

Major primary air pollutants gases are sulphur dioxide, oxides of nitrogen particularly NO_2, HF, HCl, chlorine, ammonia, ethylene and other organic substances. Particulate air pollutants are soot, dust, fine particles of cement and various other substances. Various fertilizers, pesticides and insecticides used in aerial spray are also important air pollutants. The common sources of the pollutants, factors affecting the effect of pollutant and the injury symptoms produced in plants are discussed below.

10. Major gaseous pollutants

Sulphur dioxide (SO_2)

Sulfur dioxide is a major component in acid rain. One of the byproducts of sulfur dioxide is sulfuric acid, and both can be extremely damaging to plants that are exposed to these chemicals. Exposed leaves can begin to lose their color in irregular, blotchy white spots. Some leaves can develop red, brown or black spots. When the pigments in enough tissue are damaged or killed, plants can begin to lose their leaves. Crop output is greatly reduced and growth can be stunted. This is especially noticeable in young plants.

It is the most important and common air pollutant produced in huge amounts in combustion of coal and other fuels in industrial and domestic use. It is also produced during smelting of sulphide ores. Major sources of sulfur dioxide are coal-burning operations, especially those providing electric power and space heating. Sulfur dioxide emissions can also result from the burning of petroleum and the smelting of sulfur containing ores.

SO_2 effects increase in high humidity, windy conditions, in the early morning, in the deficiency of K and Cl_2 and excess of sulphur in the soil. It interacts with ozone, NO_2 and HF. The nature of interaction depends on the relative proportion of gases. The impact of SO_2 decreases in low soil moisture, low temperature, deficiency of nitrogen, sulphur and phosphorus and sometimes in excess of nitrogen also.

In angiosperms, young leaves and in conifers, needles are most sensitive to SO_2 pollution. In general, seedlings are more sensitive than older plants. The effect of the gas usually decreases with age of the plant and lesser morphological and physiological symptoms appear in older plants.

Injury symptoms: The gas is a strong reducing agent. In low concentration, it is oxidized and used in protein synthesis of the plant. However, in high concentration, it causes swelling of thylakoids and interferes with electron transport chain. In SO_2 pollution, plants show initial reduction of photosynthesis and increased respiration. The gas reduces stomatal opening and thus causes general water stress in plants. SO_2 replaces oxygen in cellular materials and changes their nature. It affects structural proteins in the cell membrane and thus changes the membrane permeability. High concentration of the gas causes accumulation of sulphydril and decrease of sulphides in plants. SO_2 interferes with amino acid metabolism and reduces the synthesis of proteins and enzymes. It stimulates the oxidation of PGA and increases the pentose phosphate cycle activity. It reduces the level of keto acids, ATP, sucrose and glutamate in plants and increases the level of glucose, fructose and glycolate. It inactivates many enzymes either by breaking their S-S bonds or by changing their stereo structure. In lichens, the gas induces photooxidation in the phycobiont part. Most common visible symptom of SO_2 injury is water-soaked appearance of leaves which later become necrotic changing into brown spots. Color and shape of necrotic spots is characteristic in different species and NO_2 concentrations. In some species, characteristic intraveinal chlorosis is caused. In general, SO_2 pollution results in abscission of older leaves and tip necrosis in flower and sepals.

Sulfur dioxide enters the leaves mainly through the stomata (microscopic openings) and the resultant injury is classified as either acute or chronic. Acute injury (fig. 4) is caused by absorption of high concentrations of sulfur dioxide in a relatively short time. The symptoms appear as 2-sided (bifacial) lesions that usually occur between the veins and occasionally along the margins of the leaves. The colors of the necrotic area can vary from a light tan or near white to an orange-red or brown depending on the time of year, the plant species affected and weather conditions. Recently expanded leaves usually are the most sensitive to acute sulfur dioxide injury, the very youngest and oldest being somewhat more resistant.

Fig. 4. Acute sulfur dioxide injuries to raspberry [26].

Chronic injury is caused by long-term absorption of sulfur dioxide at sub-lethal concentrations. The symptoms appear as a yellowing or chlorosis of the leaf, and occasionally as a bronzing on the under surface of the leaves. Different plant species and varieties and even individuals of the same species may vary considerably in their sensitivity to sulfur dioxide. These variations occur because of the differences in geographical location, climate, stage of growth and maturation. The following crop plants are generally considered susceptible to sulfur dioxide: alfalfa, barley, buckwheat, clover, oats, pumpkin, radish, rhubarb, spinach, squash, Swiss chard and tobacco. Resistant crop plants include asparagus, cabbage, celery, corn, onion and potato. Plants damaged by sulfur dioxide can be as far as 30 miles from its source, but the most severe damage, defoliation and discoloring is typically found within five miles. For some plants, it can take exposure of only four hours to suffer damage. A wide variety of plants are vulnerable, from alfalfa and carrots to crab apple and fir trees.

Nitrogen dioxide (NO_2)

NO_2 mostly affects the leaves and seedlings. Its effects decrease with increasing age of the plant and tissue. Conifers are found to be more sensitive to this gas during spring and summer than in winters. Older needles are more sensitive to the gas than young ones.

Injury symptoms: The gas causes formation of crystalloid structures in the stroma of chloroplasts and swelling of thylakoid membrane. As a result the photosynthetic activity of the plant is reduced.

Most common visible injury symptoms are chlorosis in angiospermic leaves and tip burn in conifer needles. In angiosperms, most of the species produce water-soaked intraveinal areas that later become necrotic. Tip burn is common symptom in bracts, sepals and awns.

Fluorides

Fluirides in general, are accumulated in the plant tissues over long times. They are first accumulated in the leaves and then are translocated towards tips and margins of the leaves. The injury symptoms are produced only after a critical level of fluoride is attained. Due to such accumulation over long times, flurides generally and HF particularly can induce injury at very low atmospheric concentrations. Critical concentration for fluoride injury is 0.1 ppm for several days. Toxicity of particulate fluorides depends upon the particle size, their solubility and humidity of the atmosphere.

HF gas is much lighter than air and so can cause damage in plants even at a distance of 30 km from the source. It is a hygroscopic gas and forms acidic cloud near the source. Generally the impact of HF pollution increases with humidity and excess of P in soil while decreases in low temperature, drought and deficiency of N and Ca in the soil. In some species, impact of HF has been reported to decrease with excess of N and Ca in the soil.

In most of the species, recovery from moderate fluoride injury can occur within few days if exposure to pollutant stops. However, some highly sensitive species e.g. pine and spruce can never recover fully. HF generally affects immature leaves in angiosperms and needles in conifers.

Injury symptoms: Fluorides combine with metal components of proteins or inhibit them otherwise and thus interfere with the activity of many enzymes. As a result the cell wall composition, photosynthesis, respiration, carbohydrate synthesis, synthesis of nucleic acids and nucleotides and energy balance of the cell are affected. In the leaves subjected to HF exposure, endoplasmic reticulum is reduced, ribosomes are detached from ER, number of ribosomes is reduced and mitochondria become swollen. Chlorophyll synthesis and cellulose synthesis are inhibited. Activities of UDP-glucose-fructose transglucosylase, phosphoglucomutase, enolase and polyphenol oxidase are reduced. On the other hand activities of catalase, peroxidase, pyruvate kinase, PEP-carboxylase, glucose-6-phosphate dehydrogenase, cytochrome oxidase and pentose phosphate pathway are stimulated. In conifer needles common visible injury symptoms are chlorosis later turning into red/brown discolouration, tip burn later turning into necrosis of whole needle, formation of sharply defined red/purple bands between healthy and injured tissue. Similar symptoms are common in angiospermic leaves also. In addition, the angiospermic leaves in many species also show zonation of necrotic areas, leaf cupping, curling of leaf edges and ragged leaf margins. In sepals, petals, bracts and awns, water-soaked margins and later tip and marginal necrosis are observed. Fluorides absorbed by leaves are conducted towards the margins of broad leaves (grapes) and to the tips of monocotyledonous leaves (gladiolus).

Little injury takes place at the site of absorption, whereas the margins or the tips of the leaves build up injurious concentrations. The injury (fig. 5) starts as a gray or light-green water-soaked lesion, which turns tan to reddish-brown. With continued exposure the necrotic areas increase in size, spreading inward to the midrib on broad leaves and downward on monocotyledonous leaves.

Fig. 5. Fluoride injuries to plum foliage [26].

The fluoride enters the leaf through the stomata and is moved to the margins where it accumulates and causes tissue injury. Note, the characteristic dark band separating the healthy (green) and injured (brown) tissues of affected leaves. Studies of susceptibility of plant species to fluorides show that apricot, barley (young), blueberry, peach (fruit), gladiolus, grape, plum, prune, sweet corn and tulip are most sensitive. Resistant plants include alfalfa, asparagus, bean (snap), cabbage, carrot, cauliflower, celery, cucumber, eggplant, pea, pear, pepper, potato, squash, tobacco and wheat.

Chlorine (Cl_2)

Older plants are more sensitive to chlorine than seedlings. The age of tissue has little effect on the sensitivity and older as well as young tissues are almost equally afected by chlorine pollution.

Injury symptoms: Chlorine injury symptoms can appear from 18 hours to 8 days after exposure. In most plant species, recovery from chlorine injury can occur 3 to 4 days after exposure is stopped. Chlorine injury symptoms are quite variable in different species. Most common visible symptoms in conifers are chlorosis, tip burn and necrosis is needles. In angiosperm leaves, marginal or intraveinal necrosis, water-soaked appearance, leaf cupping and abscission are common.

Hydrogen chloride (HCl)

The HCl injury can be caused to plants even at a distance of 800 meter from the source. Like fluorides, the chloride from HCl is accumulated in the leaves and translocated towards their margins and tips. Symptoms of HCl injury appear after a critical concentration is reached, usually between 24 and 72 hours after the exposure.

Impact of HCl pollution decreases with increase in humidity, deficiency of Mg and excess of Ca. Mature plants are more sensitive to HCl than seedlings. Similarly, young fully expanded leaves are more sensitive than immature unexpanded leaves.

Injury symptoms: Most common visible injury symptoms in conifer needles are red or brown discolouration and tip burn. In angiosperm leaves, common symptoms are intraveinal water-soaked streaks, yellow or brown necrosis, tip necrosis, bleached areas around the necrosis and shot-holing. Tip burn, necrotic stipple and discolouration in sepals and petals are also observed.

Ammonia (NH_3)

Impact of ammonia on plants generally increases with humidity and decreases with drought. Effect of darkness on ammonia sensitivity is highly variable among species. Some species are more sensitive to low concentrations of ammonia than to its high concentration. Age of tissue has little effect on sensitivity and both young and old tissues are equally sensitive to ammonia.

Injury symptoms: Most common visible symptoms in conifers are black discolouration, usually sharply bordered tip burn and abscission of needles. In angiosperm leaves, common symptoms are water-soaked appearance later turning black, intercostal necrosis, slight marginal and upper surface injury, glazong/bronzing of upper surface, desiccation and abscission. Ammonia injury to vegetation has been observed frequently in Ontario in recent years following accidents involving the storage, transportation or application of anhydrous and aqua ammonia fertilizers. These episodes usually release large quantities of ammonia into the atmosphere for brief periods of time and cause severe injury to vegetation in the immediate vicinity. Complete system expression on affected vegetation usually takes several days to develop, and appears as irregular, bleached, bifacial, necrotic lesions. Grasses often show reddish, interveinal necrotic streaking or dark upper surface discolouration. Flowers, fruit and woody tissues usually are not affected, and in the case of severe injury to fruit trees, recovery through the production of new leaves can occur (fig. 6). Sensitive species include apple, barley, beans, clover, radish, raspberry and soybean. Resistant species include alfalfa, beet, carrot, corn, cucumber, eggplant, onion, peach, rhubarb and tomato.

Fig. 6. Severe ammonia injuries to apple foliage and subsequent recovery through the production of new leaves following the fumigation [26].

Organic gases (Ethylene)

Ethylene injury symptoms develop in plants only in exposure to high concentrations and take several days to develop. After exposure to the gas is stopped, level of recovery is variable in different species. Generally, younger plant parts recover but older parts do not. Much 'acute' damage to plants is caused on the fringes of polluted area or by a steady leakage of gas in low concentration.

Injury symptoms: In injuriously high concentrations of ethylene, growth of plants is stopped. In low concentrations, growth abnormalities appear. In conifers, yellow tips in needles and abscission of branches and cones are common. In angiosperms, common symptoms are epinasty or hyponasty, loss of bark, abscission of leaves and flowers, premature flower opening and fruit ripening. Ethylene affects the growth hormones and regulatory process that takes place in the plant and results in a number of outward manifestations of infection. Leaves can begin to curl and die; ethylene causes the leaves of plants to curl down and fold under as they shrivel and are stuck with necrosis. On flowering plants, buds can stop opening or flowers can begin to show signs of discoloration or die and drop sooner than expected. Even in more resistant plants like evergreen conifers, growth of the plant will be stunted, needles will be small and few pine cones will be produced. Plants such as peach trees, marigolds, blackberries and tomatoes are extremely vulnerable to damage from exposure to ethylene.

11. Minor gaseous pollutants

Hydrogen sulphide (H_2S)

Plants show wilting on exposure to this gas but the symptoms develop after about 48 hours. No injury occurs below the exposure of 40 ppm for 4 hours.

Carbon monoxide (CO)

Like ethylene this gas produces epinasty, chlorosis and abscission. However, concentration of over 1000 times that of ethylene is needed to produce same degree of damage. No injury to plants occurs below exposure of 100 ppm for 1 week.

Bromine (Br_2) and Iodine (I_2)

Studies show these gases are highly toxic to plants. HI and I_2 are readily absorbed and accumulated by plants producing visible injury symptoms similar to those of SO_2. Injury occurs at exposure of 0.1 ppm for 18 hours.

Common injury symptoms of bromine in angiosperms are necrosis of leaf margins, leaf tips and tendrils; brown discoloration and black spots later spreading to entire leaf. In conifers, yellow/white needle tips or red/brown discoloration later becoming grey/brown are common symptoms.

Mercury vapors (Hg)

Unlike other pollutants, flowers are more sensitive to Hg than leaves. Injury symptoms usually appear within 24 hours of Hg exposure but often go on increasing up to 5 days.

Common injury symptoms due to Hg-vapors pollution are abscission of oldest leaves, interveinal necrosis, chlorosis around veins, flower abscission, loss of petal colors, buds remaining closed and later becoming necrotic, blackening of stamens, pistils and peduncles.

Particulate pollutants

Different types of solid particulate materials are also important air pollutants. Each of these affects the plants in characteristic manner. Some common particulate air pollutants have been discussed below.

Cement-kiln dust

In generals, plants having hairy surface of leaves trap more dust and are, therefore, damaged more than the plants with shiny leaf surface. The cement dust forms crusts on the surface of leaves, twigs and flowers. This inhibits gaseous exchange from the surfaces of plant parts. Such crust on the leaves also inhibits light penetration and consequently reduces photosynthesis. Such crusts are especially thicker in conditions of dew, mist or light rains. In dry conditions, dust blowing with wind is highly abrasive and damages the cuticle of leaves. Cuticle is also damaged due to alkalinity of cement dust. Due to damaged cuticle plants become more susceptible to infection by pathogens.

Lime and gypsum

Lime and gypsum deposited on the soil from the air, these change the pH of the soil and thus affect the nutrient availability to plants. Such deposition usually causes appearance of various nutrient deficiency symptoms in the plants. Lime and gypsum are less adhering as compared to cement-kiln dust. However, these are also trapped and deposited on the surface of plant parts particularly the leaves with hairy surfaces and produce injury symptoms similar to cement dust. Lime and gypsum particles blowing with wind are also highly abrasive for cuticle.

Soot

Soot deposited on the surface of leaves may be washed away by rains so its damage may be reduced. However, in bright sunlight and high temperature, the damage is increased.

Soot deposited on the surface of leaves inhibits light penetration, increased surface temperature due to absorption of heat and clogging of stomata. The result of these is reduced gaseous exchange, reduced photosynthesis and general weakening of the plant growth. Necrotic spots also develop in many species due to soot deposition.

Magnesium oxide

Deposited on the soil these compounds can soon increase the soil pH to levels injurious to plants. Deposition of these substances on the soil prevents germination of seedlings. The seedlings that are able to emerge usually have yellow/brown tips of leaves and their roots are stunted. In areas of heavy pollution, composition of the vegetation changes completely.

Boron

Severe injury to plants is observed even at a distance of 200 meters from the source and mild injury may be observed up to 500 meters in all the directions from the source.

Impact of boron pollution is more severe on older leaves than on younger leaves. Boron is also accumulated in the leaves and produces injury symptoms quite similar to fluoride pollution.

Chlorides of sodium, potassium and calcium Injury symptoms produced by these chlorides in plants are very similar to those produced by SO_2 and fluoride pollution.

Sodium sulphate dust can cause necrosis of leaves of the plants. The damage increases in moist condition.

Pesticides, insecticides and herbicides

A large variety of such chemicals are sprayed on the crops these days. These substances may drift with wind to nearby areas. Generally, these chemicals are deposited on the soil and form important soil pollutants. However, in frosty conditions when crops and other plants damaged by early frost are quite susceptible to foliar spray of these chemicals, these may also be injurious air pollutants. Injury symptoms vary with the plant species and the type of chemical. Generally, the symptoms are produced on foliage and are quite similar to those produced when these substances act as soil pollutants.

12. Secondary pollutants and plants

Many of the primary pollutants under specific environmental conditions may interact with each other and produce secondary environmental pollutants or certain complex environmental conditions that are injurious to plants. Such secondary pollutants and pollution conditions are discussed below.

Photo-oxidants

PAN (Peroxyacetylnitrate-$CH_3CO.O_2.NO_2$)

The common visible symptoms of exposure to PAN are chlorosis and necrosis in leaves. It also interferes with photosynthesis, respiration and absorption and synthesis of carbohydrates and proteins. It inhibits photorespiration, NADP reduction, carbon dioxide fixation, cellulose synthesis and the enzymes associated with photosynthesis and respiration.

Ozone (O_3) is released into the atmosphere from the burning of fossil fuels and is one of the most harmful pollutants to plants. It can be carried for long distances and is readily absorbed as a part of the photosynthetic process. Plants exposed to large amounts of ozone can develop spots on their leaves. These spots are irregular and often tan, brown or black. Some leaves can take on a bronze or red appearance, usually as a precursor to necrosis. Depending on the concentration of ozone in the environment, plants can show different amounts of discoloration before the leaves begin to die.

Studies by the National Crop Loss Assessment Network show that ozone in the environment also has a detrimental effect on crop production. While crops such as cotton, soybeans and other dicots are more sensitive than monocot crops, all crops sampled over the decades-long studies show significant loss of productivity when exposed to ozone. Cotton crops show significantly less yield when exposed to levels of ozone in the atmosphere. Middle aged leaves and young plants are more sensitive to ozone. This pollutant interacts with SO_2, NO_2, PAN and heavy metals in complex manner.

Common symptoms of ozone pollution are yellowing, flecking and blotching in leaves, premature senescence and early maturity. It interferes with pollen formation, pollination, pollen germination and growth of pollen tubes. Increase in the level of RNA, starch, polysaccharides and number of polysomes is observed in ozone pollution. Ozone stimulates respiration, inhibits oxidative phosphorylation and changes membrane permeability. In some species, it inhibits the synthesis of glucon and cellulose and reduces the level of reducing sugars, ascorbic acid and ATP while in other species the effect is opposite to it. The impact of ozone on plants increases with humidity and decreases with drought, darkness, low temperature, high soil salinity, deficiency of soil phosphorus and excess of soil sulphur. Throughout the growing season, particularly July and August, ozone levels vary significantly. Periods of high ozone are associated with regional southerly air flows that are carried across the lower. Localized, domestic ozone levels also contribute to the already high background levels. Injury levels vary annually and white bean, which are particularly sensitive, are often used as an indicator of damage. Other sensitive species include cucumber, grape, green bean, lettuce, onion, potato, radish, rutabagas, spinach, sweet corn, tobacco and tomato. Resistant species include endive, pear and apricot.

Fig. 7. Ozone injuries to soybean foliage [26].

Ozone symptoms (fig. 7) characteristically occur on the upper surface of affected leaves and appear as a flecking, bronzing or bleaching of the leaf tissues. Although yield reductions are usually with visible foliar injury, crop loss can also occur without any sign of pollutant stress.

Conversely, some crops can sustain visible foliar injury without any adverse effect on yield. Susceptibility to ozone injury is influenced by many environmental and plant growth factors. High relative humidity, optimum soil-nitrogen levels and water availability increase susceptibility. Injury development on broad leaves also is influenced by the stage of maturity. The youngest leaves are resistant. With expansion, they become successively susceptible at middle and basal portions. The leaves become resistant again at complete maturation

Ground-level ozone causes more damage to plants than all other air pollutants combined. This web page describes the ozone pollution situation, shows classical symptoms of ozone injury and shows how ozone affects yield of several major crops.

Ozone enters leaves through stomata during normal gas exchange. As a strong oxidant, ozone (or secondary products resulting from oxidation by ozone such as reactive oxygen species) causes several types of symptoms including chlorosis and necrosis. It is almost impossible to tell whether foliar chlorosis or necrosis in the field is caused by ozone or normal senescence. Several additional symptom types are commonly associated with ozone exposure, however. These include flecks (tiny light-tan irregular spots less than 1 mm diameter), stipples (small darkly pigmented areas approximately 2-4 mm diameter), bronzing, and reddening.

Ozone symptoms usually occur between the veins on the upper leaf surface of older and middle-aged leaves, but may also involve both leaf surfaces (bifacial) for some species. The type and severity of injury is dependent on several factors including duration and concentration of ozone exposure, weather conditions and plant genetics. One or all of these symptoms can occur on some species under some conditions, and specific symptoms on one species can differ from symptoms on another. With continuing daily ozone exposure, classical symptoms (stippling, flecking, bronzing, and reddening) are gradually obscured by chlorosis and necrosis.

Studies in open-top field chambers have repeatedly verified that flecking, stippling, bronzing and reddening on plant leaves are classical responses to ambient levels of ozone. Plants grown in chambers receiving air filtered with activated charcoal (CF) to reduce ozone concentrations do not develop symptoms that occur on plants grown in non-filtered air (NF) at ambient ozone concentrations. Foliar symptoms shown on this web site mainly occurred on plants exposed to ambient concentrations of ozone (either in NF chambers or in ambient air).

Yield Loss Caused by Ozone

Field research to measure effects of seasonal exposure to ozone on crop yield has been in progress for more than 40 years. Most of this research utilized open-top field chambers in which growth conditions are similar to outside conditions. The most extensive research on crop loss was performed from 1980 to 1987 at five locations in the USA as part of the National Crop Loss Assessment Network (NCLAN). At each location, numerous chambers were used to expose plants to ozone treatments spanning the range of concentrations that occur in different areas of the world. The NCLAN focused on the most important agronomic crops nationally

The strongest evidence for significant effects of ozone on crop yield comes from NCLAN studies [18] (fig. 8). The results show that dicotyledonous species (soybean, cotton and peanut) are more sensitive to yield loss caused by ozone than monocot species (sorghum, field corn and winter wheat).

Particulate Matter

Particulate matter such as cement dust, magnesium-lime dust and carbon soot deposited on vegetation can inhibit the normal respiration and photosynthesis mechanisms within the leaf. Cement dust may cause chlorosis and death of leaf tissue by the combination of a thick crust and alkaline toxicity produced in wet weather. The dust coating (fig. 9) also may affect the normal action of pesticides and other agricultural chemicals applied as sprays to foliage. In addition, accumulation of alkaline dusts in the soil can increase soil pH to levels adverse to crop growth.

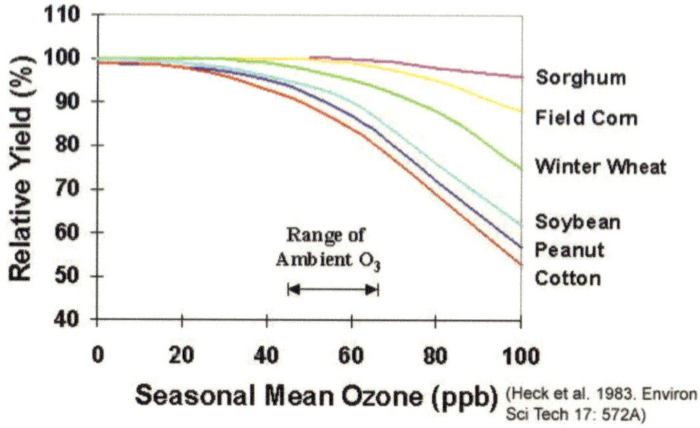

Fig. 8. Effect of ozone on yield of crops [18]. cotton image by arklite06 from

Fig. 9. Cement-dust coating on apple leaves and fruit. The dust had no injurious effect on the foliage, but inhibited the action of a pre-harvest crop spray [26].

13. Effect of atmospheric pollutants on vegetation monitoring system, why forestry monitoring system?

Because the crop plants are mostly annual plants they can not show the long-term effects produced by air pollutants. Therefore to monitor the effects of air pollution are recommended the trees, the changes in forest structure highlight the harmful effects of different air pollutants.

The evident decline of the health state of the forest in Europe since the beginning of the 1980 due to the negative impact of air pollution were illustrated by numerous publication from this period (see litt.). In the efforts to obtain objective and comparable data concerning the health of the European forests were developed a common methodology for the assessment of the forest state under the influence of air pollution. This network is known under the short term ICP- Forest (International Cooperative Programme for the investigation of the trans-boundary Pollution Influence on the Forests).

The poor health status of the forests in Central Europe concerns all the Europe. The pictures of the forests on large area were dominated by tree with defoliated crowns and an increasing rate of the death trees (fig 10). The assessment of the causes of the "new damages"- neuartiges waldschaden, in germ- is not easy because the symptoms of the decline were different from the symptoms of the damages caused by natural (biotic and abiotic) and anthropic causes.

Under the umbrella of ICP Forest Programme, were developed and implemented an European network of plots for the assessment of the parameters of the trees crowns condition known as Level I plots. The grid of the European Level I plots were established at 16 * 16 km, arranged on a transnational unified grid over Europe. In comparison with the national grids used by each country the obtained data were relevant for the evaluation of the forest health state at European level.

After 1996 were put in function the Level II monitoring plots used for the intensive monitoring and collection of comparable data related to the changes in forest ecosystems which are directly connected to specific environment at factors such as atmosphere pollution and acid deposition. Such data can help in a better understanding at the relation causes and effects in the forests decline.

Fig. 10. General aspects of silver fir crowns affected by decline in the border of the northern Carpathians (Forest District Solca)

The monitoring results contribute to the scientific basis of air pollution control policies of UN/ECE and the European Commission (EC). Fifteen years of monitoring forest condition and two decades of forest damage research have shown, however, that the discussion of recent forest damage must not be confined to the effects of air pollution alone. The comprehensive monitoring programme corresponds to the complex interrelations between natural and anthropogenic factors in forest ecosystems. Infrastructure and data of the programme are thought to be relevant for other processes of international forest policies, e.g.

those on biodiversity, climate change and sustainable forest management. In this respect the monitoring pursues the objectives of Resolution SI of the Strasbourg, Resolution HI of the Helsinki and Resolution, L2 of the Lisbon Ministerial Conference on the Protection of Forests in Europe, and contributes to global forest policies such as the United Nations Forum on Forests (ICP- Report 2007).

The monitoring results obtained each year are summarized in annual Executive Reports. The methodological background and detailed results of the individual surveys are described in Technical Reports (www.icp-forests.org).

Methodology for the crown health condition assessment of forests

The state of health of forest trees can be determined by assessing the foliage loss. With a little practice, this can be accurately estimated by the foresters or other trained personnel. The development of forest damage can be traced through repeated assessments of the same trees.

Loss of needles or leaves should be assessed after sprouting in spring or early summer and before broadleaves and larch display autumn coloration, at best in July and August. Evergreen conifers (fig. 11-18) may also be assessed in their winter state as long as they are free of snow. Assessments should be made under good light conditions in good weather: rain and fog render assessments inaccurate. Leaf or needle loss is estimated for the entire crown.

The crown is considered to reach from the peak of the tree to, the lowest strong green branch forming part of the crown as such; epicormic shoots on the stem are not considered, while those in the crown are.

Fig. 9.-12. Defoliation in % of Oak crowns [27].

Fig. 13.-16. Defoliation in % of Beech crowns [27]

A forest tree can spread its crown to a greater or lesser extent depending on the room available within the stand. Consequently, spatial conditions must be considered in crown assessment; that is, the maximum foliage that each tree could possibly produce must be taken as a basis. The photo series (fig. 19 - 26) depicts trees of the upper storey with well developed crowns enjoying optimum light conditions. It is therefore applicable to trees of the middle and lower strata only to a limited extent. Foliage loss may be determined by comparing the tree under consideration with the corresponding photo series. The appearance of the crown is matched with one of the photos and the foliage loss estimated to a degree of 5 percent accuracy. Assessments should be made with field-glasses from a distance of at least one tree-length. Field-glasses permit precise identification of bare branches and twigs and discoloration. In subsequent surveys it is important that the tree always be observed from the same side; this should either be marked on the tree itself or noted in terms of compass direction. Leaf or needle loss due to known causes, e.g., hail, lightning, whipping, insect attack, etc., should not be included but separately inventoried [27].

14. Political background and objectives of ICP forests

The objectives and the strategy of ICP Forests are based on the draft long-term strategy and the work plan for the effects-oriented activities of the Working Group on Effects (WGE) of

Fig. 19.-22. Defoliation in % of Oak crowns [27].

Fig. 23.-26. Defoliation in % of Beech crowns [27].

CLRTAP (Convention on Long-range Transboundary Air Pollution). The draft long-term strategy of WGE specifies the following long-term aims to which all ICP are expected to contribute:

Assessment of knowledge on

- The present status, long-term trends and dynamics, and the degree and geographical extent of the impact of air pollution, particularly, but not exclusively, its long range trans-boundary impact
- Exposure-response relationships for agreed air pollutants;
- Critical loads, levels and limits for agreed air pollutants;
- Interactive effects of air pollution and climate change on forest ecosystems
- Moreover, the long-term strategy of WGE specifies the following long-term priorities of special relevance to ICP Forests:
- Derivation of exposure-response functions for chemical and biological effects of air pollutants including investigation of nutrient nitrogen, acidifying compounds and ozone effects on ecosystem functions and on biodiversity, including combinations with other stresses (e.g. climate change and land use practices);
- Further development of models and mapping procedures, particularly for effects of nitrogen and ozone on the environment and for the description of dynamic processes of damage and recovery (acidification, eutrophication, heavy metal accumulation) by including to a larger extent biological effects;
- Evaluation of environmental benefits of air pollution control policies.
- In order to meet the information needs of the Working Group on Effects of Atmospheric Pollution (WGE), ICP Forests pursues the following two main objectives:
- Objective 1: A periodic overview on the spatial and temporal variation of forest condition in relation to anthropogenic and natural stress factors (in particular air pollution) by means of European-wide and national large-scale representative monitoring on a systematic network.
- Objective 2: A better understanding of the cause-effect relationships between the condition of forest ecosystems and anthropogenic as well as natural stress factors (in particular air pollution) by means of intensive monitoring on a number of selected permanent observation plots spread over Europe and to study the development of important forest ecosystems in Europe.

These objectives imply in accordance with the long-term priorities of WGE contributions to calculations of critical loads and levels and the assessment of their exceedances. They imply also dynamic modeling of the response of forest ecosystems to deposition scenarios expected for the future. Additional insight is gained by compiling available studies from the National Focal Centers (NFCs) and from related programmes inside and outside of Convention on Long-range Trans-boundary Air Pollution.

15. Strategy of ICP forests

Monitoring activities

In order to meet its data generation and reporting obligations, ICP Forests employs data collection at two levels.

- Large-scale monitoring (Level I) provides a periodic overview of the spatial and temporal variation in a range of attributes related to forest condition. Level I plots, national forest

inventory (NFI) plots, and other related inventory plots may be combined when appropriate, feasible and necessary, according to defined and agreed procedures.
- Intensive monitoring (Level II) is carried out on plots installed in important forest ecosystems.

These plots are dedicated to in-depth investigation of the interactive effects of anthropogenic and natural stress factors on the condition of forest ecosystems.

Quality assurance and control

All monitoring activities are harmonized by ICP Forests among the participating countries and are laid down in this Manual. This ensures a standard approach for data collection and evaluation and can form the nucleus for a future common European forest monitoring programme. A consistent quality assurance approach is applied within the programme covering the set up of methods, data collection, submission and investigation as well as reporting. Quality assurance and control is supervised by the Programme Coordinating Group through its Quality Assurance Committee. A set of Expert Panels cares for data quality assurance within the specific surveys and for the further development of monitoring methods and standards. This includes field checks, inter-calibration courses, laboratory ring tests, and data validation.

Data evaluation and reporting

A range of monitoring variables is required to meet the information requirements of Convention on Long-range Trans-boundary Air Pollution and other international institutions. The Programme Coordinating Group and the Expert Panels are responsible for a data evaluation and reporting approach which takes the medium term work-plan of Working Group on Effects of Atmospheric Pollution into account. International and national data from other programmes and institutions should be included in combined analysis. The main topics for data analysis are:

- Forest condition
- Effects on forest ecosystems from
- Acidity and nitrogen
- Ozone

Contributions in the fields of

- Climate change
- Biodiversity

Trends in deposition and their interactive effects on the adaptation and vulnerability of forest ecosystems are evaluated. This includes spatial and temporal changes and cause-effect relationships with special emphasis on critical loads and their exceedances. Dynamic models and transfer functions derived from suitably selected intensive monitoring plots are used to investigate the effects of climatic factors and greenhouse gases on forest ecosystems and applied to the large scale monitoring plots. These models are validated against measured data collected at the plots. Furthermore, data gathered at the plots are used in an integrated manner to investigate the carbon sequestration potential of forests, ozone fluxes to forests and contribute to assess status and trends of forest biodiversity at the pan-European level.

Results

The integrative monitoring approach of ICP Forests using the Level I and Level II networks provides robust data on the health and stability of forests. This facilitates an understanding of the effects of deposition on the role and functioning of forest ecosystems in protecting soils and water. Furthermore the programme surveys can contribute to the understanding and forecast of climate change effects on forests and can be used to supply information on the sequestration of carbon and are going to provide information on forest biodiversity as an integral part of forest ecosystems. Results are published via reports and a website (www.icp-forests.org).

ICP Forests aims to provide periodic overviews on the spatial and temporal variation of forest condition in relation to man-made and natural stress factors (particularly air pollution); to contribute to a better understanding of the cause-effect relationships between the condition of forest ecosystems and man-made and natural stress factors (particularly air pollution); and to study the development of important forest ecosystems in Europe.

More specifically, to support harmonized forest monitoring by linking existing and new monitoring mechanisms at the national, regional and EU level (tab. 2); to collect quantitative and qualitative forest data related to climate change, air pollution, biodiversity, and forest condition; and to contribute information on sustainable forest management to the Ministerial Conference on the Protection of Forests in Europe.

Survey	Number of plots		Assessment frequency
	installed	Data submitted for 2007	
Crown condition	836	462	Annually
Foliar chemistry	904	200	Every two years
Soil condition	615	0	Every ten years
Soil solution chemistry	302	169	Continuously
Tree growth	811	70	Every five years
Deposition	657	353	Continuously
Ambient air quality (active)	84	27	Continuously
Ambient air quality (passive)	254	167	Continuously
Ozone induced injury	114	43	Annually
Meteorology	265	191	Continuously
Phenology	186	58	Several times per year
Ground vegetation	777	67	Every five years
Litterfall	262	105	Continuously
Remote sensing	National data		Preferably at plot installation

Table 2. Surveys and number of plots for Level II monitoring. The variation in assessment frequency results in different numbers of plots with data submission for the different surveys (after www.icp-forests.org)

Conclusions after 25 years of forest monitoring at European level

For 25 years, forest condition has been monitored by ICP Forests in close cooperation with the European Commission. The system combines an inventory approach with intensive monitoring. It provides reliable and representative data on forest ecosystem health and vitality and helps to detect responses of forest ecosystems to the changing environment. The data collected so far provide a major input for several international programmes and initiatives, such as the Convention on Long-range Trans-boundary Air Pollution and the Ministerial Conference for the Protection of Forests in Europe.

Participating countries	Forest area (x 1000 ha)	% of forest area	Grid size (km X Uin)	No. of sample plots	No. of sample trees	Defoliation of all species by class (aggregates), national surveys		
						0	1	2 - 4
Albania	1063	37	no survey in 2009					
Andorra	18		16 X 16	3	73	60.3	32.9	6.8
Austria	3878	46.2	no survey in					
Belarus	7921	38.2	16 X 16	409	9620	27,7	63.9	8.4
Belgium	700	23.1	42/82	122	2858	30,7	49.1	20.2
Bulgaria	3699	33.3	42/82/162	159	5560	29,6	49.3	21.1
Croatia	2061	36.5	16 X 16	83	1991	37.2	36.5	26.3
Cyprus	298	32.2	16 X 16	15	362	3,0	60.8	36.2
Czech Republic	2647	33.6	82/162	133	5284	11.7	31.5	56.8
Denmark	527	12.2	72/162	16	384	69,0	25.5	5.5
Estonia	2213	49.1	16 X 16	92	2202	44.3	48.5	7.2
Finland	20150	66.3	162* 24 X 32	886	7182	58,2	32.7	9.1
France	15840	28.9	16 X 16	500	9949	28.7	37.8	33.5
Germany	11076	31	162*42	424	10376	36,4	37.1	26.5
Greece	2034	19.5		89	2098	42.2	33.5	24.3
Hu ngary	1904	22.5	16 X 16	78	1872	54,8	26.8	18.4
Ireland	680	6.3	16x 16	30	599	69.9	17.5	12.5
Italy	8675	28,8	16 X 16	257	6966	24.5	39.7	35.8
Latvia	3162	49	8x8	340	8036	17	69.2	13.8
Liechtenstein	8	50	no survey in 2009					
Lithuania	2150	32.9	8X8 / 16X16	983	5961	18.6	63.7	17.7
Luxembourg	89	34.4			no survey in	2009		
FYR of Macedonia					no survey in	2009		
Rep. of Moldova	318	9.4	2x2	622	13676	43.1	31.7	25.2
The Netherlands	334	9.6			no survey in	2009		
Norway	12000	37.1	32/ 92	1622	9332	43.1	35.8	21
Poland	9200	29.4	16 X 16 ,	1923	38460	24.1	58.2	17.7
Portugal	3234	36.4	no survey in 2009					
Romania	6233	26.1	16 X 16	227	5448	44.1	37	18.9
Russian Fed.	809090	73.2		365	11016	80	13.8	6.2
Serbia	2360		16 X 16/4x4	130	2765	68.1	21.6	10.3
Slovak Republic	1961	40	16x 16"	. 108	4049	9.3	58.6	32.1
Slovenia	1099	54.2	16 X 16	44	1056	18.2	46.4	35.5
Spain	11588	30.9	16 X 16	620	14880	17.8	64,4	17,7
Sweden	28300	69	Varying	3217	7097	59,9	25.1	15,0
Switzerland	1186	28.7	16 X 16	48	1040	32.3	49.4	18.3
Turkey	21389	27.5	16 X 16	563	12290	25.1	56,2	18.7
Ukraine	9400	15.4	16 X 16	1483	34498	66.4	26.8	6.8
United Kingdom	2837	11.7	no .survey in 2009					
Total	**1011322**		**Varying**	**15591**	**236980**			

Table 3. Forest surveys and defoliation classes for all tree species in European countries (2009). Results of national surveys as submitted by National Focal Centres (after www.icp-forests.org)

In the early 1980s, a dramatic deterioration in forest condition was observed in Europe and this initiated the implementation of forest condition monitoring under Convention on Long-range Trans-boundary Air Pollution. Today, the monitoring results indicate that, at the large scale, forest condition has deteriorated far less severely than was feared at that time. Stress factors like insects, fungi and weather effects have been shown to affect tree health. The drought in the Mediterranean region in the mid-1990s and the extremely warm and dry summer across large parts of Europe in 2003 led to increased levels of defoliation as a natural reaction of trees to this type of stress. The programme has also reported on acidifying deposition which is regionally correlated with defoliation and on atmospheric inputs that are accentuating other stress factors. In the past three years there has been little change in the mean levels of defoliation for the main European tree species. However, long-term trends show more deterioration than improvement (tab. 3).

The health status of forest trees in Europe is monitored over large areas by surveys of tree crown condition. Trees that are fully foliated are regarded as healthy. The Ministerial Conference on the Protection of Forests in Europe uses defoliation as one of four indicators for forest health and vitality.

- In 2009, crown condition data were submitted for 7193 plots in 30 countries. In total, 136 778 trees were assessed. This constitutes the programmer's largest number of plots for which annual data were submitted.
- In 2009, 20.2 % of all trees assessed had a needle or leaf loss of more than 25 % and were thus classified as either damaged or dead (fig. 27). This represents no change relative to 2008.
- Of the main tree species, European and sessile oak had the highest levels of damaged and dead trees, at 31.8 %.
- There were no significant changes in crown condition over the past ten years on two-thirds of the plots, but deterioration prevailed on the remaining third.
- In 2009, a fifth of the 136 778 trees studied were considered damaged or dead
- Trends vary between species, with European and sessile oak the most frequently damaged species. However, both have shown some recovery over the past five years. The health of Norway spruce and Scots pine has improved over the past 18 years. Defoliation in common beech, Holm oak and maritime pine has increased.
- There has been no significant change in tree health on most plots monitored over the past ten years. Defoliation increased on 24.4 % of plots monitored and decreased, indicating an improvement in crown condition, on only 14.9 % (fig. 28).
- Over the past 18 years there has been a clear improvement in crown condition for Scots pine and a slight improvement for Norway spruce. European and sessile oak have shown the highest mean defoliation over the past decade.
- Defoliation peaked after the extremely dry and warm summer in 2003 and has been slowly recovering since 2007. Defoliation of common beech peaked in 2004, while Holm oak showed a sharp deterioration in crown condition in the mid-1990s and again in 2005. Unfavorable weather conditions are thought to be responsible for these trends. There was a reasonably consistent increase in defoliation of maritime pine up to 2005, followed by a short period of recovery after which crown condition again deteriorated in 2009 [28], [29], [30].

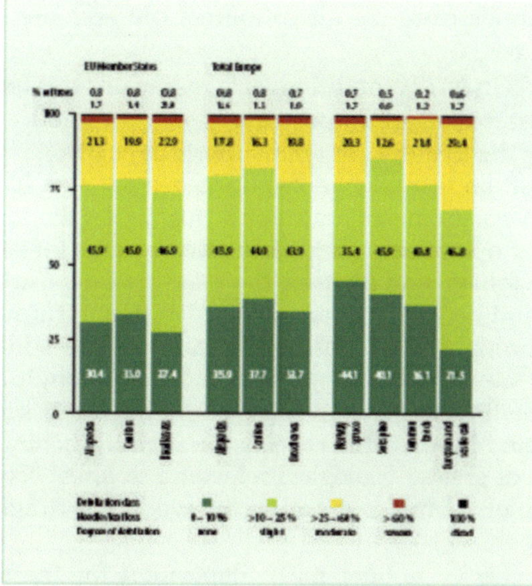

Fig. 27. Extent of defoliation for the main European tree species. Total Europe and EU, 2009. (after www.icpforests.org)

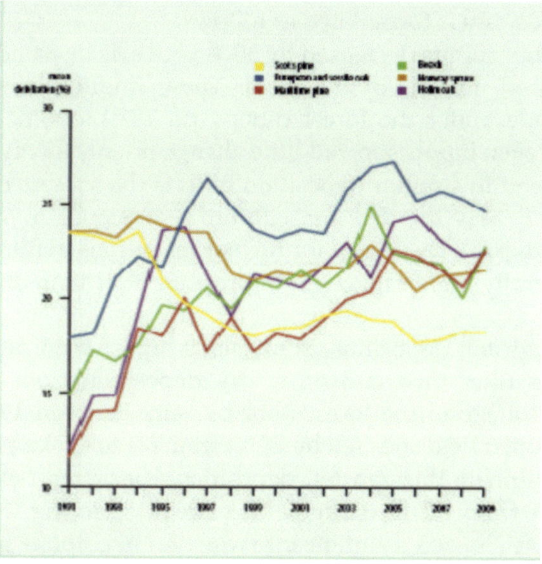

Fig. 28. Mean percentage defoliation for the most frequent tree species in European forests (after www.icpforests.org)

- Defoliation is an indicator of tree health and vitality that can be easily monitored over large areas and which reacts to many different factors, including climatic conditions and weather extremes as well as insect and fungal infestations.
- Defoliation represents a valuable early warning system for the response of the forest ecosystems to change – this is particularly relevant as climatic extremes are predicted to occur more frequently in the relatively near future.

- Deposition of pollutants from the air can affect soil and site conditions and thus the condition of forest trees.
- The status and trends in forest condition vary regionally and for different species. Local conditions may differ from the European average.

Conclusions concerning the dynamic of atmospheric deposition

ICP Forests began deposition measurements on intensive (Level II) monitoring plots in the latter half of the 1990s. Measurements are carried out within the forest stands (through fall deposition) and in nearby open fields (bulk deposition). In the forest canopy, some elements can be leached from the foliage and increase the measured deposition load, whereas others are taken up by leaves and needles and are so not detected in through fall. Bulk deposition is mostly lower than through fall deposition because of the additional deposition loads filtered from the air by the forest canopy. Thus, neither through fall deposition nor bulk deposition is equal to the total deposition received by the forest stands. However, through fall deposition is presented here as this reflects the inputs reaching the forest floor and so these measurements are of greater ecological relevance to forest ecosystems than open field measurements. On the plots, samples are collected weekly, fortnightly or monthly and are analyzed by national experts.

After intensive quality checks, annual mean deposition for the years 1998 to 2007 was calculated for plots with complete data sets. Slopes of plot wise linear regressions of deposition over time were tested for significance. Plot-specific means were calculated for the period 2005 to 2007.

The most relevant trends cam is formulated as follows:

- Mean annual sulphur inputs decreased by 30 % between 1998 and 2007, with significant reductions measured on half of the plots. These findings are based on deposition measurements made under the forest canopy on 157 plots located mostly in central Europe. Mean nitrogen inputs showed little change or only a very small decrease.
- The downward trend in sulphur deposition reflects the success of the clean air policies under the UNECE and the EU for sulphur emissions. In contrast, the nitrogen deposition data indicate a clear need for further reductions in nitrogen emissions.
- Deposition is generally higher on central European plots than on plots in northern and southern Europe.
- On average, through fall deposition in forests is higher than deposition on open field sites because trees filter dust and other dry deposition from the air which is then washed from the foliage to the forest floor by rain. Between 1998 and 2007, sulphate deposition on the open field sites fell by 26 %; from 6.1 to 4.5 kg per hectare per year.
- The decrease in sulphate through fall deposition (measured below the forest canopy) was higher at 34 %; from 10.0 to 6.6 kg per hectare per year (fig. 29).
- About half the plots showed a significant reduction in sulphur inputs over the 10-year study period. The data are mean values from around 150 measurement stations located mainly in central Europe.
- Mean nitrogen deposition within the forest stands fluctuated (for nitrogen measured as nitrate and ammonium) and few plots showed significant changes in through fall deposition.
- Slight decreases in mean nitrogen deposition at the open field plots were observed (fig. 30). The deposition data show the success of the clean air policies in Europe for sulphur emissions, and show the need for further reductions in nitrogen emissions [31], [32], [33].

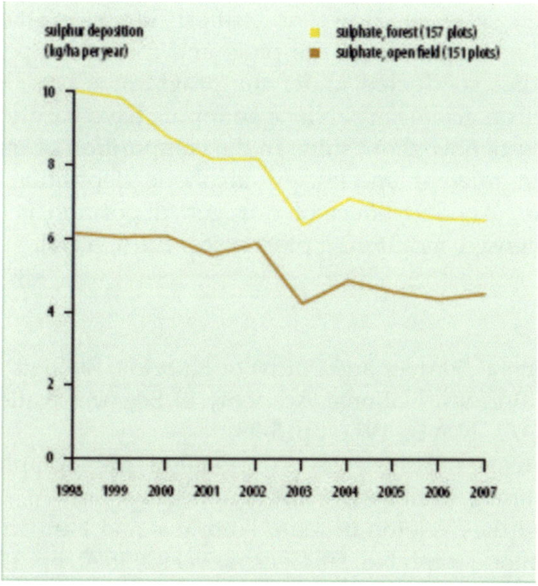

Fig. 29. Development of mean deposition of sulphate from 1998 to 2007. The forest canopy alters pollutants from the air. Inputs within the forest stands are higher than in the open field. In 2003 there was less precipitation and thus less deposition. (after ww.icpforests.org)

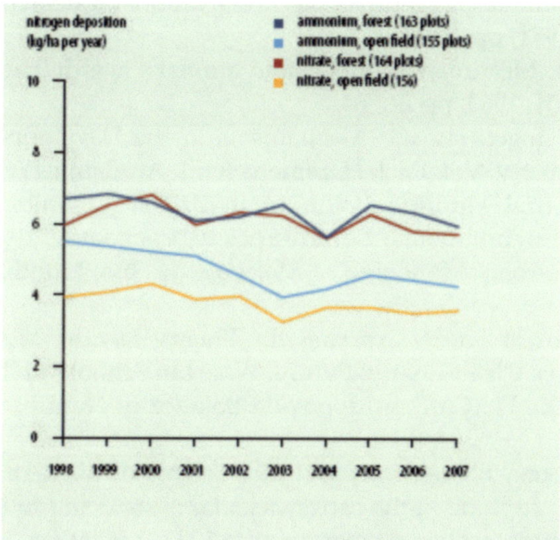

Fig. 30. Development of mean plot deposition of nitrogen compounds plots) from 1998 to 2007. Some reduction are visible in open field measurements. There was little change in deposition for the forest stands over the10 years of observation. (after www.icpforests.org)

Atmospheric deposition has been the specific focus of the programme since its inception. Current evaluations show decreasing sulphur inputs on about 50% of around 150 intensive monitoring plots since 1998, which is a result of clean air policies under the LRTAP

Convention and EU legislation. However, critical limits in the soil water are still substantially exceeded on a quarter of the plots and indicate a potential threat to forest vegetation. Earlier studies conducted under the programme have shown that the risk of storm damage is higher on acidic soils. Nitrogen inputs have hardly changed over the past ten years and the data sets now show shifts in the composition of forest ground vegetation towards more nitrogen tolerant species. Atmospheric deposition is a driver for these changes in biodiversity. Another effect of nitrogen deposition is increased tree growth which was found on intensive monitoring plots across Europe [34].

16. References

[1] Stein, L., Environmental Sources and Forms of Fluoride, Biologic Effects of Atmospheric Pollutants - Fluorides, National Academy of Sciences-National Research Council, Washington, D.C., N.A.S., 1971, pp. 5-28.

[2] Shupe, J. L., Fluorosis of Livestock, Air Quality Monograph No. 69-4, American Petroleum Institute, New York, 1969.

[3] Zipkin, I., Effects on the Skeleton of Man, Fluorides and Human Health, World Health Organization monograph No. 59, Geneva, WHO, 1970, pp. 185-201.

[4] Zipkin, I., Air Pollutants Affecting the Performance of Domestic Animals, Agricultural Handbook No. 380, U.S. Dept. of Agriculture, revised 1972, 109 pp.

[5] Zipkin, I., W. A. Lee, and N. C. Leone, Amer. J., Air pollutants and fluorides effects, Pub. Hlth. , 1957, 47:848.

[6] Suttle, J. W., Effects of Fluoride on Animals, Biological Effects of Atmospheric Pollutants - Fluorides, National Academy of Sciences-National Research Council, Washington, D.C., N.A.S., 1971, pp. 133-162.

[7] Hobbes, C. S., G. M. Merriman, Fluoride and animal's health, Univ. of Tenn. Agri. Exp. Sta. Bull. no. 351, 1963, pp 24:964.

[8] Johnson, L. C., Histogenesis and Mechanisms in the Development of Osteofluorosis, Fluorine Chemistry, Vol. IV, J. H. Simons (ed.), Academic Press, N.Y., 1965.

[9] Greenall, L., Industrial Fluoride Pollution in British Columbia, Canadian Scientific Pollution and Environmental Control Society, Vancouver, 1971.

[10] Thompson M., Carbon Monoxide - Molecule of the Month, Winchester College, UK, 2009.

[11] Robert U., Edward H., 2009, Crossing the Energy Divide: Moving from Fossil Fuel Dependence to a Clean-Energy Future, Wharton School Publishing, pp.36.

[12] Weinstock, B.; Niki, H., Carbon Monoxide Balance in Nature, Science 176 (32), 1972, pp. 290.

[13] Li L., Hsu, A. Moore, PK, Actions and interactions of nitric oxide, carbon monoxide and hydrogen sulphide in the cardiovascular system and in inflammation--a tale of three gases!, Pharmacology & therapeutics 123 (3), 2009, pp. 386–400.

[14] McNeil, Donald G. Jr., In Raising the World's I.Q., the Secret's in the Salt, New York Times, 2006.

[15] Pacyna E. G, Pacyna J. M, Steenhuisen F, Wilson S., Global anthropogenic mercury emission inventory for 2000". Atmos Environ 40 (22), 2006, pp. 4048.

[16] Solnit R., Winged Mercury and the Golden Calf, Orion Magazine, Retrieved 2007-12-03, 2006.

[17] Maprani A. C., Al T. A., MacQuarrie K. T., Dalziel, J. A., Shaw, S. A.; Yeats, Ph. A., Determination of Mercury Evasion in a Contaminated Headwater Stream, Environmental Science & Technology 39, 2005, pp. 1679.
[18] Heagle, A. S., Ozone and crop yield. Annu. Rev. Phytopathol. 27, 1989, pp. 397–423.
[19] Krupa, S. V., Gruenhage, L., Jaeger, H.-J., Nosal, M., Manning, W. J., Legge, A. H., and Hanewald, K., Ambient ozone (O_3) and adverse crop response: A unified view of cause and effect. Environ. Pollut. 87, 1995, pp. 119–126.
[20] Bowler, C., Van Montagu, M., and Inze, D., Superoxide dismutase and stress tolerance. Annu. Rev. Plant Physiol. Plant Mol. Biol. 43, 1992, pp. 83–116.
[21] Elstner, E. F., and Osswald, W. Mechanisms of oxygen activation during plant stress. Proc. R. Soc. B Biol. 102B, 1994, pp. 131–154.
[22] Taiz L., and Zeiger, E., Plant physiology, SINUAER, 2006,
[23] Aono, M., Kubo, A., Saji, H., Tanaka, K., and Kondo, N., Enhanced tolerance to photooxidative stress of transgenic Nicotiana tabacum with high chloroplastic glutathione reductase activity. Plant Cell Physiol. 34, 1993, pp. 129–135.
[24] McKersie, B., Bowley, S. R., Harjanto, E., and Leprince, O., Water-deficit tolerance and field performance of transgenic alfalfa overexpressing superoxide dismutase. Plant Physiol. 111, 1996, pp. 1177–1181.
[25] Schwartz, S. E., Acid deposition: Unraveling a regional phenomenon. Science 243, 1989, pp. 753.
[26] Heather G., Air Pollution on Agricultural Crops, Factsheet, No. 85-002, 2003,
[27] Muler, E., Stierlin, H., Water-deficit tolerance and field performance of transgenic alfalfa over expressing super-oxide dismutase. Plant Physiol. 111, 1990, pp. 1177–1181.
[28] Barbu, I., Cercetările de ecofiziologie forestieră şi fenomenul de uscare a răşinoaselor din Bucovina. Referat prezentat la Conferinţa naţională de fiziologia plantelor, ASAS Bucureşti, 1985.
[29] Barbu, I., Tipuri de uscare la bradul din România. Lucrare prezentată la Conferinţa a III-a de Ecologie Arad. Ziridava. Vol. XVI, 1986.
[30] Barbu, I., Moartea bradului - simptom al degradării mediului. Ed. Ceres, Bucureşti, 1991.
[31] Geambaşu, N., Seceta şi fenomenul de uscare a bradului din unele păduri din Bucovina. Revista Pădurilor nr. 2, 1988.
[32] Hildebrandt, G., Inventur und Beobachtung aktueler Waldschaden auf Landesebene. In: Schriftenreihe der Forstlichen Facultat Freiburg. AFZ nr. 12, Forstliche Hochschulwoche, 1983, pp. 148
[33] Hutte, P., Vom Waldsterben bedrohte Walder Polens. AFZ nr. 33, 1986, pp. 819.
[34] Spiecker, H. et al., Growth trends in European Forests. EFI Research Report nr. 5. Berlin, Springer, 1996.
1983 : Belastung durch Luftschadstoffe in der BRD., AFZ nr. 51/52, p. 1 385 - 1 386
1988 : Forest decline in Europa in 1987 attributed to an pollutants,
UNITED NATIONS, Economic and Social Council Economic Commission for Europe, Sesiunea 46, 11-14 oct. 1988
2007, 2008, 2009, United States Environmental Protection Agency (EPA) Report.
www.cleartheair.nsw.gov.au
(www.icp-forests.org).

www.icp-forests.org/Reports.htm
www.cleartheair.nsw.gov.au/science_and_research/cause_and_effect/what_effects_air_quality_.aspx
www.windows2universe.org/milagro/effects/wildlife_forests.html
http://5e.plantphys.net/article.php?ch=e&id=262

13

Phytoremediation on Air Pollution

Hua Yang and Yanju Liu
Beijing Center for Physical and Chemical Analysis
China

1. Introduction

Air pollution has been becoming a necessary evil with rapid industrialization and urbanization around the world, after it results in kinds of human health problems, such as ophthalmic, respiratory and cardiovascular diseases (Brunekreef & Holgate, 2002; Giles et al., 2011; Gudmundsson, 2011; Jamrozik & Musk, 2011; Miller et al., 2007; Nandasena, 2010). The direst threat posted by air pollution may be its hard controlling caused by its strong flowability. Either could it be spread from one source location to a larger region, even the whole planet, and the sweeping radiation pollution in air originated from Fukushima in Japan is unfortunately in this case, or conversely diluted with changes of climatic conditions (Sample, 2007).

Among all types of treatments of contaminants, including microbial bioremediation, phytoremediation stands out for its benefits yielded from self-maintaining, soil stabilization and other advantages to meet greater public approval (Doty et al., 2007). And different phytoremediation process is responsible for specific pollutant. Air pollutants can be divided into anthropogenic and natural pollutants according to their sources, or primary and secondary pollutants, which stem from reactions of primary pollutants, when taking production process into account (UNEP, 2004). However, physical, chemical and biologic pollutants belong to three general categories as usually discussed in air pollution treatment, and particulate matters, organic and inorganic chemicals, and microorganism were referred to the above categories respectively. This chapter will start from conception of phytoremediation, current state of phytoremediation of air pollutants, such as particulate matters, inorganic and organic pollutants, and a case study from recent authors' research on phytoremediation of benzene and toluene of indoor air will be finally presented.

2. Phytoremediation

Phytoremediation is a way to mitigate environment pollutions, such as in air, water and soil pollution in virtue of plants, more often than not, combined with their associated microorganisms. This concept has been widely applied to treat pollutants in soil and water (Cunningham & Ow, 1996; Schröder et al, 2002).

2.1 Principles of phytoremediation

Briefly, for phytoremediation, roles from roots to shoots are played in phytostabilization, rhyzodegradation, phytoextraction, phytodegradation, phytovolatilization and phytofiltration

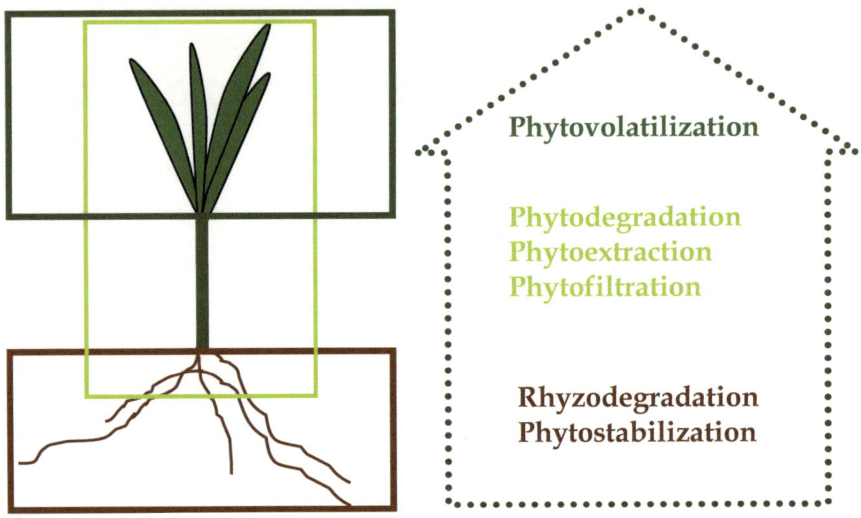

Fig. 1. Different parts of the plants are responsible for different processes of phytoremediation.

(Morikawa & Erkin, 2003) (see Fig. 1.). It is the process for phytostabilization that contaminants were sequestrated, but not degraded, in the root zones. The narrow region of 1-3 mm from root surface is rhyzosphere, where there are much more profuse microorganisms than in bulk soil. Root secretions and rhyzospheric microorganisms are joint-agents to degrade the pollutants in soil in the process of rhyzodegradation. Phytoextraction is with regard to extracting pollutants from soil and translocating them from the roots to shoots by plants. Phytodegradation is to transform and/or degrade pollutants, which enter the plants from soil, water or air. Phytovolatilization means the pollutants are volatile out of the stoma of the plants and degraded by hydroxyl radicals, otherwise they will become air pollutants. It is phytofiltration that particulate matters in air are reduced through the surface of plants (Singh & Tripathi, 2007).

2.2 Characteristics of phytoremediation

2.2.1 Main advantages

Phytoremediation is proved to be helpful to reduce pollutants in soil, water and air; it is energy-economizing and cost-effective compared with those physical or chemical methods since it is solar-energy driven cleanup technology; it is a preference for people as its aesthetic value and less environment disruption, which letting nature take its course.

2.2.2 Primary disadvantages

Plants are also living things, and there exists tolerance limits for the toxicity of pollutants; it is a kind of "leave it to chance" thing since it is too natural to disrupt the environment: optimum concentrations of pollutants for remediation varying with species of plants, intermediates possibly leading to second pollutants, time-consuming when considering growth period of plants.

2.3 Examples of phytoremediation of air pollutants

2.3.1 Particulate matters

Vegetation has been used to shield dust in many countries, and the process is called phytofiltration. A 8-m wide green belt may reduce 2 to 3 times of dust fall (Novoderzhikina et al., 1966 as cited in Singh & Tripathi, 2007). For phytofiltration, morphological characteristics, such as orientation of leaf on the main axis, size and shape, surface nature, the presence or absence of trichomes and wax deposition, are factors to trap or capture dust from ambient air.

2.3.2 Inorganic pollutants

NO_X

NO_X are air pollutants, and they are one of precursors of photochemical reaction. They can reach plant system through wet/dry deposition to foliar or roots. Plant species, plant age, NO_X concentration and other environmental conditions are factors which influence the leaf penetration of NO_X. After NO_2 enter into the plant, most of them are metabolized to organic compounds, e.g. amino acid, through nitrate assimilation pathway. The enzymes in plants, such as nitrate reductase, nitrite reductase or glutamine synthetase may play an important role in this process. Genetic engineering could turn plants into magic sink for NOx by making the enzymes overexpressed.

SO_2

Nearly 70% SO_2 in the atmosphere originates from fossil fuel combustion. SO_2 enters into plants mainly through stoma, and can be utilized in a reductive sulfur cycle in plants. They are changed into SO_4^{2-} or SO_3^{2-} in cell walls. Adenyl-5-phosphosulphate, carrier protein, carrier protein with bound sulphite and carrier protein with bound sulphide are the intermediates in the reduction pathway. The final products are cysteine or other organic compounds.

2.3.3 Organic pollutants

Formaldehyde

Formaldehyde, a ubiquitous air pollutant, is so harmful that it is classified as a mutagen and carcinogen. In the 1980's, an NASA's research in the USA revealed that low level of formaldehyde in air could be removed by plant leaves alone, while higher concentrations of the toxic chemical can be filtered by activated carbon firstly and the plant roots and associated microorganisms degrade and assimilate remained chemicals (Wolverton,1988). In cell-culture experiments (Giese,1994), when the concentration of formaldehyde is low enough (8.5mg · m^{-3}), spider plants (*Chlorophytum comosum* L.) shoot can metabolize it to organic acids, amino acids, free sugars, lipids and cell-wall components. In soybean (*Glycine max* L.) cell cultures, serine and phosphatidylcholine are the major metabolic products for formaldehyde. NAS^+-dependent formaldehyde dehydrogenase and glutathione-dependent formaldehyde dehydrogenases have been isolated from cell-suspension cultures of some plants. It appears that formaldehyde can be oxidized and then degraded in C1 metabolism in phytodegradation.

Benzene and toluene

As members of VOCs (Volatile Organic Compounds, VOCs), benzene is genotoxicity and carcinogenicity while toluene is a neurotoxic chemical (Pariselli et al., 2009). Benzene and toluene have been reported to be removed from air and be assimilated by plants (Porter, 1994; Ugrekhelidze, 1997). Porter (1994) suggested at the lower toluene exposure, a fairly modest amount light change was credited with significant increase or decrease of removal rate and the plant, the only photoresponsive organisms in the system, may be responsible for the phenomenon. Furthurmore, hydroxylation is considered to be the first step of the aromatic ring cleavage of benzene and toluene in higher plants (Ugrekhelidze, 1997). Besides, a substantial role of microorganisms in the growth medium in removing benzene and toluene was also reported by previous researches (Orwell et al., 2004; Orwell et al., 2006; Wood et al., 2002; Wood et al., 2006).

Two compartments Model

Foliage-air exchange is believed to be the primary route of plant uptake particularly for semi-volatile and volatile organics. Two-compartment model has been used to explain atmosphere-foliage bioaccumulation phenomenon, namely a relatively fast initial uptake followed by a period of slower uptake (Keymeulen, 1995; Mackay et al, 2006). In the model, the leaf was divided into two compartments according to its structure. The first compartment can be referred to the non-living plant cuticle, where physico-chemical sorption of airborne lipophilic compounds can occur. The second compartment is within the leaf interior, where further sorption and metabolism of organic compounds may exist.

3. Removal of benzene and toluene from indoor air by ornamental houseplants—a case study (Yang, 2010)

3.1 Indoor air pollution

Indoor air pollution in urban environments has become a significant health concern, as city dwellers often spend over 90% of their time indoors (USEPA, 1987; Abbritti & Muzi, 1995; American Lung Assoc., 2001). "Sick building syndrome" (SBS) or "building-related illness" is partly attributed to the indoor chemicals, particularly in air-conditioned buildings (Burge et al., 1987; Mendell and Smith, 1990; Carpenter, 1998; Brasche et al., 1999; Carrer et al., 1999; Jones, 1999; Sundell, 2004). As indoor pollutants, volatile organic compounds (VOCs) with 300 species having been detected, should not be negligible (Orwell et al., 2004), after the cocktail of VOCs might lead to additive effects on human health (Wolkoff, 1995; Weschler and Shields, 1997; Pariselli *et al.*, 2009). Thus, removing these chemical mixtures from indoor air merits serious consideration. Many experimental studies have investigated the effects of plants on VOCs (Cape, 2003). This part presents a study on applicability of phytoremediation of benzene and toluene from indoor air.

3.2 Experiments design

Indoor ornamental plants were purchased from flower markets for experiments. Plants were well watered and allowed to drain for 24 hours before fumigation.

A series of experiments were designed as follows (see Fig. 2.):

1. **Plants-Screening experiments**

This experiment aimed to acquire plants which are capable of effective-removing binary mixture of benzene (0.26 mg · m^{-3}) and toluene (0.3 mg · m^{-3}). For this purpose, initial screening experiment and replication experiment for validation were carried out in sequence. Allowing for vast amounts of species of plants, only one individual for every species of plants was fumigated in the initial screening experiment. After some ineligible ones could inevitably have slipped into candidates in the initial screening experiment, further three replicates for those species which showed relative high removal rates were

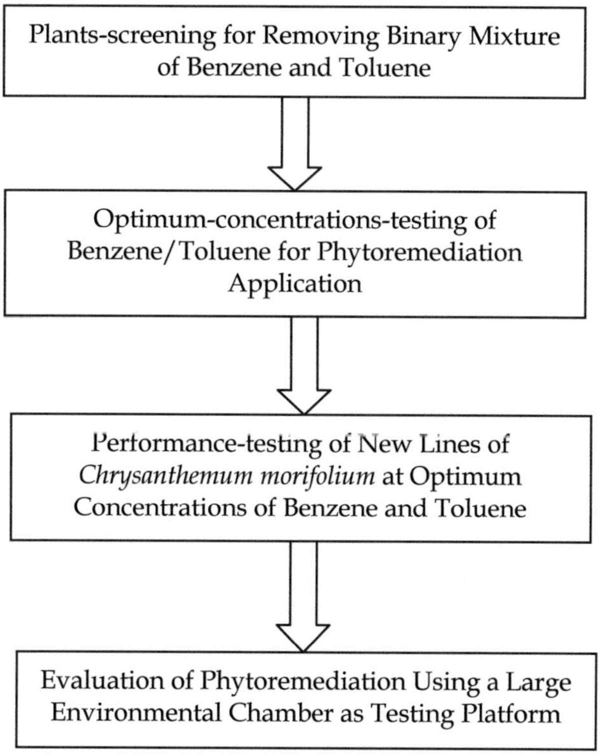

Fig. 2. Processes of the case study on Removal of Benzene and Toluene from Indoor Air by Ornamental Houseplants

established in replication experiment to confirm the removal effects and to quantitatively estimate the rate and efficiency of phytoremediation.

2. **Fumigating-experiments at gradient concentrations**

The gradient concentrations (Table 1.) were set up to investigate optimum concentrations of benzene and toluene for phytoremediation application. The plants were chosen from those plants with relatively high removal efficiency from plants-screening experiment.

3. **Evaluating-experiments for new lines of *Chrysanthemum morifolium***

It is one object in the experiments to evaluate removal performance of new lines of *Chrysanthemum morifolium* at the optimum concentrations of benzene and toluene acquired in the experiments mentioned above. *Chrysanthemum morifolium*, a Chinese traditional flower, has been so popular they are available everywhere in the country. The phytoremediation on indoor air pollution is believed to be easy to be promoted, if *Chrysanthemum morifolium* demonstrates ideal removal effects.

The other object is to establish a large environmental chamber as a testing-platform to evaluate phytoremediation on indoor air pollution.

Gradient concentrations ($mg \cdot m^{-3}$)	
Benzene	Toluene
0.26	0.31
0.64	0.76
1.3	1.5

Table 1. Gradient concentrations of benzene and toluene set up in the fumigating experiments

3.3 Fumigation system

A set of eight cylindrical, Plexiglas chambers (41 cm in diameter, 70 cm in height, ca. 0.09 m^3 in volume) were lined with Teflon film to reduce possible absorption by Plexiglas material (see Fig. 3.). Stainless steel or Teflon Pipes were used to carry gasses in order to minimize the loss of benzene/toluene. Simultaneously, compressed purified air and standard mixture of benzene and toluene entered a cylindrical, stainless steel mixing chamber. The flow rate of air and benzene/toluene was regulated by Float Flow Meter and Mass Flow Meter respectively, which were calibrated by Bubble Flow Meter before use. Thus, air containing expected concentrations of benzene and toluene was continuously and evenly distributed to fumigation chambers through 8 Float Flow Meters. It took about 2 hours to achieve expected concentrations of benzene and toluene of outlet air based on the continuous measuring the benzene/toluene concentrations at the inlets and outlets of the eight empty chambers. Fig. 4. shows the fumigation system in the study.

Fig. 3. Photograph of a Plexiglas chamber housing one potted plant

Among the eight chambers, two chambers were designated randomly as controls, one holding only pots and soil similar to those used in the experimental chambers; another empty one used to monitor stability of gasses flow. And the other six chambers housed six different plant species per trial. After three hours (from 9 a.m. to 12 noon) and six hours

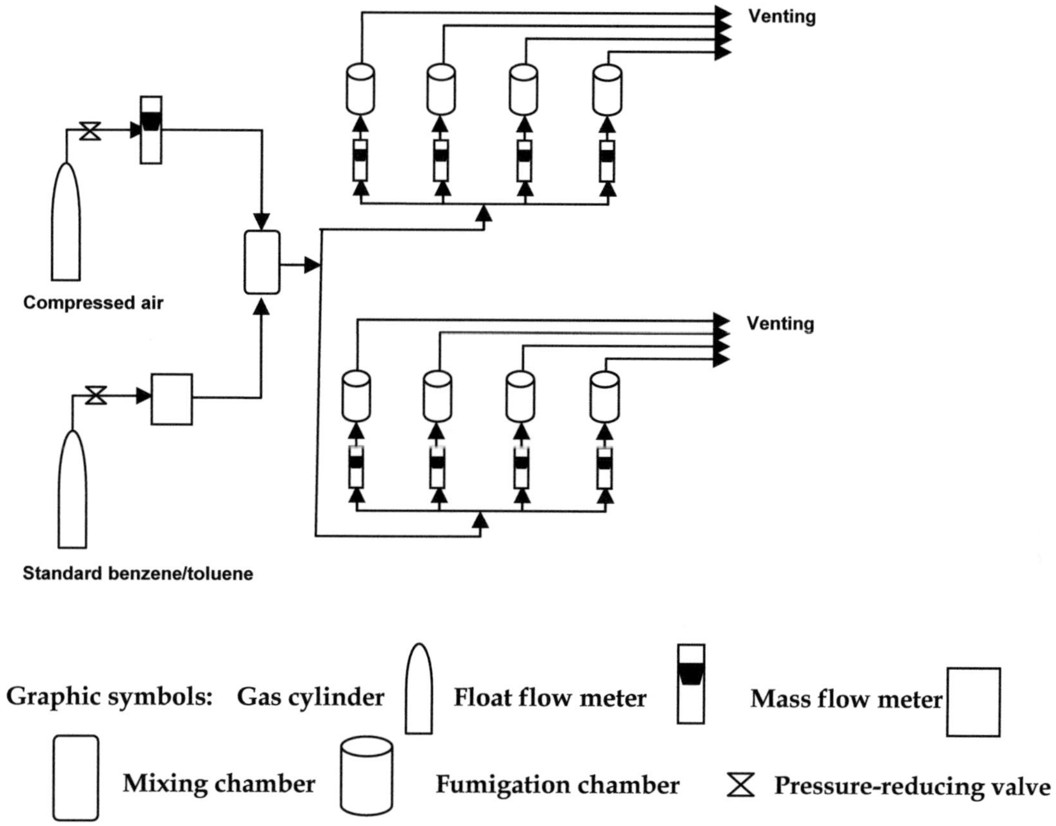

Fig. 4. Schematic diagram for fumigation system

(continue to extend from 12 noon to 3 p.m.) of exposure, gasses were sampled using a 100 mL gastight syringe for immediate analysis. The benzene/toluene removal rate was calculated as follows:

$$(C_{control} - C_{sample}) / C_{control} \times 100\% \quad (1)$$

Where C represents benzene or toluene concentration ($mg \cdot mL^{-1}$).

After sample collection, chambers were evacuated and prepared to test the next plants. The same sampling procedure with 8 empty chambers was inserted every 5 trials to test leak to guarantee uniform concentrations of benzene/toluene in all eight chambers.

14 m^3 stainless steel of environmental chamber was made to meet ASTM 1333-96 (reaproved 2002) (see Fig. 5.). Parameters inside the chamber, such as temperature (25°C), relative humidity (45%) and air exchange rate (once every four hours), were maintained during the whole testing period. In the large chamber, concentrations in three steps, namely, background of empty chamber, only chemical evaporation and plants-fumigating were analyzed. 1 mL of benzene and toluene liquid were dropped in a 90 cm-high Petri dish to evaporate. The first sampling was taken after 10 minutes of evaporation, followed by next 8

Fig. 5. The photograph of inside of the environmental chamber

samplings every one hour. Gas samples were collected from 6 air sampling ports, each of them with 15 L of sample. When plants-fumigating, four replicates of the same species of *Chrysanthemum morifolium* were placed on the four corners of the chamber floor. Ventilation and scrubbing should be conducted between fumigation experiments.

3.4 Results and discussion

1. Plants-Screening experiments

In the initial screening experiment, 94 species and cultivars were tested, among which 15 species removed more than 30% of benzene or toluene at either 3-hour or 6-hour sampling. Eight species of plants, *Rhododendron hybrids, Hosta plantaginea, Ctenanthe oppenheimiana* cv. Tricolor, *Ficus elastica, Calathea rotundifolia* cv. Fasciata, *Codiaeum variegatum, Chrysanthemum morifolium, Hemerocallis fulva,* were chosen for the following replication experiment to confirm the removal effects, since they are more easily accessible. On the other hand, fixed time dependence on removal rates was not found for all species in this phase, just as Ugrekhelidze (1997) pointed out the ability of different plants to absorb and utilize exogenous hydrocarbons varied.

In the initial screening experiment, *Crassula portulacea, Cymbidium Golden Elf.* and *Dieffenbachia amoena* cv. Tropic Snow did not display well as in previous experiment (Liu et al., 2007), less than 10% removal rate of benzene. Apart from toluene interference, fumigation fashion (kinetic/static and flow rate at inlet of each fumigation chamber), chamber size, concentration of benzene may all contribute to different removal effects

(Table 2.). In reference to studies on the effects of plants on VOCs, two main types of fumigation chambers were applied – static chambers and kinetic chambers. For a static chamber, more accessories have to be equipped with, such as thermostat and stirring devices needed to accelerate atmospheric equilibration, so that difference of concentrations at the same spot in the chamber at different time period can be evaluated effects of plants on VOCs, whereas effects of plants on VOCs can be determined by the difference of concentrations at inlet and outlet of a kinetic chamber with continuous gas-flow through (e.g. Liu et al., 2007).

For replication experiment, removal rates of 8 species of plants of benzene or toluene at either sampling (3-hour or 6-hour) were showed to be more than 30% in the replication experiment. Furthermore, for the convenience of comparison, the removal rates of benzene and toluene was normalized for leaf area (m^{-2}). For most of 8 species of plants, the removal rates of benzene after 6-hour fumigation increased compared with those after 3-hour fumigation. The case of toluene reversed. A competitive reduction might exist when both benzene and toluene penetrating into leaves were under metabolic control, but combined removal remained essentially unchanged (Porter, 1994). This could be used to elaborate the reversal trend of removal rate with time showed by benzene and toluene.

Chrysanthemum morifolium displayed an excellent performance in removing both benzene and toluene at twice samplings in replication experiment. So it was chosen for further Evaluating-experiment.

2. Fumigating-experiments at gradient concentrations

From those 15 species which removed more than 30% of benzene or toluene at 3-hour or 6-hour sampling in the initial screening experiment, six species of plants such as *Rhododendron hybrids, Ficus elastica, Codiaeum variegatum, Hemerocallis fulva, Euphorbia pulcherrima and Cymbidium sinense* were chosen to test removal capacities at gradient concentrations of benzene and toluene (Yang et al., 2011) (see Table 1). The removal rates at different condition varied. At 0.64 mg m^{-3} of benzene and/or 0.76 mg $m^{-3}$of toluene, better removal effects were showed, and considering the household application, it is better to arrange some species of plants to achieve the best removal effects taking account of concentrations, fumigation time and pollutants.

3. Evaluating-experiment for new lines of *Chrysanthemum morifolium*

Not only did *Chrysanthemum morifolium* show an excellent performance in removing both benzene and toluene according to the results showed in plants-screening experiments, but also it is a kind of symbolic Chinese traditional flower. So it was chosen for evaluating-experiment. The new lines of *Chrysanthemum morifolium* were provided by Institute of Vegetable and Flower, the Chinese Academy of Agricultural Sciences. They are bred via anther culture and radiation mutation from breeding parents of *Chrysanthemum* with spray forms. It was the first performance-testing in China to remove gaseous benzene-toluene using new lines of *Chrysanthemum morifolium* (Yang et al., 2010).

While selective removal of benzene or toluene was not significantly showed in nine lines of *Chrysanthemum*, most of them removed more toluene than benzene. Removing rates varied significantly with lines. Samples from 6-hour of fumigation suggested that above 30% of removal rates of both benzene and toluene were reached by lines of the number 2, 4, 5, 9, in the meanwhile removal rate of toluene from line of the number 7 dropped to 0 from 13% at 3-hour of fumigation (Yang et al., 2010). From the results of removal amount, lines of the number 2, 4, 5, 8, 9 should be taken for application performance testing in real world considering better performance-testing results.

Chrysanthemum morifolium and *Calathea rotundifolia* cv. Fasciata were chosen for performance evaluation in the large environmental chamber. After 10 minutes of evaporation, 10 mg m^{-3} of benzene and 8.5 mg m^{-3} of toluene were detected in the chamber. At first three hours, benzene and toluene were reduced significantly when fumigating *Chrysanthemum morifolium* compared with only chemical fumigation. It suggested that phytoremediation on indoor air pollution should be an effective way to remove organic pollutants. Besides, the environmental chamber is stable in performance, low in adsorption and background value of pollutants, convenient and safe in operation so that it could provide a testing platform for evaluating phytoremediation of gaseous pollutants.

Plant species	Fumigation fashion	Chamber size	VOCs member	Concentration level	Removal rate
Crassula portulacea Cymbidium Golden Elf. and *Dieffenbachia amoena* cv. Tropic Snow	Kinetic fumigation (0.5 L/min at inlet of each chamber)	Cylindrical; 40 cm in diameter, 60 cm in height,	Benzene	150 ppbv	More than 20% after 2-h fumigation
Epipremnum aureum	Static fumigation	Cubic; 60 × 60 × 60 cm	Benzene	25 ppmv	Less than 10% (Average removal rate in initial dose) after 3-h fumigation

Table 2. Fumigation methods and results in previous studies (Liu et al., 2007; Orwell et al., 2004)

4. Future perspective

As a technology, phytoremediation on air pollution is still in its infancy, and study on the applicability of phytoremediation is scarce. First, it is necessary that appropriate testing and evaluating platform should be established for different application. Static and kinetic fumigation fashion, fumigation box and large environmental chamber should be adopted according to goals of experiments. Second, pollutants-removing mechanism needs further studying from morphology to metabolism, from plant alone to associated microorganisms. Third, based on mechanism studies, prediction models can be established and accuracy be improved. In the meanwhile, the study of optimum concentrations of pollutants depending on species of plants should not be neglected for the technology promotion.

5. Conclusion

Phytoremediation is an eco-friendly, cost-effective way to remove pollutants from environment. Phytostabilization, rhyzodegradation, phytoextraction, phytodegradation, phytovolatilization and phytofiltration are main principles of phytoremediation. As physical and chemical air pollutants, phytoremediation of particulate matters, NO_X, SO_2, formaldehyde, benzene and toluene have been reported. Moreover, the authors studied applicability of phytoremediation from aspects of plants-screening, optimum concentrations and evaluation system. For all this, the study on the applicability of phytoremediation still needs work for future technology promotion.

6. Acknowledgements

Authors thank National Natural Science Funds (Grant No. 40875082) for supporting this project.

7. References

Abbritti, M.C. & Muzi, G. (1995). Indoor Air Quality and Health Effects in Office Buildings, In: *Proceedings of Healthy Buildings `95, An International Conference on Healthy buildings in a Mild Climate*. Maroni, M. (Ed). Vol. 1, pp. 185-195, University of Milano and International Centre for Pesticide Safety, Milan, Italy

American Lung Association. (2001). When you can't breathe, nothing else matters. In: *Air Quality*. Available from www.lungusa.org/air/

ASTM 1333-96. (Reapproved 2002). Standard Test Method for Determining Formaldehyde Concentrations in Air and Emission Rates from Wood Products Using a Large Chamber, pp. 1-12

Brasche, S.; Bullinger, M.; Gebhardt, H.; Herzog, V.; Hornung, P.; Kruppa, B.; Meyer, E.; Moreld, M.; Schwab, R.; Mackensen, S.; Winkens, A. & Bischof ,W. (1999). Factors Determining Different Symptom Patterns of Sick Building Syndrome Results From a Multivariate Analysis. In: *Proceedings of Indoor Air `99, The 8th International Conference on Indoor Air Quality and Climate*. Vol.5, pp. 402—407, Edinburgh, UK

Brunekreef, B & Holgate, S. T. (2002). Air Pollution and Health. *Lancet*, 2002; Vol.360, No.9341, (October 2002), pp. 1233–1342, ISSN 1470-2045

Burge, S.; Hedge, A.; Wilson, S.; Harris, B.J. & Robertson, A. (1987). Sick Building Syndrome: a Study of 4373 Office Workers. *Annals of Occupational Hygiene*, Vol.31, pp. 493-504, ISSN 0003-4878

Cape, J.N. (2003). Effects of Airborne Volatile Organic Compounds on Plants. *Environmental Pollution*, Vol.122, No.1, (March 2003), pp. 145—157, ISSN 0269-7491

Carpenter, D.O. (1998). Human Health Effects of Environmental Pollutants: New Insights. *Environmental Monitoring and Assessment*, Vol.53, No. 1, (October 1998), pp. 245—58, ISSN 0167-6369

Carrer, P.; Alcini, D.; Cavallo, D.; Visigalli, F.; Bollini, D. & Maroni, M. (1999). Home and Workplace Complaints and Symptoms in Office Workers and Correlation with Indoor Air Pollution. In: *Proceedings of the 8th International Conference on Indoor Air Quality and Climate*, Vol.1, pp. 129-134, Edinburgh, UK

Cunningham, S.D. & Ow, D.W. (1996) Promises and Prospects of Phytoremediation, *Plant Physiol*. Vol.110, (March 1996), pp. 715-719, ISSN 0032-0889

Doty, S. L.; Andrew, J. C.; Moore, A. L.; Vajzovic, A.; Singleton, G. L.; Ma, C.P.; Khan, Z; Xin, G.; Kang, J. W.; Park, J. Y.; Meilan, R.; Strauss, S. H.; Wilkerson, J.; Farin, F. & Strand, S. E. (2007). Enhanced Phytoremediation of Volatile Environmental

Pollutants with Transgenic Trees. *Proceedings of the National Academy of Sciences*, Vol.104, No.43, (October 2007), pp. 16816–16821, ISSN 0027-8424

Giese, M.; Bauer-Doranth, U.; Langebartels, C. & Sandermann H. (1994). Detoxification of Formaldehyde by the Spider Plant *(Chlorophytum comosum* L.) and by Soybean *(Clycine max* L.) Cell-Suspension Cultures. *Plant Physiol.* Vol.104, No.4, (April 1994), pp. 1301-1309, ISSN 0032-0889

Giles, L.V.; Barn, P.K.N.; Romieu, I.; Mittleman, M.A.; Eeden, S.; Allen, R.; Carlsten, C.; Stieb, D.; Noonan, C.; Smargiassi, A.; Kaufman, J.D.; Hajat, S.K.T.; & Brauer, M. (2011). From Good Intentions to Proven Interventions: Effectiveness of Actions to Reduce the Health Impacts of Air Pollution. Environmental Health Perspectives, Vol.119, No.1, (January 2011), pp. 29-36, ISSN 0091-6765

Gudmundsson, G. (2011). Respiratory Health Effects of Volcanic Ash with Special Reference to Iceland. A review. *The Clinical Respiratory Journal*, Vol.5, pp. 2–9, (January 2011), ISSN 1752-6981

Jamrozik, E. & Musk, A.W. (2011). Respiratory Health Issues in the Asia–Pacific Region: An Overview, *Respirology*, Vol.16, No.1, (January 2011),pp. 3–12, ISSN 1440-1843

Jones, A.P. (1999). Indoor Air Quality and Health. *Atmospheric Environment*, Vol.33, No. 28, (December 1999), pp. 4535-4564, ISSN 1352-2310

Keymeulen, R. ; Schamp, N. & Langenhove, H.V. (1995). Uptake of Gaseous Toluene in Plant Leaves: A Two Compartment Model, *Chemosphere*, Vol. 31, No.8, (October 1995),pp. 3961-3975, ISSN 0045-6535

Liu, Y.J.; Mu, Y.J.; Zhu, Y.G.; Ding, H. & Arens, N.C. (2007). Which Ornamental Plant Species Effectively Remove Benzene From Indoor Air? *Atmospheric Environment*, Vol.41, pp. 650-654, ISSN 1352-2310

Mackay, D.; Foster, K.L.; Patwa, Z. & Webster, E. (2006). Chemical Partitioning to Foliage: The Contribution and Legacy of Davide Calamari, *Environ Sci & Pollut Res*, 2006, Vol.13, No.1, (January 2006), pp. 2-8, Vol. 30, pp. 786-791, ISSN 0944-1344

Mendell, M.J. & Smith, A.H. (1990). Consistent Pattern of Elevated Symptoms in Air-conditioned Office Buildings: A Re-analysis of Epidemiological Studies. *American Journal of Public Health*, Vol.80, No.10, pp. 1193-1199, ISSN 0090-0036

Miller, K.A.; Siscovick, D.S.; Sheppard, L.; Shepherd, K.; Sullivan, J.H.; Anderson, G.L.; & Kaufman, J.D. (2007). Long-Term Exposure to Air Pollution and Incidence of Cardiovascular Events in Women. *The new england journal of medicine,* Vol. 356, No. 5, (February 2007), pp. 447-458, ISSN: 0954-7762

Morikawa, H. & Erkin Ö. C. (2003). Basic Processes in Phytoremediation and Some Applications to Air Pollution Control. *Chemosphere* , Vol.52, No.9, (September 2003), pp. 1553–1558, ISSN: 0045-6535

Nandasena, Y. L. S.; Wickremasinghe A. R. & Sathiakumar N. (2010). Aira rPollution and Health in Sri Lanka: a Review of Epidemiologic Studies. *BMC Public Health,* Vol.10, (June 2010), Available from http://www.biomedcentral.com/1471-2458/10/300)

Orwell, R.L.; Wood, R.A.; Tarran, J.; Torpy, F. & Burchett, M.D. (2004). Removal of Benzene by the Indoor Plant/Substrate Microcosm and Implications for Air Quality. *Water, Air, and Soil Pollution*, Vol.157, (September 2004), pp. 193-207, ISSN: 0049-6979

Orwell, R.L.; Wood, R.A.; Burchett, M.D.; Tarran, J. & Torpy, F. (2006). The Potted-plant Microcosm Substantially Reduces Indoor Air VOC Pollution: II. Laboratory Study. *Water, Air, and Soil Pollution*, Vol.177, (November 2006), pp. 59-80, ISSN 0049-6979

Pariselli, F.; Sacco, M.G.; Ponti, J.; & Rembges, D. (2009) Effects of Toluene and Benzene Air Mixtures on Human Lung Cells (A549). *Experimental and Toxicologic Pathology*, Vol.61, No. 4, (July 2009), pp. 381-386, ISSN: 0940-2993

Porter, J.R. (1994). Touene Removal From Air by Dieffenbachia in a Closed Environment. *Adv Space Res*, Vol.14, No.11, (November 1994), pp. 99—103, ISSN 0273-1177

Sample, I. (April 2007). Urban Air Pollution More Dangerous than Chernobyl, In: *The Guardian*, Available from http://www.guardian.co.uk/science/2007/apr/03/uknews.pollution

Schröder, P.; Harvey, P.J.; & Schwitzgébel, J.P. (2002). Prospects for the Phytoremediation of Organic Pollutants in Europe, *Environ Soi & Pollut Res*, Vol.9, No.1, pp. 1-3, ISSN 0944-1344

Singh, S.N & Tripathi, R D. (2007). *Environmental Bioremediation Technologies*, Springer, ISBN 9783540347934, Berlin, Germany

Sundell, J. (2004). On the History of Indoor Air Qulity and Health. *Indoor Air*, Vol.14, No. 7, pp. 51-58, ISSN 0905-6947

Ugrekhelidze, D.; Korte, F., & Kvesitadze, G. (1997). Uptake and Transformation of Benzene and Toluene by Plant Leaves. *Ecotoxicology and Environmental Safety*, Vol.37, No. 1, (June 1997)pp. 24-29, ISSN: 0147-6513

UNEP. Malé Declaration on Control and Prevention of Air Pollution and its likely Transboundary Effects for South Asia,(March 2004), pp. 9-16, Available from http://www.rrcap.unep.org/male/manual/national/02chapter2.pdf

USEPA. (1987). The Total Exposure Assessment Methodology (TEAM) Study. In: *EPA600/S6-87/002, USA*. Available from http://www.epa.gov/ncepihomS

Weschler, C.J. & Shields, H.C. (1997). Potential Reactions Among Indoor Air Pollutants. *Atmospheric Environment*, Vol.31, No.21, pp. 3487—3495, ISSN 1352-2310

Wolkoff, P. (1995). Volatile Organic Compounds Sources, Measurements, Emissions and the Impact on Indoor Air Quality. *Indoor Air*. Vol. Suppl. 3, pp. 1-73, ISSN: 0905-6947

Wolverton, B.C. (July 1988). Foliage Plants For Improving Indoor Air Quality. In: *National Foliage Foundation Interiorscape Seminar*, pp. 29-45, National Aeronautics and Space Administration, John C. Stennis Space Center, Hollywood, FL, United States

Wood, R.A.; Orwell, R.L.; Tarran, J.; Torpy, F. & Burchett, M. (2002). Potted-plant/Growth Media Interactions and Capacities for Removal of Volatiles from Indoor Air. *Journal of Horticultural Science & Biotechnology*, Vol.77, No.1, (January 2002), pp. 120-129, ISSN 1462-0316

Wood, R.A.; Burchett, M.D.; Alquezar, R.; Tarran, J.; Orwell, R.L., & Torpy, F. (2006). The Potted-plant Microcosm Substantially Reduces Indoor Air VOC Pollution: I. Office Field-study. *Water, Air, and Soil Pollution*, Vol.175, No.1-4, (September 2006), pp. 163-180, ISSN: 0049-6979

Yang, H. (2010). *Removal of Benzene/Toluene from Indoor Air by Ornamental Plants and its Application and the Mechanism*, Postdoctoral Research Report, Zhejiang University and Beijing Academy of Science and Technology

Yang, H.; Liu, Q.Y.; & Liu, Y.J. (2011). Removal effects of benzene/toluene in indoor air using potted-plants and its application analysis, *Environmental Chemistry*, ISSN 0254-6108

Yang, H.; Liu, Z.Y.; Ge, H.; Yang, S.H.; Ge, W.Y. & Liu, Y.J. (2010). Performance-Testing in Removing Benzene-Toluene Binary Gas Using New Lines of *Chrysanthemum*. *Northern Horticulture*, No. 226, (October 2010), pp.5-8, ISSN 1001-0009

14

Combustion Emissions Analysis of Wood-Based Waste Processing-Materials

Han Chien Lin
Department of Forest Products Science and Furniture Engineering, College of Agriculture, National Chiayi University
Taiwan

1. Introduction

The pursuit of economic development by the human has come at a great cost to the animals and plants of this planet, and has damaged the quality of our air and water. In the last 100 years more problems have come to threaten the environment, such as vanishing forests, acid rain, the burning of fossil-fuels and global warming. Currently, researchers are focusing on so-called "eco-materials" that make effective and environmentally friendly use of natural resources, such as the recycling of industrial wastes, the development of new environmentally friendly materials, etc. The evaluated items are included the frontier, amenity and adjustment of materials (Okuma, 1998). Besides, Life Cycle Analysis (LCA), started in the early 1970s, is a method that considers energy and raw material consumption, different types of emissions and other important factors related to a specific product's entire life cycle. The Society of Environmental Toxicology and Chemistry (SETAC) published the widely accepted series of guidelines and definitions of LCA, and the ISO 14040-14043 is considered the LCA standard (Bras & Román, 2006). Recycling of wood products, as one item of the LCAs, generally includes two kinds of methods. The first type is recycling of discarded forest products or resource-recovered (renewable) materials, which means to develop new wood-based materials in accordance with the principle of cascade utilization (Akiyama, 1998). The second type is the waste disposal options for spent various wood materials, including demolition, combustion and/or landfilling (Jang, 1997; Kobayashi et al., 2000). However, discarded the wood-based waste processing-materials, especially wood or/and wood-based materials that have been processed with different types of adhesives, coatings and preservatives, are extremely difficult to be recycled as a raw material, because these materials for wood products used cannot presently be used as a raw material again (Lin et al., 2001), and the waste stage has a very significant impact on the life cycle assessment results (Esser et al., 2000).

As we know, carbon dioxide (CO_2) is one of elemental compounds in the atmosphere (air) and a factor in photosynthesis and absorption. CO_2 is also the best-known greenhouse gas because it is capable of increasing the average global temperature depending on the sensitivity of the atmosphere (such as the concentration of CO_2) to greenhouse forcing (Allaby, 2000; Aplin et al., 2000). Carbon monoxide (CO), one of tasteless, colorless and non-odorous gases, is produced primarily by car exhaust and incomplete combustion of organic materials. Hence, CO can easily be inhaled and combined with the hemoglobin in blood, which prevents oxygen (O_2) transportation in the blood and causes intoxication effects such as headache, giddiness, tiredness and buzzing in the ears etc (Lin, 1992). Sulfur dioxide (SO_2) is one of the precursors for acid rain because not only a sulfurous acid (H_2SO_3) reaction is formed when SO_2 is compounded as mist in the atmosphere, but also SO_2 has a chemical reaction with O_2 and forms sulfur trioxide (SO_3). SO_3 then becomes liquid SO_3 because its' dew point is 22 °C. At the dew point a sulfuric acid (H_2SO_4) reaction is formed. When inhaling SO_2 at about 30 to 40 ppm, the body will feel giddy and experience breathing difficultly (Lin, 1992; Aplin et al., 2000). Nitrogen oxide (NOx) is so-called "thermal induced compound" because it has two reactions during combustion. The first reaction is from oxidization in which a nitric (N) substance after combustion is formed with the numerator of O_2. The second reaction involves the reaction of O_2 and N_2 from the air during the combustion process (Li et al., 1993; Jang, 2000; Takabashi et al., 1995).

On the other hand, the thermal behavior (pyrolysis, thermal decomposition or combustion) of every combustible material or product produces a toxic smoke atmosphere. This toxic smoke atmosphere, in sufficiently high concentrations, presents a hazard to exposed animals and humans. Discarded woods treated with any of these above mentioned wood-based waste processing-materials from the combustion are therefore need to be disposed of in an environmentally acceptable manner (Ibach, 1999; Lin et al., 2001; Cooper, 2001; Lin et al., 2006). The thermal composition of these wood-based waste processing-materials during combustion, including such things as, decomposition temperature, compounds emitted, total amount of volatiles evolved (Kercher & Nagle, 2001) and the amount of char formed (residual elements) (Lin et al., 2007) is need to be investigated. This is due to the fact that some of the materials are not organic compounds, and their precise molecular composition may vary. In addition, the thermal decomposition of all combustible materials produces a toxic smoke. Combustion toxicology indicates adverse effects produced from exposure to fire-generated toxic species (Hartzell, 1996), but they are not included in the wood recycling industry, basically because of technical problems and regulatory concerns, and the same applies to waste wood management (Falk, 1997). In many cases the final treatment for used wood products is destruction by the combustion (Jang, 1997). The emitted gases from burning the wood-based waste processing-materials, such as CO_2, CO, SO_2 and NOx, and so on are discharged and contribute to environmental pollution.

To provide an understanding of the emitted gases and the fundamental thermal decomposition, as well as to analyze the evolved species and the char of wood-based waste processing-materials, including wood adhesives, wood coatings and wood preservatives, the focus for this chapter is divided into three categories. The first category involves using a Flue-gas Analyzer from the exit of a 45° flammability testing cabinet as specified in the Continuous Emission Monitoring (CEM) technique; the emission gas concentrations during the combustion are analyzed. The second category involves the study of thermal decomposition, and examines the effectiveness of the thermal properties of these materials, using Thermogravimetric Analysis (TGA). In the third category two kinds of spectrometric techniques are applied. One applies Thermogravimetric Analysis with Infrared Spectrometry (TGA-IR) to monitor the effectiveness, on an analytical scale, of these processing-materials and also to examine the emissions of certain compounds (evolved species) in each particular analysis. The others apply, Fourier Transform Infrared Spectroscopy (FTIR), Energy Dispersive X-ray Spectrometry (EDX), and Elemental Analysis (EA) etc, to examine certain residual elements (organic and inorganic materials) of the char, before/after thermal decomposition. The main goal of this chapter is to realize fundamental knowledge of the thermal decomposition of the wood-based waste processing-materials, while at the same time understanding an air pollution reference for gas emissions, evolved species and a disposal-end practice (of the residual elements) reference for the char.

2. Analysis methods of emission gas, thermal decomposition and evolved species

2.1 Continuous emission monitoring

Detecting flue-gas emissions is required by strict environmental legislation. Combustion better detects the flue-gas emission value for analyzing waste products (Jang, 2000). Continuous Emission Monitoring (CEM) techniques are utilized widely for detecting gaseous emissions such as CO, NOx, SO_2 and O_2 and CO_2 concentration at the chimney exit of furnaces to determine the concentrations of various pollutants (Jang, 2000). This portion is to introduce the emission gas concentrations during the combustion of wood-based waste processing-materials by using the CEM technique. A Flue-gas Analyzer (MSI, 2000 W. Germany) was used to examine the gas emitted of wood adhesives, wood coatings and wood preservatives from the exit of a 45° flammability testing cabinet (with reference to CNS 7614 A3125, NCYU handmade). The fuel used was liquid propane. The gas flow direction during the combustion was determined using a gas-measuring probe (MSI 2000 type B). The design and methods are shown in Fig. 1.

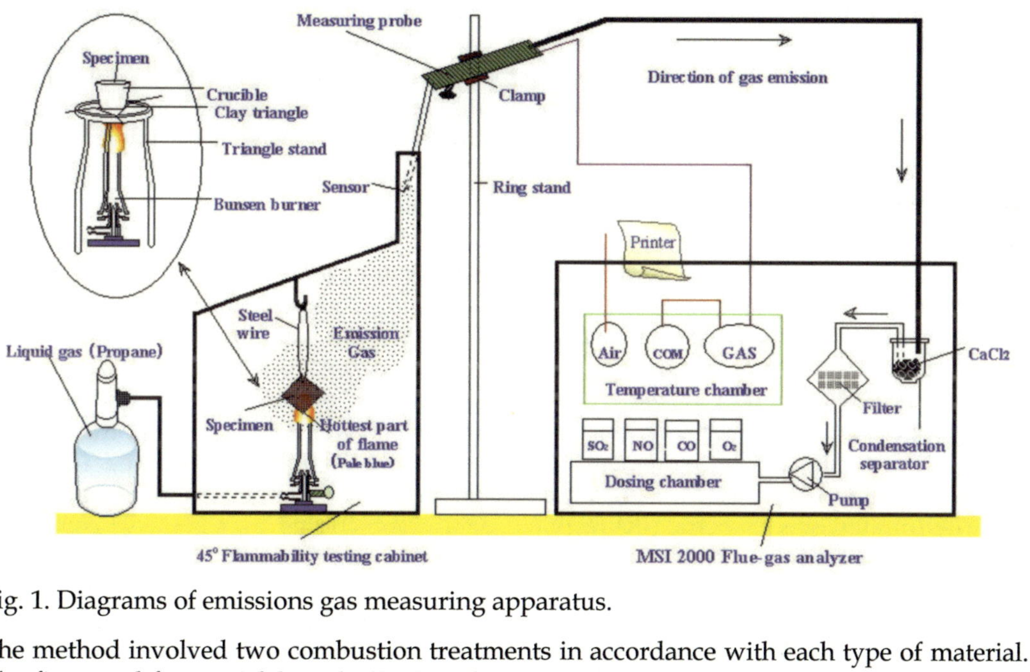

Fig. 1. Diagrams of emissions gas measuring apparatus.

The method involved two combustion treatments in accordance with each type of material. The first used for particleboards (PBs) with 4 types of wood adhesives, coatedwoods with three types of wood coatings and the coatings on a stainless steel, as well as preservative-treated woods with 2 types of wood preservatives. The specimen is hung inside the test cabinet with a steel wire and burned with a Bunsen burner adjusted to (pale blue flame)

about 850 to 980 °C. For the second treatment, the adhesives, the preservatives or wood particles was placed into a crucible on a clay triangle with a triangle stand, respectively. This method is displayed in the oval figure at the upper left in Fig. 1. The flame temperature was verified using a temperature-element apparatus (SUNTEX, ST-54) for both methods.
The combustion measureed method is described as follow. The first cycle run in the Flue-gas Analyzer involved no combustion with a specimen in the test cabinet to be certain that the measured value of the emission gas was zero and the gas temperature was room temperature. The material was ignited and then examined for 10 cycles continuously. A measured cycle for the emission gas concentrations and the emission gas temperature during specimen combustion was about 40 sec. The maximum experimental value was obtained and compared with the value from each cycle. The CO, SO_2 and NOx gases were determined according to parts per million (ppm). The O_2 and CO_2 concentrations were defined as the volume percent (VOL %) of the air in the emission exit. The gas temperature was measured simultaneously.

2.2 Thermogravimetric analysis

Thermogravimetric analysis (TGA) is a common technique for measuring the weight change of a material as a function of temperature. Throughout this chapter, TGA was carried out using a Perkin-Elmer TGA 1, fully supported by computer-controlled software options, from Perkin-Elmer Thermal Analysis Systems, for control and data handling. The wood-based waste processing-materials, including wood adhesives and wood preservatives, approximately 5.0-10.0 mg in weight, were placed into an open experimental sample pan, and suspended from a sensitive microbalance. A furnace surrounded each material sample to provide accurate heating from 50 to 850 °C, while the measurement was taking place. In order to isolate the pyrolysis step, considered the most important step in the burning process for materials or products, a nitrogen atmosphere was selected for TGA analysis (Hsu et al., 2000). All runs were purged in nitrogen atmosphere, at a flow rate of 50 mL/min, with several heating rates of 5, 20 and 40 °C/min. The thermal behavior for the wood-based waste processing-materials was then analyzed.

2.3 Thermogravimetric analysis with infrared spectrometry

TGA, described in the previous section, is a quantitative technique, which does not identify evolved compounds. To understand the evolved species produced by the pyrolysis of the wood-based waste processing-materials, analysis was carried out using a TGA (Perkin-Elmer Pyris 7 TGA model), linked to a secondary method of analysis, Fourier-Transform Infrared Spectroscopy (using a Perkin-Elmer FTIR), which was able to identify the evolved species during the TGA measurements (Statheropoulos& Kyriakou, 2000). The TGA and FTIR (TGA-IR) were connected using a transfer line with a continuous scanner in a gaseous cell. The Infrared spectrometric (IR) response data was recorded by a personal computer. The time-based software obtained and analyzed the evolved species as they occurred. The samples, weighing about 5.0 mg, were placed in an open pan and suspended from a sensitive microbalance in the Pyris 7 TGA model. The conditions were set the same as for the Perkin-Elmer TGA 1, described above, for heating and nitrogen flow rates. The evolved species analysis was performed on the TGA-IR. However, the IR response obtained for various evolved species using the TGA-IR appeared about 10 sec later, when compared to the TGA traces. In order to discuss the criteria and limitations of both TGA-IR and TGA for each particular analysis, and to ensure accurate results, it was necessary to carefully study and compare the results obtained between the pyrolysis temperature (obtained from TGA) and the decomposition time of the evolved species from the IR response (obtained from TGA-IR).

3. Emission gas, thermal decomposition and evolved species

3.1 Wood adhesives

3.1.1 Emission gas of various adhesive types of particleboard

For particleboard (PB) with various adhesives, the particles used were shaving-type wood particles (involved import mixer wood particles 85 % and dismantling material 15 %) from a PB plant in Japan, the Dantani Corporation. Urea formaldehyde resin (UF, Oshika Resin-180), melamine-urea formaldehyde copolymer resin (MUF, PB-1230), phenol formaldehyde resin (PF, PB-1310) and diphenylmethane diisocyanate adhesives (MDI, PB-1605), from the Oshika Shinko CO. LTD, Japan, were used to project as the manufacture the PBs. The preparation and manufacture of PB please see the reference (Lin et al., 2001).

The results of the relationships between emission gas temperatures, the concentration of O_2 and CO_2 and the combustion time for PB manufactured with 4 types of adhesives are shown in Fig. 2.

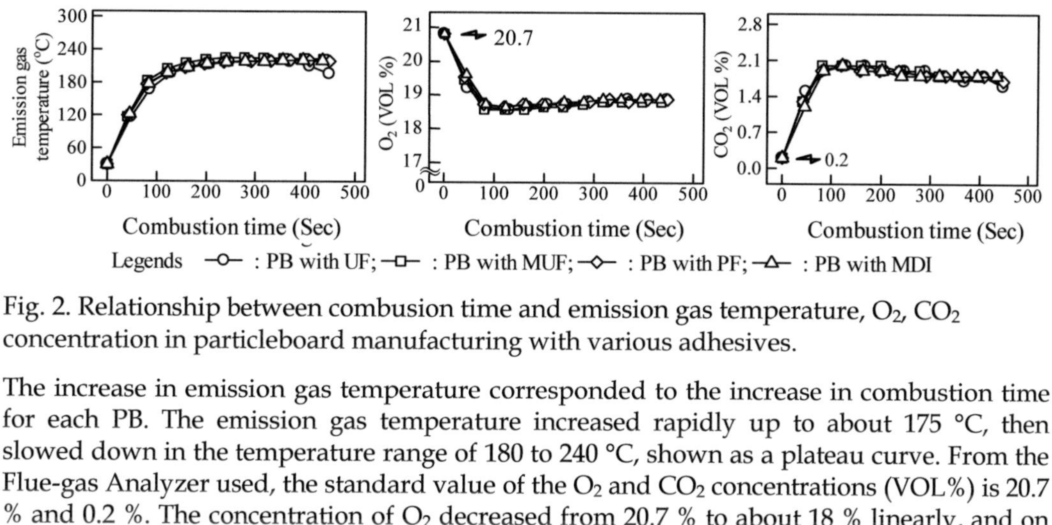

Fig. 2. Relationship between combusion time and emission gas temperature, O_2, CO_2 concentration in particleboard manufacturing with various adhesives.

The increase in emission gas temperature corresponded to the increase in combustion time for each PB. The emission gas temperature increased rapidly up to about 175 °C, then slowed down in the temperature range of 180 to 240 °C, shown as a plateau curve. From the Flue-gas Analyzer used, the standard value of the O_2 and CO_2 concentrations (VOL%) is 20.7 % and 0.2 %. The concentration of O_2 decreased from 20.7 % to about 18 % linearly, and on the contrary the concentration of CO_2 increased from 0.2 % to about 2.0 %. Both O_2 and CO_2 approached the shape of a plateau curve and had a close relationship during combustion. It was indicated that the influence of adhesive types for manufacturing the PB on the gas temperature and either of the concentration of O_2 or CO_2 during the combustion was insignificantly.

It is well known that PB is basically manufactured from particles and adhesives. In order to understand the above-mentioned PB which produced serious pollution gases, the combined results for emission gases for these PB, 4 types of adhesives, and wood particles are shown in Fig. 3. The peak for CO gas was found at about 100 sec during combustion and then the decreasing tendency slowed down to the end of combustion treatment for all of the materials. The CO had the highest point at about 100 sec for about 20 ppm and the decreasing tendency was reached quickly after 100 sec. In other words, the influence of different type adhesives on CO was insignificantly during the combustion. For the adhesives, MDI was more significant on the emission of CO for about 70 ppm (at about 100 sec) during the combustion than the others because it was considered that the reason of incomplete combustion in which MDI had the higher resin solids, 96.3 % (Lin et al., 2001). The result showed that CO was also produced from the wood particles during the combustion at about 100 sec for about 16 ppm. This is attributed to the particles from the wood and a carbon-hydrogen chemical compound.

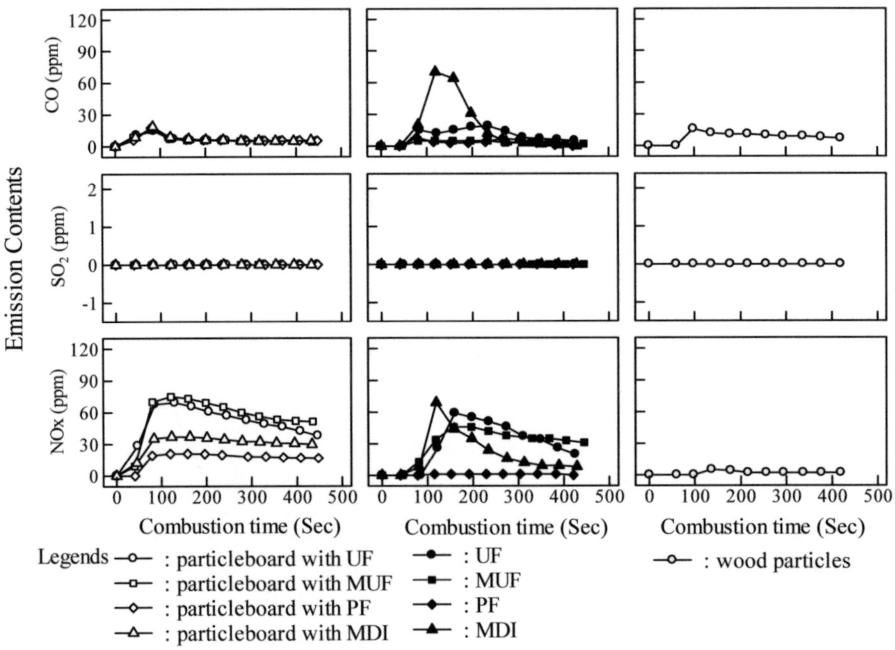

Fig. 3. Relationship between combustion time and emmission contents (CO. SO_2, Nox) in particleboard manufacturing with various adhesives, various adhesives and wood particles.

The emission SO_2 did not produce in the emission gas during the combustion for the PB with various adhesives and the adhesives because the sulphur (S) content was not included in 4 types of adhesive from the typical molecular structure (Chen, 1996; Lin et al., 2004). It is suggested that PB and wood adhesives was not the origin of the SO_2 in environment when it was concerned with final disposal (combustion).

The results showed that the PB and the adhesives were all produced the emission gas, NOx, but it for the wood particles was insignificantly during the combustion. It is considered that NOx is a main pollution origin. The reason for NOx occurring is because N is included in both wood particles and adhesives. Either the O_2 and N_2 reaction from the air and particles or the fuel, propane, produced the N residues. The typical molecular structure UF, MUF and MDI, in addition to PF, was involved a nitric-base that was easily to produce the oxidization with the numerator of O_2 during the combustion (Lin et al., 2004). Moreover, the NOx emissions for PB had the same tendency in which the peak point was about 100 sec from the start of combustion with a decreasing tendency with the combustion time from 100 sec to the end of the experiment. The quantities of emission gases, NOx, for the PB after about 100 sec were followed by MUF (70 ppm), UF (67 ppm), MDI (35 ppm), and PF (20 ppm) in that order. MDI produced the highest quantities of NOx at about 100 sec. It is inferred that the ignition point or/and heat decomposition of the adhesive might influence the combustion time and the quantity of emission gas (Lin et al., 2001).

As the results, the emission contents of CO, SO_2 and NOx from the combustion of PB and adhesives were found that the PB manufactured with PF adhesive had the least pollution gas during the combustion than the others. It is suggested that PF is the best adhesives with the least environmental impact to use for manufacturing PB, when considering the final treatment by the combustion.

3.1.2 Thermal decomposition of various adhesives

To comprehend the thermal behavior (pyrolysis, thermal decomposition) of the four types of adhesives, the specimens were measured by TGA, after being air-dried at ambient temperatures. Decomposition profiles were obtained while heating at a rate of 20 °C/min in nitrogen, between 50 °C (initial temperature) and 850 °C (end temperature). The relationships of TGA and DTG curves between temperature and weight loss, and the derivative weight for each adhesive specimen, UF, MUF, PF and MDI, are shown in Fig. 4.

As expected, the four types of adhesives typically showed gradual weight loss. The thermogram, resulting from the analysis of UF (solid line) and MUF (dotted line), showed that there were two subtle changes (shoulders) at temperatures of 50 to 390 °C, suggesting that both had almost identical initial (the first sharp step) and secondary decomposition temperatures, because UF and MUF are copolymers. The adhesive specimen weight for PF (dash line) was significantly less at temperatures between 50 °C and 500 °C in the TGA tests; the weight loss for MDI (dash-dot-dash line) was significant at temperatures from

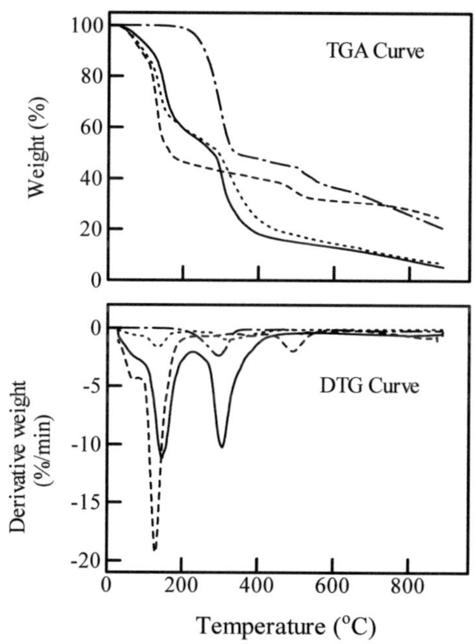

Fig. 4. TGA and DTG curves for 4 types of adhesives (UF, MUF and PF and MDI) in nitrogen at 20°C/min.

200 to 400°C. This indicates that the chemical structure, in the molecular chain, or chemical composition, of PF and MDI is different (Chen, 1996; Lin et al., 2004). The results also found that MDI exhibited the highest initial decomposition at a temperature of 297°C and better overall thermo-stability than the other types of adhesives. PF and MDI provided a better yield of char at about 24.88 and 20.61 wt%, respectively, at 850°C. This showed that theirthermal behavior was better than that of UF (5.40 wt%) or MUF (6.99 wt%), as the proposed action mechanism is based on the charred layer acting as a physical barrier, which slows down heat, and results in mass transfers between the gas and condensation phases (Camino et al., 1993, Hornsby, 1996 and Zhu & Shi, 2001).

3.1.3 Evolved species of various adhesives

Figure 5 shows the Infrared spectrometric (IR) response of the four types of adhesives in the TGA-IR measurements. The change in absorbance (concentration of evolved species %) corresponded to the increase in decomposition time for the four adhesives. The IR response of all adhesives in the TGA-IR tests showed that the absorbance of UF, MUF and MDI was higher than that of PF. This indicated that the concentration of evolved species of PF was less than that of MDI, and less than that of the urea copolymer series adhesives, UF and MUF. Of them all, PF was considered to be the best adhesive with the least environmental impact. A similar conclusion was reached in Fig. 3 and the previous research (Lin et al., 2001), which indicated that PF was the most environmentally friendly adhesive for manufacturing particleboard, because the emission contents (such as CO, SO_2 and NOx gases) of PF were less than those of UF, MUF and MDI, when the used PB was destroyed by combustion. Therefore, it has been suggested that PF is the most environmentally friendly adhesive, because it has the least environmental impact when used in many industrial and domestic applications, such as for structural members in furniture or architecture.

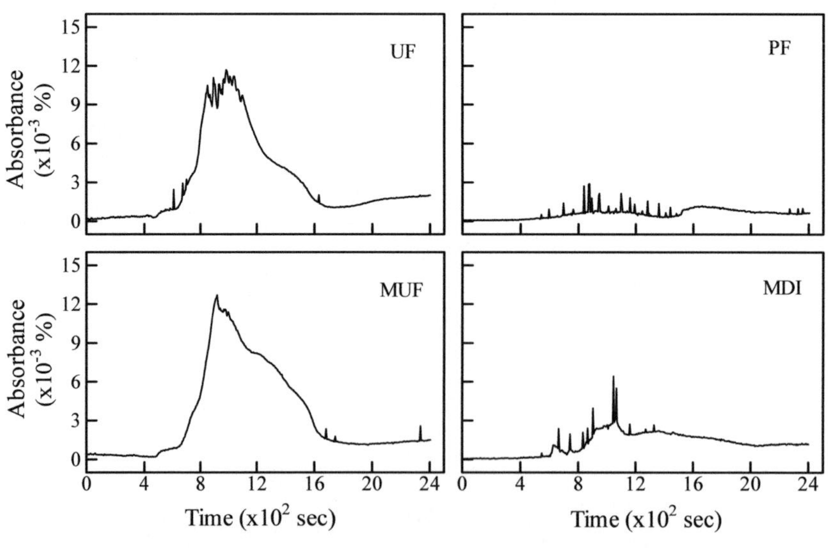

Fig. 5. IR response of 4 types of adhesives (UF, MUF and PF and MDI) from TGA - IR analysis.

In recent years, there has been much attention paid to the use of adhesives (formaldehyde copolymer compounds), because, during combustion, they emit toxic gases and corrosive smoke; exposure to these toxic and acidic fumes can cause great harm to people and can also damage costly equipment (Ladomersky, 2000). With the use of TGA-IR, it is possible to simultaneously monitor evolved species, pyrolysis temperature and concentrations formed (absorbance), as well as the evolution profile of certain compounds. Moreover, IR spectrometry is a common means of evaluating a functional base when a material undergoes degradation. The evolution of decomposition time (or pyrolysis temperature) on absorbance and wavenumbers (basically, IR spectra) for the various types of adhesive specimens (UF, MUF, PF and MDI) is shown in Fig. 6.

For TGA-IR spectra of UF, the CO_2 peak, located at 2300 and 2400 cm^{-1} was shown beginning from about 240 sec (at a temperature of about 127 °C in Fig. 4) to the end. CO_2 increased linearly to the top point of absorbance at about 960 sec (367 °C), then decreased at about 1680 sec (607 °C) and kept the same concentration of CO_2 till the end of the tests. 3545 and 3509 cm^{-1} was assigned to the N-H stretching of NH_2 moieties (Zhu & Shi, 2001) from about 360 sec (167 °C) to the end. A higher concentration of N-H stretching was seen between 840 and 1440 sec (327 to 527 °C). The range from 480 to 1180 sec (207 to 440 °C) showed a weak peak at 2921 and 2851 cm^{-1}, respectively, which was attributed to the C-H stretching of CH_2 moieties. Another weak peak, at 1447 cm^{-1}, was attributed to C-H stretching of CH_2 moieties, from 480 sec. The weak peaks at 1701 and 1625 cm^{-1} were assigned to the C=O (carbonyl group), stretching from 480 sec to the end. Moreover, several weak peaks at 972 or 962 cm^{-1} also emerged due to C=O stretching, during the period from 480 sec to the end. The top point of absorbance for C=O stretching at 972 or 962 cm^{-1} was found at 960 sec, which was the same as the top point of absorbance for CO_2. The results of MUF indicated that UF and MUF present the same tendency in the left side of Fig. 6, as both belong to the same copolymer urea formaldehyde series. The results for the absorbance peak and the time location (decomposition temperature) of CO_2, N-H stretching, C-H stretching and C=O stretching for MUF were the same as for the TGA-IR spectra of UF; the N-H deformation vibration of NH_2 moieties was found at 1605 and 1649 cm^{-1}, from about 360 sec (167 °C) to the end. The weak peak at 1449 cm^{-1} was attributed to C-H deformation vibration of CH_2 moieties, while at 1414 cm^{-1}, it was assigned to C-N stretching. Both of these began close to 480 sec (207 °C) to the end. A weak peak of C=O stretching, located at 1736 cm^{-1}, was also found from 600 sec to the end (247 to 850 °C). In addition, two weak peaks of triazine, located at 1365 and 1565 cm^{-1}, began from about 480 sec to the end.

For the TGA-IR spectra of PF, the weak peaks of CO_2 were also located at 2300 and 2400 cm^{-1} and began from about 600 to 1440 sec (247 to 527 °C), almost disappearing towards the end of the tests. Some of weak board absorbance at 3300 to 3600 cm^{-1} was assigned to OH groups from about 720 sec (287 °C) to 2160 sec (767 °C). A weak peak at 2980 cm^{-1} was assigned to out-of-plane bending of C-H on the benzene (aromatic) ring, from 1440 sec (527 °C) to the end. The spectrum results in this study are similar to those reported by Reghunadhan Nair et al (2001). Another weak peak at 1100 cm^{-1} was attributed to the stretching vibration of benzene C-C, from 1560 sec (567 °C) to the end. Besides, the results of MDI indicated that the two weak peaks of CO_2 were the same for UF, MUF and MDI, located at 2300 and 2400 cm^{-1} and which began from about 480 to 1560 sec (207 to 567 °C), almost disappearing towards the end of the tests. Some weak board absorbance located at 3700 cm^{-1}, was assigned to N-H stretching from about 600 sec (247 °C) to 1440 sec (527 °C). In addition, some weak board absorbance, found at 3100 cm^{-1}, was assigned to OH group,

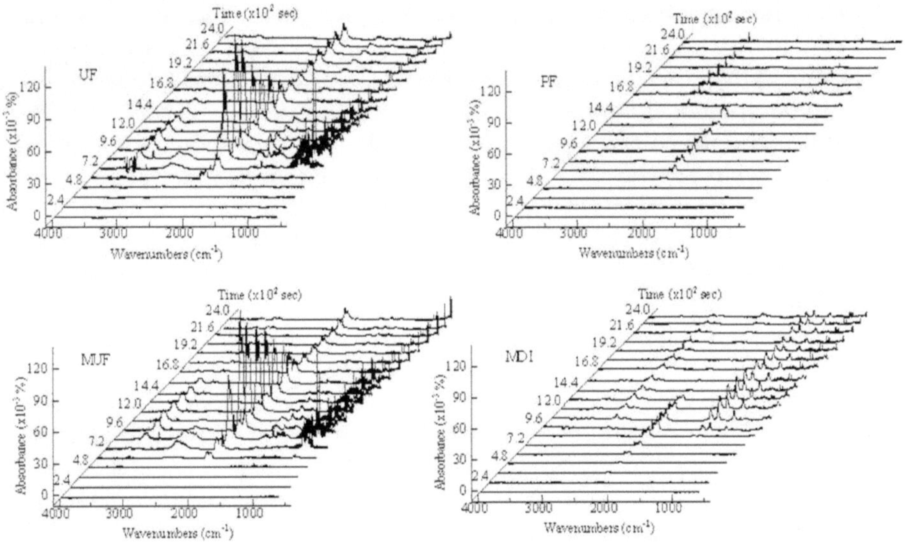

Fig. 6. TGA – IR spectra of 4 types of adhesives.

from about 960 sec (367 °C) to the end. Other weak peaks at 1624 cm^{-1} were assigned to C=O (carbonyl group) stretching from 840 sec (327 °C) to the end. Two weak peaks at 972 or 927 cm^{-1} also emerged as C=O stretching, from 1200 sec (447 °C). Results obtained also showed several weak peaks at 1114 and 1206 cm^{-1}, which were assigned to the C-NH group and N=C group (Levchik et al., 1995) from 840 sec (327 °C) to the end, respectively.

To clarify the effectiveness of the analytical scale towards the four adhesives, and to identify the evolved compounds of each particular analysis, the most probable structures for the different polymers found in these adhesives (Chen, 1996; Lin et al., 2004). UF analysis of the TGA-IR response showed that the thermal decomposition of the NH groups at a temperature of about 167 °C was initially determined as a possible degradation species. The same results were previously reported in literature (Zhu & Shi, 2001). The CH_2 and C=O groups were then degraded at about 207 °C; the final contents of the char, therefore, were practically negligible (5.40 wt%). Some of the MUF evolved species are the same as the UF evolved species, because they belong to the same series of copolymers. The TGA-IR results showed that the N-H groups were degraded at a temperature of 167 °C, while the CH_2 and triazine groups were degraded at 207 °C. In the case of PF, the thermal decomposition of the OH groups was found to be significant at 287 °C, although it could be degraded earlier, at temperatures of 65 and 129 °C; this information was obtained from the results of the TGA tests. In addition, the CH_2 groups were degraded at 527 °C and the benzene C-C at 567 °C. For MDI, the C=O and N=C groups were degraded at about 327 °C, which was analogous to the thermal decomposition of the OCN group during MDI pyrolysis, because of the resultant chemical reactions. Also, the char left at 850 °C in the TGA tests for PF and MDI was 24.88 and 20.61 wt%, respectively. The benzene (aromatic groups) structures are for PF and MDI; that is, the thermal performance was enabled when the chemical compounds included aromatic groups.

As the results, when compared to the molecular structures of the adhesives, enabled us to point out the kinds of evolved species (functional bases or the change in molecular structures) for referencing the thermo-degradation processes of the adhesives.

3.2 Wood coatings

3.2.1 Emission gas of wood coatings specimens

In the style-conscious wooden furniture industry, good finishing is perhaps the most critical factor. The sale of furniture is highly influenced by its decorative value and attractive appearance. However, the expectation of life for wooden furniture is about 17 years (Akiyama, 1998). Discarded wooden furniture are extremely difficult to recycle because the coatings on the furniture surface cannot presently be used as a raw material (particles or chips) again when it is considered with the bond quality of reuse materials, such as various PBs. In many cases the final treatment for used wooden furniture is destruction and then treats with combustion. Moreover, wood coatings are generally produced with various chemical compounds. The follow portion is to introduce the combustion emissions from wood coatings, and also to explore the modification of chemical structure using a potential application of Fourier Transform Infrared Spectroscopy (FTIR) from wood coated with various wood coatings before/after an accelerated weathering test, which is the QUV artificial accelerated-degradation (QUV degradation). Three types of commercial coatings, nitrocellulose lacquer coatings (NC lacquer, Kuo Rong Corporation in Taiwan), polyurethane coatings (PU coatings, Kuo Rong Corporation in Taiwan), and ultraviolet curing coatings (UV curing coatings, Chia Tai Corporation in Taiwan) were coated on the tangential sections of *Cryptomeria japonica* D. DON., and on a stainless steel, respectively. Before combustion, the nondegraded and degraded coating woods were analyzed using the FTIR. The data produced are to comprehend and compare the environmental pollutions from the coated materials with wood coatings before/after QUV degradation when they are discarded by combustion, and to as a basic reference to develop the environmentally friendly wood coatings hopefully. The basic properties of coatings, the finishing process and the accelerated weathering tests please see the references (Lin & Huang, 2004; Lin, 2005).

The emission gas temperature of all coated materials was rapidly up to about 160-170 °C, increased slowly up to 180-200 °C and then kept in the temperature range of 200-210 °C, shown as a plateau curve. The concentration of O_2 decreased from 20.7 % to about 18-19 % linearly, and on the contrary the concentration of CO_2 increased from 0.2 % to about 1.2-2.0 %. Both O_2 and CO_2 then approached the shape of a plateau curve as well as had a close relationship during combustion.

It is well known that wood coatings are generally combined with various chemical compounds. To understand the coated specimen produced serious pollution gases after QUV degradation, the combined results of emission contents are shown in Fig. 7. The highest emission quantity of CO and NOx for the coated specimens of NC lacquer coatings was about 198 and 102 ppm, 110 and 47 ppm for PU coatings and 90 and 9 ppm for UV curing coatings. The emission SO_2 in this experiment as expected did not produce in the emission gas during the combustion for the coated specimens before/after QUV degradation because the sulphur content was not included in 3 types of wood coatings from the typical molecular structure (Lay, 1985). It is suggested that the coatings are not the origin of the SO_2 in environment when it was concerned with final disposal (combustion). Three types of coatings still produced those pollutants (CO, SO_2 and NOx) after QUV degradation, especially for NOx gas. This is because the chemical structure of these coatings produced a series of photochemical chain reaction and the coating film separated (peeled) from the coated substrate (Chang & Chou, 1999). However, the pollutants were relatively a decreasing tendency for the coatings after QUV degradation. The above-mentioned results

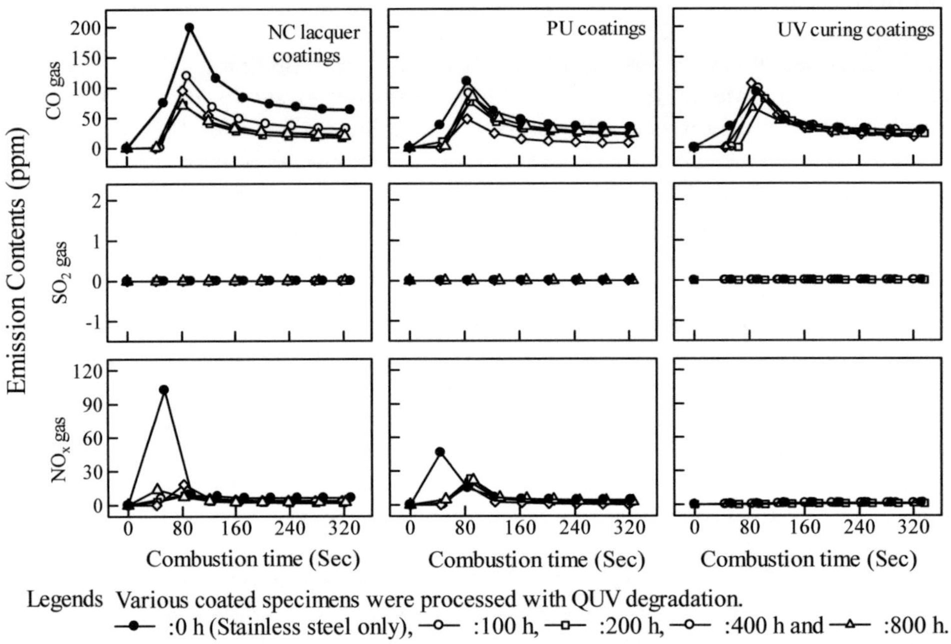

Fig. 7. Relationship between combustion time and emission contents (CO2, SO 2, NOx) for stainless steel and various coated specimens.

suggested that UV curing coatings is the most environmentally friendly coatings while the coatings are considered to discard process with combustion.

3.2.2 FTIR analysis of wood coatings

Chang (1990) reported that FTIR, possessing many unique advantages, is a powerful and potential technique for the analysis and characterization of micro-specimens of various wooden materials, coatings and coatedwood, especially with particular attention to the chemical modifications of materials after weathering and photodegradation. Moreover, It is well known that wood coatings are generally composed with various resin and chemical compounds. Basically, the manufactured principle of NC lacquer is to substitute NO_2+ for OH of CH_2OH or vascular cellulose and then became NC lacquer. In other wards, NC composed with alkyd resin and then added some of additives to become NC lacquer. To comprehend and compare the production of combustion emissions from three types of wood coatings, the coatedwoods before/after degraded at 0, 100, 200, 400 and 800 h with the QUV accelerated weathermeter were analyzed using the FTIR, shown in Fig. 8. The FTIR spectra of NC lacquer for QUV 0 h showed that the peaks at 1643 and 1273 cm^{-1} were assigned to the asymmetric NO_2 stretching vibration and symmetric NO_2 vibration of CH_2ONO_2 or vascular cellulose, and another peak at 830 cm^{-1} was assigned to N-O (nitrogen-oxygen) stretching vibration (Pouchert, 1981). Results showed that these peaks disappeared on the FTIR spectra when the NC lacquer coatedwoods after QUV degradation of 100, 200, 400 and 800h. This is because CH_2ONO_2 of NC lacquer is easily degraded by photodegradation (Chang, 1994). Therefore, the results in Fig. 7 obtained that the emission of NOx for the NC lacquer after QUV degradation decreased significantly more than control ones.

In general, the main compound of commercial NC lacquer is alkyd resin. To improve the properties of NC lacquer, it added some of additives, such as plasticizer etc. This indicated that from the FTIR spectra of NC lacquer (Fig. 8), the control coatedwood was showed a location of peak at 1720 cm^{-1}, assigned to C=O (carboxylic group) stretching vibration that was probably from alkyd resin and plasticizer, or NC degradation to become -COOH base. After QUV degradation, this peak frequency was changed its intensity and location. This is because the change in the chemical structure of NC lacquer occurred, as well as C=O produced some degraded derivatives (Chang, 1990). The results in Fig. 7 indicated that the emission of CO for the NC lacquer coatedwoods after QUV degradation decreased significantly more than control ones. The results in Fig. 8 also showed that the FTIR spectra of NC lacquer coatedwoods after QUV degradation at 100 h appeared a peak at 1015 cm^{-1}, assigned to -COOH stretching. It is indicated that this commercial NC lacquer due to QUV degradation produced a degraded derivative, -COOH base, but this degraded derivative disappeared at QUV degradation of 200, 400 and 800 h. Hence, the results in Fig. 7 showed that for the NC lacquer coatedwoods after QUV degradation at 100 h, the emission of CO decreased after 120 sec during combustion and the decreasing tendency then slowed down to the end of combustion. It is inferred that because -COOH base decomposed easily during the combustion.

FTIR spectra in the middle of Fig. 8 showed that PU coatings degraded by QUV degradation at 0, 100, 200, 400 and 800 h. The FTIR spectrum for the coatings control ones showed that the location of peak at 1716 cm^{-1} was assigned to C=O (carboxylic group) stretching vibration of urethane and peak at 1217 cm^{-1} was assigned to -N-C-O-R (Pouchert, 1981). After QUV degradation, FTIR spectra showed that both C=O produced the change in the intensity and the shift of peak frequency, and then increased its width. Chang (1990) reported that PU coatings due to photodegradation occurred yellowing, generated quinine-imides structure, on the coating film and then produced the degraded derivatives of ester or/and carboxyl. Hence, the results in Fig. 7 showed that for the PU coated specimens after various QUV degradations the emission of CO decreased significantly. Moreover, the peak at 1531 cm^{-1} was assigned to -C-NH of R-N-C-O-R′ for QUV degradation at 0 h. Results also showed that this peak disappeared on the FTIR spectra when the PU coatings coatedwoods after QUV degradation at 100, 200, 400 and 800 h. Therefore, the results in Fig. 7 obtained that the emissions of CO and NOx for the PU coatings coatedwoods decreased more than control ones.

FTIR spectra in the bottom of Fig. 8 showed that UV curing coatings coatedwoods degraded by various time of QUV degradations. Because the main compounds of UV curing coatings used are epoxy oligomer and acrylate monomer, the FTIR spectrum of the UV curing coatings coatedwoods after QUV degradation at 100 h showed that the location of peak at 1724 cm^{-1} was assigned to C=O stretching vibration and at 1604, 1505, 824 cm^{-1} was assigned to benzene ring. Moreover, the peaks at 1178 and 1175 cm^{-1} were assigned to C-H stretching vibration (Chang & Chou 1999). From the results of FTIR spectra, the UV curing coatings coatedwoods after QUV degradation at 100, 200, 400 and 800 h showed that the above-mention functional bases (group) do not change significantly, but from the emission of CO decreased (Fig. 7) the study is inferred that the benzene rings in UV curing coatings are extremely stable during the heat decomposition (combustion), but the UV curing coatings coatedwoods after QUV degradation are possible to produce the oxidation. It is inferred that the emission of CO, therefore, might decrease relatively during the combustion.

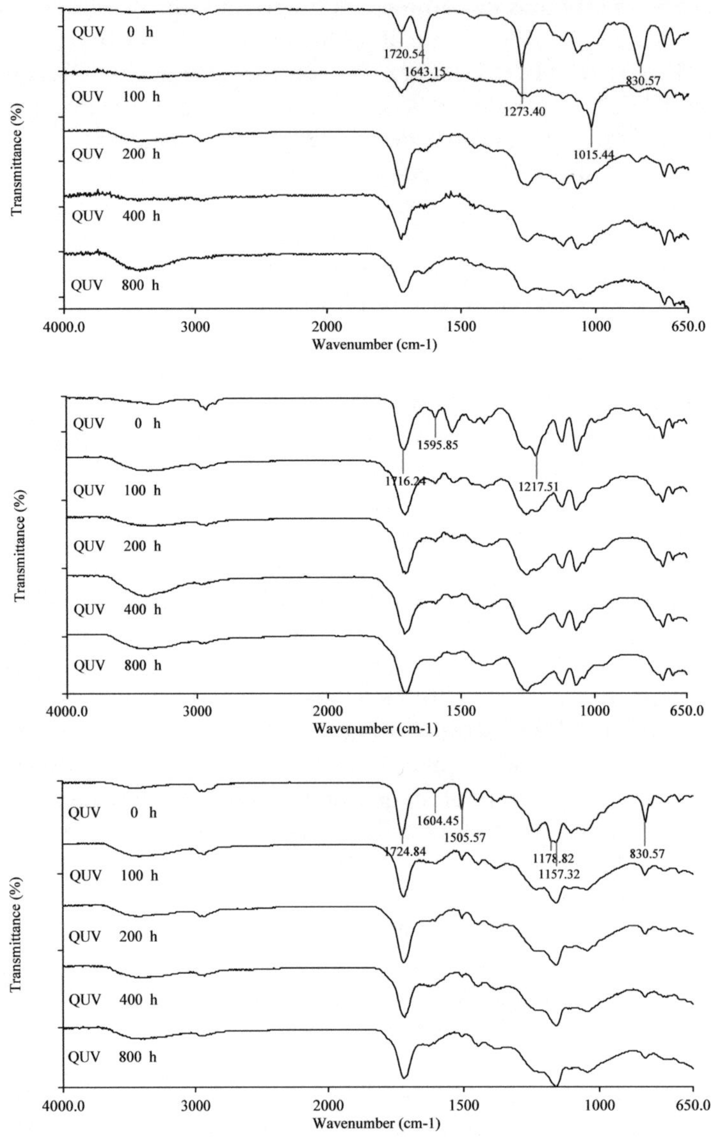

Fig. 8. FTIR spectra of NC lacquer (top of Fig), PU coatings (middle of Fig) and UV curing coatings (bottom of Fig) coatedwoods degraded by various time of QUV degradations.

To analyze the QUV degradation of coatedwoods, using FTIR showing the change in the functional bases or the molecular structures in combination with CEM analysis has led to useful results, involving the identification of emission products. Three types of coatings still produced those pollutants (CO and NOx) after QUV degradation, but the pollutants are relatively a decreasing tendency while comparing to the control ones. When analyzed to the QUV degradation of the coatings, the results enabled us to point out the kinds of functional base or the change in molecular structures for referencing the emissions of the wood coatings.

3.3 Wood preservatives

3.3.1 Emission gas of preservative-treated woods

Extra protection is needed when wood is exposed to adverse environments. This is because wood becomes subject to degradation by a variety of natural causes when left untreated in many outdoor applications. For the past few decades, wood can be protected from the attack of decay fungi, harmful insects, or marine borers by applying chemical wood preservatives with long-lasting results (Connell, 1991). Wood preservatives can be divided into two general classes: (1) oilborne (oil-based) preservatives, such as coal-tar creosote (creosote oil) and petroleum solutions of pentachlorophenol and (2) waterborne (water-based) preservatives that are applied as water solutions, such as chromated copper arsenate (CCA) and ammoniacal copper Quats (ACQ) (Ibach, 1999). However, environmental concerns have drastically changed the active ingredients of wood preservatives, resulting in restricted use (Sanders & Windom, 1980; Sanders & Riedel, 1987; Weis & Weis, 1992a, 1992b, 1996, 1999; Cooper, 2001), especially since substantial amounts of CCA or creosote oil remain in the wood for many years and the disposal of scrap wood has become an ever increasing problem (Cox, 1991; Cooper et al., 2003; Solo-Gabriele et al., 2003). Some years ago, alternative copper-based preservatives, ACQ, comprised of a combination of copper and organic biocides were developed relatively, and are being used increasingly for wood preservation in many countries.

Because wood preservatives are generally produced with various chemical compounds, it is necessary to understand the composition of the gases that are emitted when burning these preservative-treated woods, and the amount of char formed (residual metal) during combustion (Jang, 1997; Kercher & Nagle, 2001; Humphrey, 2002). The recognition of the need for abatement of air pollution has led to further interest in, and investigation of combustion as a major issue when disposing of cellulose wastes and cellulose residual materials (Helsen & Bulck, 1998). Regardless of waste minimization efforts, improved disposal-end management practices will play a key role in minimizing the impacts of disposed of CCA-treated wood in the next 25 to 40 years (Cooper et al., 2003), and of the ACQ-treated woods as well (Lin et al. 2006). To provide an understanding of the disposal of CCA- and ACQ-treated woods and the basic emitted gases generated by burning these materials, our goal in this portion is to acquire fundamental knowledge of the disposal-end practice (residual elements) reference for the char. The emissions from the discharge gas during the preservative-treated specimen combustion and the residual elements of the char after combustion are interpreted, respectively.

One of the studies used *Taiwania cryptomerioides* Hay (Taiwania) as a specimen to be treated with chromated copper arsenate (CCA) or ammoniacal copper quats (ACQ) preservatives. CCA and ACQ-treated woods, the preservatives themselves as well as their individual ingredients during combustion were investigated (Lin et al. 2007). Basically, both woods treated with CCA or ACQ are considered to be a type of cellulose material (Humphrey, 2002). The recognition of the need to reduce air pollution has led to further interest in, and the investigation of the feasibility of combustion to dispose of cellulose wastes and residual materials (Helsen & Bulck, 1998). The results of the emission contents for Taiwania and the two types of treated woods are shown in Fig. 9.

The peak for CO gas for the samples of Taiwania (T) was found to be 40 sec with 116.0 ppm after the beginning of combustion, after which it showed a decreasing tendency until it became a plateau curve until the end of the combustion time,. However for the CCA specimens a maximum value of CO was obtained at about 115 sec with 108.0 ppm, and for the ACQ specimens the measurement of the CO gas increased linearly to about 80 sec with 159.3 ppm from the start of the combustion. After that, both became a plateau curve and remained so until the end of the combustion time. This indicated that the emission of CO gas not only was concerned with the ignition time (Ladomersky, 2000; Lin et al., 2001), but also because it resulted in complete combustion at a different combustion time due to the different types of preservatives.

In middle of Fig. 9, The results also found no SO_2 being emitted gas by any of the specimens during combustion, but the specimens whether or not they were impregnated with preservatives all produced NOx gas (the bottom of Fig. 9). Although it was insignificant for the amount of wood and CCA specimens during the combustion, the peak of NOx gas for the ACQ specimens was about 26-26.5 ppm at about 75 to 115 sec of the combustion time. NOx is considered the main source of pollution from burning wood treated with ACQ. The reason for NOx occurring from the burning of preservative-treated wood is because N is included in both the wood (Shirokae & Samezima, 1996; Kercher & Nagal, 2001; Humphrey, 2002) and in the preservatives (Lin et al., 2006). Either the O_2 or the N_2 reaction from the air and the samples or the fuel, propane, produced the N residues (Lin, 1992; Takabashi et al., 1995). It is inferred that the ignition point and/or the heat decomposition of the preservatives may influence the combustion time and the quantity of emission gases, NOx.

To understand which preservative-treated wood produced the more serious pollution, the combined results of the emission contents from CCA and ACQ preservatives and their main ingredients of chromium trioxide (CrO_3), cupric oxide (CuO), arsenic pentoxide (As_2O_5), and N-alkyl benzyl dimethyl ammonium chloride (BKC) are shown in Fig. 10.

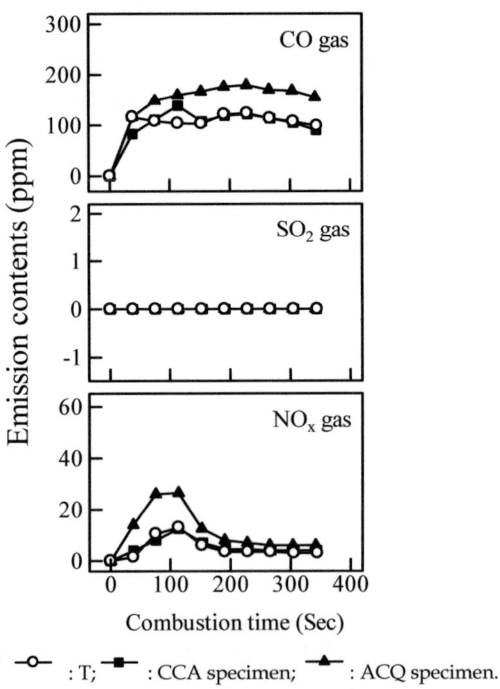

Fig. 9. Relationship between combustion time and emission contents for various experimental specimens.

The above-mentioned situation of incomplete combustion for wood particles (Taiwania) produced CO during combustion. This is attributed to the fact that the particles are wood and are a carbon-hydrogen compound. (Shirokae & Samezima, 1996; Helsen & Bulck, 1998). The same tendency was evident for the CCA and ACQ specimens (Fig. 10). In other words, the influence of different types of preservatives on the production of CO was insignificant during combustion. However, among the ingredients of the preservatives, AS_2O_5 (157.0 ppm) and BKC (85.5 ppm) were more significant on the emission of CO during combustion than the others. As expected there was no SO_2 in the emission gases in this experiment during the combustion of the wood, the preservatives or their ingredients because there was no sulphur content in either of the 2 types of preservatives (Nicholas & Schultz, 1994). It is suggested that CCA and ACQ are not the source of the SO_2 in the environment. The results also showed that the ingredients of the preservatives all produced NOx in the emission gases. However, during the combustion of the wood particles, CCA-3, AS_2O_5, CrO_3, and CuO, produced less NOx than the ACQ specimens and ACQ. Of these materials, ACQ and BKC were more significant for the emission of NOx during combustion than the others chemicals (ingredients). The maximum NO_x of the ACQ and ACQ specimens was about 23.5-26.5 ppm, and was about 32.0 ppm for BKC. This is because the main combinations of ACQ are BKC, and in addition nitrogen (N) is also included in the BKC (Nicholas & Schultz, 1994). This indicates that NOx is a main source of pollution in wood with ACQ preservatives.

Moreover, wood is considered an ecological material (Takahashi et al., 1995), but it is necessary to determine the thermal composition of these preservatives during combustion,

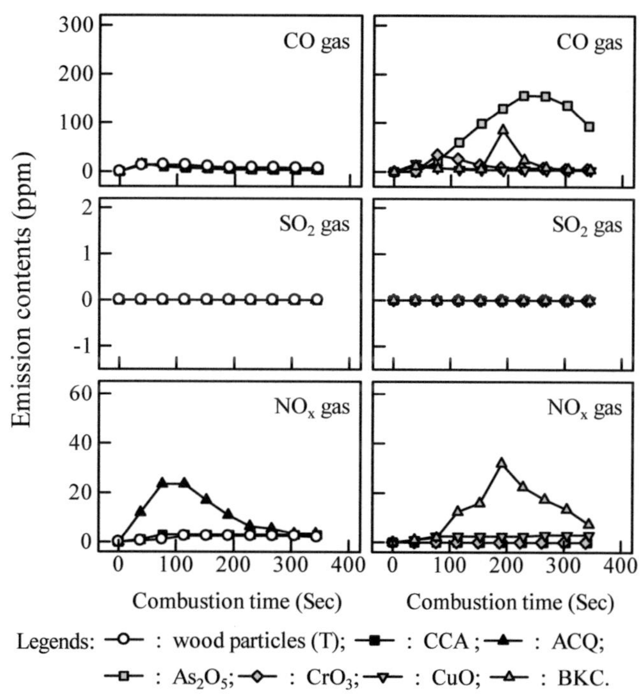

Fig. 10. Relationship between combustion time and emission contents for various wood preservatives and the main components of CCA and ACQ.

including such things as, decomposition temperature, compounds emitted, total amount of volatiles evolved (Kercher & Nagle, 2001) and the amount of char formed (residual elements) (Lin et al., 2007). This is due to the fact that both CCA and ACQ are not organic compounds, and their precise molecular composition may vary. To clarify the residual elements (inorganic and organic materials) of discarded woods treated by either CCA or ACQ, the main ingredients, the wood preservatives, their make-up and the char before/after combustion are analyzed with an Energy Dispersive X-ray Spectrometer (EDX), and are analyzed after combustion using an Elemental Vario CHNS/O analyzer (EA, Germany). The result for the carbon (C), nitrogen (N), sulfur (S), and hydrogen (H) elements was then determined. The EDX results are shown in Table 1.

The main elements of Taiwania (untreated specimens), both preservative woods and preservatives were C, 56.35 % for the untreated specimens, 47.96 % for the CCA specimens, and 50.24 % for the ACQ specimens. For the preservatives, CCA was 4.30 %, and ACQ was 3.74 %. For both of their ingredients, AS_2O_5 was 3.25 %, CrO_3 was 1.50 %, CuO was 3.76 %, but BKC was 89.35 %. After combustion, the relative proportion of C was higher, 89.42 % for the untreated specimens, 60.42 % for the CCA specimens, 77.89 % for the ACQ specimens, and 2.70 % for CCA, and 22.13 % for ACQ. In addition AS_2O_5 was 1.47 %, CrO_3 was 1.33 %, CuO was 14.02 %, and BKC was unmeasured (trace amount). The results of element O showed that it had relatively decreased for all samples after being burned and the amount of chromium (Cr) for the CCA specimens was 7.38 %, for CCA it was 34.99 %, and the CrO_3 was 64.15 % after being burned. Three of them were obviously more than the relative amount of As (Arsenic, 4.08 % for the CCA specimen, 11.89 % for CCA, and 36.92 % for AS_2O_5), and Cu (copper, 5.18 % for the CCA specimen, 16.69 % for CCA-3, and 58.39 % for CuO). This is because the Cr from the char of the CCA preservatives is a stable element (Hirata et al., 1993; Kercher & Nagle, 2001; Cooper et al, 2003; Helsen & Bulck, 2005; Lee et al., 2005). The char of the ACQ specimens left an amount of inorganic metal elements, Cu (2.67 %). As the results (Table 1) indicated, the main char of the ACQ and CuO was Cu, 50.14 and 58.39 % which was almost unable to be combusted toward the end of the combustion (Kercher & Nagle, 2001; Lee et al., 2005; Lin et al., 2006).

Element	*Taiwania cryptomerioides* Hay (wt %)		CCA Specimen (wt %)		CCA (wt %)		As_2O_5 (wt %)		CrO_3 (wt %)		CuO (wt %)		ACQ Specimen (wt %)		ACQ (wt %)		BKC (wt %)	
	before	after	before	after	before	after	before	after	before	after	before	after	before	after	before	after	before	after
C	56.35	89.42	47.96	60.42	4.30	2.70	3.25	1.47	1.50	1.33	3.76	14.02	50.24	77.89	3.74	22.13	89.35	–[1)]
O	43.45	8.09	49.43	20.55	59.78	33.50	28.59	49.23	28.14	34.52	24.69	21.89	49.30	7.17	29.94	26.38	–	–
Cr	–	–	0.08	7.38	7.41	34.99	–	–	70.36	64.15	–	–	–	–	–	–	–	–
Cu	–	–	1.75	5.18	16.29	16.69	–	–	–	–	71.55	58.39	0.45	2.67	65.76	50.14	–	–
As	–	–	0.72	4.08	10.21	11.89	68.16	36.92	–	–	–	–	–	–	–	–	–	–

[1)] Race amount (can not be measured).

Table 1. EDX analysis of various experimental samples before and after the combustion

Furthermore, the results of EA are shown on Table 2. In addition to the high level of carbon (C), the amount of N (1.42 %) was higher for ACQ than that for the other specimens, and the BKC was almost burned out (could not be measured). This inferred that the ACQ specimen could greatly produce NOx gases (Fig. 10) during combustion. The result of the EA analysis inferred that the ACQ specimen could produce NOx gases during combustion, because the amount of N (1.42 %) was higher for ACQ than for the other specimens, and the BKC was almost burned out. The above results suggest that by using EDX and EA to analyze the residual elements in char, the results can provide experimental data for referencing the char of preservative-treated woods in CEM tests.

Element	*Taiwania cryptomerioides* Hay	CCA specimen)	CCA	As_2O_5	CrO_3	CuO	ACQ specimen	ACQ	BKC
C	79.58 (0.03)	75.38 (0.54)	0.03 (0.01)	– 1)	87.04 (0.17)	63.73 (0.08)	80.93 (0.20)	4.51 (0.11)	–
N	0.28 (0.04)	0.15 (0.01)	0.05 (0.01)	–	0.97 (0.05)	0.03 (0.02)	0.33 (0.01)	1.42 (0.01)	–
S	1.17 (0.62)	0.60 (0.04)	0.36 (0.02)	–	3.22 (0.02)	0.53 (0.17)	0.54 (0.08)	0.41 (0.07)	–
H	3.27 (0.03)	2.90 (0.04)	0.19 (0.01)	–	2.68 (0.20)	0.36 (0.59)	3.13 (0.03)	0.55 (0.03)	–

1) Race amount (can not be measured).

Table 2. EA analysis of various experimental samples after the combustion

3.3.2 Thermal decomposition of wood preservatives

To comprehend the thermal properties (thermal decomposition, decomposition step, and char) of the three types of preservatives, the preservatives samples were measured by TGA, after having been oven-dried. All runs were purged in an air atmosphere, at a flow rate of 50 mL/min, with 2 kinds of heating rate, at 5 °C/min (as a slow-heating regime) or at 40 °C/min (fast-heating regime), in order to isolate the decomposition step. That is to say, the slow-heating and fast-heating regimes were applied to compare the differences in their thermal properties and their char. The relationships of the TGA and the DTG curves between temperature and weight loss, and the derivative weight for each preservatives sample, creosote, CCA and ACQ, are shown in Fig.11.

The results obtained that regardless of the heating rate, the three types of wood preservatives typically showed gradual weight loss and the decomposition temperature of each of the preservatives for the heating rate at 5 °C/min was lower than that at 40 °C/min. Regardless of different heating rate, the char (wt %) of the CCA was more than that of the ACQ, and the amount of char for the creosote was lower than that for either of the water-based preservatives. Hirata et al. (1993) reported that Cr and Cu from the CCA preservatives were largely present in the char and that arsenic (As) was easily volatilized with the increase in temperature during combustion. Some researches (Kercher & Nagle, 2001; Cooper et al., 2003; Helsen & Bulck, 2005) stated that in the TGA tests, when CCA preservatives were analyzed with air, AS_2O_5 was formed in a gaseous state and that it disappeared at a temperature of over 600 °C. In addition, CuO produced Cu_2O when the temperature reached about 800 °C both in air atmosphere, CO was then further volatilized and only Cu remained. Also, Cr (III) is a stable element with an unproductive thermal decomposition (Lin et al., 2006). In other words, it is evident that after combustion the Cr and Cu from CCA are remained. The results of ACQ show that the weight loss drastically decreases in air at either heating rate. The composition temperature for the heating rate of 5 °C/min was lower than that of 40 °C/min. These results are the same as reported previously (Lin et al., 2006). ACQ preservatives, a combination of copper and organic biocides, indicated that the main residual element was Cu (Lin et al., 2007), that is, the other elements were volatilized with the increase in temperature during combustion, such as BKC, one of the main compounds used in this study for ACQ. Therefore, the char of CCA was higher than that of ACQ (the top of Fig. 11), based on the different heating rates (5 or 40 °C/min) set.

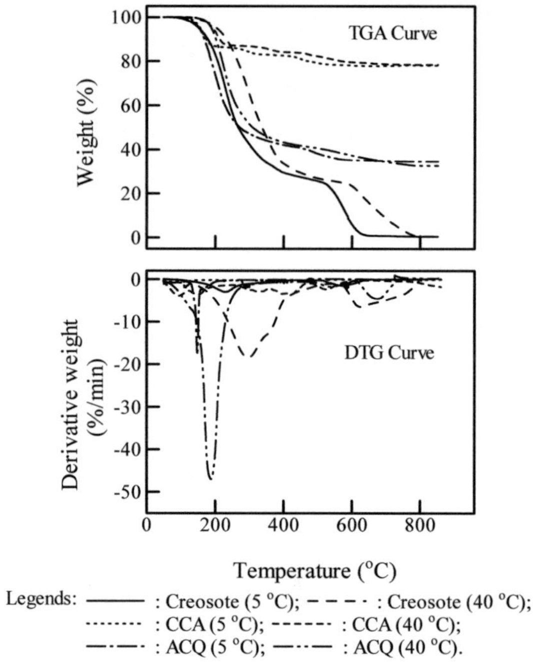

Fig. 11. TGA and DTG curves for 3 types of wood preservatives (Creosote, CCA and ACQ) in air at 5 or 40°C/min.

3.3.3 Evolved species and char of wood preservatives

Because the thermal decomposition of all combustible materials produces a toxic smoke, and combustion toxicology indicates adverse effects produced from exposure to fire-generated toxic species (Hartzell, 1996), the understanding of the evolved species and the char of wood preservatives is analyzed. Figure 12 shows the infrared spectrometric (IR) response of the three types of preservatives for both heating rates, at 5 °C/min and 40 °C/min in the TGA-IR measurements.

The change in absorbance (concentration of evolved species, %) corresponds to the increase in decomposition time for the three preservatives. Comparing the IR response of all preservatives at a heating rate of 5 °C/min, the absorbance levels for creosote were higher than for CCA and ACQ. This indicated that the concentration of evolved species of water-based preservatives was less than those of oil-based ones. Of them all, CCA was inferred to be the best preservatives with the least environmental impact, if the disposal process of the preservatives was combustion and considering the results of air pollution only. It is indicated that CCA is the better environmentally friendly preservatives for wood protection, because the emission contents of CCA are less than those of ACQ, considering their environmental impact only during combustion.

Furthermore, the right side of Fig. 12 also shows the IR response of the three types of wood preservatives with a heating rate of 40 °C/min in the TGA-IR measurements. The change in concentration of evolved species corresponds to the increase in decomposition time for the three preservatives. The IR response of ACQ showed a greater peak at the end than that of creosote and CCA. This suggests that at 40 °C/min of the heating rate, the concentration of evolved species for ACQ is higher than that for either creosote or CCA. Comparing the IR

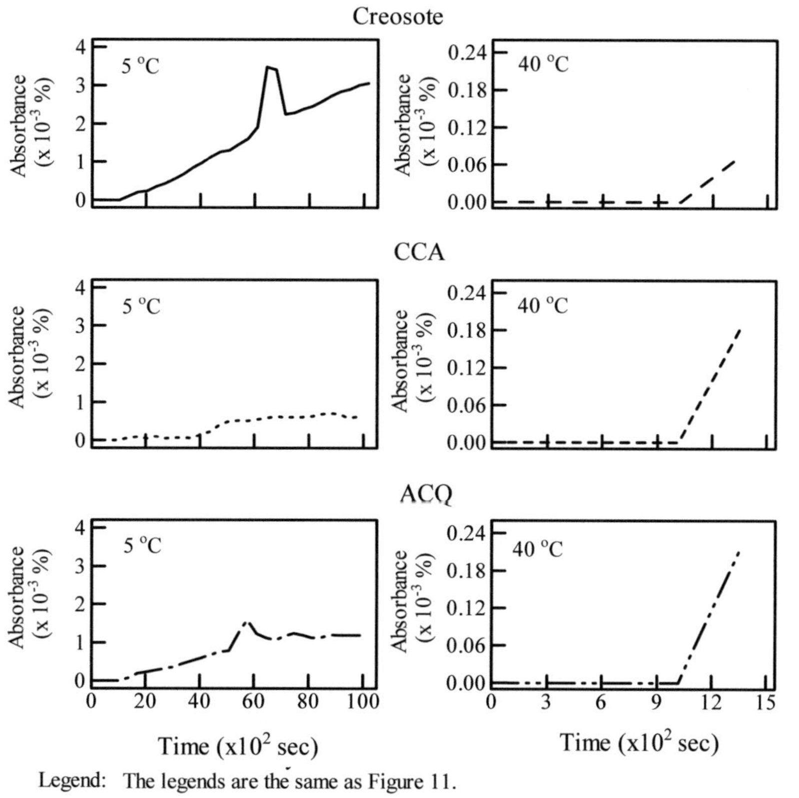

Fig. 12. IR response of 3 types of wood preservatives in air at 5 or 40°C/min from TGA-IR analysis.

response of all preservatives at both heating rates, the absorbance levels at 5 °C/min for oil-based preservatives were higher than for water-based one, but at 40 °C/min the results were reversed. This indicates that for wood preservatives the concentration of evolved species is related to the heating rate, because of the variation in decomposition steps.

At present much attention is being given to the use of wood preservatives, because of the fact that during combustion they emit toxic gases and produce a corrosive smoke. Exposure to these toxic emissions is harmful to humans and in addition can cause damage to expensive equipment (Ladomersky, 2000; Lin et al., 2007). With the use of TGA-IR, it is possible to simultaneously monitor evolved species, decomposition temperature and the concentrations formed (absorbance), as well as the evolution profile of certain compounds. Moreover, IR spectrometry is a common means of evaluating a functional base when a material undergoes degradation. The evolution of decomposition time (or temperature) on absorbance and wavenumbers (basically, IR spectra) for three types of preservatives is shown in Fig. 13. These results indicate that from the TGA-IR response the evolved species for the slow-heating regime (5 °C/min) were less than those for the fast-heating regime (40 °C/min), even for the different preservatives.

For the creosote, the CO_2 peak, located at 2296 and 2356 cm^{-1}, started showing from about 3600 sec (at a temperature of about 349 °C in the TGA tests) to the end for a heating rate of 5 °C/min and was steady from beginning to the end for a heating rate of 40 °C/min. The other CO_2 peak was located at 617 cm^{-1} for the heating rate of 5 °C/min, and was from about 2400 sec (249 °C) to the end, but at 666 cm^{-1} for a heating rate of 40 °C/min the other CO_2 peak was from about 400 sec (82.5 °C) to the end. In addition, the peaks at 1585 and 3780 cm^{-1} were due to the N-H stretching of NH_2 moieties (Zhu and Shi, 2001) from about 1200 sec (149 °C) to the end for the heating rate of 5 °C/min. A higher concentration of N-H stretching was seen between 6000 and 6600 sec (549 to 599 °C). For the heating rate of 5 °C/min the range from 5400 to 9000 sec (499 to 799 °C) showed a slight peak at 718 cm^{-1}, which was attributed to the groups of - CH_2 - rocking and another slight peak, at 3670 cm^{-1}, was attributed to the OH group, from about 1800 sec (199 °C) to the end. However, for the heating rate of 40 °C/min, some of the weak peaks, at 3741, 1794, 1316 and 814 cm^{-1}, were due to the OH group, C=O (carbonyl group) stretching, the group of C-NO_2 stretching and N-H deformation vibration of NH_2 moieties, all ranging from the beginning of the combustion to the end.

For CCA, these results indicate that the fast-heating regime presented many more evolved species. The absorbance peaks and the time location (including the decomposition temperature) of CO_2 at 2296 and 2356 cm^{-1} were from about 2400 sec (249 °C) to the end for the heating rate of 5 °C/min, but from the beginning to the end for the heating rate of 40 °C/min. C=O stretching at 1725 cm^{-1} was from about 1200 sec (149 °C) to the end for the heating rate at 5 °C/min and from the beginning to the end for the heating rate of 40 °C/min. One of the lower peaks was from the beginning to the end for the heating rate of 5 °C/min and from about 200 sec (176 °C) to the end for the heating rate of 40 °C/min.

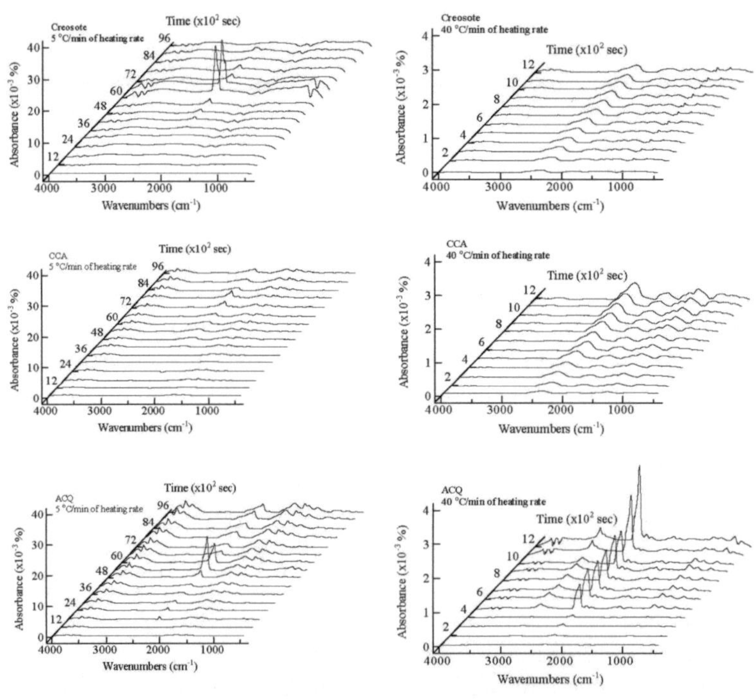

Fig. 13. TGA-IR spectra of 3 types of wood preservatives in air with 5 or 40°C/min.

This is attributed to N-H stretching at 3780 cm^{-1}. Other low peaks for the groups of - CH_2 - rocking at 710 to 718 cm^{-1} from the beginning to the end showed up in the slow and faster-heating regimes. Except those evolved species, several weak peaks, for the heating at a rate of 40 °C/min only, at 1805, 1437, 1300, 932 and 817 cm^{-1} emerged due to =CH_2, OH groups, C-NO_2 stretching, -OH out-of plane wagging, -NH_2 moieties etc.

For ACQ at the heating rate of 5 °C/min (bottom left in Fig. 13), the peaks of CO_2 were located at 2296 and 2356 cm^{-1} and began from about 1800 sec (199 °C) to the end of the tests, and the highest absorbance showed at 5400 sec (499 °C). The other CO_2 weak peak, located at 666 cm^{-1}, was from about 2400 sec (249 °C) to the end and a higher concentration of CO_2 was also seen at that point (5400 sec.). Another wavenumber at 3747 cm^{-1} was assigned to the N-H stretching of NH_2 moieties (Zhu and Shi, 2001) from about the beginning to the end and the highest point was located at the end. For the heating rate of 40 °C/min, the CO_2 peak was located at 2296 and 2356 cm^{-1} from the beginning to about 1210 sec (850 °C). The range from about 400 sec (310 °C) to the end showed two weak peaks of absorbance at 3609 cm^{-1} and 3747 cm^{-1}, which were attributed to the N-H stretching of NH_2 moieties and the OH groups. In addition, several weak peaks, for the heating at a rate of 40 °C/min only, at 1731, 1385, 1273, 1075, 930, and 716 cm^{-1} were due to C=O stretching, -CH_3, N-NO_2, C-O and -OH out-of plane wagging, and groups of - CH_2 - rocking etc. from about 400 sec (310 °C) to the end, during the period of the TGA-IR tests.

From the above results, it is suggested that using TGA-IR, to analyze the thermal behavior of wood preservatives, in combination with TGA analysis, has led to useful results, involving the identification of pyrolysis products.

4. Conclusion

To realize fundamental knowledge of the thermal decomposition of the wood-based waste processing-materials, including wood adhesives, wood coatings and wood preservatives, while at the same time understanding an air pollution reference for emission gases, evolved species and a disposal-end practice (of the residual elements) reference for the char. This chapter is mainly focused on examining the emission gas concentration and the fundamental thermal decomposition, as well as analyzing the evolved species and the char of wood-based waste processing-materials. The following results were obtained.

By using the CEM, the larger or greater these materials are, the greater the emission contents and the longer the combustion time. The concentration of O_2 was negatively and CO_2 was positively related to the emission gas temperature. Both O_2 and CO_2 had a close relationship during combustion. The main pollutants during combustion were from the ingredients of these materials, and the incomplete combustion was one of main variable factors related to the quantities of emission content (CO) and change O_2 and CO_2 concentration in the air. SO_2 was not produced during combustion for these materials because their ingredients did not have a sulfur-base. The emission contents of CO, SO_2 and NOx were found that for the adhesive, the PB manufactured with PF adhesive had the least pollution gas during the combustion than the others. This suggested that PF is the best adhesive with the least environmental impact to use for manufacturing PB. For wood coatings and wood preservatives, the emission contents were still produced after having been accelerated weathering degradation, but they were relatively a decreasing tendency. Furthermore, the results of FTIR, when analyzed to the degradation of the coatedwoods for referencing the emissions of the coatings, indicated that the change of various functional bases or/and

molecular structures were occurred. The result of the EA indicated that prior to combustion, the amount of nitrogen (N) is higher for the ACQ-treated woods than that for the CCA ones; therefore, the NOx gases during combustion were higher for the ACQ-treated woods.
By using the TGA, the results of wood adhesives found that MDI exhibited the highest initial decomposition and better overall thermo-stability than the other types of adhesives. PF and MDI provided a better yield of char and their thermal behavior was better than that of UF or MUF. The thermal properties of CCA- or ACQ-treated wood were similar to that of untreated wood in either nitrogen or air atmosphere, but the pyrolysis temperature for the slow-heating regime was lower than that of fast-ones. Regardless of different atmosphere or heating rate, the char of the CCA-treated wood and the CCA was more than that of the ACQ ones and the ACQ. The EDX analysis of both-treated woods indicated that the relative proportion of carbon was higher after thermal decomposition. Chromium (Cr) from the char of CCA is a stable element, and the amount of Cr is obviously more than that of Arsenic (As) and copper (Cu), and the char of ACQ left a high amount of Cu.
By using the TGA-IR, the IR response of the adhesives showed that the absorbance of UF, MUF and MDI was higher than that of PF; this indicated that the concentration of evolved species of PF was less than that of MDI, and less than that of UF and MUF. These results, when compared to the molecular structures of the adhesives, enabled us to point out the kinds of evolved species (functional bases or the change in molecular structures) for referencing the thermo-degradation processes of the adhesives. The preservatives showed that considering the decomposition step (thermal time or temperature) at different heating rates, the results obtained showed that at the slow-heating regime, the concentrations of CO_2 produced by creosote and ACQ were higher than that of CCA, but at the fast-heating regime, the concentration of CO_2 for ACQ was higher than that of creosote and CCA. Moreover, these results, when compared to the chemical ingredients of the preservatives and EDX analysis, enabled us to point out the kinds of evolved species for referencing the thermo-decomposition processes of the preservatives before/after time degradation.
The use of CEM for investigating the emission gases, TGA for analyzing the thermal properties of the wood-based waste processing-materials, in combination with TGA-IR, FTIR, EDX and EA analysis, has led to useful results, regarding the identification of thermal decomposition products (evolved species and char).

5. Acknowledgment

The author is grateful to Dr. Yasuhide Murase, Professor Emeritus of Kyushu University, Japan for his research assistance, and offers my sincere appreciation to the National Science Council, Taiwan through Grant No. NSC 94-2313-B-415-008 for experimental financial support.

6. References

Akiyama, T. (1998). *Wood Recycles*, Sunchoh Publication, ISBN 4-88282-186-9 C3052, Japan, pp. 7-29

Allaby, M. (2000). *Basics of Environmental Science 2ND*, Routledge, ISBN 9780203147085, Australia, pp. 44-50

Aplin, G.; Beggs, P.; Brierley, G.; Cleugh, H.; Curson, P.; Mitchell, P.; Pitman, A. & Rich, D. (2000). *Global Environmental Crises Second Edition*, Oxford, ISBN 0-19-550827-0, Australia, pp. 30-60

Bras, B. & Román, F. (2006). An Introduction to Life Cycle Analysis/Assessment (LCA), In: *LCA intro*, Available from http: www.srl.gatech.edu/education/ME4171/LCA_intro .ppt

Camino, G.; Costa, L. & Luda, M. P. (1993). Mechanistic aspects of intumescent fire retardant system. Makromol. Chem., *Macromol. Symp.*, Vol.74, pp. 71-83, ISSN1022-1360

Chang, S. T. (1990). Chemical Characterization of Woody Materials and Coating Using Fourier Transform Infrared Spectroscopy, *Q. Jour. Chin. For.*, Vol.23, No.1, pp. 87-103, ISSN 0578-1345

Chang, S. T. (1994). Study on the Lightfastness and the Weatherability of Finishes on Wood Surface, *Forest Productions Industries, Taiwan ROC.*, Vol.13, No.2, pp. 275-294, ISSN 0254-6523

Chang, S. T. & Chou, P. L. (1999). Photo-Discoloration of UV-Curable Acrylic Coatings and the Underlying Wood, Polymer Degradation and Stability, Vol.63, pp. 435-439, ISSN 0141-3910

Chen, C. M. (1996). *Wood adhesives*, National Institute for Compilation and Translation, Taiwan, ROC, pp. 280-286

Connell, M. (1991). Industrial Wood Preservatives - The History, Development, Use, Advantages and Future Trends. In: *Thompson R, ed. The chemistry of wood preservation*, pp. 16-34, Cambridge, Royal Society of Chemistry

Cooper, P. A. (2001). Analysis of Consumer Lumber Waste Management Options, *Final report prepared for Wood Preservation Strategic Options Process*, Manufacturers / Treaters Steering Committee and Consumer Waste Lumber Working Group

Cooper, P.; Ung, T.; Kazi, F. & Qi, D. (2003). Two Approaches of CCA Treated Wood: Extraction for Recycling and Wood Cement Composites, *Presented at the AWPA Annual Meeting*, Boston, MA

Cox, C. (1991). Chromated Copper Arsenate, *J. of Pesticide Reform*, Vol.11, No.1, pp. 2-6

Esser, P.; Eggels, P. & Voss, A. (2000). Waste Management of Wood Products in Life Cycle Assessment, *Presented at the 31st Annual IRG Meeting*, Kona, Hawaii, US, IRG/WP 00-50154

Falk, B. (1997). Opportunities for the Woodwaste Resources, *Forest Products J.*, Vol.47, No.6, pp. 17-22, ISSN 0015-7473

Hartzell, G. E. (1996). Overview of combustion toxicology, *Toxicology*, Vol.155, pp. 7-23, ISSN0300-483x

Helsen, L. & Bulck, E. V. D. (1998). The Microdistribution of Copper, Chromium and Arsenic in CCA Treated Wood and Its Pyrolysis Residue Using Energy Dispersive X-ray Analysis in Scanning Electron Microscopy, *Holzforschung*, Vol.52, No.6, pp. 607-614, ISSN 0018-3830

Hirata, T.; Inoue, M. & Fukui, Y. (1993). Pyrolysis and Combustion Toxicity of Wood Treated with CCA, *Wood Sci Technol*, Vol.27, pp. 35-47, ISSN 1435-0211

Hornsby, P. R. (1996). The application of hydrated mineral fillers as fire retardand and somke suppressing additives for polymers, *Macromol Symp.*, Vol.108, pp. 203-219, ISSN 1022-1360

Humphrey, D.G. (2002). The Chemistry of Chromated Copper Arsenate Wood Preservatives, *Reviews in Inorganic Chemistry*, Vol.2, No.1, pp. 1-40, ISSN 0193-4929

Hsu, C. K.; Cheng, K. B. & Lee, J. S. (2000). The applications of the thermal analysis techniques on polymer materials, *J. Spinning and Weaving Engineering, Taiwan ROC*, Vol.8, pp. 43-58

Ibach, R. E. (1999). *Wood Handbook - Wood as an Engineer Material. Chapter 14 Wood preservatives*, Forest Products Society, ISBN 1-892529-02-05, USA, pp. 14-2 - 14-9

Jang, Y. M. (1997). *Waste Processing*. WEN-JING Book Corp, ISBN 957-512-726-9, Taipei, pp. 49-85

Jang, Y M. (2000). *Technique of Burning Process*, WEN-JING Book Corp, ISBN 957-512-264-x, Taipei, pp. 213-221, 421-430

Kercher, A. K. & Nagle, D. C. (2001). TGA modeling of the thermal decomposition of CCA treated lumber waste, *Wood Science and Technology*, Vol.35, pp. 325-341, ISSN 0043-7719

Kobayashi, N.; Thikae, M. & Tominaga, Y. (2000). The properties and manufacture of renewable board from particleboard. In: *Abstracts of the 50th Annual Meeting of the Japan wood Research Society*, Japan Wood Research Society, Kyoto, Japan, April 3-5, 2000, pp. 285

Ladomersky, J. (2000). Emission analysis and minimization from the wood waste combustion, *Drevarsky Vyskum*, Vol.45, No.4, pp. 33-44, ISSN 0012-6136

Lee, H. L.; Shiah, T. C.; Hsu, F. L. & S. Lin, J. (2005). Effects of New Waterborne Wood Preservatives on the Durability and Thermal Properties of Wood, *Taiwan J. For Sci.*, Vol.20, No.2, pp. 139-156, ISSN 1026-4469

Levchik, S. V.; Camino, G.; Luda, M. P.; Costa, L.; Costes, B.; Henry, Y.; Morel, E. & Muller, G. (1995). Mechanistic study of thermal behavior and combustion performance of epoxy resin: I Homopolymerized TGDDM, *Polymer for Advanced Technologies*, Vol.6, No.2, pp. 53-62, ISSN 1099-1581

Lay, K. Y. (1985). *Coatings and Finishing in Practice*, Fu Han Publication, Taiwan, pp. 84-107

Li, Y. D.; Yan, W. Y. & Juang, T. S. (1993). *Property of Polymer Substance*, KOU-LIH Book Corp Taipei, pp. 325-328

Lin, H. C.; Huang, J. C.; Fujimoto, Y. & Murase, Y. (2001). Analysis of gases emitted from particleboard combustion, *Forest Products Industries, Taiwan ROC.*, Vol.20, No.2, pp. 165-174, ISSN 0254-6523

Lin, H. C. & Huang, J. C., (2004). Analysis of Combustion Emissions from Wood Coatings after QUV Degradation, *J. Agric. For., NCYU*, Vol.1, No.1, pp. 50 – 68, ISSN 1029-0206

Lin, H. C.; Ohuchi, T. & Murase, Y. (2004). Estimating Thermal Behavior and Analyzing Evolved Species of Adhesives through Thermogravimetric Analysis combined with Spectrometric Techniques, *J. of Fac. Agr., Kyushu Univ.*, Vol.49, No.2, pp. 449 – 459, ISBN 0023-6152

Lin, H. C. (2005). Potential Application of FTIR to Combustion Emissions Analysis of Degradation Coatedwood. *J. NCYU*, Vol.75, pp. 15-40, ISSN 1029-0206

Lin, H. C.; Ohuchi, T.; Murase, Y.; Shiah, T.-C.; Gu, L. T.; Lee, M. J. & Wu, Y. D. (2006). Application of TGA and EDX Analysis to Evaluate the Process of Preservative-Treated Woods, *J. of Fac. Agr., Kyushu Univ.*, Vol.51, No.2, pp. 337-344, ISBN0023-6152

Lin, H. C., Ohuchi, T. & Murase, Y. (2007). Analysis of Combustion Emissions and Char from CCA- and ACQ-Treated Woods, *J. of Fac. Agr., Kyushu Univ.*, Vol.52, No.1, pp. 91-98, ISBN 0023-6152

Lin, J. F. (1992). *Precaution of Special Chemical Substance*, Labors Department in Taiwan, pp. 105-107

Nicholas, D. D. & Schultz, T. P. (1994). Biocides That Have Potential as Wood Preservatives - An Overview. *Proceeding from the Conference sponsored by the Forest Products Society*, Georgia, USA, pp. 169-173

Okuma, T. (1998). *Wood-Based Resource Materials*. Tokyo University Publication, Japan, pp. 17-20

Pouchert, C. J. (1981). *The Aldrich Library of Infrared Spectra Edition III*, Aldrich Chemical Company INC., ISBN10-9999780024, Wisconsin, USA. pp. 233, 435-436, 1065

Reghunadhan Nair, C. P.; Bindu R. L. & Ninan K. N. (2001). Thermal characteristics of addition-cure phenolic resins, *Polymer Degradation and Stability*, Vol.73, pp. 251-257, ISSN 0141-3910

Sanders, J.G. & Windom H. L. (1980). The Uptake and Reduction of Arsenic Species by Marine Algae, *J. Environmental science*, Vol.10, pp. 555-567, ISSN 1001-0742

Sanders, J. G. & Riedel G. F. (1987). Control of Trace Element Toxicity by Phytoplankton, *Recent Adv Phytochem*, Vol.21, pp. 131-149, ISSN 9780080451251

Shirokae, S, & Samezima I. (1996). *Wood Science Series 4 Chemistry*, Kaiseisha, ISBN 978-4-906165-43-8, Japan pp. 15-19

Solo-Gabriele, H. M.; Townsend T. G. & Schert J. (2003). Environmental Impacts of CCA Treated Wood: A Summary from Seven Years of Study Focusing on the US Florida Environment, *Presented at the 34th Annual IRG Meeting*, Brisbane, Australia, IRG/WP 03-50205

Statheropoulos, M. & Kyriakou S. A. (2000). Quantitative thermogravimetric-mass spectrometric analysis for monitoring the effects of fire retardants on cellulose pyrolysis, *Analytica Chimica Acta.*, Vol.409, pp. 203-214, ISSN 0003-2670

Takabashi, T.; Suzuki, M. & Nakano, T. (1995). *Wood Science Series 5 Environment*, Kaiseisha, ISBN 978-4-906165-89-6, Japan pp. 7-15.

Weis, J. S. & Weis P. (1922a). Transfer of Contaminants from CCA-Treated Lumber to Aquatic Biota, *J Exp Mar Biol Ecol*, Vol.161, pp. 189-99, ISSN 0022-0981

Weis, J. S. & Weis P. (1922b). Construction Materials in Estuaries Reduction in the Epibiotic Community on Chromated Copper Arsenate (CCA)-treated wood, Mar *Ecol Prog Series*, Vol.83, pp. 45-53, ISSBN 0171-8630

Weis, J. S. & Weis P. (1996). Reduction on Toxicity of Chromated Copper Arsenate (CCA)-Treated Wood as Assessed by Community Study, *Mar Environment Res*, Vol.41, pp. 15-25, ISSN 01411136

Weis, J. S. & Weis P. (1999). Accumulation of Metals in Consumers Associated with Chromated Copper Arsenate-Treated Wood Panels, *Mar Environment Res*, Vol.48, pp. 73, ISSN 01411136

Zhu, S. & Shi W. (2001). Hyperbranched polyurethane acrylates containing phosphate as radiation curable flame-retardant coating material, *Conference Proceedings, China*, pp. 108-114

15

An Assessment of the Effectiveness of California's Local Air Pollution Controls on Agricultural Sources

C.-Y. Cynthia Lin
University of California at Davis
USA

1. Introduction

Air pollution has been recognized as a significant environmental problem in California since the early 20th century. Between 1905 and 1912, regulations were enacted by the city council of Los Angeles to regulate emissions [1]. As motor vehicle traffic increased, a new type of smog was observed and described: the "Los Angeles" or photochemical smog, as distinct from the "London" smog that resulted from coal combustion. Arie Haagen-Smit from Caltech characterized the chemistry of this smog and identified ozone as the principal oxidant in the early 1950's. Meanwhile, the first air district in the U.S. was created in Los Angeles in 1947, and was later merged with other local districts in 1977 to form the South Coast Air Quality Management District [1].

Negative effects of air pollution have been extensively documented, and include impairment of human lung function, degradation of materials, and injury to plants. In addition to affecting human health, the high ambient ozone levels found in Southern California and the San Joaquin Valley also cause yield reductions up to 30% for some crops [2].

In addition to having some of the nation's most polluted air basins, California also has the nation's most stringent set of state and local air quality standards. Although regulation has led to improvements in air quality [3], exceedances of air quality standards still take place. For example, between 1990 and 1998, the San Joaquin Valley Air Basin experienced an average of 97 days per year above the eight-hour ozone standard, while the Sacramento Valley Air Basin experienced an average of 30 days per year above the standard during the same time period [4].

Farming and livestock operations are significant sources of emissions in California, and bear the negative effects of specific air pollutants as well. Agriculture-related air pollution results from primary emissions from machinery and vehicles employed in production, chemical compounds used in the course of production, e.g. pesticides, as well as emissions from the agricultural systems themselves. For example, agricultural livestock emit nitrogen compounds such as oxides of nitrogen (NO_x) and ammonia. Vehicles used in agricultural production emit volatile organic compounds (VOCs), NO_x and carbon monoxide (CO) [5]. These emissions may lead to the formation of secondary air pollutants, such as ozone, that are deleterious to workers as well as crops [6].

This chapter evaluates whether existing air pollution control policies, particularly those targeted at agriculture, have succeeded in improving air quality, as measured by the number of exceedances of the CO and NO_2 standards. The following air pollution control policies are examined: policies and regulations for agricultural burning, visible emissions, fugitive dust, emission of particulate matter (PM) and PM precursors, emissions of nitrogen compounds, orchard and citrus heaters that release black carbon, and penalty fees.

This chapter builds upon the existing environmental economics literature on air quality, most notably the econometric analysis of the impact of federal particulate matter regulation on infant health conducted by Chay and Greenstone [7] and the study of the impact of air pollution on infant death in California by Currie and Neidell [8], in several ways. First, this paper focuses on the effects of regulation rather than on the effects of air quality. The results therefore have direct implications for policy. Second, the econometric methodology used in this paper exploits the natural variation in policy among the different air districts in California to identify the effects of these policies. Third, this paper examines multiple policies, not just one.

Results from the multivariable regressions point to mixed effects of the air pollution control policies on air quality. Agricultural burning policies and penalty fees reduce the pollution from CO. Other policies such as the prohibition on visible emission, fugitive dust, particulate matter, nitrogen and the reduction of animal matter are correlated with higher levels of CO. Regulations on orchard and citrus heaters have no significant effect on the number of exceedances of the CO and NO_2 standards.

Results of this research will lead to a better understanding of the regulations affecting air quality, and will provide insight into the appropriate development of management practice to mitigate air pollution problems.

The remainder of the chapter proceeds as follows. The next section provides background information on the state of California with respect to air quality management. Section 3 describes the data. Section 4 presents the methods and results from the econometric analysis. Section 5 concludes.

2. California's air pollution policy

The state of California is divided into fifty-eight different counties and is overseen by the Governor of California. California is divided into thirty-five air districts, which are called either Air Pollution Control Districts or Air Quality Management Districts. These air districts are responsible for controlling air pollution from stationary sources. Several air districts span the areas of many counties, while some counties belong to different air districts. Some air districts, such as the Great Basin Unified air district, trace the division lines marking the regional air basin [9].

The different air districts in California have their own set of laws and regulations regarding stationary sources. These laws and regulations are written by the authorities within the air district and are applicable to the entire air district. These laws and regulations must be at least as stringent as the standards set by the federal government [10].

The fact that each air district has its own set of laws and regulations is crucial to the analysis of the policies presented in this paper. The variation between the policies of the different air districts within the state of California provides a setting for a natural experiment. Since the different air districts all fall in the same state, they are subject to the same federal and state laws. The districts also share many other similar characteristics, for example climate, geographical location, etc. The variation in air quality over the time the different policies take place, then, must be due mostly to the policies themselves and not due to geographical location, climate or different state laws. Thus, the variation in similar policies across air districts enables one to better single out the effect of policies on air quality and health.

Among the many different laws and regulations governing each of the thirty-five air districts in California, this paper focuses on the following eight types of policy:

Agricultural burning: This policy regulates open outdoor fires used in agricultural operations in the growing of crops, the raising of animals, the disposal of agribusiness waste, or for purposes such as forest management, range improvement, irrigation system management, etc. The policy requires burning permits and imposes no-burn days.

1. *Visible emissions*: This policy provides limits for visible emissions. In many districts, emissions from agricultural operations are exempt.
2. *Fugitive dust*: The purpose of this policy is to reduce the amount of particulate matter entrained in the air as a result of anthropogenic fugitive dust sources by requiring actions to prevent, reduce or mitigate fugitive dust emissions.
3. *Particulate matter*: This policy imposes limits on particulate matter emissions.
4. *Nitrogen*: This policy imposes limits on NO_x emissions.
5. *Reduction of animal matter*: This policy requires that the gases, vapors and gas-entrained effluents from any article, machine equipment, or other contrivance used for the reduction of animal matter to be incinerated or processed.
6. *Orchard and citrus heaters*: This policy regulates orchard and citrus heaters.
7. *Penalty fee*: Stationary sources with the potential to emit regulated pollutants (including nitrogen oxides, VOCs, CO and PM10) above a certain amount need to obtain permits to operate consistent with the requirements of Title V of the federal Clean Air Act as amended in 1990. This policy requires operators of units requiring Title V permits to pay a penalty if they fail to pay for their permit within a certain number of days after it is issued.

These eight policies were chosen because they were similar in theme across the different counties in the different air districts but they either vary in whether they were implemented at all, or differ in the date of implementation. For example, the prohibition on fugitive dust applies in Amador and Imperial Counties but not in El Dorado and Monterey Counties. Of Amador and Imperial Counties, the dates of implementation of the prohibition policy differ. In Amador County the policy did not take effect until the year 2000, while the policy was implemented in Imperial County prior to 1994 [11].

The similarity of theme in the different policies chosen and the difference in the details of each of the policy add to the quality of the analysis. Since the policies are similar in theme but are different in details, comparisons can be made and the effect of each policy can be examined. For example, once controls are taken into account, differences between air quality in Amador County and El Dorado County can be said to be attributable to the prohibition on fugitive dust which is effective in Amador County but not in El Dorado County. Likewise, the differences in air quality in Amador County and Imperial County can be attributed to the differences in the date of implementation of the prohibition on fugitive dust.

3. Data description

We use annual county-level data from 1980-2000.

The policy variables used in this paper are constructed from the California Air Resources Board's online database of state and county laws and regulations concerning air quality in the different air districts [11]. For each of the eight policies chosen, dummy variables for whether or not the policy is in place for each county for each year since 1980 to 2000 were constructed. Summary statistics for the policy variables are presented in Table 1.

Policy variable	# obs	mean	s.d.
Visible Emissions	1178	0.698	0.459
Fugitive Dust	1178	0.338	0.480
Particulate Matter	1178	0.690	0.463
Nitrogen	1178	0.397	0.490
Reduction of Animal matter	1178	0.518	0.500
Orchard and citrus heaters	1178	0.296	0.457
Agricultural Burning	1537	0.856	0.351
Penalty Fee	510	0.720	0.450

Table 1. Summary statistics for policy variables

To measure air quality, this paper focuses on two agriculture-related air pollutants: CO and nitrogen dioxide (NO_2). For each pollutant, data on the number of days per year exceeding the state standards as compiled by the California Air Resources Board are used. In particular, for CO, the number of days exceeding the state's 1-hour standard for CO of 20 parts per million (ppm) and the number of days exceeding the state's 8-hour standard for CO of 9.0 ppm are used. For NO_2, the number of days exceeding state's 1-hour standard for NO_2 of 0.18 ppm is used. Summary statistics for the air quality variables are in Table 2. On average 0.696 days exceeded the state standard for 1-hour CO, 4.452 days exceeded the state standard for 8-hour CO, and 0.542 days exceeded the state standard for 1-hour NO_2. The number of days exceeding the state standard for all three measures of air pollution has decreased significantly over time.

Air quality variable	# obs	mean	s.d.	min	max	trend
# days exceeding the state standard for:						
CO, 1-hour	1187	0.696	3.283	0	43	-0.999* (0.014)
CO, 8-hour	1187	4.452	14.631	0	133	-0.524* (0.063)
NO_2, 1-hour	1069	0.542	3.273	0	42	-0.089* (0.015)

*Note: The "trend" column gives the coefficients on year of the regression of air quality on time and a constant. The standard errors are given in parentheses under the coefficients. * indicates the coefficient is significant to the 5% significant level.*

Table 2. Summary statistics for air quality variables

The socio-economic data used in this paper are obtained from the U.S. Bureau of Health Professions' Area Resource File. Socio-economic data used in this paper include population, and the percentage of county area occupied by farms, and per capita income. Summary statistics for these variables are presented in Table 3. Both population and per capita income have a significant upward trend over time, while there is no significant trend over time for the percentage of county area occupied by farms.

Variable	Obs	Mean	Std. Dev.	Min	Max	Trend
Population (thousand people)	1102	291.97	971.64	0.011	9519.34	45.78* (5.25)
Percentage of area that are farms	1218	378.3276	259.4809	0	922	0.011 (1.115)
Per capita income	1160	17164.32	6748.864	6807	57982	804.273* (21.071)

Table 3. Summary statistic for socio-economic variables

4. Econometric analysis

In order to identify the effects of the chosen air quality policies on air quality, multivariable regressions that exploit the natural variation in policy among the different air districts in California are used.

It is possible that the correlation between the regulations and air quality reflect some omitted characteristics such as socio-economic status that are correlated with both regulations and air quality. To address the possibility of omitted variables bias, we use a method similar to that used by Currie and Neidell [8]: we include a rich set of covariates and employ county fixed effects to capture any unobserved characteristics of counties that are constant over time. The control variables used are year, population, population density, per capita income, and acres of farmland. The Hausman test was used to determine whether controlling for fixed effects was more appropriate than controlling for random effects.

The natural variation of policy among air districts mitigates the potential endogeneity of regulation, since two neighboring counties that may share similar characteristics and that may have similar levels of pollution prior to the implementation of a particular policy, all else equal, may still belong to different air districts.

The econometric model is:

$$pollution_{it} = \beta_0 + \beta_1 policy_{it} + x_{it}'\beta_2 + u_i + \varepsilon_{it} \quad (1)$$

where $pollution_{it}$ is the value of the pollution variable (number of days exceeding the state's 1-hour standard for CO, the number of days exceeding the state's 8-hour standard for CO, or the number of days exceeding state's 1-hour standard for NO_2) in county i in year t; $policy_{it}$ is a dummy variable denoting whether or not the particular policy under consideration was in place in county i in year t; x_{it} is a vector of controls (year, population, population density, per capita income, and percentage of land area that are farms); and u_i is a county fixed effect.

The results for the effects of policy on air pollution are presented in Table 4. Because the results of the Hausman test favored the fixed effects model for most of the regressions, only the fixed effect results are reported.

Policy variable	*Dependent variable is # days exceeding the state standard for:*		
	CO, 1-hour	CO, 8-hour	NO_2, 1-hour
agricultural burning	-0.777*	-3.013*	-0.788
	(0.226)	(1.123)	(0.306)
p-value (Pr >F)	0.00*	0.00*	0.00*
R^2	0.017	0.014	0.030
# obs	626	626	553
visible emission	0.632*	3.013*	0.138
	(0.238)	(1.178)	(0.351)
p-value (Pr >F)	0.00*	0.00*	0.00*
R^2	0.000	0.003	0.012
# obs	660	660	561
fugitive dust	0.643*	3.725*	0.102
	(0.258)	(1.272)	(0.364)
p-value (Pr >F)	0.00 *	0.00*	0.00*
R^2	0.004	0.004	0.015
# obs	660	660	561
particulate matter	0.632*	3.013*	0.138
	(0.238)	(1.178)	(0.351)
p-value (Pr >F)	0.00*	0.00*	0.00*
R^2	0.001	0.004	0.012
# obs	660	660	561
nitrogen	0.580*	2.424*	0.392
	(0.171)	(0.846)	(0.236)
p-value (Pr >F)	0.00*	0.00*	0.00*
R^2	0.003	0.000	0.015
# obs	660	660	561
reduction of animal matter	0.545*	2.626*	0.020
	(0.234)	(1.158)	(0.334)
p-value (Pr >F)	0.00*	0.00*	0.00*
R^2	0.001	0.000	0.014
# obs	660	660	561
orchard and citrus heaters	0.058	-0.114	0.045
	(0.225)	(1.111)	(0.337)
p-value (Pr >F)	0.00*	0.00*	0.00*
R^2	0.001	0.000	0.013
# obs	660	660	561
penalty fee	-0.221*	-4.495*	0.014
	(0.091)	(0.756)	(0.017)
p-value (Pr >F)	0.00*	0.00*	0.75
R^2	0.043	0.146	0.001
# obs	376	376	378

*Notes: Standard errors are in parentheses. The controls are year, population, population-density, per capita income, percentage of land-area that are farms, and county fixed effects. * indicates the coefficient is significant to the 5% significant level.*

Table 4. The effects of policy on air quality

The results point to mixed effects of the chosen air pollution control policies on air pollutants. An agricultural burning policy significantly reduces the number of days exceeding the state standard for both 1-hour and 8-hour CO. Agricultural burning policies also reduce the number of days exceeding the state standard for NO_2, but the effect is not significant at a 5% level.

The results also suggest that having prohibitions on visible emission, fugitive dust, particulate matter, nitrogen and the reduction of animal matter seem to be significantly correlated with increasing pollution from both 1-hour and 8-hour CO. These policies also increase pollution from NO_2, but the effect is not significant at a 5% level.

The regulations on orchard and citrus heaters has no significant effect on the number of days exceeding the state standard for CO or NO_2.

A penalty fee has a significant negative effect on the number of days exceeding the state standard for both 1-hour and 8-hour CO, but no significant effect on the number of days exceeding the state standard for NO_2.

None of the policies examined had a significant effect on the number of days exceeding the state standard for NO_2.

According to the results, the regulations that were most effective in improving air quality were the regulations on agricultural burning and the penalty fees for noncompliance with the standards.

5. Conclusion

This chapter evaluates whether existing air pollution control policies, particularly those targeted at agriculture, have succeeded in improving air quality, as measured by the number of exceedances of the CO and NO_2 standards. The following air pollution control policies are examined: policies and regulations for agricultural burning, visible emissions, fugitive dust, emission of particulate matter (PM) and PM precursors, emissions of nitrogen compounds, orchard and citrus heaters that release black carbon, and penalty fees.

Results from the multivariable regressions point to mixed effects of air pollution control policies on air quality, as measured by the number of exceedances of the CO and NO_2 standards. Agricultural burning policies and penalty fees reduce the pollution from CO. Other policies such as the prohibition on visible emission, fugitive dust, particulate matter, nitrogen and the reduction of animal matter are correlated with higher levels of CO. Regulations on orchard and citrus heaters have no significant effect on the number of exceedances of the CO and NO_2 standards.

The regulations that were most effective in improving air quality were the regulations on agricultural burning and the penalty fees for noncompliance with the standards.

Results of this research will lead to a better understanding of the regulations affecting air quality, and will provide insight into the appropriate development of management practice to mitigate air pollution problems.

6. Acknowledgment

Kanittha Tambunlertchai provided excellent research assistance. I thank Gary Chamberlain and John Karlik for discussions. I received financial support from the Giannini Foundation for Agricultural Economics, an EPA Science to Achieve Results graduate fellowship, a National Science Foundation graduate research fellowship, and a Repsol YPF - Harvard Kennedy School Pre-Doctoral Fellowship in energy policy. All errors are my own.

7. References

[1] Jacobson, M.Z. (2002). Atmospheric Pollution. Cambridge University Press.
[2] Hall, J.V.; Winer, A.M.; Kleinman, M.T.; Lurmann, F.W.; Brajer, V. & Colome, S.D. (1992). Valuing the health benefits of clean air. *Science*, Vol.255, pp. 812-817.
[3] Lin, C.-Y.C.; Jacob, D.J. & Fiore, A.M. (2001). Trends in exceedances of the ozone air quality standard in the continental United States, 1980-1998. *Atmospheric Environment*, Vol. 35, pp. 3217-3228.
[4] California Air Resources Board [ARB]. (2006). Top 4 summary: Top 4 measurements and days above the standard, September 2006, Available from: http://www.arb.ca.gov/adam/cgi-bin/db2www/adamtop4b.d2w/start)
[5] Canadian Environmental Protection Agency [EPA]. (2004). Environment Canada, July 2004, Available from: http://www.ec.gc.ca/environment_e.html
[6] Winer, A.M.; Olszyk, D.M. & Howitt, R.E. (1990). Air quality impacts on California agriculture, 1990-2010, In: *Agriculture in California: On the Brink of a New Millennium, 1990-2010*, 89-112, University of California Agricultural Issues Center.
[7] Chay, K. & Greenstone, M. (2003). Air quality, infant mortality, and the Clean Air Act of 1970. NBER Working Paper Series (Working Paper No. 10053), Cambridge, MA.
[8] Currie, J. & Neidell, M. (2005). Air pollution and infant health: What can we learn from California's recent experience? *Quarterly Journal of Economics*, Vol.120, No.3, pp. 1003-1030.
[9] California Air Resources Board [ARB]. (2004). Organizations in the California Air Resources Board, July 2004, Available from: http://www.arb.ca.gov/html/org/org.htm
[10] U.S. Environmental Protection Agency [EPA]. (1993). The plain English guide to the Clean Air Act, July 2004, Available from: http://www.epa.gov/oar/oaqps/peg_caa/pegcaain.html
[11] California Air Resources Board [ARB]. (2004). Laws and regulations database, July 2004, Available from: http://www.arb.ca.gov/html/lawsregs.htm

16

Air Pollution Control in Municipal Solid Waste Incinerators

Margarida J. Quina[1], João C.M. Bordado[2] and Rosa M. Quinta-Ferreira[1]
[1]*Research Centre on Chemical Processes Engineering and Forest Products*
Department of Chemical Engineering, University of Coimbra
[2]*Department of Chemical and Biological Engineering, IBB, Instituto Superior Técnico*
Portugal

1. Introduction

Municipal solid waste (MSW) remains a major problem in modern societies, even though the significant efforts to prevent, reduce, reuse and recycle. At present, municipal solid waste incineration (MSWI) in waste-to-energy (WtE) plants is one of the main management options in most of the developed countries. The technology for recovering energy from MSW has evolved over the years and now sophisticated air pollution control (APC) equipment insures that emissions comply with the stringent limits established in developed countries. This chapter shows the role of incineration in WtE processes in the ambit of MSW management, giving an overview of the MSWI technologies and APC devices used for cleaning the gaseous emissions. The main focus is on the key air pollutants, such as dioxins and furans. At the end, the impact of emission on health risks is also briefly considered.

2. Contribution of MSWI in modern solid waste management systems

The waste hierarchy in force in European Union, Directive 2008/98/EC, and in other developed countries sets out the following options for waste management: prevention, re-use, recycling, other recovery (e.g. energy recovery) and disposal. Indeed, nowadays modern systems embrace in general different methodologies aiming as much as possible to achieve sustainable global solutions. Life Cycle Assessment (LCA) tools have been used to assess the potential environmental burdens of different waste management strategies, from environmental, energetic and economic point of view. These calculations have shown that landfilling, even if gas is recovered and leachate is collected and treated, should be avoided, due to the fact that resources in the waste are inefficiently utilised (Sundqvist, 2005). Environmental sound alternatives include incineration, material recycling, anaerobic digestion or composting.

Incineration is a combustion process at high temperature that allows rather complete oxidation of solid wastes, liquids or gases. Combustion systems may be very complex involving simultaneous coupled heat and mass transfer, chemical reaction and fluid flows. A global equation for representing combustion of wastes in air, may take the following form (Jenkins et al., 1998):

$$C_{x1}H_{x2}O_{x3}N_{x4}S_{x5}Cl_{x6}Si_{x7}K_{x8}Ca_{x9}Mg_{x10}Na_{x11}P_{x12}Fe_{x13}Al_{x14}Ti_{x15} + n_1\,H_2O + n_2\,(1+e)(O_2+3.76N_2) \rightarrow n_3\,CO_2 + n_4\,H_2O + n_5\,O_2 + n_6\,N_2 + n_7\,CO + n_8\,CH_4 + n_9\,NO + n_{10}\,NO_2 + n_{11}\,SO_2 + n_{12}\,HCl + n_{13}\,KCl + n_{14}\,K_2SO_4 + n_{15}\,C + \ldots \quad (1)$$

It is important to note that the empirical formula represented in Eq. (1) is incomplete since it includes only 15 elements and a real waste may contain a lot more, some of them found in traces; the molar indices x1 to x15 can vary widely; n_1 corresponds to the moisture in waste; n_2 is related with the amount of air (considered as a binary mixture of O_2 and N_2) used in the combustion; (1+e) is the excess of air in relation to the stoichiometric amount, usually ranges from 1.2 to 2.5 (depending on whether the fuel is gas, liquid or solid) (BREF, 2006); n_3 to n_{15} correspond to the stoichiometric coefficients of the different species that can be found as reaction products, among many others that can be released in the emissions. If the incinerated material is represented by a simpler formula, like $C_uH_vO_wN_xS_y$, then the combustion equation may be simplified and represented by Eq. (2)

$$C_uH_vO_wN_xS_y + (u+v/4-w/2+y)\,O_2 \rightarrow u\,CO_2 + v/2\,H_2O + x/2\,N_2 + y\,SO_2 \quad (2)$$

In the scope of thermal treatments of solid wastes, Fig. 1 shows the difference in terms of pyrolysis, gasification and incineration by taking into account the amount of air present.

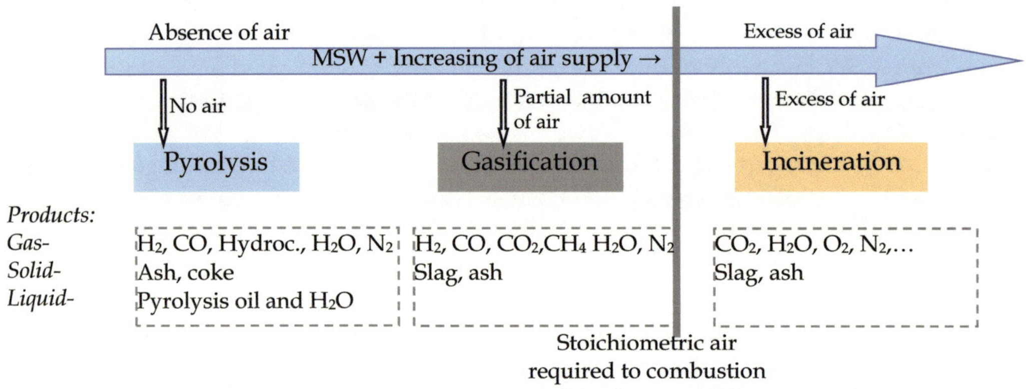

Fig. 1. Classification of the thermal technologies for treating MSW (based on DEFRA, 2007).

These thermal processes correspond to very different technologies in the way of treating waste and energy recovery. In incineration, the energy is released through the oxidation reactions, and its recovery occurs directly from the gases formed.

At present, municipal solid waste incineration (MSWI) in waste-to-energy (WtE) has confirmed to be an environmentally friendly solution and a common alternative to landfilling, while allowing recovery of a large part of the energy contained in MSW. In practice, MSWI has several advantages and disadvantages as reported in Table 1. Nevertheless, the main problems associated to these processes are probably the large volume of gaseous emissions which may pose environmental health risks (Moy et al., 2008) and hazardous solid wastes that remain after incineration as fly ash or air pollution control (APC) residues (Quina et al., 2008a,b).

MSW is generated by households and other similar wastes in nature and composition, which in general is collected and managed by or on behalf of municipal authorities, and includes materials such as paper, plastics, food, glass and household appliances. Fig. 2 shows typical composition of MSW usually associated to these waste streams, based on Gentil et al. (2009), and information reported by environmental agencies from Portugal (APA) and from USA (EPA) for the reference year 2009.

Advantages	Disadvantages
- handle waste without pre-treatment - reduce landfilling demand for MSW - reduce waste volume by 90% - reduce waste weight by 70% - possibility of recovering energy (electricity or heat) - if well managed, low air pollution is released - destroy potential pathogens and toxic organic contaminants - can be located close to the centre of gravity of MSW generation - reduce cost of waste transportation - require minimum land - stack emissions are odour-free - reduce organic materials mainly to CO_2 instead CH_4 and other VOC	- originate hazardous waste (APC residues), that requires safe disposal - originate slags (bottom ashes) - originate huge volume of flue gases - high investment and operating costs - high maintenance costs - require skilled staff - require suitable composition for autocombustion - negative public perception (so far)

Table 1. Advantages and disadvantages of municipal solid waste incineration.

According to Eurostat data for EU-27 State Members, MSW produced in 2008 was on average about 524 kg per capita, but it is possible to find values between 800 kg in Denmark to 300 kg in the Czech Republic (Eurostat 2010). Globally, in 2008, the EU-27 countries produced the huge amount of 259 Mt of MSW, whereas 221 Mt was accounted for in the EU-15. Figs. 3-4 depict the way that MSW stream has been treated in various countries, and in particular Fig. 3 shows the evolution in the EU-27 from 1995 to 2009 taken into account landfill, incineration, composting and recycling. It is important to note that, in 2009, about 20% of waste was incinerated, which correspond to 50.9 Mt. Considering that the average lower calorific value (LCV) should not be less than 7 MJ/kg of waste, in order to occur a chain of reactions able to self-supporting combustion, and assuming that in Europe the LCV

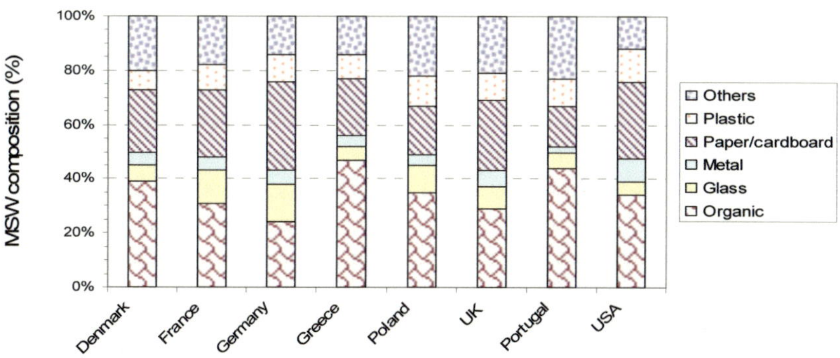

Fig. 2. Composition of MSW (based on Gentil et al., 2009, Portugal APA, US EPA).

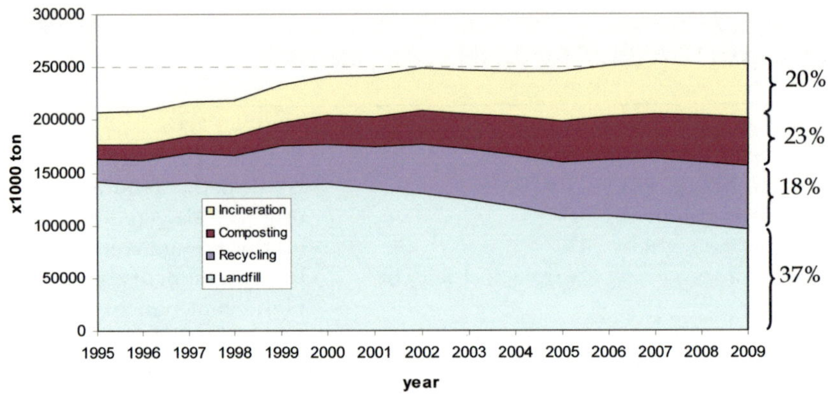

Fig. 3. Contribution of landfill, incineration, composting and recycling in EU-27 State Members (Source: based on Eurostat databases).

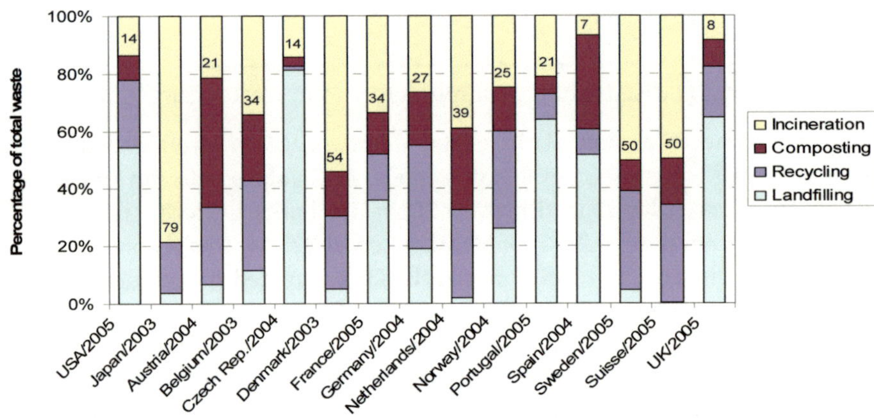

Fig. 4. Contrasting MSW management practice in selected countries, in landfills, incineration, composting and recycling (Source: based on OECD statistic databases).

is in the range of 9-13 MJ/kg (Worl Bank Report, 1999), the combustion of 50.9 Mt led to an enormous amount of energy available for recovery. Fig. 3 points out that landfilling has been gradually decreasing since 1995, and in 2009 its contribution accounts for 37%. According to Fig. 4, Japan is the country where incineration has higher contribution (79%) and in Europe, countries such as Denmark (54%) and Sweden (50%) have the highest rates.

By taking into account the information from BREF (2006) for waste incineration, Table 2 summarizes the number and total capacity of the existing incinerators in 17 European countries.

It is important to note that these numbers may vary according to the source of information used, and the year of reference. According to DEFRA (2007), in 2000, about 291 incineration sites with energy recovery located in 18 Western European countries, processed about 50

Country	Number of MSWI	Capacity Mt/year	Country	Number of MSWI	Capacity Mt/year
Austria	5	0.5	Luxembourg	1	0.15
Belgium	17	2.4	Portugal	3	1.2
Denmark	32	2.7	Spain	9	1.1
Finland	1	0.07	Sweden	30	2.5
France	210	11.7	Netherlands	11	5.3
Germany	59	13.4	UK	17	3.0
Greece	0	0	Norway	11	0.65
Ireland	0	0	Switzerland	29	3.3
Italy	32	1.7			

Table 2. Number and total capacity of the existing incinerators in 17 European countries.

million ton of waste and 50 TWh of energy recovered (40 million ton of oil equivalents). According to Directive 2008/98/CE, a formula is indicated, Eq. (3), to clarify when the incineration of MSW is energy-efficient and may be considered a recovery operation. Indeed, the energy efficiency must be equal or above 0.6 or 0.65 depending on the installation permitted before or after 31 December 2008, respectively.

$$\text{Energy efficiency} = (E_p-(E_f+E_i))/(0.97x(E_w+E_f)) \quad (3)$$

where E_p is the annual energy produced as heat (multiply by 1.1) or electricity (multiply by 2,6), GJ/year, E_f the annual energy input to the system from fuels contributing to the production of steam (GJ/year); E_w the annual energy contained in the treated waste calculated using the net calorific value of the waste (GJ/year); and E_i the annual energy imported excluding E_w and E_f (GJ/year). A corrective factor of 0.97 is introduced to accounting for energy losses due to radiation and bottom ash. It is worthwhile to refer that high efficiency is not easy to reach only through production of electricity. Hot water usage should be considered also, whenever feasible at the location.

3. Municipal solid waste incinerators and air pollution control technologies

Different technologies can be applied to MSW including mass burning with travelling grate, rotary kilns, modular-two stage combustion and fluidised bed (BREF, 2006). In Europe, grate incinerators are used in more than 90% of the installations and in the specific case of fluidised bed, MSW has to be pre-treated. The incineration technology used for MSW has been changing over the last 10 to 15 years, mainly driven by legislation requirements, which has forced low emission limits to air. According to Directive 2000/76/EC, a 'incineration plants' *correspond to any stationary or mobile technical unit dedicated to the thermal treatment of wastes with or without recovery of the combustion heat generated. This includes the incineration by oxidation of waste as well as other thermal treatment processes such as pyrolysis or gasification in so far as the substances resulting from the treatment are subsequently incinerated.* This description comprises the site and the entire incineration plant including:

- waste reception and handling (storage, on site pre-treatment facilities),
- combustion chamber (waste-fuel and air-supply systems),
- energy recovery (boiler, economizer, etc.),
- facilities for clean-up gaseous emissions,

- on-site facilities for treatment or storage of residues and waste water, stack,
- devices and systems for controlling incineration operations, recording and monitoring incineration conditions.

These areas may be distributed as indicated in Fig. 5, which represents a scheme of a typical mass burning MSW incinerator (IAWG, 1997).

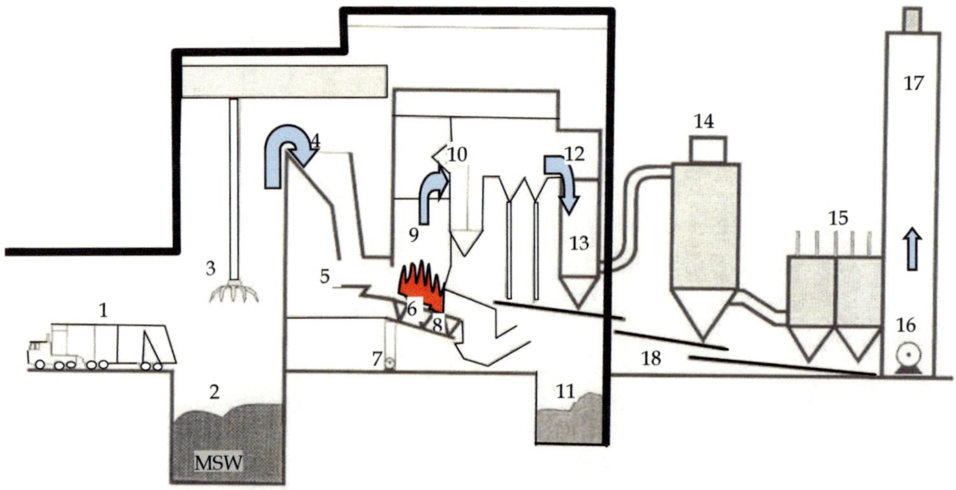

1- waste collection vehicle	7- forced-draft fan	13- economiser
2- waste storage pit	8- undergrate air zone	14- dry scrubber
3- waste handle crane	9- furnace	15- fabric filter baghouse
4- feed hopper	10- boiler	16- induced-draft fan
5- feeder	11- bottom ash bunker	17- stack
6- grate	12- superheater	18- APC residues conveyor

Fig. 5. Simplified scheme of a MSW incinerator (adapted from IAWG, 1997).

Considering the diagram of Fig. 5, a brief description of the mass flow into the incinerator is given below. MSW is in general delivered in trucks (1) and discharged into the storage pit "as-received" (2), in enough amounts for providing a continuous feeding material to the WtE plant. Then, waste is randomly picked up through a handle crane (3), and dropped into the feed hopper (4). The waste flows through the feeder (5) onto the moving grate (6) where combustion takes place. The plant should be controlled in order to optimize the combustion conditions, to ensure, as much as possible, complete carbon burn-out, and for this the residence time on the grate is usually no more than 60 min. The forced-draft fan (7) forces primary air through undergrate air zone (8) into the furnace (9), in order to supply oxygen to promote oxidation reactions, e.g. Eq (1). The primary air is in general taken from the storage pit (2) to lower the air pressure and eliminate most odour emissions from the storage area. Although it is not represented in Fig. 5, a secondary air supply system is common in the furnace, to guarantee turbulence of flue gases (secondary-air) and to ensure complete combustion. About 10-20% (v/v) of flue-gas is recirculated as secondary air. The reactions involved in this process are exothermic and release a high amount of energy that is carried over by the flue gases as heat. Indeed, for example, the upper calorific values of MSW in Germany are usually in the range of 7-15 MJ/kg (BREF, 2006). Energy recovery occurs mostly in boiler (10), superheater (12) and economizer (13). The burned-out bottom ashes are normally quenched and transported to a storage bunker (11). In most of the incinerators, the bottom ashes are transported on conveyors and ferrous metals sorted, and thus at the same time metals recycling and improvement of the slag properties take place. Slag is partly vitrified and can be handled as non-hazardous or special waste in many countries. The huge amount of gases produced during combustion contains air pollutants harmful for the environment that must comply with the stringent regulatory limits. Thus, depending on the desired cleaning degree, different air pollution control (APC) systems may be used. As an example, in Fig. 5, a dry scrubber (14) and fabric filters (15) are used. In these units, APC residues are produced and further transported through a conveyor (18) for a silo (not represented). Most of the modern incinerators treat APC residues before disposed of in monofills. Finally, by using induced-draft fan (16), the cleaned flue gas is released via the stack. Concerning air pollution, it is extremely important to note that combustion includes very fast reactions (fractions of seconds) that take place in gas phase, and self-supporting combustion is possible if heat value of the waste and oxygen concentration is sufficient. Thus, grate length should ensure the phases indicated in Fig. 6.

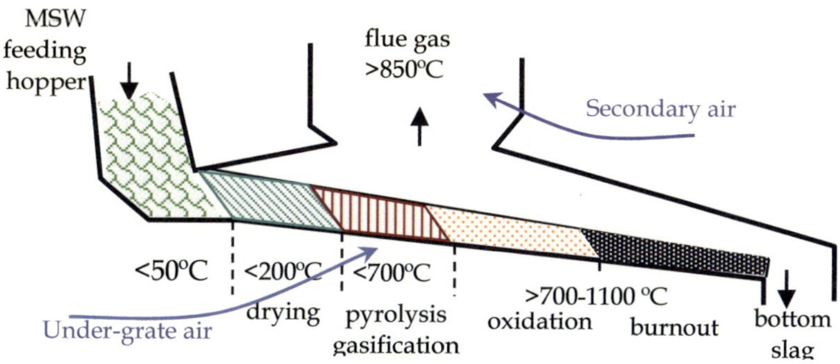

Fig. 6. Phases and temperatures on the grate of a MSWI (adapted from IAWG, 1997).

In practice, water and volatile content of MSW are evolved in the initial phase of drying at temperatures until the range of 200ºC, where no oxygen is required. The next phase corresponds to pyrolysis and gasification of the organic materials, in which organic compounds are transferred to the gaseous phase. After that, in the oxidation, the combustible gases react with oxygen releasing heat and other lower weight molecules. In fully oxidative conditions, reactions are almost complete and the major gases are water, nitrogen, carbon dioxide and oxygen. It is very important to note that those phases overlap in spatial and temporal sense. Nonetheless, some in-furnace technical measures (e.g. air distribution and furnace design) can be taken to influence those stages in order to reduce pollutants in gaseous emissions (BREF, 2006). European legislation imposes a minimum gas phase combustion temperature of 850ºC and at least 2 s for the residence time.

In general, MSW incineration plants operate 24 h/day and close to 365 days/year. Availability of the plant is very often over 98% and this imposes several redundancies of equipment and on-operation maintenance procedures. Fig.7 summarizes the main inputs and outputs of MSW incineration process, where it is indicated that 1 t of waste originates nearly 300 kg of bottom ashes, 30 kg of APC residues, and the rest is emitted as flue gas. The air flow rate is a control variable that is adjusted depending on the characteristics of the stack gases. However, in general about 4000 to 4500 m^3/t of air is required to guarantee fully oxidizing atmosphere (IAWG, 1997). The flue gas volume originated in MSWI is dependent on technology, in particular concerning the existence of flue gas recirculation. However, in literature there are indications that these values may be in the range of 4500 to 6000 Nm^3/ton of waste (Achternbosch and Richers, 2002; BREF, 2006). Even though large local variations can be observed, typically 1 ton of MSW allows energy recovery in the range of 400-700 kWh of electricity and additional 1205 kWh as heat (BREF, 2006). In general, liquid effluents can occur from: APC devices mainly if wet systems are used, the collection and storage of bottom ash, the water/steam cycle, sanitary areas, rainwater, and cooling water. However, if re-circulation is maximized reduced amounts of wastewater are produced. Chemical reactants used in incinerator plants may be ammonium hydroxide (NH_4OH) or ammonia as gas (NH_3) for reducing NO_x in furnace, neutralizing agents (e.g. $Ca(OH)_2$) and adsorption materials (e.g. activated carbon). Typical consumptions rates per ton are 0.8 kg of NH_4OH, 8 kg of $Ca(OH)_2$ and 0.5 kg of activated carbon.

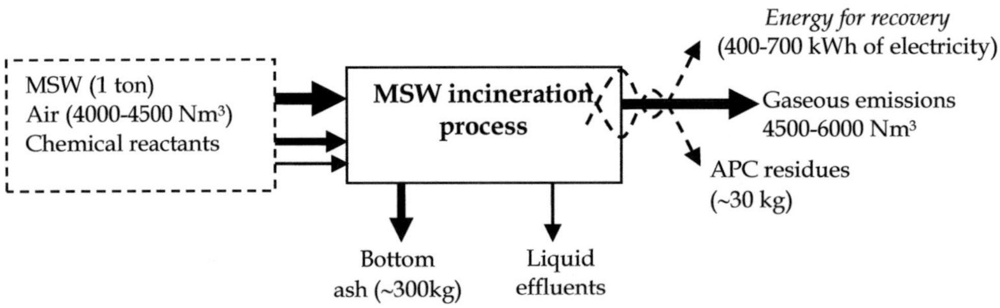

Fig. 7. Main inputs and outputs of MSW incinerators.

3.1 Gas cleaning systems for waste incinerators

The gas mixture leaving the incineration furnace has three main types of components that have to be removed to the possible extent before the exit at the stack:

- Fly ash which is composed by particles pneumatically transported by the gaseous flow;
- Acids and acids precursors, such as sulphur dioxide, nitrogen oxides, hydrochloric acid;
- Dioxins and analogues that are compounds formed by radical recombination with structures such as polychloro dibenzodioxins and the respective furan analogues.

The hot gas mixture leaving the furnace, exchanges heat at the surface of the heat exchanger vertical tubes, inside which the high or medium pressure steam is generated and before entering the cleaning systems, part of this gas is diverted through a booster to be injected in the cameras below the moving grate of the furnace. This gas recycling is essential in terms of overall energy recovery and also very important to promote one easier control of the stoichiometric excess of oxygen in the furnace. The remaining effluent gas mixture has to be cleaned by several unit operations in the gas cleaning systems.

3.2 The main targets

During the heat exchange at the surface of the boiler tubes, the effluent gas is cooled and some additional solid formation occurs, increasing the particulate fraction. This raw gas effluent of the boiler system has to be cleaned up to the environmental standards applied at the location where the incinerator is installed. The main ranges for the concentrations in the raw gas and the mandatory final emission limits most commonly used are included in

Table 3 together with the calculated efficiency required for the cleaning of each type of polluting agent. This high removal efficiency required, indicated in the Table 3, imposes the use of multiple systems for the gas cleaning, usually in sequence.

Pollutant	Concentration in raw gas from boiler (mg/Nm^3)	Max admissible at exhaust (mg/Nm^3)	Removal efficiency required (%)
Fly Ash	1500 - 2000	10	99,9
HCl	300 - 2000	10	> 99
SO_2	200 - 1000	5	99.5
NO_x	200 - 500	70	86%
HF	2 - 25	1	96%
Hg	0.2 – 0.8	0.01	99%
Cd + other metals	2 -15	0.05	>99.5
Dioxins (ng I-TEQ/Nm^3)	0.5 – 5	0.1	98%

Table 3. Required efficiency for flue gas cleaning systems.

3.3 Unit operations for gas cleaning

A large number of unit operations based on primary separation processes can be used for the gas cleaning of the flue gas generated in waste incineration systems. In Table 4 for each type of flue gas pollutant, a combination of unit operations is indicated with the respective typical range of reduction. The well designed sequence of gas cleaning methods allows for a drastic reduction of pollutants as stated by the waste incineration BREF Table 5 (adopted with comments from Table 5.2 of the BREF, 2006).

Pollutant	Process Steps	Reduction (%)
SO_x	Wet scrubber or dry multicyclone	50 - 90
HCl	Wet scrubber or semi-dry	75 - 95
NO_x	Selective catalytic reduction	10 - 60
Heavy metals	Dry scrubber + electrostatic precipitator	70 - 95
Fly ash *	Electrostatic precipitator + fabric hose filter	95 - 99.9
Dioxins & Furans	Activated carbon + fabric hose filter	50 - 99.9

*Very often the fly ash surface has adsorbed other pollutants such as dioxins and heavy metals

Table 4. Gas cleaning processes and typical range of specific pollutant reduction by combination of unit operations.

3.4 Separation of fly ash and activated carbon

Fly ash generated at power plants where the composition of the fuel is reasonably constant, is very often collected and used as raw material for the production of Portland cement. Fly ash generated at waste incinerators is usually contaminated with heavy metals and other dangerous substances and have to be treated as a hazardous residue, requiring inertization before the disposal is controlled landfill.

Substance	Non-Continuous samples	Half-hour mean value	Daily mean value	Notes
Dust		1-20	1-5	Lower levels achieved with fabric filters as hose bags filters
HCl		1-50	1-8	Use of wet processes preferred
SO_2		1-150	1-40	Use of wet processes preferred
NOx with SCR		40-300	40-100	Additional energy demand and costs
NOx with SCR		30-250	120-180	At high raw gas NOx levels, NH_3 slip to be taken into account, preferred method in conjunction with wet processes
TOC		1-20	1-10	Optimum combustion conditions are required
CO		5-100	5-30	Optimum combustion conditions
Hg	<0.05	0.001-0.03	0.001-0.02	Input control, carbon-based adsorption processes
PCDD/PCDF (ng ITEQ/Nm^3)	0.01-0.1			Optimum combustion conditions, temperatures controls to reduce synthesis, carbon-based adsorption processes.

Table 5. Attainable emission levels for waste incineration facilities (excerpt from Table 5.2 of BREF (2006) "Operational emission levels associated with the use of BAT" for air pollutants expressed in mg/Nm^3).

The activated carbon, in powder form, is very often used to adsorb organic pollutants such as dioxins and furans, as it will be detailed later on, this powder is collected together with the fly ash (small systems) in a specific stage of cleaning. The main type of equipments used for the removal of solid particles of the fly ash and activated carbon are: cyclones, electrostatic precipitators, and fabric hose filters, Table 6. Construction details, as well as operating modes have been described extensively (Niessen, 2002) and therefore it was considered more relevant to focus the detail here on the range of applicability for the fly ash separation.

3.4.1 Cyclones

Cyclones are rather efficient for the removal of solid particles with average diameter over 100 µm from gaseous flows. The design of cyclones was reviewed elsewhere (Morcos, 1989; Lee and Huffman, 1996; Amutha Rani et al., 2008) and since for their construction steel or stainless steel can be used the range of permissible operating temperatures is rather wide. In gas cleaning of flue gas from incinerators cyclones are very often used as primary separators followed by other separating units designed for the retention of particles of the lower size present in the fly ash.

3.4.2 Electrostatic precipitators (ESP)

Nearly one century ago, Dr. Frederick Cottrell introduced the concept of particle separation by the application of an electric field. Overall efficiency of ESP depends mainly on the plates and rapper design, and collection zones have to be carefully designed to assure the adequate thickness of laminar boundary layer, in order to prevent reentrainment into the gas stream the aggregate formed with the collected particles. Typical gas velocities inside the precipitation zone are always below 1 m/s and very often below 0.5 m/s. Electrostatic precipitation, as many other process equipment items in an incineration plant have to be adjusted to steady state conditions but with the control system prepared to modulate under unbalance to peak conditions. Dynamic control is often required and the fine control of efficiency is of utmost importance (Bordado and Gomes, 1999).

Equipment	Typical efficiency for fly ash	Typical efficiency for AC	Typical pressure drop range	Maximum operating temperature	Range of particles sizes
Cyclones	up to 80%	up to 50%	10 to 1000 Pa	1300 °C	≥ 20 μm
ESP	up to 99%	up to 80%	50 to 300 Pa	450 °C	0.08 to 20 μm
Fabric hose filters	up to 99%	up to 99%	500 to 2000 Pa usually with a booster fan	240 °C	0.04 to 50 μm

Table 6. Fly ash and activated carbon (AC) separation.

3.5 Separation of acids

The acids present in the flue gas, such as HCl and HF and the precursor of acid SO_2 can be separated by different processes: dry process (with the use of a solid adsorbent); semi dry (with the use of a spray absorber); wet process (with aqueous solutions).

Criteria	Wet FGT (W)	Semi-Wet FGT (SW)	Dry line FGT (DL)	Dry Sodium bicarbonate FGT (DS)
Air emissions performance	+	0	-	0
Residue production	+	0	-	0
Water consumption	-	0	+	+
Effluent production	-	+	+	+
Reagent consumption	+	0	-	0
Ability to cope with inlet variations of pollutant	+	0	-	0
Plume visibility	-	0	+	+
Process complexity	-	0	+	+
Costs - capital	generally higher	medium	generally lower	generally lower
Costs operational	medium	generally lower	medium	generally lower

Note

+ : means that the use of the technique generally offers an advantage in respect of the assessment criteria considered

0 : the technique offers no significant advantage or disadvantage in respect of the assessment criteria considered

- : means that the use of the technique generally offers a disadvantage in respect of the assessment criteria considered

Table 7. An example assessment of some IPPC relevant criteria that may be taken into account when selecting between wet/semi-wet/dry flue gas treatment (FGT) options.

As solid adsorbents, the dry lime (calcium hydroxide) is more commonly used but dry sodium bicarbonate can also be an option. A comparison of the specific processes for acids separation is qualitatively presented in Table 7. Each one of these alternatives has nowadays several variants and refinements but it was considered out of the scope of this review the fine detail of the multiple available alternatives.

3.6 Nitrogen oxides reduction

Nitrogen oxides content in flue gas are usually reduced by two reactive processes: SNCR (selective non-catalytic reduction) and SCR (selective catalytic reduction). Other chemical process focused on the reduction of NO_x formation during the combustion have been extensively studied at laboratory and pilot plant level, and may are based on radical quenchers that minimize N_2 oxidation by radical reactions. However most, if not all those processes, have a large negative impact on combustion efficiency and consequently they cause an increase content of VOC and namely PAH in the flue gas.

The SNCR process uses ammonia (NH_3) as reactant or for smaller systems urea ($CO(NH_2)_2$) as reducing agents, with direct injection into the furnace. At high temperature, urea decomposes with ammonia formation, Eq.(4), and the main overall reduction reactions take place between 850 °C and 1050 °C according with the schemes Eq.(5)-(6):

$$CO(NH_2)_2 + H_2O \rightarrow 2\,NH_3 + CO_2 \qquad (4)$$

$$4\,NO + 4\,NH_3 + O_2 \rightarrow 4\,N_2 + 6\,H_2O \qquad (5)$$

$$6\,NO_2 + 8\,NH_3 \rightarrow 7\,N_2 + 12\,H_2O \qquad (6)$$

The SCR process (Jungtten et al., 1988) uses a solid catalyst usually in a fixed bed and operates between 200 as 400 °C, in presence of ammonia. This range of operating temperature and the fact that catalysts deactivate in presence of strong acid media, imposes the SCR module to be installed after particulate material, as well as acidic components are already removed from flue gas. The more representative overall reactions taking place at the surface of the catalyst are indicated by Eq. (7)-(9). A comparison of the main features SNCR and SCR processes is presented in Table 8.

$$4\,NO + 4\,NH_3 + O_2 \rightarrow 4\,N_2 + 6\,H_2O \qquad (7)$$

$$6\,NO + 4\,NH_3 \rightarrow 5\,N_2 + 6\,H_2O \qquad (8)$$

$$2\,NO_2 + 8\,NH_3 + O_2 \rightarrow 3\,N_2 + 6\,H_2O \qquad (9)$$

3.7 Carbon monoxide and volatile organic compounds

The increase of carbon monoxide (CO) and/or volatile organic compounds (VOC) content in the flue gas is a strong indication of inappropriate burning conditions in the furnace. Several adjustments by the control systems can be adopted, but the more common are:

i. Increase of raw air inlet to the furnace;
ii. Reduction of flue gas recycling to the furnace;
iii. Slight increase of pressure below the grid.

Both i) and ii) will increase the stoichiometric excess of air in the furnace, allowing for a more complete oxidation of the wastes, and of the respective volatiles, such as thermal

	SNCR	SCR
Advantage	Lower investment cost Lower corrosion problems	More efficient
Disadvantage	Limited efficiency	Higher investment cost Higher pressure drop Requires higher O_2 excess
Efficiency for NO_x reduction	Up to 70% Typical 30 to 60%	Up to 85% Typical 50 to 80%

Table 8. SNCR versus SCR processes for NO_x reduction.

degradation products. For wastes with a rather high content of halogens or phosphorus (for example in presence of increased content of flame retardants and/or PVC), the adoption of i) and specially ii) will be very effective, but increase of stoichiometric oxygen over 10% usually results in an overall energy recovery efficiency reduction due to energy consumption to heat-up the unavoidable N_2 inlet. The adoption of iii) is especially effective for wastes with high moisture content, resulting in a more effective lifting of the waste material from the moving grate. It is important to refer that measure iii) can only be adopted with the so called high pressure drop grates in which the grate perforations and gaps are rather small (for example VON ROLL type grates).

Separation of carbon monoxide and volatile organic compounds present in the flue gas is not an easy task. One of the very few specific processes, worth to refer, is the catalytic oxidation in heterogeneous converter. In general terms, additional O_2 is injected in the flue gas stream and this mixture is further oxidized over a fixed bed of a strong oxidation catalyst such as platinum non woven felts.

For rather small waste incinerators, and when the waste has one high content of PVC disposables (e.g. hospital waste) this oxidation step is included in the gas cleaning system, usually close to the boiler since the oxidation is more effective at high flue gas temperatures. For larger systems such as ones used for mass burning of municipal wastes this systems is very seldom included in the design due to the very high specific investment cost and also due to the additional loss of energy due to boosting requirements to compensate the pressure drop.

3.8 Moisture condensation in gas cleaning systems

Moisture condensation is one of the major problems that can occur in gas cleaning systems. The occurrence of condensation causes drastic corrosion problems as well as fly ash aggregation as a mud or "paste" and induces the malfunction of several gas cleaning steps mainly the hose fabric filters. Both in the design as well as in the operation of incineration systems, to avoid moisture condensation is mandatory and in case of continuous occurrence the life span of the overall plant will be reduced as well as the average availability due to more frequent cleaning and maintenance requirements.

Incidence on operating costs is therefore also significant, and although several mitigating measures can be adopted from the control point of view, the best may be of course to prevent its occurrence by a careful design of the plant. Not only the overall water balance has to be considered for steady and dynamic conditions but careful estimation of the dew point of the flue gas in different critical points must be performed.

The main contributions for the total water present in the flue gas come from:

i. Moisture in the inlet air;

ii. Moisture vaporized from the wastes;
iii. Water formed in the combustion reactions;
iv. Water vaporized in the flue gas stream in the wet or semi-wet cleaning steps;

Rather high values of i) and ii) are expected to occur simultaneously in rainy days or if snow accumulates on the waste at the collection points, and therefore active prevention of condensation is of utmost importance at the design stage.

The flue gas temperature, from the outlet of the boiler until the stack, is becoming lower, partly due to energy recovering systems, and therefore the probability of the occurrence of condensation becomes higher toward the last stages of the gas cleaning system.

Re-heating of the flue gas, to avoid condensation can be achieved in practice by different processes. The more common ones are:

i. Injection of limited flow of hot flue gas boosted stream in the main stream: this is very effective system of re-heating, but it has the disadvantage that hot flue gas from the boiler is more contaminated, and the overall cleaning efficiency is slightly reduced;
ii. By heat exchanging with hot flue gas: in this case the two streams are kept separated and the only drawback is the increased pressure drop thought the heat exchanger and the higher investment cost.

In both cases the heat-up to, say 2 to 5 degree centigrade over the higher dew point temperature that can occur is usually effective to prevent condensation since the pressure profile is, as a rule, very stable. To prevent condensation, the re-heating is very often considered in the design at least in two points of the cleaning system:

i. Immediately before the exhaust stack to avoid the formation of "plume" caused by condensation during cooling by the cold air;
ii. Before the hose fabric filters to avoid clogging by mud formation in the inside wall of the filter.

The use of corrosion resistant materials and special coatings where condensation is more prone to occur, is indeed a good design practice, but to prevent the occurrence is certainly also of major importance.

3.9 Emerging technologies for gas cleaning

3.9.1 Oil and emulsion scrubbing

Dioxins and polyaromatic hydrocarbons (PAH) have a very limited solubility in aqueous solutions and therefore their removal by aqueous wet scrubbers is rather limited. A small depletion is observed due to condensation in the rather cold water solution, as well as in the sludge of solid particles at which surface they remain adsorbed, but due to their marked lipophilic character they tend to float and eventually be removed from the scrubbing solution by the gas flow.

High boiling oils, partly unsaturated proved to be an efficient scrubbing media, as well as their respective oil-in-water emulsions stabilized by non-ionic surfactants with HLB between 7 and 10 (Encyclopedia of emulsion technology, 1988). The oil emulsion retains mostly dioxins, furans and PAH and it is considered a good practice to dispose the emulsion as soon as total content of those pollutants reaches 0.1 mg/L. The design of the scrubbing systems considers a buffer volume of emulsion that allows for the emulsion exchange up to 4 times per year in the worst operating conditions. Emulsion preparation vessel and transfer pump is therefore to be included in the auxiliary equipment.

Disposal of the contaminated emulsion includes an emulsion breaker, liquid-liquid separation and absorption of the oil phase in wood particles or other lipophilic absorber, such as open cell polyurethane (Clark et al., 2008) or phenolic foam, and the feeding of the solid material for the furnace for further energy recovery.
A separate burner for the oil into the furnace can also be installed, but operators of waste incinerators do not like to feed liquids or oils to the furnace for a large number of operational problems that can occur.

3.9.2 Dedusting and particle aggregation

Although very efficient for over 99% of the particles the electrostatic precipitators and the fabric hose filters are not effective for nanoparticles as well as for the smaller sub-micron particles. Flue gas exhaust at the stack, although very clean has however a small content of nanoparticles. Recent concerns on the physiological effect of respirable nanoparticles induced research on their study and on methods of avoiding their presence in the exhaust flue gas. Improving particle aggregation and deducting seems the more promising alternative and intense research is being conducted in that direction. So far it was proved that very fine dust and nanoparticles can be effectively aggregated by water dispersible polymers with controlled water compatibility (Nunes et al., 2011).

4. Main pollutants released during MSWI and the European legislation

The main direct impacts caused by incineration are emissions to air, residues production (bottom ash and APC residues), emissions to water, energy production/consumption, material consumption, noise and vibration, fugitive emissions (including odour mainly from waste storage), storage/handling/processing risks of wastes. Among these, emissions to air are really significant, since the incineration of 1 ton of MSW produces a huge volume of flue-gas. Some measurements revealed that 4500-6000 Nm^3/ton (at 11% O_2) are usually emitted. Depending on the technology, operating conditions and the composition of waste incinerated, diverse pollutants are formed and emitted in flue gas:

- particulate matter - with a broad distribution size;
- acids and other gases- HCl, HF, HBr, HI, SO_2, NO_X, NH_3, ...;
- heavy metals - Hg, Cd, Tl, As, Ni, Pb, Sb, Se, Sn, Zn, ...;
- carbon compounds - CO, VOC, PCDD/PCDF, PCB, PAH,...;

In this scope, measures for reducing pollutant precursors in wastes, such as chlorine and volatile heavy metals, may have positive effects. In practical terms, redesign products and packages could be an important aspect for reducing emissions. With respect to heavy metals, it is important to avoid batteries, leather, pigments, cans, and for reducing chlorine, wastes such as plastics (PVC) must be avoided.
Since the early 1970s Europe has been strongly committed to the environment protection of air. In this ambit, it should be mentioned the Directive 2000/76/EC of the European Parliament and of the Council, on the incineration of waste that aims "*to prevent or to limit as far as practicable negative effects on the environment, in particular pollution by emissions into air, soil, surface water and groundwater, and the resulting risks to human health, from the incineration and co-incineration of waste.*" This Directive, known as Waste Incineration Directive (WID), states that continuous measurements of NO_x, CO, total dust, TOC, HCl, HF and SO_2 should be carried out, and at least twice a year for heavy metals (Cd, Tl, Hg, Sb, As, Pb, Cr, Co, Cu, Mn, Ni, V) and dioxins and furans. In fact, WID imposes very restrictive limit values for the

main air emission pollutants. For example, the emission limit value allowed to the total concentration of dioxins and furans calculated using the concept of toxic equivalence is 0.1 ng I-TEQ/Nm³. Table 9 shows the air legal limits values indicated in WID, as well as the usual ranges that can be observed before treatment of flue-gas (crude flue-gas).

	Before treatment (BREF, 2006)	Ref.1	After treatment (BREF, 2006)	Legal limits*
Dust (mg/Nm³)	1000-5000	2000-10000	0.1-4	10
CO (mg/Nm³)	5-50		2-45	50
TOC (mg/Nm³)	1-10		0.1-5	10
HCl (mg/Nm³)	500-2000	400-1500	0.1-6	10
HF (mg/Nm³)	5-20	2-20	0.01-0.1	1
SO_2 (mg/Nm³)	200-1000	200-800	0.2-20	50
NO+NO_2 (mg/Nm³)	250-500	200-400	20-180	200 #
NO (mg/Nm³)	<40			not defined
Hg (mg/Nm³)	0.05-0.5	0.3-0.8	0.0002-0.05	0.05
Cd,Tl (mg/Nm³)	<3	3-12	0.0002-0.03	0.05
Other heavy metals (Sb, As, Pb, Cr, Co, Cu, Mn, Ni, V)	<50		0.0002-0.05	0.5
Dioxins and Furans (ng I-TEQ/Nm³)	0.05-10	<1-5	0.0002-0.08	0.1
CO_2	5-10%			not defined
H_2O	10-20%			not defined

Ref.1 - Achternbosch and Richers (2002), *set on Directive 2000/76/EC, #for nominal capacity ≥ 6 ton/h or new incineration plants.

Table 9. Concentrations of gaseous emissions before and after treatment and legal limits.

To prove compliance with the emission values set in WID, measured concentrations, C_M, shall be standardised (indexed with S), C_S, at the following conditions: 273 K, 101.3 kPa, 11 % O_2, and dry gas (M_S=0) by using the following equation:

$$C_S = C_M \frac{21 - O_{2,S}}{21 - O_{2,M}} \frac{100 - M_S}{100 - M_M} \frac{T_M}{T_S} \frac{P_S}{P_M} \tag{10}$$

where oxygen concentration ($O_{s,M}$), moisture (M_M), temperature (T_M) and pressure (P_M) represent the measured conditions.

The emissions of pollutants such as HCl, HF, SO_2, NO_x and heavy metals depend mostly on the waste characteristics and air pollution control devices. Furnace technology determine to a great extent CO, VOC and NO_x. The total dust emitted is determined upon APC devices. Fig. 8 shows a mass balance in respect to some pollutants per 1 ton of MSW incinerated.

4.1 Dust

During the incineration process, part of the waste is incombustible and removed from the incinerator as solid wastes, usually referred as bottom ash or slag, APC residues and very low quantity of residual particles in the flue gas. The matrix compounds of these particles are mainly aluminium and silicon oxides. Concerning air pollution from MSWI, the most import fraction is the finest inorganic particles that remain after the combustion of the

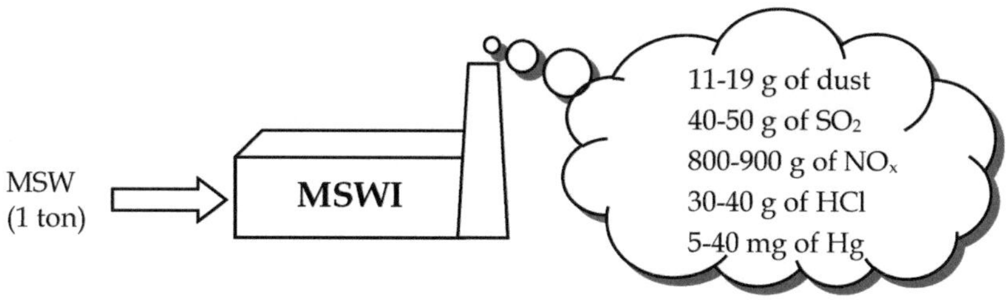

Fig. 8. Mass balance in respect to some pollutants in stack gases.

waste and pass with the flue gas out of the furnace. Some of these particles settle as boiler ash and are removed in the hoppers. A very important aspect concerning air pollution is the fact that as flue gas cools diverse gaseous compounds (some pollutants) condense on the particle surfaces. Since high concentration of HCl is present in the boiler, the condensate phases may include diverse metals chlorides such as $ZnCl_2$, $PbCl_2$ and $CdCl_2$ (World Bank Report, 1999).

Particulate matter is often grouped into categories based on their diameters (e.g. $PM_{2.5}$ with a diameter less than 2.5 μm). For example, $PM_{2.5}$ may persist in the atmosphere for long periods, travelling long distances and they are capable to penetrate deep into the respiratory tract. Their toxicity is normally correlated with the respective chemical composition.

Air pollution control (APC) residues produced in incineration plants are classified as hazardous solid waste mainly due to its high alkaline pH, high content in volatile heavy metals and soluble salts, and trace amounts of very toxic compounds such as dioxins and furans (Quina et al 2008b). The current APC disposal methods in use are very limited, and in the near future they are likely to become unsustainable, due to stringent environmental regulations. Development of new and safer inertization processes will be a must in the future. When APC residues are formed in devices with $Ca(OH)_2$ injection, reaction products such as CaOHCl, CaF_2, $CaSO_3$ or $CaSO_4$ may be present in considerable concentration, as result of HCl, HF and SO_2 gases neutralization. Often activated carbon is also added which also ends up in APC residues.

4.2 Carbon monoxide

Carbon monoxide (CO) is a gas that can be found in the flue-gas as a product of the incomplete combustion of organic compounds that is measured on-line and usually used to check the incineration efficiency. In MSWI it is usual to assume that when CO concentration is low in flue-gas then the gas burn out has high quality, and the TOC emissions are low in those conditions. CO arises in combustion chambers when there is not enough O_2 for full oxidation or if temperature is not high enough to guarantee full reaction to CO_2. Carbon monoxide oxidation to carbon dioxide occurs after some time of its release to the atmosphere (BREF, 2006).

4.3 Volatile organic compounds

Volatile organic compounds (VOC) are products of incomplete combustion and may cover a wide range of compounds, since they can include carbon chains or rings that have high vapour pressure (e.g. 0.27 kPa at 250 °C). When combustion operates at high efficiency, very

low concentrations of organic compounds are expected into air emissions. Indeed, in general, flue-gas from MSWI only contains trace quantities of volatile organic compounds (VOC) that are quantified on-line in the flue-gas as total organic carbon (TOC). Commonly speciation of VOC may include compounds such as methane, ethane, propane, butane, pentane, hexane, heptane, ethylene, benzene, toluene and xylene (BTX), ethylbenzene, acetylene, formaldehyde, acetone, etc. Methane (CH_4) is not formed as long as oxidative conditions are observed in WtE process. In fact, CH_4 may only arise in the waste bunker, due to anaerobic processes during long periods of storage.
Beyond the organic compounds referred above, traces of polycyclic aromatic hydrocarbons (PAH), polychlorinates biphenyls (PCB) and dioxins (PCDD/PCDF - discussed in section 4.10) may occur in flue-gas, mostly adsorbed into particles.

4.4 Acid chloride and fluoride

Chlorine and fluorine may be present in MSW in significant quantities. The main sources of chlorine are plastics such as PVC, salty food waste and other inorganic chlorides. Fluorine can arise into plastics such as PTFE, fluorinated textiles and other inorganic fluorides. Under the furnace conditions chlorine and fluorine are mostly converted into acid hydrogen halides, HCl and HF, and part of these may react to form metal chlorides.
In general, HCl and HF are removed by the same technology, by adding neutralization agents, for instance lime ($Ca(OH)_2$), sodium hydroxide (NaOH), or calcium carbonate ($CaCO_3$). The following reactions show the separation of acid pollutants with $Ca(OH)_2$

$$2\,HCl + Ca(OH)_2 \rightarrow CaCl_2 + 2\,H_2O \qquad (11)$$

$$2\,HF + Ca(OH)_2 \rightarrow CaF_2 + 2\,H_2O \qquad (12)$$

These reactions may entail complex mechanisms due to various gas/solid phases involved, and some studies refer that instead $CaCl_2$ the main phase formed is CaOH (Quina et. al. 2008b). Gaseous emissions concentrations are in general in the range of 0.1 -6 mg/Nm^3 for HCl and 0.01 -0.1 mg/Nm^3 for HF, and thus below the regulatory limits.

4.5 Sulphur dioxide

Sulphur dioxide is a reaction product of S-compounds contained in the waste with oxygen, and its concentration in the flue-gas is proportional to the amount present in the waste:

$$C_xH_yS + w\,O_2 \rightarrow CO_2 + H_2O + SO_2 \qquad (13)$$

Regardless of whether in organic or organic forms, most of the sulphur oxides end up in stack gases as SO_2. According to BREF (2006), in spite of the regulatory threshold of 50 mg/Nm^3, most of the installations reveal emissions lower than 20 mg/Nm^3. This pollutant can be removed by the reactions:

$$SO_2 + Ca(OH)_2 \rightarrow CaSO_3 + 1/2\,H_2O \qquad (14)$$

It is important to note that SO_2 is a highly reactive gas with short half life indoors. Although it is a known respiratory irritant and bronchoconstrictor, its effects seem limited to patients with asthma and bronchitis, while sensitivity to exposure varies widely (Katsouyanni et al., 1997). In some cases, SO_2 is highly correlated with fine particles, and thus these pollutants may be surrogates for each other. Sulphur dioxide is also known as highly toxic for Koch's bacillus and thus, the incidence of tuberculosis is recognized as lower in areas of SO_2 release.

4.6 Nitrous dioxides

Nitrogen oxides, NO_x, are well known acidic and global warming gases, and during incineration process three main different mechanisms can lead to their formation. Indeed, part of the air nitrogen and part of the nitrogen contained in the waste (or fuel) may be oxidized to NO_x. The nitrous oxides produced from air are known as *thermal* NO_x, and the ones originated from waste as *fuel* NO_x. These reactions can significantly occur for temperatures above 1300 °C and whenever oxygen is not a limiting reagent. In WtE plants, thermal NO_x is often much greater than fuel NO_x. The formation of nitrogen oxides through radical reactions with organic compounds is known as *prompt* NO_x, and has little contribution in WtE incineration plants.
Legislation set a regulatory limit of 200 mg/Nm^3, and to comply with this limit, primary and secondary measures may be taken. The main primary techniques can involve both over supply air and excessive high furnace temperatures. Secondary measures may demand injection of reduction agents such as ammonia (as 25% aqueous solution) or urea, for promoting SNCR indicated in Eq. (5)-(6) (section 3.6).
The reduction of NO_x can attain 85% at temperature of 1000°C, and thus some NH_3 can also be released in the flue-gas (BREF, 2006). When urea is used as reducing agent in SNCR, a little N_2O may arise. Nitrous oxide (N_2O) is usually correlated with CO, since it is formed under low oxygen concentrations, but it is not included in NO_x emissions. Alternatively, SCR by using specific catalysts can be integrated in the APC area, after acid neutralization and dedusting. Depending on the catalyst, SCR reactions can take place in the range of 180 to 450°C. Stack emissions typically exhibit 95% of NO and 5% of NO_2 (BREF, 2006). Specific emission of NO_x may be in the range of 800-900 g/ton of MSW.

4.7 Mercury

Mercury is by far the most thermally mobile metal, highly toxic, and at 357 °C it is all volatilized into the flue gas. Although restriction in uses of Hg, some peak loads of MSW with batteries, electrical devices, thermometers and specific wastes may lead to incompliance with the regulatory threshold of 0.05 mg/Nm^3.
The oxidation state of Hg depends mostly on the characteristics of the flue-gas, but in general, the most significant forms in MSWI are Hg^0 and $HgCl_2$. When the HCl concentration in flue-gas is higher than SO_2 (a reduction agent), mercury is mainly in the form $HgCl_2$, which is easier to remove from the gas phase than Hg^0.
For this element, the primary measures are those which avoid as much as possible to introduce in the furnace contaminated wastes that contain Hg. Nevertheless, secondary techniques are usually implemented in order to comply with the regulatory limit at stack emissions. The lowest emissions of Hg are observed when sulphur doped activated carbon is used. Some analysis showed that wet acidic scrubbers can serve as a sink for Hg if it occurs in chloride form (BREF, 2006). The amount of Hg in MSW is in the range of 0.5- 5 g Hg/ton MSW, and from this less than 2% goes to bottom ashes, 70-90% is released into APC residues, and less than 10% is emitted to air emissions, which correspond to 0.0002-0.05 mg/Nm^3 (IAWG, 1997).

4.8 Cadmiun and tallium

Cadmium is an environmental problematic toxic element that may arise in air emissions of MSWI due to its thermal mobility. Although some countries restrict applications of Cd, frequent sources in MSWI are electronic devices (including accumulators), paints, Ni-Cd batteries, and cadmium-stabilised plastics. Under the conditions usually observed in the furnace, Cd is mainly converted into $CdCl_2$. The APC devices normally lead to a concentration of Cd in APC residues, and less than 1% is released to the atmosphere. The reported range is usually 0.0002-0.03 mg/Nm^3 (11% of O_2) (BREF, 2006).
The thallium amount in municipal solid waste is virtually non-existent and very often below any detection limit.

4.9 Other heavy metals

During the main stages of incineration, combustion temperatures attained in the furnace determine the evaporation degree of heavy metals and most of inorganic salts. Indeed, besides Hg, Cd and Tl also Sb, As, Pb, Cr, Co, Cu, Mn, Ni and V must be under control in stack emissions. These are very toxic metals since they are carcinogenic or may provoke respiratory damages. Most of these metals are in flue-gas as oxides or chlorides, which condense onto dust particles as the gases cool down downstream of the boiler. Therefore, it is expected that the part of these metals that vaporize in furnace, probably ends as condensed phases in APC residues. Thus, according to measurements in diverse European incinerators, the limit of 0.5 mg/Nm^3 is usually complied (BREF, 2006). Lead and chromium are currently the most problematic heavy metals found in APC residues that lead to a classification as hazardous waste (Quina et al, 2008b).

4.10 Dioxins

Polychlorinated dibenzo-p-dioxins (PCDD) and polychlorinated dibenzofurans (PCDF) are commonly referred as *dioxins* and include a set of tricyclic aromatic persistent organic pollutants (POPs), that are ubiquitous in the environment and may provoke significant risk to human health (WHO, 1989). Recently, Hites (2011) presented a short but very interesting overview and history about dioxins, where it is referred that PCDD/PCDF is now a well-known environmental contaminant family. In fact, depending on the position of chlorine in the benzene ring, 210 congeners can be formed. Fig. 9 (a)-(b) shows the generic structure of dioxins and furans, where chlorine may be attached at positions m=1- 4 and n=6-9, and thus, it is possible to form 75 dioxin and 135 furan congeners. The most toxic dioxin is 2,3,7,8-tetrachlorodibenzo-*p*-dioxin (2,3,7,8-TCDD) and it has been considered as molecule characterized with the lowest lethal dose to 50% of a population (LD_{50}). For example, for guinea pigs, LD_{50} is only 0.6 µg/kg. PCDD/PCDF are in general very lipophilic, solid at room temperature, with low volatility and insoluble in water. Thus, their dispersion in atmosphere is likely to occur mainly in particulate matter.

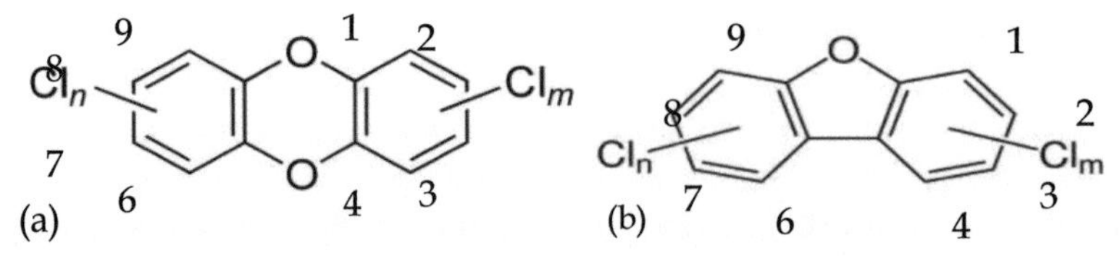

Fig. 9. Generic structure of (a)- polychlorinated dibenzo-*p*-dioxin (PCDD), (b)- polychlorinated dibenzofurans (PCDF).

PCDD/PCDF are considered unintentionally produced persistent organic pollutants (POPs), in respect to which the Stockholm Convention obliges Parties to decrease the total release. According to UNEP Chemical (2005) these by-products are formed in various processes and activities (e.g. waste incineration, metal and mineral production, heat and power generation, transport, open burning, production and use of goods) or they can be introduced into processes as contaminants in raw materials (e.g. disposal of wastes). In 1977, Olie and co-workers showed that PCDD/PCDF are in trace quantities in fly ashes and flue gases of MSW incinerators (Olie et al., 1977). Since then, it is well known that dioxins are formed in trace quantities in incineration processes (UNEP Chemical, 2005). McKay (2002) reviewed methods to minimise the formation of these very toxic compounds in MSWI. Indeed, in addition to PCDD/PCDF originally present in the furnace feedstock, three formation theories should be considered:

- precursor compounds such as chlorinated hydrocarbons already in the MSW feed or formed in furnace by reacting rapidly with other groups (e.g. chlorobenzene) ;
- *de-novo* synthesis of smaller and relatively innocuous chemical molecules combining together in the low temperature range;
- radical recombination with prevalence of the structures with longer life-span (more symmetrical and more stable).

The maximum formation rate of dioxins is in the range 300–400 °C, and the knowledge of this fact associated with the information about the reaction mechanisms have been the basis of designing combustion systems (McKay, 2002). Therefore, two strategies have been adopted for complete destruction in combustion and for minimize their formation. In this scope, it is very important that the combustion temperature should be above 1000 °C, the residence time should exceed 2 s, and high turbulence in the combustion chamber should be promoted (Reynolds number higher than 50,000). As an end-of-pipe treatment, it should be assured that a very rapid gas cooling from 450 to 250 °C is able to occur. Currently, semi-dry lime scrubbing and bag filters coupled with activated carbon injection have been playing an important role in prevention or minimisation of dioxins in stack gas emission to the environment (McKay, 2002). In addition, after the regulatory limit of 0.1 ng I-TEQ/Nm^3 entered into force, MSWI is no longer a significant sources of dioxins. In fact, some nineties studies refer that MSWI seem to be the most important source of PCDD/PCDF to air emissions (Olie et al., 1998). However, the emissions of MSWI of units that comply with the limit established in Directive 2000/76/EC are no longer so important (Abad et al., 2003; Lee et al., 2007). Indeed, measurements for MSWI plants, equipped with up-to-date APC systems showed that average concentration is around 0.048 ng I-TEQ/Nm^3, and the monitored values are all in the range 0.01-0.08 ng I-TEQ/Nm^3 (Abad et al., 2006). Table 10 shows the concentration of the seventy congeners that should be taken into account to calculate the equivalent toxicity (I-TEQ), from four different references (Ref1-Ref4).

At this point it is important to note that there is evidence that some additional treatment systems for flue gas may result in a dramatic decrease in dioxins concentrations emitted in stack flue gas, having been reported reductions of 99.98% (Lee et al., 2007). On the other hand, while the concentrations are very low, these chemicals are highly toxic. In 1997, the International Agency for Research on Cancer declared the 2,3,7,8-TCDD to be carcinogenic to humans (Abad et al., 2006). PCDD/PCDF are among the most widely studied organic chemical and there are a large number of publications on the toxicological effects (WHO 1989; Van Den Berg et al., 1994; Mukerjee, 1998). The effect of dioxin on humans is associated with increased risk of chloracne and hyperpigmentation, changed liver function

Congener	Ref1 (pg/Nm³)*	Ref2 (ng/Nm³)	Ref3 (ng/Nm³)	Ref4 (ng/Nm³) 1996	2000
2,3,7,8-TeCDD	0.34 ± 0.23	0.00259	0.0003	1.29	0.007
1,2,3,7,8-PeCDD	3.53 ± 2.23	0.00886	0.0004	15.41	0.026
1,2,3,4,7,8-HxCDD	5.92 ± 3.13	0.0109	0.0010	21.10	0.032
1,2,3,6,7,8-HxCDD	27.0 ± 16.6	0.0299	0.0010	48.88	0.085
1,2,3,7,8,9-HxCDD	11.2 ± 6.83	0.0155	0.0000	55.28	0.074
1,2,3,4,6,7,8-HpCDD	204.7 ± 104.4	0.213	0.0000	227.66	0.533
OCDD	403.6± 198.4	0.340	0.0000	251.24	0.665
2,3,7,8-TeCDF	9.25 ± 14.2	0.0190	0.0000	94.10	0.029
1,2,3,7,8-PeCDF	12.6 ± 8.1	0.0222	0.0000	10.81	0.010
2,3,4,7,8-PeCDF	30.0 ± 15.3	0.0388	0.0010	81.04	0.028
1,2,3,4,7,8-HxCDF	25.4 ± 13.1	0.0306	0.0000	121.76	0.052
1,2,3,6,7,8-HxCDF	46.3± 24.1	0.0350	0.0010	69.21	0.050
2,3,4,6,7,8-HxCDF	86.4± 46.9	0.0516	0.0010	10.47	0.003
1,2,3,7,8,9-HxCDF	21.4 ± 13.9	0.00446	0.0010	135.83	0.102
1,2,3,4,6,7,8-HpCDF	145.0± 93.5	0.110	0.0000	227.06	0.231
1,2,3,4,7,8,9-HpCDF	52.7 ± 34.9	0.0240	0.0000	68.01	0.040
OCDF	219.1 ± 210.0	0.0906	0.0000	198.91	0.345
ΣPCDD/PCDF	1305	1.05	0.0067	1638.1	2.312
I-TEQ (PCDD/DF)	46.9 ± 22.5	0.0511	0.0016	111.39	0.086

Ref.1- Abad et al. (2006); Ref.2 -Chen et al. (2008); Ref.3- Lee et al. (2007); Ref.4- Meneses et al (2004);

*-average value ± standard deviation.

Table 10. Concentration of PCDD/PCDF in air emissions from MSWI.

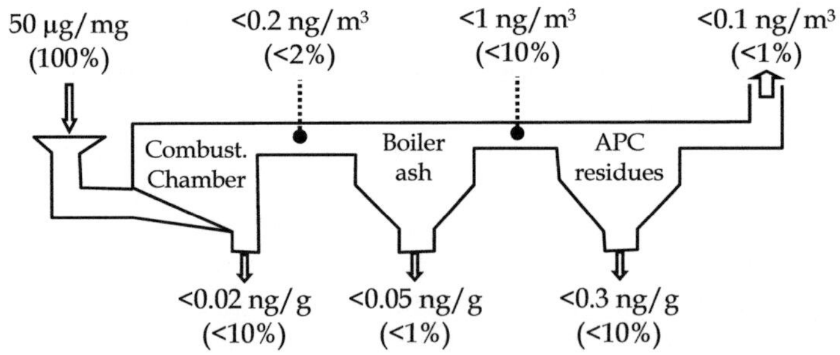

Fig. 10. PCDD/PCDF mass balance in modern MSW incinerator plant (adapted from McKay, 2002).

and lipid metabolism, changes in activities of various liver enzymes, depression of the immune system, and endocrine-and nervous-system abnormalities. It is a potent teratogenic and fetotoxic chemical in animals (Mukerjee, 1998). Indeed, 2,3,7,8-TCDD seems to be a very potent promoter of carcinogenesis liver of rat, and populations occupationally exposed to dioxins have increased incidences of soft-tissue sarcoma and non-Hodgkin's lymphoma.

Fig. 10 shows a representative mass balance in a modern MSWI unit, where it is implicit that overall the incineration process is now a sink of dioxins. In general, if high technology combustion is combined with sophisticated APC devices an emission factor of 0.5 µg I-TEQ/ton of MSW can be assumed to air, 15 µg I-TEQ/ton to fly ashes and 1.5 µg I-TEQ/ton to bottom ashes (UNEP, 2005).

At the end, it is very important to note that the 2,3,7,8-TCDD concentration in human tissue is decreasing by about a factor of 7 over a 25-year period (Hites, 2011).

4.11 Green house gases

Green house gases enclose carbon dioxide (CO_2), methane (CH_4), nitrous oxide (N_2O), hydrofluorocarbons (HFC) perfluorocarbons (PFC) and sulphur hexafluoride (SF_6). Specific emissions of CO_2 usually observed in MSWI are in the range of 0.1 to 1.7 ton/ton of MSW. The organic compounds of MSW are biomass and fossil wastes, and so the fraction of CO_2 from fossil must be taken into account for the climate changes. Concerning methane, since the furnace is kept at high oxidative conditions, no emissions of CH_4 are expected in stack gases. Some methane can be formed during long periods of waste storage and under anaerobic conditions. In the cases where primary air is supplied from the storage bunker, the CH_4 is oxidized to CO_2 and H_2O.

In Table 11 a summary of the most important techniques/technologies is indicated for the reduction of the main incineration pollutant described above (BREF, 2006).

5. The impact on health of emissions to air from MSWI

The activity of MSW incineration may pose environmental problems at various levels, concerning either ambient contamination provoked mainly by gaseous emissions or coming from the landfilling of APC residues after being submitted to solidification/stabilization treatment. Highly toxic PCDD/PCDF or harmful heavy metals such as Cd, Pb and Cr can be spread in the surroundings of incinerators plants and landfill disposal facilities. Ultimately, population health living nearby may be adversely affected by carcinogenic and non-carcinogenic compounds so that assessing their risk on the potential increase of human diseases is truly advisable. In this ambit some literature studies have addressed this theme.

Cangialosi et al. (2008) described the health impact associated to the gaseous emissions of a municipal solid waste incinerator in Italy, based on an atmospheric dispersion model involving pollutants content and human exposure level. Food intake revealed to be the major threat when compared with air inhalation and dermal contact, being, however, the carcinogenic (dioxins, furans, Cd) and non-carcinogenic (Pb and Hg) contamination values lower than those accepted. Therefore, their final conclusions supported that no relevant health problems may be endorsed to modern MSWI systems. Volatile organic compounds and bioaerosols possibly attributed to a MSWI plant in Spain (Vilavert et al., 2011) were checked and potential risks for human health were not identified once pollutants concentrations were similar to those observed in other regions. In a comparative study, Moy et al. (2008) reported that the landfill impact in human health through inhalation is five times higher than the one associated to the incineration operation, with cancer risk for both alternatives considered as acceptable when analysing data from New York City, whereas non-cancer health problems showed to be unacceptable. Concerning landfill sites for disposal of air pollution control (APC) residues from municipal solid wastes in UK, health risks for direct and indirect human exposure were also assessed with dispersion models by

Macleod et al. (2006) revealing that irrelevant danger was associated to inhalation while tolerable daily soil intake (TDSI) could be possibly surpassed due to the high deposition levels of those residues. Investigations about the impact of harmful PCDD/PCDF released

Pollutant	Techniques
Particles	Electrostatic precipitators Wet electrostatic precipitators Condensation electrostatic precipitators Ionization wet scrubbers Fabric filters Cyclones and multi-cyclones
Acid gases (HCl, HF, SO_x, …)	Wet-scrubber Semi-dray scrubber (e.g. suspension of lime) + bag filter Dry-scrubber (e.g. lime or sodium bicarbonate)
Direct desulphurisation	Injection of adsorbents (e.g. calcium compounds) directly into the incineration chamber
Oxides of nitrogen (NOx)	Primary techniques: air and temperature control, flue-gas recirculation, Secondary techniques: Selective Non- Catalytic Reduction (SNCR) and Selective Catalytic Reduction (SCR).
Hg	Primary techniques: separate collection, restrictions of receipt contaminated wastes Secondary techniques: scrubber by adding oxidants, activated carbon, furnace coke or zeolites
Other heavy metals	Converted into no-volatile oxides and deposited into fly ash, all techniques referred to remove particles can be applied. Activated carbon injection into scrubbing units.
Organic carbon compounds	Adsorption on activated carbon. SCR used for NOx. Catalytic bag filters Static bed filters Rapid quenching of flue-gas
Greenhouse gases (CO_2, N_2O)	All techniques used for NOx. Increase energy recovery efficiency
APC residues	Treated (e.g. solidification/stabilization and disposed of) Thermal treatment (vitrification, melting, sintering) Extraction and separation Chemical stabilisation
Bottom ash	Separation of metals Screening and crushing Treatment using ageing conditions High temperature slagging rotary kilns

Table 11. Most important techniques for the reduction of the main incinerator pollutants.

from a MSWI in Korea (Oh et al., 2006) showed that those emissions affected more directly atmosphere quality, with furan levels in air closer to the emitted values than those found in soil. In order to deepen consequences of environment contamination of MSWI, Cordier et al. (2004) analysed birth malformations in the population surrounding 70 incinerators in France. Though congenital problems were not significantly more relevant in those groups, higher incidence of some anomalies associated to renal, urology, cardiology and skin pathologies could be ascertained to persistent pollutants accumulated from the old incineration methodologies and road traffic pollution as well. The effect resulting from the application of more restrictive environment EU legislation regarding MSWI emissions was considered by Meneses et al. (2004), by determining PCDD/Fs concentrations in air, soil and vegetation, and their danger for human health, having concluded that both pollutants emissions and cancer risk was reduced after those measures implementation. Other contamination sources were, nevertheless, pointed out since pollutants emission levels diminished more significantly than health problems after the installation of new cleaning gas systems. Also, the replacement of old systems (electrostatic precipitator) by new technologies (semi-dry scrubber, fabric filters and charcoal injection) in Spain (Abad et al., 2003) revealed a significant decrease on dioxins emissions spread in stack gas, bottom ash and air pollution control residues, where the higher dioxin content was found. Heavy metal deposition around a MSWI area in USA showed quite different pollution concentrations during fall and winter (Feng et al., 2000). Methodologies envisaging minimisation of dioxins released from MSWI processes were reviewed by McKay (2002) accounting for three possible sources, furnace feedstock, precursor compounds or innocuous molecules forming dioxins by heterogeneous catalytic reactions or homogeneous thermal reactions. Finally, it is worthy to refer that in September 2009, the Health Protection Agency from UK analysed the on-going research about health impact of municipal solid waste incinerators and conclusions about the low influence in humans diseases of modern technologies were justified by the fact that emissions are nowadays well controlled, imparting, therefore, low contribution to global air pollution.

6. Conclusion

At present, municipal solid waste incineration (MSWI) in waste-to-energy (WtE) is considered to be an environmentally friendly technology, while allowing recovery of a large part of the energy enclosed in MSW. Indeed, MSWI is one of the main management options in most of the developed countries. Nonetheless, the main problems associated to these processes are the large volume of gaseous emissions which may pose environmental health risks (if not well cleaned) and the hazardous fly ash or air pollution control (APC) residues that remain after incineration. The main air pollution formed in MSWI that are nowadays under regulatory observation are: dust, CO, TOC, HCl, HF, SO_2, NO_x, Hg, Cd, Tl, Sb, As, Pb, Cr, Co, Cu, Mn, Ni, V and PCDD/PCDF.

However, it is very important to note that air pollution induced by MSWI has been strongly reduced by some up-to-date sophisticated air pollution control equipments. In fact, process engineering of the gas cleaning systems can still be extended to lower emission limits if the environmental legislation so requires. It is only a question of overall cost of operation per ton of incoming waste to be incinerated.

Some studies that analysed in detail the health impact associated to the gaseous emissions of MSWI have revealed that no relevant health problems may be associated to modern MSWI systems.

7. References

Abad, E. Martinez, K. Caixach, J. Rivera, J., (2006). Polychlorinated dibenzo-p-dioxins, dibenzofurans and 'dioxin-like' PCBs in flue gas emissions from municipal waste management plants, *Chemosphere*, 63, pp. 570–580.

Abad, E., Caixach, J., Rivera, J. (2003). Improvements in dioxin abatement strategies at a municipal waste management plant in Barcelona, *Chemosphere*, 50, pp. 1175–1182.

Achternbosch, M., Richers, U. (2002). *Materials flows and investment costs of flue gas cleaning systems of municipal solid waste incinerators*, Forschungzentrum Karlsruhe GmbH, ISSN 0947-8620, Karlsruhe, Germany.

Amutha Rani, D., Boccaccini, A.R., Deegan, D., Cheeseman, C.R. (2008). Air pollution control residues from waste incineration: Current UK situation and assessment of alternative technologies, *Waste Management*, 28, 2279-2292.

Bordado, J.C.M., Gomes, J.F.P. (1999). Practical determination of the efficiency of electrostatic precipitators. *Powder and Bulk Engineering*, 2 , 37-42.

BREF (2006). *Integrated Pollution Prevention and Control – Reference Document on the Best Available Techniques for Waste Incineration*, European Commission.

Cangialosi, F., Intini, G. , Liberti, L., Notarnicola , M., Stellacci, P., (2008). Health risk assessment of air emissions from a municipal solid waste incineration plant - A case study, *Waste Management* , 28, pp. 885–895.

Chen, C.-K., Lin, C., Lin, Y.-C., Wang, L.-C., Chang-Chien, G.-P. (2008). Polychlorinated dibenzo-*p*-dioxins/dibenzofuran mass distribution in both start-up and normal condition in the whole municipal solid waste incinerator, *Journal of Hazardous Materials*, 160, pp. 37–44.

Clark, J.W., Bordado, J.C.M., Correia, C.S.R., Correia, N.D.S., Oil absorption foam, WO2008043545, Appl. 17April 2008.

Cordier, S., Chevrier, C., Robert-Gnansia, E., Lorente, C., Brula, P., Hours, M. (2004). Risk of congenital anomalies in the vicinity of municipal solid waste incinerators, *Occupational and Environmental Med*icine, 61, pp. 8–15.

DEFRA (2007). *Incineration of Municipal Solid Waste*, Report of Department for Environment, Food & Rural Affairs, UK.

Directive 2000/76/EC of The European Parliament and the Council, of 4 December 2000, on the incineration of waste.

Directive 2008/98/EC of The European Parliament and the Council of 19 November 2008 on waste and repealing certain Directives.

Encyclopedia of emulsion technology, Ed. Paul Becher, Marcel Dekker, New York (1988).

Eurostat (2010). Environmental statistics and accounts in Europe, Eurostat Statistical books, European Commission. <http://www.epp.eurostat.ec.europa.eu>

Feng, X., Melander, A.P., Klaue, B., (2000). Contribution of municipal waste incineration to trace metal deposition on the vicinity, *Water, Air, and Soil Pollution,* 119, pp. 295–316.

Gentil, E., Clavreul, J., Christensen, T.H. (2009). Global warming factor of municipal solid waste management in Europe, *Waste Management and Research,* 27, pp. 850-860.

Hites, R.A. (2011). Dioxins: an overview and history, *Environment Science and Technology*, 45, pp. 16-20.

IAWG (1997). *Municipal Solid Waste Incinerator Residues*, The international ash working group, Studies in environmental Science 67, Elsevier, The Netherlands.

Jenkins, B.M., Baxter L.L., Miles Jr., T.R., Miles T.R. (1998). Combustion properties of biomass, *Fuel Processing Technology*, 54, pp. 17–46.

Jungten, H. Richter, E. Knoblauch K., Hoang-Phou T., (1988) Catalytic NOx reduction by ammonia and carbon catalysts, *Chemical Engineering Science*. 43, 419-428.

Katsouyanni, K. , Touloumi, G., Spix, C., Schwartz, J., Balducci, F., Medina, S., Rossi, G., Wojtyniak, B., Sunyer, J., Bacharova, L., Schouten, J. P., Ponka, A., Anderson, H. R. (1997) . Short term effects of ambient sulphur dioxide and particulate matter on mortality in 12 European cities: results from time series data from the APHEA project, *British Medical Journal* , 314, pp. 1658-1663.

Lee, C.C., Huffman, G.L. (1996). Medical waste management/incineration, *Journal of Hazardous Materials*, 48, 1-30.

Lee, S.-J., Choi, S.-D., Jin, G.-Z., Oh, J.-E., Chang, Y.-S., Shin, S. K., (2007). Assessment of PCDD/F risk after implementation of emission reduction at a MSWI, *Chemosphere*, 68, pp. 856–863.

Macleod, C., Duarte-Davidson, R., Fisher, B., Ng, B., Willey, D., Shi, J.P., Martin, I., Drew, G., Pollard, S. (2006). Modeling human exposures to air pollution control (APC) residues released from landfills in England and Wales, *Environment International*, 32, pp. 500–509.

McKay, G. (2002). Dioxin characterisation, formation and minimisation during municipal solid waste (MSW) incineration: review, *Chemical Engineering Journal*, 86, pp. 343-368

Meneses, M., Schuhmacher, M., Domingo, J.L. (2004). Health risk assessment of emissions of dioxins and furans from a municipal waste incinerator: comparison with other emission sources, *Environment International*, 30, pp. 481–489.

Morcos, V.H. (1989). Energy recovery from municipal solid waste incineration - a review. *Heat Recovery Systems and CHP*, 9, 115-126.

Moy, P. Krishnan, N. Ulloa, P. Cohen, S. Brandt-Rauf, P.W. (2008). Options for management of municipal solid waste in New York City: A preliminary comparison of health risks and policy implications, *Journal of Environmental Management*, 87, pp. 73–79.

Mukerjee, D. (1998). Health impact of polychlorinated dibenzo-p-dioxins: A critical review, *Journal of the Air & Waste Management Association*, 48, pp. 157-165.

Niessen, W.R. (2002). *Combustion and incineration processes*. 3rd Ed., Marcel Dekker Inc., New York, USA.

Nunes, A. Duarte, A. Bordado, J., PCT Patent, 2011.

Oh, J.-E., Choi, S.-D., Lee, S.-J., Chang, Y.-S., (2006). Influence of a municipal solid waste incinerator on ambient air and soil PCDD/Fs levels, *Chemosphere*, 64, pp. 579–587.

Olie, K., Addink, R., Schoonenboom, M., (1998). Metals as catalysts during the formation and decomposition of clhorinated dioxins and furans in incineration processes, *Journal of the Air & Waste Management Association*, 48, pp. 101-105.

Olie, K., Vermeulen, P., Hutzinger, O., (1977). Chlorodibenzo-p-dioxins and chlorodibenzofurans are trace components of fly ash and flue gas of some municipal incinerators in The Netherlands, *Chemosphere*, 6, pp. 455-459.

Quina, M.J., Bordado, J.C.M., Quinta Ferreira, R.M., (2008a) Treatment and Use of Air Pollution Control Residues from MSW Incineration: an Overview, *Waste Management*, 28, pp. 2097-2121.

Quina, M.J., Santos, R.C., Bordado, J.C.M., Quinta Ferreira, R.M. (2008b). Characterization of air pollution control residues produced in a municipal solid waste incinerator in Portugal, *Journal of Hazardous Materials,* 152, pp. 853-869.

Sundqvist, J.O. (2005). *How should municipal solid waste be treated – a system study of incineration, material recycling, anaerobic digestion and composting,* Report of Swedish Environmental Research Institute.

UNEP Chemicals (2005). *Standardized Toolkit for identification and quantification of dioxin and furan releases.* Genebra, Switzerland.

Van Den Berg, M., Dejongh, J., Poiger, H., Olson, J.R. (1994). Toxicokinetics and metabolism of polychlorinated dibenzo-*p*-dioxins (PCDDs) and dibenzofurans (PCDFs) and their relevance for toxicity. *Critical review in toxicology,* 24, pp. 1–74.

Vilavert, L., Nadal, M., Figueras, M. J., Kumar, K. , Domingo, J.L. (2011). Levels of chemical and microbiological pollutants in the vicinity of a waste incineration plant and human health risks: temporal trends, *accept for publication in Chemosphere.*

WHO (1989). *Polychlorinated dibenzo-para-dioxins and dibenzofurans.* Geneva, World Health Organization, (Environmental Health Criteria, No. 88).

World Bank Report (1999). *Municipal Solid Waste Incineration,* Technical Guidance Report of World Bank, Washington D.C.

Part 4

The Impact of Air Pollution in the Economy and Development Goals: Case Studies

17

Indoor Air Pollution in Mexico

Iván Nelinho Pérez Maldonado, Lucia Guadalupe Pruneda Álvarez, Fernando Díaz-Barriga, Lilia Elisa Batres Esquivel, Francisco Javier Pérez Vázquez and Rebeca Isabel Martínez Salinas
Laboratorio de Toxicología, Facultad de Medicina, Universidad Autónoma de San Luis México

1. Introduction

General health conditions in Mexico have improved considerably over the last 50 years. This is clearly seen in life expectancy at birth. Between 1950 and 2000, the country added 25 years to this indicator. However, the respective differences among the states show the inequality prevailing in the country (SSA, 2005).

The differences found per region are an important fact to consider in understanding the Mexican situation. In the modern and industrialized northern region, the population is concentrated mostly in urban zones. The southern region is clearly traditional, unindustrialized and with a high indigenous population living in small, dispersed rural communities. This explains why the health indicators for the northern region are similar to those of developed countries, while the same indicators for the southern region are similar to those of developing countries with social and economic problems.

In 2002, there were 43,719,756 persons under 19 years of age in Mexico, representing just over 40 percent of the total population. The birth rate per 1,000 inhabitants has decreased from 45 in 1960 to 17 in 2000, while the child mortality rate in the first year of life decreased from 19.1 per 1,000 inhabitants in 1998 to 16.78 per 1,000 inhabitants in 2002. The leading causes of death also changed radically over the past 50 years. Transmissible diseases and congenital illnesses were displaced as the primary causes of death by non-transmissible diseases and injuries. In the same period, the percentage of deaths due to intestinal infections decreased by a factor of 14 (from 14.3 percent to 1 percent), while deaths from heart disease quadrupled (from 4 percent to 16 percent; SSA, 2005).

Another fact pointing to a clear epidemiological transition in the country is the changing pattern of causes of mortality among children under one year of age. Previously, infant deaths were concentrated in the postnatal period (between one month and one year of age), mostly caused by acute respiratory infections and diarrhea. Presently, infant deaths are concentrated in the first 28 days of life, due to perinatal causes and congenital anomalies requiring high-technology intervention (SSA, 2005).

Common infections and congenital events continue to cause major harm to health in highly marginalized groups. Over the last 10 years, deaths from diarrhea in children under five years of age have decreased by 85 percent. However, there are southern states with mortality rates from diarrhea above 40 per 100,000 inhabitants under five years of age, i.e., five times higher than the rates found in the more developed northern states. The same is

found with deaths from acute respiratory infections, another clear example of the persistently lower quality of life (SSA, 2005).
Despite the major progress in health systems, the problems of poverty, social inequality, marginalization, the lack of services, and environmental air, water and soil pollution constitute important factors associated with a poor quality of life for a high percentage of the child population, primarily in the country's rural areas.
Air pollution is a generalized problem in Mexico's major metropolitan areas. However, current demographic growth, industrial concentration, greater numbers of vehicles, increased fuel consumption and inadequate urban mobility patterns have caused the problem to increase in other areas, such as medium-size cities.
Indoor air pollution caused by the burning of wood or charcoal used for cooking or heating constitutes a public health problem with repercussions for the population under five years of age and women of reproductive age, especially in the country's marginalized areas. In 1990, one of every three Mexicans (91 percent of rural inhabitants and 11 percent of urban inhabitants) used wood for cooking. In 1993, 25.6 million persons were estimated to use wood as household fuel, decreasing to 17.2 million inhabitants in 2000.

2. Biomass as energy source

The use of solid fuels for cooking and heating is likely to be the largest source of indoor air pollution on a global scale. Nearly half the world continues to cook with solid fuels such as wood, crop residues, agricultural wastes, and animal dung, with wood being the most commonly used (Bruce et al. 2000). When used in simple cooking stoves, these fuels emit substantial amounts of toxic pollutants. In households with limited ventilation (as is common in many developing countries), exposures experienced by household members, particularly women and young children who spend a large proportion of their time indoors, have been measured to be many times higher than World Health Organization (WHO) guidelines and national standards (Bruce et al. 2000; Smith 1987). In this regard, the number of annual deaths in developing countries associated with domestic biomass combustion is estimated at 1,849,000 (SSA, 2005).
In México currently, biofuels represent about 8% of the total energy demand (SE, 1998), 46% of residential energy use (Figure 1) and more than 80% of the energy demand in the rural sector (Masera, 1996b). The three main types of biofuels used in the country are: bagasse, which is used in the sugar cane industry, fuelwood and charcoal. Fuelwood is by far the dominant woodfuel, with charcoal being used mainly in street industries and for barbecues. Total fuelwood use accounts for 3 times the total commercial timber legally harvested in the country (Masera, 1996a). In this regard, in 1990, one of three Mexicans used fuelwood for cooking, including 91 percent of rural residents and 11 percent of urban residents. It is estimated that 25.6 million people used this fuel in their homes in 1993 and that by 2000 this number had declined to 17,256,471. Fuelwood is also used in many small (cottage) industries, like pottery making, "tortilla" making, brick making, and others. The use of fuelwood is concentrated in rural areas and small towns. Fuelwood markets are mostly of a local nature. Most fuelwood comes from forest areas (including here all degraded lands and semi-arid forests); many of the species used are of no commercial value, and the use of agricultural residues and dung is not widespread (Masera, 1996a; Masera et al, 1997). Among the states in México with the highest use of fuelwood are Oaxaca and Chiapas, where it is estimated that 50–60 percent of the

population uses this type of fuel (Figure 2). The general pattern is that a higher proportion of people are exposed to fuelwood and charcoal in the southern part of the country. These are largely rural states with some of the poorest populations; thus, exposure to fuelwood-related pollutants is more prevalent here.

The heaviest biomass fuel usage by household is in southern and north central Mexico, where utilization may approach 90 to 100 percent in some localities (Figure 3). These are largely rural states with some of the poorest populations.

The majority of households using solid fuels in México burn them in open fires or simple stoves that release most of the smoke into the home (Figure 4). The resulting indoor air pollution is a major threat to health, particularly for women and young children, who may spend many hours close to the fire. Furthermore, the reliance on solid fuels and inefficient stoves has other, far-reaching consequences for health, the environment, and economic development (Reddy et al., 1997; WHO, 1997).

3. Indoor air pollution

The current dominant pattern of household fuelwood use presents several problems. People depend mostly on

open fires, leading to very high indoor air pollution levels, particularly for women and children. Fuelwood combustion emits a complex mixture of organic compounds and gases, which include carbon monoxide (CO), nitrogen and sulphur oxides (NOx and SOx), aldehydes, polycyclic aromatic hydrocarbons (PAHs), volatile organic compounds (VOCs), chlorinated dioxins, breathable particulate matter (PM) with diameters < 10 microns (PM10),

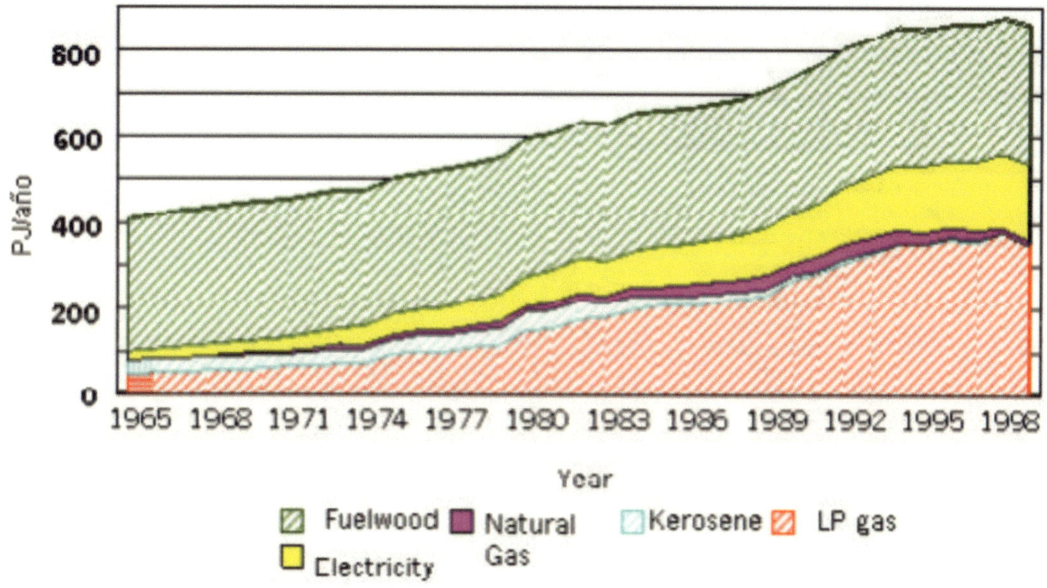

Fig. 1. Residential Energy Use in Mexico 1965-1998. **Source:** Masera O.R, Drigo R., and Trossero M.A. 2003. Woodfuels integrated supply/demand overview mapping. Universidad Autónoma de México, FAO-EC Partnership Programme. Food and Agriculture Organization of the United Nations.

and free radicals (Albalak, 2001; Mishra, 2003; Table 1). Moreover, wood users rely on simple and rustic stoves such as open "three-stone" fires and mud, clay, or metal stoves that result in incomplete and inefficient combustion (Reddy et al., 1997; WHO, 1997).

Compound	Major indoor sources
Carbon monoxide	Fuel/tobacco combustión
Nitrogen oxides	Fuel combustión
Sulphur oxides	Coal combustion
Aldheydes	Furnishing, construction materials, cooking
Polycyclic aromatic hydrocarbons	Fuel/tobacco combustión, fumes from food being cooked.
Volatile and semi-volatile organic compounds	Fuel/tobacco combustión, consumers products, Furnishing, construction materials, fumes from food being cooked.
Chlorinated dioxins	Fuel/tobacco/ coal combustión
Breathable particulate matter	Fuel/tobacco combustión, cleaning, fumes
free radicals	Indoor chemistry

Table 1. Toxic compounds generated during fuelwood combustion.

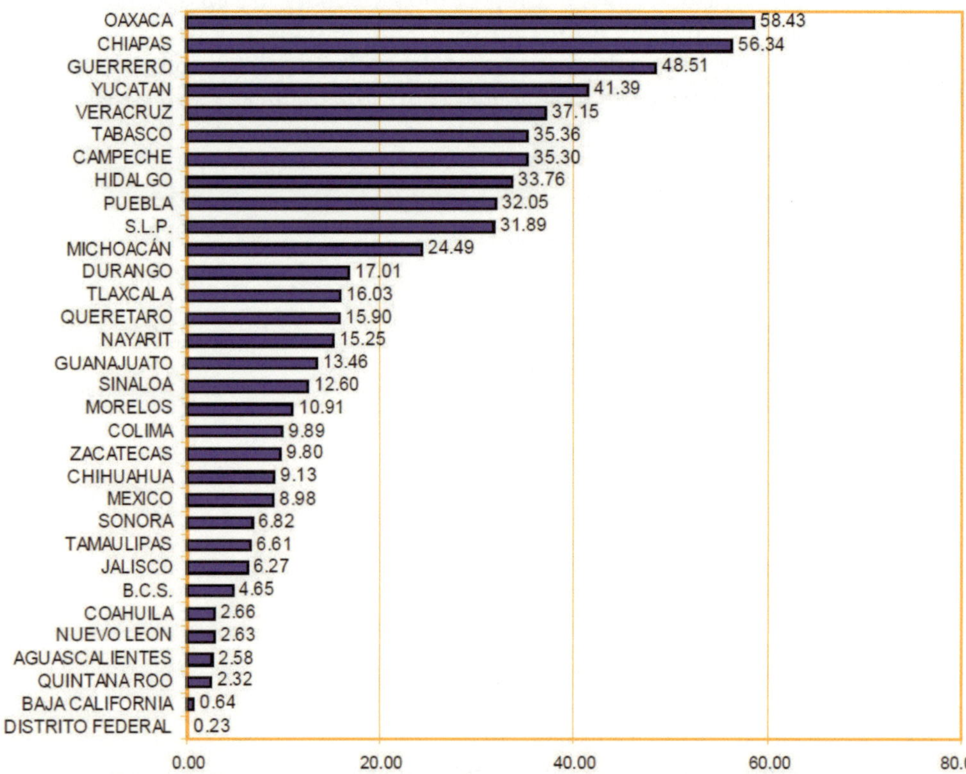

Fig. 2. Percentage of Mexico's General Population Exposed to Biomass Smoke, by Region, 2000. Source: Diagnóstico Nacional de Salud Ambiental y Ocupacional 2002. Dirección General de Salud Ambiental.

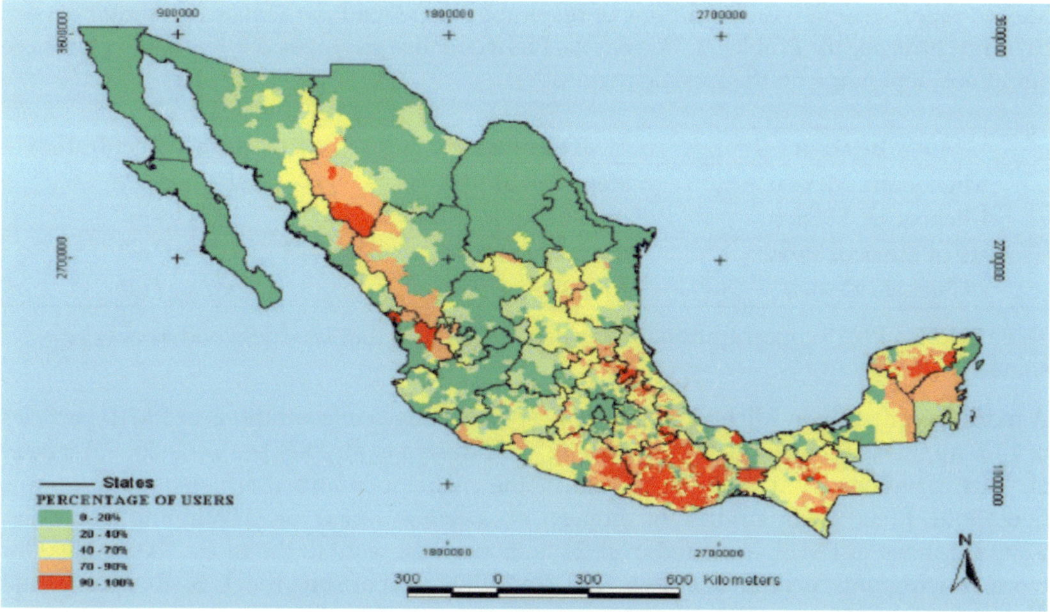

Fig. 3. Percentage of Fuel Wood Users, at the Municipal Level, in Mexico, 2000. **Source:** Masera O.R, Drigo R., and Trossero M.A. 2003. Woodfuels integrated supply/demand overview mapping. Universidad Autónoma de México, FAO-EC Partnership Programme. Food and Agriculture Organization of the United Nations.

Fig. 4. Traditional wood-burning cookstove.

For example, studies carried out in rural households belonging to different regions of Mexico report average concentrations of respirable suspended particulates ranging between 537 and 1020 μg/m³ (Table 2). Per capita fuelwood use averages 2.0 kg/day, with large variations depending on the specific region.

Study location	Reference	PM10 mean concentration
Michoacán, México	Masera et al. 2007	1020 μg/m³
Michoacán, México	Saatkamp et al. 1999	995 μg/m³
State of México, México	Brauer et al. 1996	768 μg/m³
Chiapas, México	Riojas, et al. 2001	537 μg/m³

Table 2. Mean PM10 concentration in Mexican households that use fuelwood as energy source.

A more recent study in Michoacán, México report average concentrations of PM10 particles of 1020 μg/m³ with maxim levels of 4230 μg/m³ (Masera et al. 2007; see Table 2). Saksena et al. 2004, have compiled data on several of the main pollutants associated with various household fuels from studies of homes in a wide range of developing countries. Concentrations of PM10, averaged over 24-hour periods, were in the range 300 to 3,000 (or more) micrograms per cubic meter (μg /m³). By comparison, the U.S. Environmental Protection Agency's annual air pollution standard for PM10 is 50 μg/m³, one to two orders of magnitude lower than levels seen in many homes in developing countries.

With use of biomass, CO levels are generally not as high in comparison, typically with 24-hour averages of up to 10 parts per million (ppm), somewhat below the World Health Organization (WHO) guideline level of 10 ppm for an eight hour period of exposure. Much higher levels of CO have been recorded, however. For example, a 24-hour average of around 50 ppm was found in Kenyan Masai homes (Bruce et al. 2002), and one Indian study reported carboxyhemoglobin (COHb) levels similar to those for active cigarette smokers (Behera et al. 1988). In this regard our group, demonstrated in a rural community in San Luis Potosí, Mexico that use biomass as the principally energy source, mean COHb level of 4.93% and moreover, 55% of those people presented COHb levels above 2.5% considered a safe level (Kleinman, 2000). Moreover, this 55% of people had geometric mean COHb levels of 8.39% (Torres-Dosal et al. 2008). In this study also was monitored exposure to PAH´s in urine of people using 1-hydroxypyrene (1-OHP) as exposure biomarker to PAH´s. Because PAH exposure occurs as a mixture of compounds, and because pyrene is almost always found in this mixture, pyrene and its metabolite 1-hydroxypyrene (1-OHP) are considered appropriate surrogate markers of total PAH exposure (Jacob and Seidel, 2002). Results show that the geometric mean level of 1-OHP ranged between 1.1 to 17.8 μmol/molCr (mean ± SD; 6.71 ± 3.58 μmol/molCr; (Torres-Dosal et al. 2008). The levels of 1-OHP found in this study are higher than the mean urinary 1-OHP concentrations from people throughout the world (range: 0.03–0.76 μmol/mol creatinine; Levine, 1995; Zhao et al., 1992). These values are also higher than levels observed in people living in traditional houses from rural districts in Burundi (1.50 μmol/mol creatinine, range: 0.26–15.62), that use wood as the principal fuel (Viau et al., 2000). We don't know the type of wood burned in Burundi, but it is possible that the difference in exposure between Mexico (our study) and Burundi is the type of wood used. It has been shown that different emission rates of PAHs and other compounds are dependent on fuel and burning conditions (Jenkins et al., 1996; Zhang and

Smith, 1999; Mcdonald et al., 2000). In a more recent study performed in México, our group evaluated PAHs exposure in children in different residential categories (sites included in the study are recognized for their industrial activity, indoor wood combustion, waste disposal and brick manufacturing using different materials as fuel sources). We performed a random sampling in nine communities in Mexico (Table 3). The participant group consisted of 65 children living in communities with brick furnaces (communities of San Vicente and Tercera in San Luis Potosí (SLP), state); 105 children who live in houses where firewood is the domestic fuel (communities of Ramonal in Quintana Roo state, Ventanilla in Oaxaca state, Victoria in Chiapas state and Tancuime in San Luis Potosí, state); 32 children living next to a municipal landfill in San Luis Potosi city with waste combustion (community of Milpillas) and 56 children living in areas next to highways with moderate and high vehicular trafficc (communities of El Centro and Domingo in San Luis Potosí, state, respectively). The highest levels of 1-OHP in this study were found in Victoria, Chiapas (mean, 4.4 ± 3.7 μmol/mol creatinine; 5.9 ± 5.1 μg/L), in this community, children are exposed to biomass combustion. Interestingly, the communities that also are using biomass combustion as Victoria (i.e., Ramonal, Ventanilla and Tancuime) had children with urinary 1-OHP mean levels, similar to those found in Victoria (Martínez-Salinas et al. 2010). When grouped by exposure scenario, (a) moderate vehicular traffic; (b) heavy vehicular traffic; (c) fumes from a municipal landfill; (d) fumes from brick kilns and (e) indoor air pollution by biomass combustion, the highest levels of urinary 1-OHP were found in children exposed to indoor air pollution (approximately one order of magnitude higher than the other scenarios; Table 4). Children living in communities with brick kiln industry, children living in Milpillas (landfill) and children living in Domingo (heavy vehicular traffic) were the next communities in order (Table 4), leaving the community of El Centro (moderate vehicular traffic; Table 4) at the end (Martínez-Salinas et al. 2010). Moreover, Jongeneelen (2001) proposed a benchmark guideline for occupational exposure to PAHs, taking into account urinary 1-hydroxypyrene levels. Following this guideline, the reference value as a 95th percentile in non occupational exposed controls is 0.24 μmol/mol creatinine and 0.76 μmol/mol creatinine for non-smokers and smokers, respectively (first level). A no-biological-effect-level of 1-hydroxypyrene in urine for exposed workers was fixed at 1.4 μmol/mol creatinine. It is the lowest reported level at which no genotoxic effects were found and therefore the estimate for the second level of the benchmark guideline. Finally, two reference values for the third level were proposed 2.3 μmol/mol creatinine and 4.9 μmol/mol creatinine, in two types of industry, coke ovens and primary aluminum production, respectively, and it was designated occupational exposure limit (OEL). When our results in children were compared with this guideline for adult workers, only in El Centro did we find a low risk condition; in the rest of the communities, an important percentage of children were found at risk (Table 5). It is important to note that the guideline values proposed by Jongeneelen (2001) are derived for workers and for adult populations. Thus, our results are more important in terms of public health as we studied children in non-occupational scenarios.

Finally, taking account the above results mentioned assessed in children, our group performed a study to evaluated 1-OHP levels in women (aged 15-50 years) in two exposure scenarios: 1) People living in an indigenous community in La Huasteca in San Luis Potosi state that use fuelwood as the principally energy source and 2) People living in the city of San Luis Potosi that use gas LP as the principally energy source. As expected the levels of 1-OHP in urine were higher in people that use fuelwood (table 6).

Community	Characteristics
Victoria, Chiapas	Rural community with fuelwood combustion
Ventanilla, Oaxaca	Rural community with fuelwood combustion
Ramonal, Quintana Roo	Rural community with fuelwood combustion
Tancuime, SLP	Rural community with fuelwood combustion
San Vicente, SLP	Community with 50 brick kilns using different materials as fuel source
Milpillas, SLP	Community living next to municipal landfill in the city of San Luis Potosí
Tercera, SLP	Community with 75 brick kilns using different materials as fuel source
Domingo, SLP	Community in San Luis Potosí state exposed to heavy vehicular traffic
Centro, SLP	Urban community in the city of San Luis Potosí with moderate vehicular traffic

Table 3. Characteristics of studied sites

Residential Group	Mean ± SD 1-OHP (µmol/mol creatinine)
moderate vehicular traffic	0.08 ± 0.2
heavy vehicular traffic	0.20 ± 0.2
fumes from a municipal landfill	0.30 ± 0.4
fumes from brick kiln	0.35 ± 0.3
indoor air pollution by biomass combustion	2.26 ± 2.8

Table 4. Urine concentrations of 1-OHP in children by exposure scenario in México.

4. Health effects associated with indoor air pollution

The adverse effects on respiratory health of products of incomplete solid-fuel combustion are summarized in Table 7, which also includes some of the known or proposed mechanisms of damage. Exposure to solid fuel smoke can be lifelong, beginning before birth and early infancy, and continuing during adulthood, especially in women, who are traditionally charged with the task of cooking. Exposure is longer in cold communities that require fire-related heating, and may adversely impact lung growth and development, both directly and through an increase in lung infections. Indoor air pollution from indoor burning of solid fuels has been associated with an increased risk of several diseases and health conditions (Table 7). In general, studies are scarce, and show varied health outcomes. Moreover, they commonly lack a quantitative exposure assessment, and rely instead on qualitative or semi-quantitative indicators, such as the use of open fire indoors. The amount of time that children and/or women spend in proximity to fires is the crucial determinant of the health impact of indoor air pollution (Barnes et al. 2005). For other health outcomes, the adverse effects of exposure to solid fuel smoke from coal or biomass is expected, as from exposure to tobacco smoke, but information is lacking or scarce about other consequences such as low birth weight and adverse perinatal outcomes (stillbirth), among others (WHO, 2007) For example, biomass smoke in Guatemalan women has been shown to increase diastolic blood pressure (McCracken et al. 2007) According to World Health Organization estimates, worldwide exposure to solid fuel smoke produces 1.6 million deaths yearly, 693 000 due to COPD and 910 000 due to acute lower respiratory infections (ALRI), as well as 38.5 million disability adjusted life years (DALYs), most due to ALRI, being the eighth overall cause of DALYs in the world and the eleventh cause of death (Smith et al. 2004). This is likely an underestimation, as only diseases with a strong evidence base, i.e., COPD, ALRI, and lung cancer from coal burning, are considered.

Residential Group	<0.24 μmol/mol creatinine (%)	0.24-1.39 μmol/mol creatinine (%)	1.4-2.3 μmol/mol creatinine (%)	>2.3 μmol/mol creatinine (%)
moderate vehicular traffic	95.0	5.0	0.0	0.0
heavy vehicular traffic	53.0	47.0	0.0	0.0
fumes from a municipal landfill	66.0	32.0	2.0	0.0
fumes from brick kiln	51	47.0	2.0	0.0
indoor air pollution by biomass combustion	7.0	14.0	25.0	54.0

Table 5. Percentage of children in each range of 1-OHP urinary levels.

Community	Mean ± SD 1-OHP (μmol/mol creatinine)
La Huasteca in San Luis Potosi state	3.97 ± 5.51
City of San Luis Potosí	0.44 ± 0.70

Table 6. Urine concentrations of 1-OHP in women (aged 15-50 years) in San Luis Potosi state.

In México information regarding health effects associated with indoor air pollution is scarce. However, in a study conducted in the rural village of Solis, composed of 13 small communities, located 200 km northwest of Mexico City and at an elevation of 2,600 m above sea level. Women in Solis have cooked with wood, crop residues, and corn cobs as fuel, but now some of them also use natural gas either alone or supplemented with biomass fuel. When compared with those cooking with gas, current use of a stove burning biomass fuel was associated with increased reporting of phlegm (27 vs. 9%) and reduced FEV1/FVC (79.9 vs. 82.8%). Levels of FEV1 were 81 ml lower and cough wasmore common (odds ratio, 1.7;

95% confidence interval, 1.0–2.8) in women from homes with higher PM10 concentrations. All women found with moderate airflow obstruction were cooking with biomass stoves (Regalado et al. 2006). Similar results were found by Romieu et al. 2009.
In order to study other health effects associated with indoor air pollution, our group evaluated blood carboxyhemoglobin (% COHb) and DNA damage in blood cells (with comet assay) in a community in San Luis Potosi state exposed to wood smoke (Torres-Dosal et al. 2008). Since carbon monoxide (CO) is probably the most important single contaminant emitted during combustion of wood (Viau et al., 2000), A marker use to indirectly assess CO exposure is carboxyhemoglobin (COHb) that reflects binding of CO to the hem portion of hemoglobin. A concentration of COHb <2.5% is currently considered safe (Kleinman, 2000). The lowest level of COHb, at which adverse effects are observed, ranges from 2.9 to 3% (Estrella et al., 2005). Moreover, COHb concentrations >5% are associated with effects, such as neurobehavioral function, impaired visual function, task performance, and maintaining alertness (WHO, 1999; EPA, 2000; Raub and Benignus, 2002). All studied individuals had a geometric mean COHb level of 4.93% and 55% of those people presented COHb levels above 2.5% considered a safe level. While, DNA damage in people exposed was 5.8 ± 1.3 of Tail Moment. In this study also was monitored 1-OHP in studied individuals and significant positive correlation was obtained between urinary 1-OHP and DNA damage in blood cells. Our data suggest that DNA damage in lymphocytes is at least partially related to exposure to wood smoke, as a relationship was found between DNA damage in lymphocytes and levels of 1-OHP in urine of people studied (Table 8).

Compound	Potential health effect
Breathable particulate matter	Wheezing, exacerbation of asthma, respiratory infections, chronic bronchitis, and chronic obstructive pulmonary disease.
Carbon monoxide	Low birth weight, increase in perinatal deaths
Polycyclic aromatic hydrocarbons	Lung cancer, cancer of mouth, nasopharynx and larynx.
Nitrogen oxides	Wheezing, exacerbation of asthma, respiratory infections, reduced lung functions in children.
Sulphur oxides	Wheezing, exacerbation of asthma, Exacerbation of chronic obstructive pulmonary disease, cardiovascular disease.

Table 7. Adverse effects on respiratory health associated with indoor air pollution

5. Interventions programs

Far from a relatively simple problem with fixed technical solutions, the "biofuel problématique" in developing countries has turned out to be a very complex issue with multiple implications. The patterns of household biofuel use are very heterogeneous, as are also the people, the environment and the cultures that depend on these fuels to solve their essential cooking needs. The uses of energy in the home—for example, for cooking and keeping warm and as a focus of social activities—have important attributes that are specific to the locality, culture, and individual households and are often associated with established traditions and deeply held beliefs. Encouraging the use of cleaner and more efficient energy technologies by populations that are among the poorest in the world has not been easy, but recent years have seen progress being made with respect to suitable technology that meets the needs of households and with respect to the development of supportive policy.

There is limited experience with improved wood-burning cookstove programs in Mexico (Olguín, 1994; Arias and Cervantes, 1994; Navia, 1992; Dutt et al., 1989). The government launched a large-scale effort at the beginning of the '80s, with poor results, and basically abandoned the field (Vargas, 1990). In recent years, a number of initiatives have been started in different regions of Mexico, particularly within the states of Michoacan, Chiapas, Oaxaca and San Luis Potosí, working with a diversity of stove designs and dissemination schemes (RETA, 2004; Cayetano, 1997). Most groups still work with massive Lorena-type improved stoves, but there are groups now disseminating "Justa" and portable "Rocket" stoves in southern Mexico. Most of the improved cookstove programs are local (village- level) or micro-regional in scope, they are generally part of larger initiatives directed at restoration of local forests or conservation of biodiversity, or they are part of the portfolio of well-established peasant organizations. Efforts are still concentrated on the construction of stoves with little or no follow-up; stoves are entirely or mostly subsidized; there is still emphasis on stoves being constructed by users themselves. Some institutions promoting these stove programs have become more aware of users' priorities and needs, resulting in higher stove acceptance rates.

	Before program intervention	**After program intervention**
% COHb (mean ± SD)	4.9 ± 4.3	1.0 ± 0.2
1-OHP (μmol/mol creatinine; mean ± SD)	6.7 ± 3.6	4.8 ± 3.3
Tail moment (comet assay; mean ± SD)	5.8 ± 1.2	2.4 ± 1.3

Table 8. A risk reduction program using biomarkers of exposure (COHb and 1-OHP) and a biomarker of effect (DNA damage).

In this regard, several intervention programs have been implemented. For example, the Grupo Interdisciplinario de Tecnología Rural Apropiada (GIRA) has recently disseminated 4,000 improved Patsari cookstoves, most of them in the Purépecha region of Michoacán state, Mexico. In paired comparisons in a subset of kitchens in a single community before and after installation of an improved Patsari cookstove, 48-hour average kitchen concentrations of carbon monoxide (CO) and fine particulate matter (PM2.5) were reduced by 66 % (n = 32) and 67 % (n = 33), respectively (Masera et al. 2007). The results are excitants; however health concerns are now a major focus of the new initiatives. A randomized study was conducted in the Central Mexican state of Michoacan. Households were randomized to receive the Patsari stove or keep their traditional open fire. A total of 552 women were followed with monthly visits over 10 months to assess stove use, inquire about respiratory and other symptoms, and obtain lung function measurements. Women who reported using the Patsari stove most of the time compared with those using the open fire had significantly lower risk of respiratory symptoms (relative risk [RR], 0.77; 95%confidence interval [CI], 0.62–0.95 for cough and RR, 0.29; 95% CI, 0.11–0.77 for wheezing) adjusted for confounders. Similar results were found for other respiratory symptoms as well as for eye discomfort, headache, and back pain (Romieu et al 2009). Our group performed a study to evaluate a risk reduction program using biomarkers of exposure (COHb and 1-OHP) and a biomarker of effect (DNA damage) in a community in San Luis Potosí state. An initial census survey of the study area was used to identify homes with open fires stoves, indoor soot adhered to roofs and internal walls and dirt floors. Thus, for the intervention we offered 1) removal of indoor soot adhered to roofs and internal walls, 2) paving dirt floors, and 3) an improved stove (the Patzari stove),which is constructed using sand, clay and cement; it also has a metal chimney that expels the smoke outdoors (Masera et al., 2007; Figure 5). The complete 3-stage risk reduction program was applied in all houses studied. We applied a questionnaire to confirm that the improved stove was well accepted by people in the community. The program was introduced in ten indigenous houses during August 2005; we assessed exposure to carbon monoxide, PAHs (1-OHP) and DNA damage previously and one month after the intervention program. Table 8 shows the geometric mean COHb level in all subjects studied before and after intervention. Before intervention all individuals had a geometric mean COHb level of 4.93% and 55% of those people presented COHb levels above 2.5% considered a safe level. Moreover, this 55% of people had geometric mean COHb levels of 8.39%. However, in all the studied individuals the levels of COHb were reduced to below 2.5% (mean level 1.0%) one month after intervention. Urinary 1-OHP was used to assess the exposure to PAHs (Table 8). Results show that the geometric mean level of 1-OHP before intervention was significantly higher in studied people (6.71±3.58 μmol/molCr) than after the program (4.80±3.29 μmol/ mol Cr). The amount of DNA damage in the studied people is depicted in Table 8. When compared, DNA damage in people exposed before the intervention was higher (5.8±1.3 of Tail Moment) than when the program was introduced (2.8±0.9 of Tail Moment). When the studied population was divided by gender or age groups, we did not observe differences in Tail Moment. Usually, risk reduction programs for indoor air pollution caused by biomass combustion only include the installation of a stove with a chimney (Chapman et al., 2005; Smith-Sivertsen et al., 2004; Khushk et al., 2005; Naeher et al., 2000). Those programs were evaluated either through a decrease of indoor air pollution (Riojas-Rodriguez et al., 2001; Naeher et al., 2000; Khushk et al., 2005; Zuk et al., 2007) and/or by a reduction in the incidence of diseases related with this kind of contamination (Riojas-Rodriguez, et al., 2001; Chapman et al., 2005; Boy et al., 2000; Smith-Sivertsen et al., 2004). We believe that our study represents an improvement in this area, as we installed a stove with a metal chimney that expels smoke outdoors, but also indoor soot adhering to roofs and internal walls was removed, and dirt floors were paved (it has been shown that biomass burning might be the major origin of PAHs in rural soil; Zhang et al., 2006). It is important to mention that soot is a sink of several chemicals generated during combustion (Jonker and Koelmans, 2002a, 2002b) and moreover, an extremely slow desorption of PAHs from soot has been demonstrated (Jonker et al., 2005). Therefore, although we do not have information regarding the importance of each stage in risk reduction, the elimination of soot sources (in roofs, walls and dirt floor) is a relevant action considering that this material is a sink of hydrophobic compounds such as PAHs and Dioxins. Children and adults can be exposed to this material, by inhalation (indoor air particles), ingestion (dust particles) or direct dermal exposure. This work was a pilot study and although with the limitation in sample size, the improvement in exposure (1-OHP, and COHb) and effect (comet assay) was so homogeneous that the State Government of San Luis Potosi, using the precautionary principle has expanded the program to different communities. We are following our studies in these sites.

Cookstove dissemination programs are evolving from "projects" narrowly targeting stove construction and fuelwood savings to more integrated "program" approaches looking at providing a set of health, environmental and socio-economic benefits. For them to be successful, systemic and interdisciplinary approaches are needed, including work on technology innovation, users' needs and priorities, market development and innovative financing. Strong and sustained efforts need to be devoted to monitoring stove performance in the field and to ensure the sustainability of the achieved benefits. Being more complex and ambitious, the new approaches come with a new set of challenges. More than ever, an intense cross-fertilization among groups, both North-South and South-South, is urgently needed. Donors and governments need to realize that sustained and longterm efforts rather than the typical two- to three-year projects are needed for setting up the type of integrated approaches described above. There is a need to support continuous technology innovation, getting research institutions to work hand in hand with local organizations on new stove designs that are more robust, cost-effective and clean. Participatory research is also needed to better understand users' priorities and needs and also to better assess the dynamics of multiple fuel use. Innovative and creative financial mechanisms are required that help users

Fig. 5. Patsari wood-burning cookstove.

overcome the investment costs of improved cookstoves. Users – specifically women -- need to have a strong say in program design and implementation. Finally, to be successful in the medium and long term, cookstove programs need to look at the overall policy context and integrate cookstove programs within larger sustainable rural development strategies. The challenge is substantial, but fostering the current partnership and networks on household energy use and learning from each other's experience will surely pave the way to a larger number of success stories in the near future.

Characteristics of Patsari stoves: The Patsari is oriented to mass dissemination; it is built with the help of a mould, and includes several custom-made pieces. The size and geometry of the combustion chamber (primary furnace), the tunnels and the secondary furnaces are designed to increase heat transfer to the pot. Metal flat pans (or comales) are tightly adjusted to the furnaces to prevent smoke leakage. The stove comes with a metal chimney, which is inserted into the stove body with the help of a custom-made metal base built in the stove. Currently, two models of Patsari, one with a single entry for fuelwood and a second with two entries, are being disseminated. In the latter case, both a ceramic comal and a metal comal are used in the primary furnaces (Figure 6).

Fig. 6. Design of Patsari wood-burning cookstove.

6. Conclusion

Indoor pollution will continue to be an essential field for health studies and interventions, because exposure to varied indoor substances will likely increase in coming years. Better studies dealing with genetic susceptibility to indoor pollutants, their carcinogenic effect and their impact on lung growth, lung development and, later on, lung aging, are also required. To answer several of these questions, longitudinal studies are required. A formal evaluation of improved stove programs from many viewpoints is also essential to improve guidance for countries and communities as they implement their own programs. Although local adaptation of programs will always be required, improved biomass stoves will likely be more common, with better community acceptance, reduced burden on forests, and increased spare time for homemakers.

7. Reference

Albalak, R. (2001). Indoor respirable particulate matter concentrations from an open fire, improved cookstove, and LPG/open fire combination in a rural Guatemalan community. Environ Sci Technol. 35:2650–5.

Arias, T.; & Cervantes, V. (1994). Las estufas de barro ahorradoras de leña ¿Una tecnología apropiada para la región de la Montaña de Guerrero?, Programa de Aprovechamiento Integral de Recursos (PAIR)-UNAM, México.

Barnes, B.; Mathee, A.; & Moiloa, K. (2005). Assessing child time-activity patterns in relation to indoor cooking fires in developing countries: a methodological comparison. Int J Hyg Environ Health. 208: 219–225.

Behera, D.; Dash, S.; & Malik, S. (1988). "Blood Carboxyhaemoglobin Levels Following Acute Exposure to Smoke of Biomass Fuel." Indian Journal of Medical Research 88: 522–524.

Boy, E.; Bruce, N.; Smith, K.R.; & Delgado, H. (2000). Fuel efficiency of an improved wood-burning stove in rural Guatemala: implication for health, environment and development. Energy Sustain Dev. 2:21–29.

Brauer, M.; Bartlett, K.; Regalado-Pineda, J.; & Pérez-Padilla, R. (1996). Assessment of particulate concentrations from domestic biomass combustion in rural Mexico. Environ.Sci. Tech, 30(1): 104-109.

Bruce, N.; Pérez-Padilla, R.; & Albalak, R. (2000). ''Indoor air pollution in developing countries: a major environmental and public health challenge''. Bulletin of the World Health Organization, 78(9): 1078-1092.

Bruce, N. G.; Bates E.; Nguti, R.; Gitonga, S.; Kithinji, J.; & Doig, A. (2002). "Reducing Indoor Air Pollution through Participatory Development in Rural Kenya." In Proceedings of 9th International Conference on Indoor Air Quality and Climate, Monterey, CA, 590–95.

Cayetano, H. (1997). Curso taller de promoción y construcción de estufas rurales en la comunidad de Santa Cecilia Lalana, Asesoría Técnica a Comunidades Oaxaqueñas (ASETECO), Oaxaca, Oax., Mexico.

Chapman, R.S.; He, X.; Blair, A.E.; & Lan, Q. (2005). Improvement in household stoves and risk of chronic obstructive pulmonary disease in Xuanwei, China: retrospective cohort study. BMJ. 331:1050–1056.

Dutt, G.S.; Navia, J.; & Sheinbaum, C. (1989). Cheranátzicurin: tecnología apropiada para cocinar con leña. Ciencias. 15: 43-47.

EPA, U. S. Environmental Protection Agency. (2000). Air quality criteria for carbon monoxide. EPA 600/P-99/001F. Washington, D.C. Office of Research and Development, U. S. Environmental Protection Agency.

Estrella, B.; Estrella, R.; Oviedo, J.; Narváez, X.; Reyes, M.T.; Gutiérrez, M.; & Naumova, E.N. (2005). Acute respiratory diseases and carboxyhemoglobin status in school children of Quito, Ecuador. Environ Health Perspect. 113:607–611.

Jacob, J.; & Seidel, A. (2002). Biomonitoring of polycyclic aromatic hydrocarbons in human urine. J Chromatogr B Analyt Technol Biomed Life Sci 778:31–47.

Jenkins, BM.; Jones, A.D.; Turn, S.Q.; & Williams, R.B. (1996). Emission factors for polycyclic aromatic hydrocarbons from biomass burning. Environ Sci Technol. 30:2462–2469.

Jongeneelen, F.J. (2001) Benchmark guideline for urinary 1-hydroxypyrene as biomarker of occupational exposure to polycyclic aromatic hydrocarbons. Ann Occup Hyg. 45:3–13

Jonker, MT.; Hawthorne, S.B.; & Koelmans, A.A. (2005). Extremely slowly desorbing polycyclic aromatic hydrocarbons from soot and soot-like materials: evidence by supercritical fluid extraction. Environ Sci Techno. 39: 7889–7895.

Jonker, M.T.; & Koelmans, A.A. (2002a) Sorption of polycyclic aromatic hydrocarbons and polychlorinated biphenyls to soot and soot-like materials in the aqueous environment: mechanistic considerations. Environ Sci Technol. 36: 3725–3734.

Jonker, M.T.; & Koelmans, A.A. (2002b). Extraction of polycyclic aromatic hydrocarbons from soot and sediment: solvent evaluation and implications for sorption mechanism. Environ Sci Technol. 36: 4107–4113.

Khushk, W.A.; Fatmi, Z.; White, F.; & Kadir, M.M. (2005). Health and social impacts of improved stoves on rural women: a pilot intervention in Sindh, Pakistan. Indoor Air. 15: 311–316.

Kleinman, M.T. (2000). Carbon monoxide: evaluation of current California air quality standards with respect to protection of children. Prepared for California Air Resources Board, California Office of Environmental Health Hazard Assessment. Irvine, CA: University of California Irvine; [Departament of Community and Environmental Medicine].

Levine, J.O. (1995). First international workshop on hydroxypyrene as a biomarker for PAHexposure inman—summary and conclusions. Sci Total Environ. 163:164–168.

McCracken, J.P.; Smith, K.R.; Diaz, A.; Mittleman, M.A.; & Schwartz J. (2007). Chimney stove intervention to reduce long-term wood smoke exposure lowers blood pressure among Guatemalan women. Environ Health Perspect. 115: 996–1001.

Martínez-Salinas, R.I.; Leal, M.E.; Batres-Esquivel, L.E.; Domínguez-Cortinas, G.; Calderón, J.; Díaz-Barriga, F.; & Pérez-Maldonado, I.N. (2010). Exposure of children to polycyclic aromatic hydrocarbons in Mexico: assessment of multiple sources. Int Arch Occup Environ Health. 83:617–623.

Masera, O. (1996a). Deforestación y degradación forestal en México. Documentos de trabajo N°19, Grupo Intrediciplinario de Tecnología Rural Apropiada (GIRA A.C.) Pátzcuaro, México.

Masera, O. (1996b). Uso y Conservación de Energía en el Sector Rural: El caso de la leña. Documentos de Trabajo N°21, Grupo Interdisciplinario de Tecnología Rural Apropiada (GIRA A.C.). Pátzcuaro, México.

Masera, O.; Navia, J.; Arias, T.; & Riegelhaupt, E. (1997). Patrones de Consumo de Leña en Tres Micro-regiones de México: Síntesis de Resultados. Proyecto FAO/MEX/TCP/4553(A). Grupo Intrediciplinario de Tecnología Rural Apropiada (GIRA A.C.). Pátzcuaro, México.

Masera, O.R; Drigo, R.; & Trossero, M.A. (2003). Woodfuels integrated supply/demand overview mapping. Universidad Autónoma de México, FAO-EC Partnership Programme. Food and Agriculture Organization of the United Nations.

Masera, O.; Edwards, R.; Armendáriz-Arnez, C.; Burrueta, V.; Johnson, M.; Rojas-Bracho, L.; Riojas-Rodríguez H.; & Smith, KR. (2007). Impact of Patsari improved cookstove on indoor air quality in Michoacán, México. Energy Sustain Dev. 9:25–36.

Mcdonald, J.D.; Zielinska, B.; Fujita, E.M.; Sagebiel, J.C.; Chow, J.C.; & Watson, J,G. (2000). Fine particle and gaseous emission rates from residential wood combustion. Environ Sci Technol. 34:2080-2091.

Mishra, V. (2003). Effect of indoor air pollution from biomass combustion on prevalence of asthma in the elderly. Environ Health Perspect. 111:71-78.

Naeher, L.P.; Smith, K.R.; Leaderer, B.P.; Mage, D.; & Grajeda, R. (2000). Indoor and outdoor PM2.5 and CO in high- and low-density Guatemalan villages. J Expo Anal Environ Epidemiol. 10: 544-551.

Navia, J. (1992). Estufas mejoradas, programa de difusión en Cheran Atzícurin, Primera Reunión Internacional sobre energía y medio ambiente en el sector residencial mexicano, comp. J. Quintanilla.

Olguín, E. (1994). 'Evaluación y optimización del uso de la leña a nivel familiar y de pequeñas industrias rurales. Instituto de Ecología, A.C. Jalapa, Veracruz, Mexico.

Raub, J.A.; & Benignus, V.A. (2002). Carbon monoxide and the nervous system.Neurosci Biobehav Rev. 26:925-940.

Red de Tecnologías Apropiadas (RETA). (2004). Memorias del Taller de Intercambio de experiencias sobre estufas ahorradoras de leña, San Cristóbal de Las Casas, Chiapas, México, 11 pp.

Reddy, AKN.; Williams, RH.; Johansson TB. (1997) Energy after Rio: prospects and challenges. New York: United Nations Development Programme.

Regalado, J.; Perez-Padilla, R.; Sansores, R.; Paramo Ramirez, J.I.; Brauer, M.; Pare, P.; & Vedal, S. (2006). The Effect of Biomass Burning on Respiratory Symptoms and Lung Function in Rural Mexican Women. Am. J. Resp. Crit. Care Med. 174: 901-905.

Riojas, R.H.; Romano, P.; Santosburgoa, C.; &Smith, K.R. (2001). Household firewood use and the health of children and women of Indian communities in Chiapas, México. Int J Occup Environ Health, 7(1): 44-53.

Romieu I.; Riojas-Rodrıguez, H.; Marron-Mares, A.T.; Schilmann, A.; Perez-Padilla, R. & Masera, O. (2009). Improved Biomass Stove Intervention in Rural Mexico. Impact on the Respiratory Health of Women. Am. J. Resp. Crit. Care Med. 180: 649-656.

Saksena, S.; Thompson, L.; & Smith, K.R. (2004). "Indoor Air Pollution and Exposure Database: Household Measurements in Developing Countries." http://ehs.sph.berkeley.edu/hem/page.asp?id=33.

Secretaría de Energía (SE). (1998). Balance Nacional de Energía. Gobierno de México, Secretaría de Energía, México D.F.

Secretariat of Health in Mexico (SSA). (2005). Children Health and the Environment in North America. A First Report on Aviable Indicators and Measures.

Available in http://www.cec.org/Storage/27/1802_CountryReport-Mexico-CHE_en.pdf. Accessed on 18-03-2011.

Smith KR. (1987). Biofuels, air pollution, and health: a global review. New York: Plenum Press.

Smith, K.; Mehta, S.; & Maeusezahl-Feuz, M. (2004). Indoor air pollution from household use of solid fuels. In: Ezzati M, Lopez A, Rodgers A, Murray C, eds. Comparative quantifi cation of health risks. Global and regional burden of disease attributable to s elected major risk factors. Geneva, Switzerland: World Health Organization. pp 1435-1493.

Smith-Sivertsen, T.; Diaz, E.; Bruce, N.; Diaz, A.; Khalakdina, A.; Schei, M.A. et al. (2004). Reducing indoor air pollution with a randomized intervention design—a presentation of the stove intervention study in the Guatemala Highlands. Nor Epidemiol. 14: 137–143.

Saatkamp, B.D.; Masera, O.; & Kammen, D.M. (1999). Energy and health transitions in development: fuel use, stove technology, and morbidity in Jarácuaro, Mexico. Energy for Sustainable Development. IV(2): 7-16.

Torres-Dosal A, Pérez-Maldonado IN, Jasso-Pineda Y, Martínez Salinas RI, Alegría-Torres JA, Díaz-Barriga F (2008) Indoor air pollution in a Mexican indigenous community: evaluation of risk reduction program using biomarkers of exposure and eVect. Sci Total Environ 390:362–368

Vargas, F. (1990). Breve diagnóstico sobre el proyecto de estufas rurales en México, Secretaria de Agricultura y Recursos Hidraúlicos (SARH), Mecanografiado, 8 pp.

Viau, C.; Hakizimana, M.; & Bouchard, M. (2000). Indoor exposure to polycyclic aromatic hydrocarbons and carbon monoxide in traditional houses in Burundi. Int Arch Occup Environ Health. 73:331-338.

World Health Organization. (WHO. 1997). Health and environment in sustainable development. Geneva, Switzerland: World Health Organization.

World Health Organization. (WHO. 1999). Environmental health criteria 213, carbon monoxide. IPCS, International Programme on Chemical Safety. Second Ed.Geneva, Switzerland: World Health Organization; 1999.

World Health Organization. (WHO. 2007). Indoor air pollution from solid fuels and risk of low birth weight and stillbirth: report from a symposium held at the Annual Conference of the International Society for Environmental Epidemiology (ISEE), September 2005, Johannesburg. Geneva, Switzerland.

Zhang, J. (1999). Smith KR. Emissions of carbonyl compounds from various cookstoves in China. Environ Sci Technol. 33:2311–2320.

Zhang, H.B.; Luo, Y.M.; Wong, M.H.; Zhao, Q.G.; & Zhang, G.L. (2006). Distribution and concentrations of PAHs in Hong Kong soils. Environ Pollut. 141:107–114.

Zhao, Z.H.; Quan,W.Y.; Tian, D.H. (1992). The relationship between polynuclear aromatic hydrocarbons in ambient air and 1-hydroxypyrene in human urine. J Environ Sci Health. A27:1949–1966.

Zuk, M.; Rojas, L.; Blanco, S.; Serrano, P.; Cruz, J.; Angeles, F.; Tzintzun, G.; Armendariz, C.; Edwards, R.D.; Johnson, M.; Riojas-Rodriguez, H.; Masera, O. (2007). The impact of improved wood-burning stoves on fine particulate matter concentrations in rural Mexican homes. J Expo Sci Environ Epidemiol. 17:224–232.

18

Epidemiologic and Economic Impact of Natural Gas on Indoor Air Pollution in Colombia 2007

Nelson Alvis Guzmán[1], Luis Alvis Estrada[1] and
Fernando De la Hoz Restrepo[2]
[1]*Universidad de Cartagena*
[2]*Universidad Nacional de Colombia*
Colombia

1. Introduction

Colombian development since 1990, it can be describe according to evolution of three components of the Human Development Index: internal product, life expectancy and education. As for the gross domestic product, there has been show a slow but steady growth, excluding the 1999 crisis that resulted in a decrease of about four years, from which now is recovering. Life expectancy of Colombian population increased for about 5 years during 90th decade, because important epidemiologic and demographics changes, such as improvements in infant mortality and maternal health services, leading to a gradual recovery of the indicator, with some differences between genders, because of the greater impact than men suffered from the effect of violence. In education, there have had significant advances that although were hampered by the economic crisis, currently with the emphasis in social policies relation to education, there has been considerable improvement in the indicators. Although the overall picture is positive, there are still large differences between population groups, making equity a major challenge to human development in Colombia. Analyses by state, indicators are similar between Bogota and Hungary (country of high human development, 38th in the world) while the Chocó resemble Kenya (country of low human development, since 148 in the world). For Zones, most of the indicators in rural areas are lower than in urban areas and between social groups by deciles of the income distribution is unsatisfactory (Departamento Nacional de Planeacion de Colombia, [DNP], 1991).

In search of a universal sign for development, Colombia and 188 nations agreed at the Millennium Summit in September 2000 within the framework of the General Assembly of the United Nations, eight Millennium Development Goals in the long term, with coordination of UN, World Bank, IMF and OECD. Each country undertook to define national targets, which can be reached in 2015, for each of the Target (DNP, 1991).

However, there are studies that question the possible compliance of Colombia, some of the MDGs, especially related to poverty reduction: "... the scope in terms of poverty reduction is not so encouraging. Although according to international standards (PPP U.S. $ 1 and $ 2 PPP), the country would be very close to achieving the target reduction of 50%, it is clear that when the measuring stick becomes the national poverty line developments indicate that the moderate level of poverty would be only 20% lower than the 2001 (The poverty rate would rise from 55.2% to 44.2%). These results indicate that progress in other areas of society is not enough to boost the capacity of generating household income Colombians and that the strategy should focus on attacking directly additionally other determinants of poverty such as lack of access to labor markets and financial and serve the population in extreme poverty in an integrated way to break the poverty traps facing this population " (Nuñez, 2007).

In this sense, the MDGs by the different countries depends not only on the commitment of states to finance social development and undertake appropriate actions, also the contribution of civil society and the various cooperating agencies are willing to make to achieve more just societies.

Moreover, both programs targeting of social spending by the state as part of national and international, are coherent strategies for progress towards achieving better results in material development and social welfare. However, although the application of resources in development projects and / or involve social opportunities for communities to improve their capacities to address the different risks; the sole decision to implement grant programs to improve access to social services, does not mean that the results are guaranteed. Therefore, it is necessary to impact assessment procedures for specific programs and projects in order to gauge the true impact of the funds on the expected results.

2. Household natural gas as health technology

Living conditions, especially those related to health, an important influence on the economic situation and welfare of everyone in any society, which is particularly true in low income countries (where social safety nets are weak or nonexistent) and for the poor, due to the vicious circle of poverty and disease (Schultz, 1962). Health is considered a capital stock which the product is long lasting health, which can be used to obtain money or wages in the market for "goods" in the home or non-market sector (Grossman, 1972). It is well documented the relationship between health risk and poverty (World Health Organization [WHO], 2001).

On the other hand, the use of solid fuels for cooking and / or heating is the largest source of air pollution inside houses in the world. Nearly half of households in the world continue to use biomass fuels (wood, charcoal, dung and agricultural wastes). When biofuels are used, for example the stoves inefficient combustion, generate high concentrations of pollutants. Exposure to biomass smoke has been associated with low birth weight, acute respiratory infections, infant mortality and recently, anemia and mental retardation in children (Mishra, 2007). These pollutants, called solid fuel smoke (smoke) include respirable particulates, carbon monoxide, nitrogen oxide and sulfur, benzene, formaldehyde, 1,3 butadiene and polyaromatic compounds such as benzo (alpha) pyrene (Smith, 1987; Pandey, 1989). In households with limited ventilation, (both in developing countries and industrialized) exposure to such contaminants is higher in women and children, who stay longer in the house. Their measurements showed that concentrations of these pollutants were above the international standards (WHO) (Smith, 1987; Bruce, 2000).

The combustion of natural gas worldwide is classified as the cleanest among the traditional industrial fuels. In fact, emissions of particulate matter to meet the highest international standards, without having to invest in gas processing equipment. One of the great advantages of natural gas over other fuels, are the low combustion emissions, which can be seen in the following table (Innergy Soluciones Energeticas, 2007).

Fuel	MP Particulado Matter	SOX Sulfur Oxide	NOX Nitric Oxide
Gas Natural	1	1	1
Gas City	3	61	0,5
Gas	1,4	23	2
Kerosene	3,4	269	1,5
Diesel	3,3	1.209	1,5
Fuel Oil N° 5	15	4.470	4
Fuel Oil N° 6	39,4	4.433,	4
Coal	157	5.283	6
Firewood	140	13	2

Table 1. Concentrations of contaminants in relation to alternative fuels different of the NG

Most research on environmental pollution intradomicilliary has been conducted in developing countries and has been associated with the determination of pollutants and their sources (Spengler, 2001). This has been done by estimating the levels of risk factors generated by the combustion of biomass, through the determination of the exposure variable. To do this, Smith et al. 2004, proposes the use of the well established exposure-response epidemiological studies conducted outside the houses and measuring concentrations of pollutants, such as an estimate of what happens inside the houses (Smith, 2004). Another approach is to use studies in developing countries, dividing the population into categories of exposed and unexposed to smoke the product of biomass burning. However, there are gaps in data on exposure of such populations, and variability between groups exposed. In addition, this method can be affected by the differential use of fuels and different mixtures of pollutants, which could prevent establishment of only two categories.
Apart from the use of biomass fuels, another factor to consider is related to the ventilation conditions of housing and particularly the kitchen. This equation can build a better, in which exposure to smoke from solid fuels (biomass) is a function of the number of people using such fuels, multiplied by a factor of ventilation.

$$P_e \cong P_b \times F_V$$

PE = Population exposed to smoke from biofuels
PB = Population using biofuels
FV = Ventilation factor
Based on this model were estimated using solid fuels for subregions in the world (Smith, 2004). For the subregion in which America is Colombia (AMR-B), the population using solid fuels was estimated at 24.6% (CI. 18.8-30.8). This means that in Colombia about a quarter of the population could be using biofuels. Of total DALYs, it is estimated that 2.7% is attributed to diseases associated with smoke from solid fuels (2.5% in men and 2.8% in women). Of this total, 32% occur in Africa (AFR-D), 37% in Southeast Asia (SEAR-D) and 16% in China, Australia and other countries (WPR-B) (Smith, 2004). Surely, that is spread by region and intra-regionally, as far as access to clean fuels is different for each region.
The strength of evidence for each factor was determined based on the structured assessment of causality, using the criteria of Bradford Hill (Hill, 1992). Therefore the specificity, dose-response and experimental evidence is often difficult to assess environmental exposures and their health effects, we used the epidemiological evidence together with the emission

mechanism of exposure and pollution (Smith, 2000; Zodpey, 1999). There are determined three health effects for which there is clear evidence:

1. Acute Lower Respiratory Infection
2. Chronic Obstructive Pulmonary Disease
3. Lung Cancer.

Children under 5 years are the highest exposure group, although levels are unknown globally (Ezzati, 2001; Saksena, 1992). About half of the world's countries cook with solid fuels, including more than 75% of people in India, China and nearby countries, 50 to 75% of people in some countries of South America and Africa. Inadequate ventilation in homes is a common factor in developing countries, which increases the exposure, especially for women and children who stay more time inside the house. In many measurements made by the WHO, has been found that concentrations of pollutants inside homes exceed by a large margin, the measurements on the outside. Multiple studies have shown a consistent relationship between the use of solid fuels and several diseases. These tests show that smoke from solid fuels causes about 35.7% of the IRA, 22% of COPD and 1.5% of cancers tracheae, bronchi and lungs (WHO, 2002).

Health intervention is considered, all that technology into their practices creates possibilities of improving the living conditions of the individual and / or community. Natural gas has become a health intervention as far as a major impact not only the household economy, but can reduce the burden of respiratory diseases in the communities in which it is implemented, especially in children, women and elderly.

3. Estimation of economic and epidemiological impact of household natural gas - analysis plan

To perform the study combined the following studies: a) secondary data analysis (a systematic review of the literature) and analysis of databases provided by the Foundation Promigas, demographic data from DANE 2005 Census, and databases of economic statistical series of the Bank of the Republic. b) Analysis of disease burden estimated from the parameters identified in the previous studies. c) Analysis of the cost burden of illness and the estimated costs avoided then implemented the program natural gas home connections (Arredondo, 1997). d) Cost effectiveness analysis of connections of households to service natural gas home (Drummond, 2005).

The study was conducted in two phases: a) Estimating the burden of disease associated diseases biomass smoke (ARI and COPD) in the population covered by the project further at this stage to combine the results of the literature review; b) Estimating the potential economic impact of the grant program's connection to the Household Natural Gas -HNG.

The estimated population data are from field research conducted in the projected 2005 population census DANE.

We carried out an analysis of cost-effectiveness of introducing natural gas house on morbidity and mortality from respiratory diseases in children and older adults, using a simple decision that assessed the burden and costs of illness in the absence and presence of HNG.

Both share the possible scenarios in which events can be found elsewhere, for example: patients with ARF and / or IRA without COPD and / or COPD. The transition from one event to another is determined by the probability of coverage and efficiency of natural gas as a health technology. For the assessment of the model assumes that: 1) persons living in homes with natural gas which are not protected against respiratory diseases and are equally likely to suffer respiratory infections caused by other causes, 2) the effective protection that natural gas generates against respiratory diseases depends also on housing conditions (ventilation of the kitchen and independence and efficiency of the stove).

The study horizon is 5 years, for both children and adults, with a discount rate of 3% for both costs and outcomes. The parameters for the model shown in the following table.

Model Inputs	Base Case	Plausible Range Lower Limit	Plausible Range Upper Limit	Source - Reference
Demográphics				
Households covered	34.137			
Population covered	143.861	--	--	Current analysis
Population under 5 years	11.940	--	--	Current análysis
Population aged 60	14.675	--	--	Current análysis
Average monthly income per household	174	137	250	Current analysis
Monthly expenditure on fuels other than HNG	9,1	8,0	10,1	Current analysis
Average monthly expenditure without grant HNG	4,0	3,5	4,5	Current análysis
Average monthly expenditure on subsidized HNG	2,1	1,9	2,3	Current análysis
Discount rate	3%			Current análysis
Odds (cumulative incidence)				
Cumulative Incidence of IRA	12,6	11,5	13,5	(Constenla, 2007)
Proportion of hospitalization	30	25	35	(Alvis, 2008)
Proportion of hospital mortality	4	3	5	(Alvis, 2008)
Prevalence of EPOC[1]	8,9	7,3	13,4	(Caballero, 2005)
Proportion of hospitalization	30	25	35	(Alvis, 2008)
Proportion of hospital mortality	8	5	11	(Alvis, 2008)
Effectiveness of Gas to avoid events				
Effectiveness ARI	62,7	55	69	Current analysis
Effectiveness COPD	26,6	22	31	Current analysis
Costos (US$ 2005)				
Average cost of patients hospitalized*	886	867	1 325	(Constenla, 2007)
Average cost of patients Ambulatory ARI*	93	78	221	(Constenla, 2007)
Average cost of patients hospitalized COPD*	5 215	4 800	5 500	(Pérez, 2007)
Average cost of patients ambulatory COPD*	500	450	550	(Pérez, 2007)
Cost of installation of HNG	146	127	170	Current analysis
Weighting of disability				
Disability weighted for ARI	0,280			(Murray, 1996)
Disability weighted for COPD	0,334			(Murray, 1996)

* Includes direct and indirect costs ** Including installation costs and the average annual turnover of the home. Installation costs are deferred for 5 years

Table 2. Parameters for the economic assessment of the impact of the installation program Household Natural Gas

[1] The study by (Caballero, 2005) was performed in 5539 adults of 40 years or more in 5 cities in the country (Bogotá, Bucaramanga, Medellín, Cali and Barranquilla). In conclusion the prevalence of COPD in Colombia is 8.9%, with the Medellín city that was the highest percentage 13.5% and Barranquilla with the lowest 6.2%. The study assessed exposure to wood smoke and found that 60.7% of the population had ever used firewood for cooking. The highest frequency, 67.5%, was observed in Bucaramanga, followed by 62.5%, in Cali, 59.2% in Medellín, 58.1% and 56.3% in Bogota, Barranquilla. The prevalence among men and women were similar: 60.1% in men and 61.0% in women. 64.8% have used the wood for over ten years.

The outcome measures considered or result of the program for the model are: The total costs of the burden of respiratory disease associated with smoke from biomass, the costs of averted disease burden by the grant program to HNG connections, where both ambulatory (medical) and ARI in children hospital and COPD in older adults, deaths from these diseases and the years of disability-adjusted life (DALYs) averted.

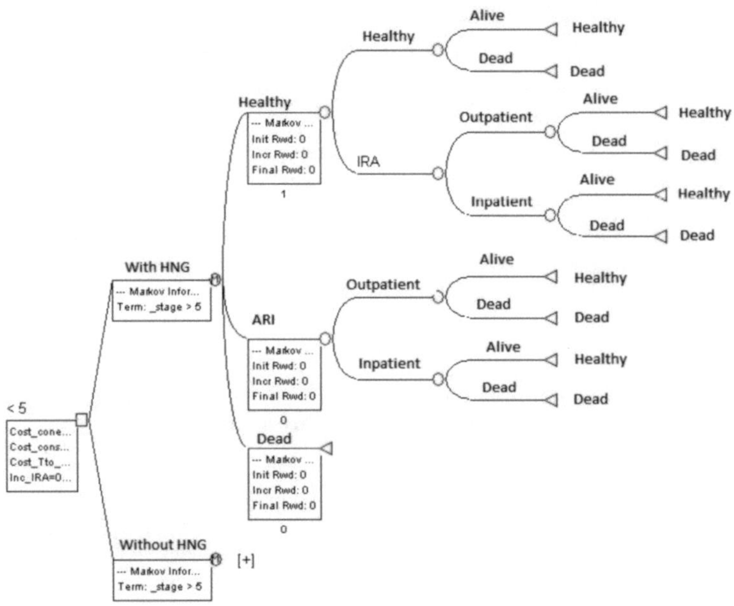

Fig. 1. Choice models for ARI.

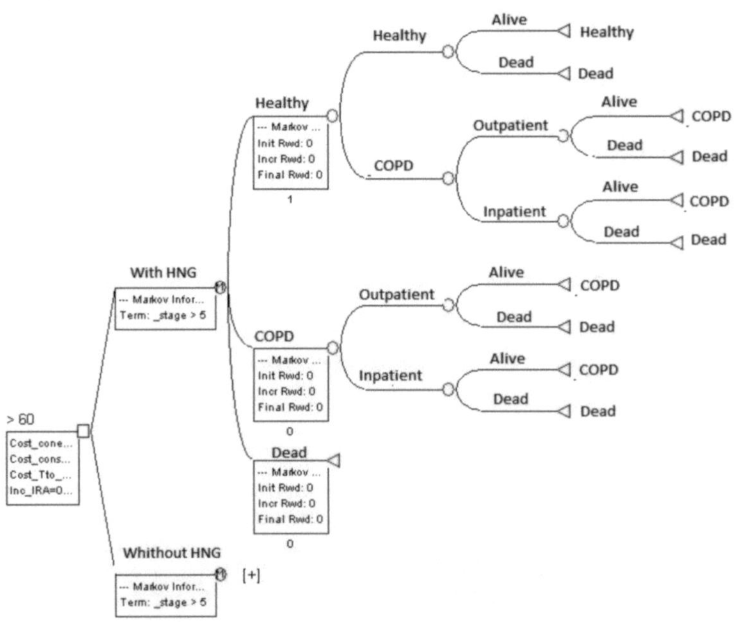

Fig. 2. Choice models for COPD

It also uses the ratio of incremental cost-effectiveness (ICER) for comparing alternatives with and without a program using the following formula:

$$\text{ICER}=\frac{\text{Connections Program Costs to HNG} - \text{Avoided costs of the burden of respiratory disease associated with smoke from biomass}}{\text{Consequently avoided (DALYs, deaths avoided, AVS)}}$$

The ICER compares the difference in cost on the difference in health outcomes in different scenarios. For the analysis of cost effectiveness of disease avoided costs are deducted from the cost of implementing the grant program and are divided between health outcomes (life expectancy gained per woman, year of life saved, deaths prevented, etc.). The costs are expressed in dollars and pesos in 2008. All future costs and benefits are discounted at an annual rate of 3%.

3.1 Sensitivity analysis

For the sensitivity analysis using different assumptions for the parameters. Moreover, to assess the profitability of the investment is estimated Internal Rate of Return of the project assuming that: a) all households covered by the project receive an average income between 137 and $ 250 a month, b) the costs of the monthly averages without subsidy bill are $ 2.3 to 1.9, c) the average monthly cost of the subsidy bill is $ 4.5 to 3.5, d) all households receive a subsidy of the cost of installation HNG of $ 146, and e) all households prior to the installation of the HNG, various fuels consumed by an average monthly cost between $ 10.1 and 8.0.Households were followed for 20 years and yields are assumed constant.

4. Burden of disease from ARI and COPD in population

In 84 localities in the absence of the facilities of HNG would occur between 6288 and 7382 outpatient cases of ARI and between two thousand five hundred and five hundred thousand hospitalizations in the 5 years of observation. This could generate between 47 and 129 deaths and between 75 thousand and 127 thousand DALYs per IRA during the period. Also be presented between 5 and 9 thousand cases of outpatient visits for COPD and hospitalizations between 1200 and 3152, just as would occur between 61 and 347 deaths from COPD and between 12 thousand and 35 thousand DALYs.
As shown in Table 3, the installation of Household Natural Gas avoided during the study period, about 4 thousand to 5 thousand cases of outpatient ARI and between 1200 and 2300 for COPD. In addition, about 1,200 hospitalizations avoided by IRA and 500 COPD. Also avoid death between 45 and 170 would represent about 45 thousand to 90 thousand DALYs.

5. Economic impact

The economic costs of the burden of disease from ARI and COPD on stage without HNG would be between US$ 10.7 and US$ 23.6 million. On stage with HNG these costs would be reduced by about 32%, see Table 4.
As for the incremental cost effectiveness of the HNG would be between minus 8 thousand (cost savings) and 18 thousand dollars to avoid a death, and betwen under US$ 16 to US$ 17 avoid a DALY. The net cost would be from under 1.4 million dollars (cost savings) and 834 thousand dollars of 2005.

	In the absence of the installation of Household Natural Gas			In the presence of the installation of Household Natural Gas		
(ARI) Events < 5 years	Base Case	Lower Limit	Upper Limit	Base Case	Lower Limit	Upper Limit
Outpatient cases	6.562	6.288	7.382	2.165	2.075	2.436
Hospitalizations	1.969	1.572	2.584	650	519	853
Deaths	79	47	129	26	16	43
DALYs	96.049	75.697	127.723	31.696	24.980	42.149
(COPD) Events > 60 years						
Outpatient cases	6.049	4.906	9.006	4.476	3.631	6.664
Hospitalizations	1.815	1.227	3.152	1.343	908	2.332
Deaths	145	61	347	107	45	257
DALYs	19.764	12.822	35.711	14.625	9.488	26.426
Total avoided events						
Outpatient cases avoided of ARI				4.068	3.899	4.577
Hospitalizations avoided for ARI				1.220	975	1.602
Deaths avoided for ARI				49	29	80
DALYs avoided for ARI				59.550	46.932	79.188
Outpatient cases avoided of COPD				1.573	1.276	2.341
Hospitalizations avoided for COPD				472	319	820
COPD Deaths				38	16	90
COPD AVAD avoided				5.139	3.334	9.285

Table 3. Burden of disease in the localities. Colombia 2008

	In the absence of the installation of Household Natural Gas			In the presence of the installation of Household Natural Gas		
(ARI) Events < 5 years	Base Case	Lower Limit	Upper Limit	Base Case	Lower Limit	Upper Limit
Outpatient cases	616.810	591.109	693.911	234.388	224.622	263.686
Hospitalizations	1.748.065	1.396.024	2.294.335	664.265	530.489	871.847
Subtotal costs ARI	2.364.874	1.987.133	2.988.246	898.652	755.111	1.135.533
(COPD) Events > 60 years						
Outpatient cases	3.024.324	2.453.063	4.502.883	2.238.000	1.815.267	3.332.133
Hospitalizations	9.279.836	6.272.482	16.119.419	6.867.079	4.641.637	11.928.370
Subtotal costs COPD	12.304.160	8.725.545	20.622.302	9.105.079	6.456.903	15.260.503
Total Cost Burden	14.669.035	10.712.678	23.610.547	10.003.731	7.212.014	16.396.037
Total costos avoided by the project				4.665.304	3.500.664	7.214.511
Total installation costs of GND				4.984.002	4.335.399	5.803.290
Cost/Deaths avoided for the period				3.682	-8.289	18.473
Cost / DALY avoided for the period				5	17	-16
Total neto costs for the period				318.698	-1.411.221	834.735

Table 4. Costs of Burden of disease in the localities. Colombia 2008

Other benefits generated by the HNG connection to households are the savings that are generated by stop using different fuel HNG which is more than twice as costly as paying the monthly bill in the event that households do not receive service subsidies from the State. In the scenario where households receive the state subsidy of these costs can be up to 4 times what the household pays each month for the service. If we assume that these households maintained this condition for about 20 years, it is possible to evaluate returns on those savings can generate the initial investment by the donor (grant connect HNG).

Ítems	Average	Lower Limit	Upper Limit
Total household income	1.096.087.341,3	865.834.808,4	1.569.325.590,3
Spending on different fuel to GND	57.009.359,7	50.367.234,2	63.652.387,1
GND spending without subsidy	25.109.209,4	21.970.558,3	28.247.860,6
GND spending with subsidy	13.182.335,0	11.926.874,5	14.437.795,4
State subsidies	11.926.874,5	10.043.683,8	13.810.065,2
Savings without subsidies	31.900.150,3	28.396.676,0	35.404.526,5
Savings with subsidies	43.827.024,8	38.440.359,8	49.214.591,7
Internal Rate of Return without subsidies	42%	37%	46%
Internal Rate of Return without subsidies	57%	50%	64%

Table 5. Financial return of the project for households for 20 years. Colombia 2008

Adding these savings to the costs avoided by the impact of the HNG on the incidence of respiratory diseases, we can assess the IRR of the project from the perspective of the donor, assuming that their interest is the achievement of maximum benefit to each household subsidized. As shown in Figure 3 the IRR is positive in both scenarios (with and without subsidies) as low as 44% during the first five years and a maximum of 78% if the period of observation is 20 years, this scenario without the subsidy cost of the bill (consumer subsidy). These numbers vary in the same scenario with allowance for consumption of between 65% and 96% when we evaluated the project to 20 years.

6. Sensitivity analysis of the results

As shown in Figure 4, a sensitivity analysis in a way to display the variables that have the greatest impact incremental effect on the cost of connecting homes to the HNG are: a) the prevalence of COPD, b) the costs of HNG installation, c) the effectiveness of the HNG to impact the prevalence of COPD, d) the proportion of hospitalizations for COPD and e) the average costs of outpatient and hospitalization ARI. In this regard, the decision to subsidize the costs of installing the HNG in poor households, allowing for an intervening variable of greatest impact for the evaluation of benefits of installing.

Fig. 3. Internal rate of return of the the installation of Household Natural Gas.

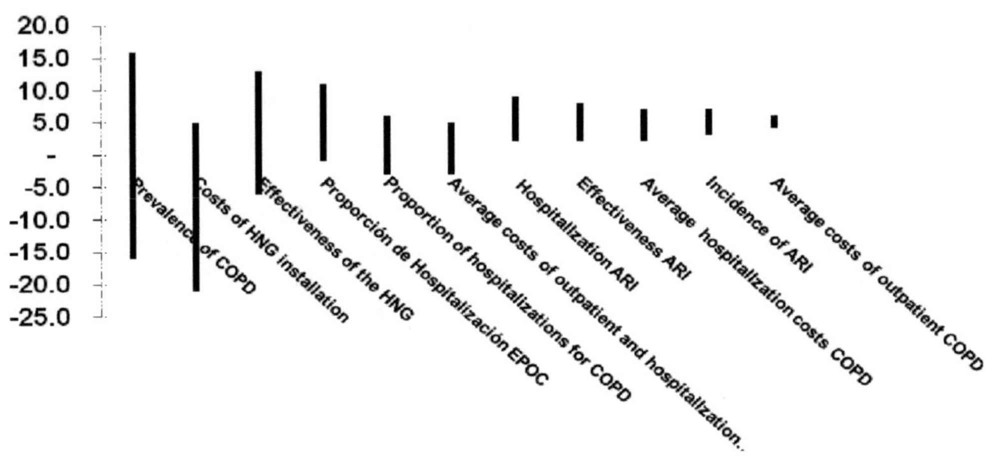

Fig. 4. Tornado Diagram of ICER of Cost/DALYs

7. Conclusion

According to the DANE census of 2005 in Colombia were 9.74 million households of which 3.92 million (40.3%) accessed the service of HNG. For the fourth quarter of 2008 the Ministry of Mines and Energy of Colombia reported 4.93 million homes connected to the HNG (Murray, 1996), which represents a growth of more than one million homes in three years and covers over 50% of all households. These households are concentrated in 462 locations in about 350 municipalities in 22 of the 33 departments in the country. Of these households, 55% belong to strata 1 and 2 (Murray, 1996). Since according to the quality of life survey 2003 (Alzate, 2006) 95.5% of rural households in Colombia belong to the socioeconomic strata 1 and 2, we can say that the coverage in the poorest and most vulnerable populations are not yet adequate.

On the other hand, this demonstrated the increased risk of ARI, COPD and lung cancer in those households who cook with biofuels (Norman, 2007). In our study, the limited information, only measure the disease burden of ARI and COPD which assumes an underestimation of the protective effects of GND, for example, about lung cancer. However, as shown, assuming the WHO criteria (WHO, 2001), the GND is a highly cost effective technology and thus its incorporation into benefits for the social development of poor and marginalized communities, is beyond dispute.

Thus, this study is a good estimator of the potential impact of universal access to GND by the poorest and most vulnerable households, which are precisely those that generate the greatest burden of disease by the pollution. This investigation provides means of support of public policy with a view to providing decision makers to mitigate or eliminate pollution as change or improve the efficiency of the stoves, fuel switching to biomass use GND generating reducing the frequency of cases of ARI, COPD and other respiratory diseases and the reorientation and improvement in the household economy (Larson, 2002).

These findings are consistent in scale, with the referenced document CONPES 3344 through 2005 which set out "guidelines for the formulation of policy on prevention and control of air pollution" (DNP, 2005) and estimated that in Colombia, According to Larsen, is generated per year 6 000 deaths, 7 400 new cases of chronic bronchitis (COPD), 13 000 hospitalizations and 255 000 visits to emergency rooms, with costs between 230 and 600 billion pesos in 2004 ie between about 100 and 270 million dollars.

In Colombia there are already incentives rule (Article 97 of Act 142 of 1994) which facilitate the implementation of the policy of subsidies: "In any case, the cost of household connection, attack and meter of stratas 1, 2 and 3 can be covered by the municipality, the department or the nation through budgetary support, but political will is an important input to achieve the goals for the reduction of pollution by smoke from biomass. However, given the high rates of return for this project (37% and 46% in cases that do not subsidize the consumption of the service and between 50% and 64% in the scenario with subsidies from the state) is very difficult to escape of these choices.

In summary we can state that: 1) Given the high rates of return of the project, both on stage and without subsidies subsidized consumption, for every $ 100 invested by the donor in the year zero of the project, households received net benefits of between 65 and 96 dollars annually. 2) Decisions to invest in subsidizing the connection of poor households to the GND are highly cost effective, to be cost saving, especially in poorer households with higher risks of respiratory diseases (health risk) and lower income (economic risk) is say the most vulnerable. So, with less than $ 20 is spent a year of life saved, which in the worst case is 8 times lower than required to save a year of life using the pneumococcal vaccine, which needs between $ 154 and $ 5252 in different countries of Latin America and the Caribbean (Constenla, 2007; Sinha, 2008). 3) If the state took the decision to subsidize both the connection and GND consumption by households more vulnerable, more could be close to achieving the MDGs, as is clearly demonstrated improvements in quality of life households and high social return on investment in this area, which is essential in the fight against poverty and misery in our country.

8. References

DNP. Departamento Nacional de Planeación.. (2005). Metas y estrategias de colombia para el logro de los objetivos de desarrollo del milenio - 2015, In : *Consejo Nacional de Política Económica y Social República de Colombia, Conpes 091*. Bogotá D.C.

Nuñez, J; Gonzalez, N. & Hernández, G. (2007). ¿Es posible alcanzar los objetivos del Milenio en Colombia? Una evaluación de estrategias de focalización y financiamiento, In : *Archivos de Economía* N° 330. Departamento Nacional de Planeación, Dirección de Estudios Económicos: Bogotá D.C., Colombia.

Schultz, T. (1962). Reflections on investment in man. *Journal of Political Economy*, Vol. 70, pp. 101-109.

Grossman, M. (1972). On the concept of health capital and the demand for health. Journal of Political Economy, Vol. 80, No. 2, (March-April 1972), pp. 223-255.

WHO, World Health Organization (2001). Macroeconomics and health: investing in health for economic development. Report of the Commission on Macroeconomics and Health. In : *Commission on Macroeconomics and Health*, Geneva.

Mishra, V. & Retherford R. (2007). Does biofuel smoke contribute to anaemia and stunting in early childhood? *Int J Epidemiol*, Vol. 36, No. 1, pp. 117-29.

Smith, K. (1987). Biofuels, air pollution and Health: a global review. Plenum, Editor. ISBN: 0-306-42519-X, New York.

Pandey, M ; Smith, K ; Boleij, J. & Wafula, E. (1989). Indoor air pollution in developing countries and acute respiratory infection in children. *Lancet*, Vol. 333, No. 8635, pp. 427-9.

Bruce, N; Perez, R. & Rachel, A. (2000). Indoor air pollution in developing countries: a major environmental and public health challenge. In: *Bulletin of the World Health Organization*, Vol. 78, No. 9, pp. 1067-1071.

Agency, U.S.E.P. (1997). Revisions to the National Ambient Air Quality Standards for Particles Matter. Federal Register. pp. 38651-38701.

Innergy Soluciones Energeticas. (2007). Ventajas Para el Medio Ambiente. Ventajas del Gas Natural, 02.10.2007, Available from: http://www.innergy.cl/ventajas1.htm.

Spengler, J ; Samet, J. & McCarthy, J. (2001). Indoor Air Quality handbook., ISBN: 0-07-445549-4. New York: McGraw Hill.

Smith, K ; Mehta, S. & Maeusezahl, F. (2004). Indoor Air Pollution from Household Use of Solid Fuels. In : *Comparative quantification of health risks: Global and regional burden of disease due to selected major risk factors*, World Health Organization: Geneva.

Hill, A. & Hill, I. (1992). Bradford Hill's Principles of Medical Statistics. *Biometrics*. Vol. 48, No. 2, pp. 659-660.

Smith, K. ; Samet, J ; Romieu, I. & Bruce, N. (2000). Indoor air pollution in developing countries and acute lower respiratory infections in children. *Thorax*. Vol. 55, No. 6, pp. 518- 532.

Zodpey, P. & Ughade, S. (1999) Exposure to cheaper cooking fuels and risk of agerelated cataract in women. *Indian Journal of Occupational and Environmental Medicine*, Vol. 3, No 4,(Oct- Dec 1999) pp. 159-161.

Ezzati, M. & Kammen, D. (2001). Quantifying the effects of exposure to indoor air pollution from biomass combustion on acute respiratory infections in developing countries. *Environmental Health Perspectives*, Vol. 109, No. 5, (May 2001) pp. 481-488.

Saksena, S .; Prasad, R.; Pal, R. & Joshi, V. (1992). Patterns of daily exposure to TSP and CO in the Garhwal, Himalaya. *Atmospheric Environment*, Vol. 26A, No. 11, (August 1992), pp. 2125-2134.

WHO, World Health Organization. (2002). Addressing the links between indoor air pollution, household energy, and human health. In : *Meeting report of the WHO-USAID global consultation on the health impact of indoor air pollution and household energy in developing countries*, Geneva.

Caballero, A; Torres, C. ; Maldonado, D. ; Jaramillo, C. & Guevara, D. (2005). Prevalencia de la Enfermedad Pulmonar Oclusiva Crónica en cinco ciudades colombianas. *Revista Médica Sanitas*. Edición Especial.,.

Alvis, N; Alvis, L. & Orozco, J. (2008). The cost-effectiveness of installing natural gas as a sanitary alternative for rural communities on the Colombian Caribbean coast burning biomass fuels. *Rev. salud pública*, Vol. 10, No. 4, (september 2008) pp. 537 - 549.

Pérez, N ; Murillo, R. ; Pinzón, C. & Hernández, G. (2007). Smoking Attributable Costs of Lung Cancer, COPD, and AMI in Colombia (A PAHO Multicentric Proyect). *Rev Colomb Cancerol*. Vol. 11 No. 4, (December 2007) pp. 241 - 249.

Arredondo, A. & Damian, T. (1997). The economic costs in the production of health services: from the cost of inputs to the cost of case management. *Salud Publica Mex*, Vol. 39, No. 2, pp. 117-24.

Drummond, M. ; O'brien, B. ; Stoddart, G. & Torrance, G. (2005). Methods for the economic evaluation of health care programs. ISBN 0198529457. Third Edition ed., Oxford Medical Publications Oxford GB:.

Murray, C. & Lopez, A. (1996). The Global Burden of Disease. Boston: Harvard University Press, .

Alzate, M. (2006). La estratificación socioeconómica para el cobro de los servicios públicos domiciliarios en Colombia ¿Solidaridad o focalización?. In : *Estudios y Perspectivas 14*, ISBN: 92-1-322968-2, Bogotá D.C..

Norman, R. ; Barnes, B. ; Mathee, A. ; Bradshaw, D. & South African Comparative Risk Assessment Collaborating Group. (2007). Estimating the burden of disease attributable to indoor air pollution from household use of solid fuels in South Africa in 2000. *S Afr Med J*, (August 2007), Vol. 97, No. 8 Pt 2, pp. 764-71.

Larson, B. & Rosen, S. (2002). Understanding household demand for indoor air pollution control in developing countries. *Soc Sci Med*, Vol. 55, No. 4, pp. 571-84.

DNP, Departamento Nacional de Planeación,. (2005). Lineamientos para la formulación de la política de prevención y control de la contaminación del aire. In : *Consejo Nacional de Política Económica y Social República de Colombia, Conpes 3344*. Bogotá D.C.

Constenla D. ; Gomez E.; De la Hoz, F.; O'Loughlin, R.;, Sinha, A.; Valencia, J. & Valenzuela, M. (2007) The Burden of Pneumococcal Disease and the Cost Effectiveness of a Pneumococcal Vaccine in Latin America and the Caribbean: A review of the evidence and a preliminary economic analysis. Available from: http://www.sabin.org.

Sinha, A. ; Constenla D. & O'Loughlin, R.; Gomez E.; De la Hoz, F. ; Valencia, J. ; Valenzuela, M. & De Qadros, C. (2008). Cost-effectiveness of pneumococcal conjugate vaccination in Latin America and the Caribbean: a regional analysis. *Rev Panam Salud Publica*, (November 2008)Vol. 24, No. 5, pp. 304-13.

19

Economic Growth and Air Pollution in Iran During Development Programs

H. Asadikia, R. Oyarhossein, S. Zare and I. Saleh
Department of Agricultural Economics,
University of Tehran
Iran

1. Introduction

Air pollution means combination of air with gases that lead to decrease in quality of the air. Pollutant elements are Carbon dioxide, Monoxide, Sulfur Dioxide and etc.
Estimated annual losses of mortality due to urban air pollution are 640 million dollars i.e. 5100 billion Rials that is equivalent 0.57 percent of GDP. The disease caused of urban air pollution creates 260 million- dollar or 2100 billion- Rial cost (0.023 percent of GDP) for the economy of Iran. (World Bank, 2005).
Grossman and Krueger (1991) has analyzed relationship between air pollution and economic growth in USA. GNP per each person, time and pollution index were input in that model. Results show that there is relationship like "U" between GNP per each person and SO_2 emission.
Lee (1996) has surveyed air pollution in Korea during 1985-92. The effects show that CO and income have relationship as "U" whereas Sulfur Dioxide, NO_2 and Ozone did not have it.
Khan (2002) has perused correlation between revenue in each family and CO, NO and Ozone in USA for the year of 1990. One of the important things is that in this research population, labor force, unemployment, literate people, workmen's, housekeeper, rental house that have an effect on pollution, were Exogenous Variables. Consequences expressed that linkage in equation between income and pollution of gasses is inverse.
Pazhoyan and etal (2007) have examined respect of economic growth and air pollution with Kuznets hypothesis in 67 countries with different income. The effect of economic growth, rural population, provision of environment, number of cars, the ratio of summation of export and import on GNP have been surveyed and results emphasized Kuznets curve for environment in those countries.
This study surveys relationship between air pollution and economic growth and also, evaluation of Economical and Social Development Programs in the field of air pollution reduction. For this purpose, the study used data for 1979-2005 from the Balance Sheet Energy (2009), Iran Department of Environment (2009), Central Bank of Iran (2009) and World Development Reports (2008) has been collected. Calculation and estimates is done by Shazam and SPSS packages.

2. Methodology

The estimation of relationship between economic growth and environmental destruction has done by the use of Grossman and Krueger model with some adjustments (Eq. (1))

$$LNP = \alpha_0 + \alpha_1 LNGDP + \alpha_3 LNGDP^2 + \alpha_4 LNcar + \alpha_5 LO + \alpha_6 LNU + \alpha_7 D_1 + \alpha_8 D_2 + \alpha_9 D_3 + \varepsilon \quad (1)$$

Where LNP: the logarithm of pollution variable (carbon dioxide emissions per capita in thousand tons), LNGDP: the logarithm of GDP per capita in Iran (billion Rials), $LNGDP^2$: the square logarithm of GDP per capita in Iran, LNCAR: the logarithmic number of cars produced and imported, LNO: the logarithm of the degree of openness of an economy (ratio of total exports and imports to GDP), LNU: logarithm of population (thousands), D_1: dummy variable for the function of first Development Program, D_2: dummy variable for the function of second Development Program, D_3: dummy variable for the function of third Development Program (in each dummy variables, the number 1 is related to the program years and the number 0 for the other years). For each of the variables some classes are considered, and then the effect of them is investigated by the find results from SPSS package. Many studies done in the field of air pollution considered carbon dioxide emissions as an index of the air pollution and this gas is one of the most important gases leading to climate change and global warming. Additionally, about 72 percent of greenhouse is related to carbon dioxide, so in this study carbon dioxide emission is a criterion for air pollution.

3. Summary and conclusion

Study data are time series, so stationary test is done and the result is available in Table (1) that shows two lags are optimal and whereas calculated value is more than critical value the model is stationary.

Error term	Description
data level	Variable
-5.835	AIC criteria
2	Lag length
3.5573*	Calculated value
-2.57	Critical value

(Source: research findings)
* 10% significance level

Table 1. Generalized
Dickey – Fuller Unit root test results

Chi-square test shows that there are significant differences in carbon dioxide per capita for various variable levels. Considering the fifth column table (2), when the population is over 60 million people, 100 percent of pollution will be more than five ton. Significant differences exist in different classes of the export and import at one percent level of and when imports increase the level of pollution also increases. It is obvious that the highest pollution level in import and export class is in 40 to 50 billion-dollar level.

variable	variables Classification	1-2 ton (Percent)	3-4 ton (Percent)	3-4 ton (Percent)	More than 5 ton (Percent)	Total	Test Results χ^2
Country population	20-30	25	0	8	0	7	**20.87
	40-60	50	50	0	0	22	
	50-60	25	50	33	0	33	
	More than 60	0	0	59	100	38	
Total export and import (million person)	20000-30000	75	25	0	0	18	**22.02
	40000-50000	25	25	50	0	34	
	50000-60000	0	37.5	16	0	19	
	More than 60000	0	12.5	34	100	29	
GDP in fixed price (billion Riala)	100000-200000	100	37.5	0	0	25	***31.16
	200000-300000	0	62.5	59	0	44	
	300000-400000	0	0	41	60	25	
	More than 400000	0	0	0	33	3	
The numbers of produced and imported cars	8000-28000	50	37.5	0	0	18	**17.54
	28000-48000	0	25	8	0	11	
	48000-68000	0	37.5	25	0	22	
	More than 68000	50	0	67	100	49	

(Source: research findings),** 5%,*** 1% significance level

Table 2. Estimated effects of different factors on carbon dioxide emissions

Considering table (3) the emissions of CO2 during various Development Programs is different. In first and second Development Programs pollution emissions is almost equal, while during third Development Program pollution increase has had significant decrease.
The highest effect on pollution is related to economic growth variable. Since increase in economic growth is as an important goal for all government, it is necessary to this reach this goal while sustainable development is considering and environmental damage is at least.
As an illustration the most important variable that has an effect on pollution is GNP after that population has significant effect that Government must control it with training them to protect the environment.
Although summation of import and export on GNP are not punctual, it explains if importing and exporting is done easier, we can hope that productions with high expenditure will be imported without harmful effect on effect on economy of the Iran.
Instead of that we can export productions with competitive advantages.
Also government must supervise the number of cars and consumption of fuel with pollution must be prohibited, rules in big city should be modified and etc.
The of third Development Program had the remarkable effect on falling the pollution. Authorities should expand it if they want to receive sustainable development I the world.

variables	coefficients	standard deviation	t statistic
The logarithm of GDP in fixed price	-7.7103***	1.729	-4.458
the logarithm of GDP square in fixed price	2.842***	0.5672	5.011
The logarithm of total vehicles number produced and imported	0.0292	0.0204	1.432
Logarithmic degree of openness in a economy	-0.2879	0.0960	0.2998
The logarithm of population	0.3372**	0.1491	2.262
dummy variable for the function of first Development Program	0.200***	0.0394	5.079
dummy variable for the function of second Development Program	0.2447***	0.0532	4.596
dummy variable for the function of third Development Program	0.0794**	0.0436	1.821
Constant coefficient	2.6196	2.249	1.165

R-SQUARE = 0.9494 R-SQUARE ADJUSTED = 0.9270
DURBIN-WATSON = 2.7128
JARQUE-BERA NORMALITY TEST- CHI-SQUARE(2 DF)=1.6833
P-VALUE=0.431

(Source: research findings),** 5%,*** 1% significance level

Table 3. Estimated logarithmic model

4. References

Central Bank of Iran. 2009.

Grossman,G. m. , A. G. Krueger. 1991 .Environmental Impacts of a North American Free Trade Agreement; National Bureau of Economic Research, NBER Working Paper. 3914.

Iran Department of Environment. 2009.

Khanna,N. 2002. The Income Elasticity of Non-point Source Air pollutants: Revisiting the Environmental Kuznets Curve; Economics Letters, 77, 387-392.

Lee, k. and M. Lee. 1996. The Evaluation of the Regional Development Considering Environment, Environmental Economic Review, Vol.5 No.1, pp. 143-188.

Pazhoyan, j. and Moradhasel,N. 2007. Surveying the effect of economic growth on tha air pollution, economic magazine, VOL.7 NO. 4, PP 160-141.

the Balance Sheet Energy. 2009.

UNDP. 2008. Human Development Report 2004 (Energy And The Environment), World Bank 2007b, Baised on DATA from IEA.

www.ttic.ir/new/ Air polution.aspa.

World Bank. 2005. Human Development Report. (www.worldbank.org)

World Development Reports. 2008.

20

Managing Air Pollution: How Does Education Help?

Gaye Tuncer Teksoz
Middle East Technical University
Turkey

1. Introduction

We aware since Tbilisi Declaration (UNESCO, 1977) that, science and technology can no doubt provide solutions to environmental problems, which probably helped to cause, nevertheless, solutions sought should not be short-term ones nor too narrowly conceived. Solutions, on the other hand, have to take into account social and cultural factors which are so often at the root of environmental problems. What is necessary is a close examination of the complex relationships between people and their environment. The equilibrium in the flow of matter and energy through natural ecosystems as well as ecosystems already modified by humanity must be re-established. In addition, models of economic growth, development, environment and culture must be reconsidered. It has become essential to look-over the lifestyles to distinguish between the essentials and luxuries for both the environment and development. This is one of the basis to advocate a holistic approach to the management of environmental problems. Therefore, the recognition of reasons, results and implications of environmental problems must be coupled with an increasing awareness of solidarity among nations. Improved management of the environment should aim reducing existed disparities as pertaining a sustainable use of natural resources and at bringing about international relations based on equity. Environmental Education (EE), therefore, has an evident role to play if the issues are to be grasped and if all concerned are to be provided with the knowledge, skills, and attitudes to modify the existing situation for the better. Building on more than 30 years of experience in environmental education, education for sustainable development (ESD) continues to highlight the importance of addressing the issues of natural resources as part of the broader agenda of sustainable development. In December 2002, the United Nations General Assembly (UNGA) adopted resolution (57/254) to put in place a United Nations Decade of Education for Sustainable Development (DESD), spanning from 2005 to 2014. Teaching society how to behave responsibly towards the environment lies at the core of ESD; the founding value of ESD is respect: for others and respect for the planet and what is provides us with. ESD wants to challenge us all to adopt new behaviours and practices to secure our future, seeks to integrate the principles, values, and practices of sustainable development into all aspects of education, in order to address the social, economic, cultural and environmental problems we face in the 21st century. Air pollution is one of the themes of environmental perspectives of the ESD. Educating about air pollution builds the skills and attitudes needed to question the way we think, the values we hold and the decisions we make in the context of sustainable development. Improving awareness about the sustainability involves issues like the impact of human activities on earth systems, control of green house gases, land and energy use, consumption patterns, pollution and transport. ESD for air pollution can be appreciated as one of the powered tools for air pollution management since it offers innovative ways of framing to make sense in people's daily lives and of translating passive awareness into active concern and behaviour change. EE and ESD for air pollution has been covered in several researches all over the world the major areas of focus being, problems on the implementation of EE, misconceptions of both teachers and students especially related to global warming and ozone layer depletion issues , attitudes towards air pollution problems and solutions and environmental literacy components and environmental responsible behaviour (Cutter, 2002; Michail et al., 2007; Dove, 1996; Gayford, 2000; Özdemir & Çobanoğlu, 2008; Summers, et al., 2000; Michail et al.,2007). The researchers also interested in the regional differences in students' and citizens' attitudes and behaviours toward air pollution issues. All this research have shown that people of all ages has positive attitudes toward air pollution issues; that is to say we all are aware of the threats of population increase, industrialization and consumption patterns on the air pollution problems but, almost none of us can make connections between these threats and individual contributions. Therefore, through ESD, we seek for ways to help people to touch the reality, so to make sustainable decisions. The main idea of this chapter therefore, is to present both reasons and results of air pollution in a wider perspective, i.e. through the point of view of EE and ESD, as well as to present education as a component for air pollution management.

2. Action and response

Hunter-gatherers live in the forest, agriculturalists live adjacent to but within striking distance of the forest, and urban-industrial men live away from the forest. Paradoxically, the more the spatial separation from the forest the greater the impact on its ecology, and the further removed the actors from the consequences of this impact! (Gadgil & Guha, 1992, p.67)

Earliest probable evidence of fire used deliberately to clear forests in the Kalambo Falls site in Tanzania points out 60,000 years before present. (Grove, 1995). Air pollution is concurred with the appearance of humans, continues to grow and as we stand still unconsciously, it will continue to intrude the life on Erath. Signals are clear and actually there are number of evidences for air pollution beginning from ancient civilizations that should not be disregarded:

When Homo sapiens first lighted fire, its smoke provided the first medium of environmental pollution. The burning of fuels for heating and cooking has contributed to indoor air pollution. The walls of caves, inhabited several thousands of years ago, are covered with thick layers of soot. The presence of smoke must have made breathing difficult and must have irritated the eyes in the confined space as well. Most of the lungs of mummified bodies from the Palaeolithic have a black tone. In the first inhabited areas smoke was not driven away (one of the practical reasons might have been protection against mosquitoes) and the people dwelling in these inner areas found shelter in the smoke (McNeill, 2001). Humans, on the other hand, seem to have been living together with this unhealthy form of air pollution for many thousands of years. Following section presents the brief history of air pollution beginning from 13th century till today, based mainly on the "Environmental History Timeline, which originally appeared in Mass Media and Environmental Conflict, a book written by Mark Neuzil and William Kovarik published by

Sage in 1996. The first web publication of the timeline was on 6/18/96, and it was expanded in 1998 and 2001 (www.radford.edu/~wkovarik/envhist).

The Roman Senate introduced a law about 2000 years ago, according to which: 'Polluting air is not allowed'. The Institutes issued under the Roman emperor Justinian in 535 AD were used as a text in law schools. Under the section Law of Things, our right to the air is clear: 'By the law of nature these things are common to mankind - the air, running water, the sea, and consequently the shores of the sea.' (Makra & Brimblecombe, 2004).

In 1257, Queen Eleanor of Provence was forced to leave Nottingham Castle for Tutbury Castle because heavy coal smoke fouls the air.

In 1306, Edward I forbidden coal burning in London, but like many attempts to regulate coal burning, it has little effect.

Between 1560 and1600, rapid industrialization in England leaded to heavy deforestation and increasing substitution of coal for wood. In 1590, Queen Elizabeth was "greatly grieved and annoyed" by coal smoke in Westminster Palace.

In 1661 John Evelyn wrote "Fumifugium, or the Inconvenience of the Air and Smoke of London Dissipated" to propose solutions for London's air pollution problem. These include large public parks and lots of flowers:

"The immoderate use of, and indulgence to, sea-coal in the city of London exposes it to one of the foulest inconveniences and reproaches that can possibly befall so noble and otherwise incomparable City... Whilst they are belching it forth their sooty jaws, the City of London resembles the face rather of Mount Aetna, the Court of Vulcan... or the suburbs of Hell [rather] than an assembly of rational creatures..."

In his diary, Evelyn wrote in 1684 that smoke was so severe "hardly could one see across the street, and this filling the lungs with its gross particles exceedingly obstructed the breast, so as one would scarce breathe."

Abraham Darby of Coalbrookdale, England used coal instead of wood in 1709 for manufacturing iron. British coal production around this time was 3 million tons per year, or five times more than the rest of the world combined.

In 1775 English scientist Percival Pott found that coal was causing an unusually high incidence of cancer among chimney sweeps.

Benjamin Franklin noted in 1784 that the switch from wood to coal had saved what remained of England's forests and he urged France and Germany to do the same.

In 1795 Sir Thomas Percival, a UK physician, leaded group of doctors to supervise textile mills and recommended hours and working conditions, children were only permitted to work 12 hours per day.

Phileppe Lebon became the first to illuminate a public building with gas. The hotel Seignelay in Paris was lit using wood gas, not coal. Project was restricted and then Lebon's untimely death by robbery 1804 ended the experiments.

In 1804, impacts of smoke had begun to be felt in Pittsburgh. The smoke affected the "comfort, health and peace and harmony" of the new city. As in most other cities, the solution was to build higher chimneys.

First gas light introduced in 1812 in London. This "town gas" or manufactured gas would be used in every major US and European city, but residual coal tar would remain an environmental problem well into the 21st century.

In 1819 British Parliamentary committee expressed concern that steam engines and furnaces "could work in a manner less harmful to public health."

James Fenimore Cooper wrote "The Pioneers" in 1823, which contains the idea that humans should "govern the resources of nature by certain principles in order to conserve them.
Jean Baptiste Joseph Fourier wrote the first scientific reference to global warming in 1824, "Remarks on the Temperature of the Terrestrial Globe and Planetary Spaces", in which he proposed the theory that the sun's heat is partially trapped in the earth's atmosphere like a giant glass jar.
In 1835 Alexis de Tocqueville published Journey to England and described the industrial city of Manchester:
"Thirty or forty factories rise on the tops of the hills...six stories (high). The wretched dwellings of the poor are scattered haphazard around them. Round them stretches land uncultivated but without the charm of rustic nature.,, the fetid, muddy waters stained with a thousand colours by the factories ... Look up and all around this place and you will see the huge palaces of industry. You will hear the noise of furnaces, the whistle of steam. These vast structures keep air and light out of the human habitations which they dominate; they envelope them in perpetual fog; here is the slave, there the master; there is the wealth of some, here the poverty of most."
In 1843 House of Commons Select Committee on the Smoke Nuisance recommended all manufacturers be removed to a distance of 5 to 6 miles from city centre.
In 1848, Andrew Jackson Downing, a landscape architect, proposed creation of a 500 acre People's Park in New York, which is now known as Central Park.
Novelist Charles Dickens published his novel "Bleak House" in 1853, with an image of London as a twisted, twilight world of smoke, shadows and wraiths. Dickens wrote:
"Smoke lowering down from chimney-pots, making a soft black drizzle, with flakes of soot in it as big as full-grown snowflakes -- gone into mourning, one might imagine, for the death of the sun."
In 1860, Prof. Augustine Mouchot of Lycee de Tours, France, said: "*One cannot help coming to the conclusion that it would be prudent and wise not to fall asleep regarding this quasi security. Eventually industry will no longer find in Europe the resources to satisfy its prodigious expansion Coal will undoubtedly be used up. What will industry do then?*" Mouchot's answer was to build solar energy machines. In 1874 he built a collector with 54 square feet of reflecting surface for alcohol distillation which worked at the rate of 5 gallons a minute. The machine could also power a 1/2 hp motor and develop 75 psi of steam.
John Tyndall explained the "greenhouse effect" in 1863 in a lecture to the British Royal Society entitled "On Radiation through the Earth's Atmosphere." It was the first confirmation and extension of Joseph Fourier's idea that the earth would be much colder without its atmosphere.
First of a series of "killer fogs" in London occurred in December 1873. Over 1,150 died in three days. Similar incidents happened in 1880, 1882, 1891, 1892 and later.
In 1874 German graduate student Othmar Zeider discovered chemical formula for the insecticide DDT.
Inversions lead to another "killer fog" in London with 700 deaths in January 1880.
In 1881, Chicago became the first American city to create a local ordinance regulating smoke discharges, followed that same year by Cincinnati.
In the same year, in 1881, Norway tracked first signs of acid rain on its western coast.
English writer Edward Carpenter published "**Civilization: Its Cause and Cure**" in 1889, which later had a great influence on Mahatma Ghandi. The following is what Carpenter wrote about the town of Sheffield:
"Only a vast dense cloud, so thick that I wondered how any human being could support life in it, that went up to heaven like the smoke from a great altar. An altar, indeed, it seemed to me, wherein

thousands of lives were being yearly sacrificed. Beside me on the hills the sun was shining, the larks were singing; but down there a hundred thousand grown people, let alone children, were struggling for a little sun and air, toiling, moiling, living a life of suffocation, dying (as the sanitary reports only too clearly show) of diseases caused by foul air and want of light -- all for what? To make a few people rich!"

Clarence Kemp, "the father of solar energy in the U.S.", patented first commercial Climax Solar Water Heater in1891. By 1910, the Climax had competition, especially from the Night and Day solar hot water company, which used a secondary loop from the collector to a water tank. By 1920, over 5,000 Night and Day heaters had been sold in California. At the same time, a boom in solar hot water heaters started in Florida, where electricity was a very expensive competitor. About 15,000 units were sold by 1937.

In April, 1896, Swedish chemist Svante August Arrhenius summarized scientific opinion about the effect of carbon dioxide in the atmosphere, predicting a global temperature increase of 8 or 9 degrees Fahrenheit for a doubling of CO_2 in the atmosphere; "On the Influence of Carbonic Acid in the Air upon the Temperature of the Ground".

In 1898, Coal Smoke Abatement Society formed to pressure government agencies to enforce pollution laws in England.

In 1900, the world's leading scientists gathered in Paris to consider new elements with unusual powers discovered by Pierre and Marie Curie. Minerals like thorium, uranium and radium emitted a new kind of light, the Curies had found. A year later, Ernest Rutherford and Frederick Soddy found that thorium was turning itself into radium -- evidence of long-sought transmutation of metals. Tapping the energy within atoms would mean that a future awaited that "would bear as little relationship to the past as a dragonfly does to that of its aquatic prototype." Indeed, he said, the power would allow mankind to "transform a desert continent, thaw the frozen poles, and make the whole world one smiling garden of Eden." (Weart, 1988, p. 6).

Smoke Prevention Association of America founded in Chicago in 1907.

Air pollution lawsuit begun in Supreme Court in 1907. In various decisions through 1915, the Court decided to limit the amount of sulphur and other noxious fumes that can emerge from the Tennessee Copper Co. following a suit by the State of Georgia. The suit involved sulphur dioxide fumes from Copper Basin smelters in Tennessee that were killing forests and orchards and making people sick over the Georgia border. The state of Tennessee refused to move against the copper companies and disputed Georgia's right to interfere. Georgia sued in 1907 and won in 1915 after investigation and attempts to reduce the pollution, including a court-mandated reduction and mandatory inspections by a university professor. The majority opinion was delivered by the Chief Justice:

"It is a fair and reasonable demand on the part of a sovereign that the air over its territory should not be polluted on a great scale by sulphurous acid gas, that the forests on its mountains should not be further destroyed or threatened by the act of persons beyond its control, that the crops and orchards on its hills should not be endangered." - Georgia v. Tennessee Copper Co. and Ducktown Sulphur, Copper & Iron Co, 206 U.S. 230 (1907)

In 1908, Swedish chemist Svante Arrhenius argued that the greenhouse effect from coal and petroleum use is warming the globe. According to his calculations, doubling CO_2 would lead to average temperature increase of 5 to 6 degrees Celsius. Rather than being alarmed, Arrhenius was pleased that people in the future would *"live under a warmer sky and a less harsh environment than we were granted."* In his book World in the Making, he says that with increased CO_2 *"we may hope to enjoy ages with more equable and better climates, especially as regards the colder regions of the earth, ages when the Earth will bring forth much more abundant crops than at present for the benefit of rapidly propagating mankind."*

Glasgow, Scotland, winter inversions and smoke accumulations kill over 1,000 in 1909.

In October 1921, General Motors demonstrated car powered by 30 percent alcohol-gasoline blend.

In December 1921, General Motors researchers discovered tetraethyl lead as an anti-knock gasoline additive. Despite strong private warnings about its danger and a secret Public Health Service inquiry, the new gasoline went on sale without safety tests 14 months after it was invented, with disastrous consequences.

Leaded gasoline went on sale in February 1923 in Dayton, Ohio at a gas station.

In 1926, the first large scale survey of air pollution occurred in U.S., in Salt Lake City.

Air pollution control begun in 1928 in eastern US cities, reporting sunlight cut by 20 to 50 percent in New York City.

In 1930, Meuse River Valley killer smog incident occurred in Belgium. Three day weather inversion in this industrial valley killed 63, with 6,000 made ill.

The date 1939, October 11 was recorded as St. Louis smog episode. Smog was so thick that lamps were needed during daylight for a week.

Donora, Pennsylvania smog incident occurred in October 30-31, 1948. Twenty people died, 600 hospitalized and thousands stickered in this nationally publicized environmental disaster.

In 1948 a "killer fog" in London caused 600 deaths.

First US conference on air pollution held in 1949.

Four thousand people died in the worst of the London "killer fogs" in Dec. 4-8, 1952. Vehicles used lamps in broad daylight, but smog was so thick that busses run only with a guide walking ahead. By Dec. 8 all transportation except the subway had stopped.

In May 4, 1953, Gilbert N. Plass presented paper on global warming at American Geophysical Union. The Washington Post story (May 5) says:

*World Industry, **pouring its exhausts into the air, may be making the earth's climate warmer,** a Johns Hopkins physicist, reported here yesterday. Releases of carbon dioxide from burning coals and oils, said Dr. Gilbert N. Plass, blanket the earth's surface 'like glass in a greenhouse.' So much carbon dioxide has been released in this industrial century that the earth's average temperature is rising 1 1/2 degrees (F) a century, he said. Similar but more naturally caused changes in the air's carbon dioxide content may account for the ice ages and warm intervals in geologic time, he added... Latest experimental and theoretical calculations, he reported, show that **doubling the carbon dioxide content of the atmosphere causes surface temperatures to rise four degrees (F)** if no other changes occur. But, he added, still other earth warming factors may also be triggered by increased carbon dioxide in the air. It could cause less rainfall by its effect on the clouds and less cloud cover for the earth,' both tending to make the climate warmer and drier,' he said. Dr Plass said the newer calculations bolster the theory first proposed in 1861 that decreases in the carbon dioxide content of the earth's atmosphere caused the ice ages in geologic history. The theory, he said, has not generally been accepted because the effects 'appeared to be too small.' It appears now, he said, that **even the physicists supporting the theory underestimated the climate-changing effects of the carbon dioxide** content in the earth's atmosphere.* ("Industrial Gasses Warming Up Earth, Physicist Notes Here," Washington Post, May 5, p. 5, probably by Nate Haseltine).

New York smog incident killed between 170 and 260 in November, 1953.

Heavy smog conditions shut down industry and schools in Los Angeles for most of October in 1954.

In 1955 International Air Pollution Congress held in New York City.
Another killer smog occurred in London in 1956; 1,000 died.
British Parliament passed Clean Air Act in 1956.
World's first commercial nuclear electric power plant was opened at Sellafield in the United Kingdom in March 31, 1956.
Chelyabinsk nuclear waste explosion occurred in Kyshtym in Russia in 1957-58. Two million curies spread throughout the region, exposing to radiation over a quarter million people.
Another smog phenomenon in London caused 750 die in 1962.
A reaction took place in 1962 to Silent Spring by Rachel Carson. Some agronomists asked whether Carson is intending to starve people by banning pesticides. By 1970 DDT was banned, but other more toxic chemicals were not. Silent Spring was often seen as a turning point in environmental history because it opened a much stronger national dialogue about the relationship between people and nature.
In January 1970, General Motors president promised "pollution free" cars by 1980 and urged the elimination of lead additives from gasoline in order to allow the use of catalytic converters.
In the seventies, air pollution was cut back dramatically through use of catalytic converters on new cars that use only unleaded gasoline. But the predicted "pollution free car" proved to be imaginary.
Disasters showed the weak and fragile side of industrial technology in 1980's:
Bhopal Disaster was recorded as one, 2 decades after the "Silent Spring": Bhopal is a town in India, was similar to many others, before that night in December 3, 1984. Union Carbide Co. fertilizer plant leaked methyl isocyanide at 5 past midnight in Bhopal. 2000 dead, another 8,000 die of chronic effects, estimated 2000 casualties, 100,000 injuries, and significant damage to livestock and crops. The International Medical Commission on Bhopal estimated that as of 1994 upwards of 50,000 people remained partially or totally disabled.
"India's night of death: (Brelis, 1984).
"The first sign that something was wrong came at 11 p.m. A worker at the Union Carbide pesticide plant on the outskirts of Bhopal (pop. 672,000), an industrial city 466 miles south of New Delhi, noticed that pressure was building up in a tank containing 45 tons of methyl isocyanate, a deadly chemical used to make pesticides. At 56 minutes past midnight, the substance began escaping into the air from a faulty valve. For almost an hour, the gas formed a vast, dense fog of death that drifted toward Bhopal.
The vapour passed first over the shantytowns of Jaiprakash and Chhola, just outside the walls of the plant, leaving hundreds dead as they slept. The gas quickly enveloped the city's railway station, where beggars were huddled against the chill. In minutes, a score had died and 200 others were gravely ill. Through temples and shops, over streets and lakes, across a 25-sq.-mi. quadrant of the city, the cloud continued to spread, noiselessly and lethally. The night air was fairly cool (about 60° F), the wind was almost calm, and a heavy mist clung to the earth; those conditions prevented the gas from dissipating, as it would have done during the day.
A few hundred yards from the chemical plant, M.A. Khan, a farmer, was lying in bed when he heard several thumps at a nearby dairy farm and sensed that his own cows were milling about restlessly. He arose and went outside. Two cows were dead on the ground. A third gave out a loud groan and collapsed as Khan watched. Then the farmer's eyes began to smart painfully. He ran into the darkness. The day after, at Bhopal's Hamidia Hospital, his eyes shut tightly and tears streaming down his cheeks, Khan described his fear: "I thought it was a plague."

Fig. 1. Bhopal Disaster; cover in Time Magazine

By week's end more than 2,500 people were dead in the worst industrial disaster the world has known. At least 1,000 more were expected to die from the fumes in the next two weeks; some 3,000 remained critically ill. In all, 150,000 people were treated at hospitals and clinics in Bhopal and surrounding communities. Most of the dead had succumbed because their lungs had filled with fluid, causing the equivalent of death by drowning. Others had suffered heart attacks. The disaster struck hardest at children and old people, whose lungs were either too small or too weak to withstand the poison. A number of the survivors were permanently blinded; others suffered serious lesions in their nasal and bronchial passages…

Another disaster showing the weak and fragile side of industrial technology happened in 1986: On April 26, 1986, at 1:23 a.m. an explosion and fire occurred in Reactor Number 4 of the Chernobyl Nuclear Power Plant in the former USSR (now Ukraine), located 80 miles north of Kiev. Before engineers and scientists could get it under control, 190 tons of highly radioactive material was released into the atmosphere. The radioactive particles rained down not only on Chernobyl, but all over Ukraine, as well as the neighbouring countries of Belarus and Russia, and drifted over to other European countries such as Poland. Scientists estimate that the amount of particles released was equivalent to the effect of 20 nuclear bombs. The Chernobyl accident remains the largest peacetime nuclear disaster ever. The massive radiation killed 31 people within a short time, mostly plant workers and people close to the accident site who died of radiation sickness. As time passed it became clear that

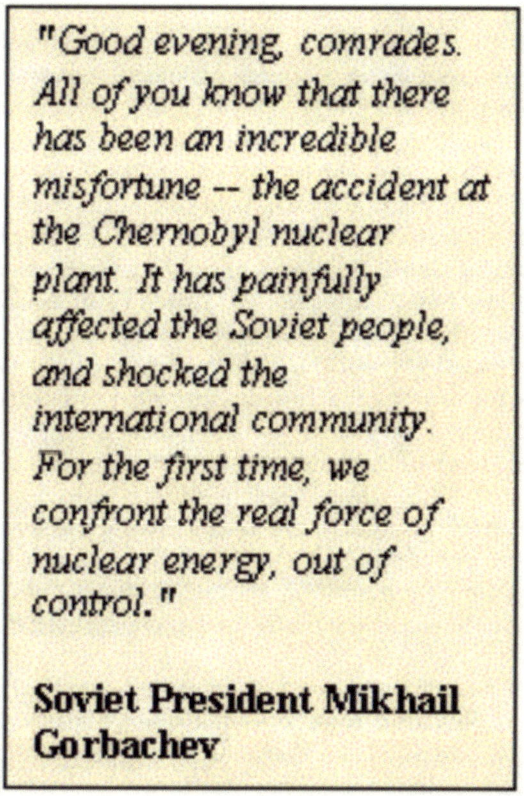
"Good evening, comrades. All of you know that there has been an incredible misfortune -- the accident at the Chernobyl nuclear plant. It has painfully affected the Soviet people, and shocked the international community. For the first time, we confront the real force of nuclear energy, out of control."

Soviet President Mikhail Gorbachev

(http://www.megaessays.com/viewpaper/99365.html)

Fig. 2. Announcement for Chernobyl Disaster

the accident had left a number of serious long-term health problems for the people who lived in the area. These health problems were made worse by the poverty, poor nutrition, and lack of medical care in the region.

In the wake of the Chernobyl disaster, in 1988 Russian scientists form the Ecology and Peace Association, electing as president S.P. Zalyghin, whose astonishing statement -- "Only the people can save nature" -- implied that the government had failed.

At the time when Zalyghin pointed out the people as the saver of the nature, UNEP was working on public awareness raising activities to reduce ozone layer lose. The story of success in ozone layer protection is presented by Stephen O. et al. (2002) and education is dedicated as one of the major contributors.

In 1985, British scientist Joe Farman published discovery of, so called, ozone hole over Antarctica and it was confirmed by US NASA satellite monitoring. Meanwhile, US EPA had begun reconsidering CFC regulations. And the United Nations Environment Program had begun negotiations under the Vienna Convention for the Protection of Ozone which leads to the 1987 Montreal Protocol. The Montreal Protocol international agreement to phase out ozone-depleting chemicals signed by 24 countries, including the US, Japan, Canada and EEC nations in the same year. The treaty calls for phasing out production and consumption of ozone depleting substances. Accordingly, beginning from March 1989, European nations begun ban on ozone - depleting chemicals.

A new report released in September 2010 by the UN, says international efforts to protect the ozone layer are a success and have stopped additional ozone losses. The joint World Meteorological Organization and U.N Environment Program report is the first comprehensive update in four years. World Meteorological Organization Research Department Director Len Barrie says the treaty is working: "It has protected us from severe ozone depletion over the past decade, global ozone, including ozone in the Polar Regions," Barrie said. "It is no longer decreasing, but not yet increasing. One phase for the ozone layer protection work realized by UNEP is related to public awareness raising activities. Apart from the technical reports, UNEP has published myriad of booklets, brochures, posters aimed at the public and the General Assembly of the United Nations designated 16 September as the International Day for the Preservation of Ozone Layer. In most countries, improved servicing was the source of early and substantial emissions reductions. One of the components of the improved servicing, on the other hand, was training. Trade and professional associations, labor unions and private companies primarily undertook training. The objective of the ozone treaties was certainly a difficult one: to persuade the entire world to give up the use of many profitable chemicals. To be persuaded were not only the governments, but also the producers of these chemicals, all major multi- national giants of industrialized countries, and thousands of industries. Behind them were the billions of consumers who wanted and needed the products that contained ozone-depleting chemicals. (Stephen, O et al., 2002).

Therefore, an important role of the success in ozone layer protection can be dedicated to education and training, believing that, re-establishment efforts of the equilibrium of natural systems that have already been distorted by human activities can be satisfactory on the condition that, they include the human itself. Any innovative technology produced to handle an air pollution problem should consider human being who will get the advantage of it.

In the book titled "Earth in Mind", Orr (1994, p.6) wrote, *"If one listens carefully, it may even be possible to hear the Creation groan every year in May when another batch of smart, degree-holding but ecologically illiterate, Homo sapiens who are eager to succeed are launched into the biosphere."* The things on which our future health and prosperity depend are in dire danger and according to Orr (1994) this is not the work of ignorant people, rather it is largely the results of work by people with degrees. Because what was wrong in their education is, it emphasized theories instead of values, concepts rather than human beings, abstraction rather than consciousness, answers instead of questions, ideology and efficiency rather than conscience. And he added (page 8), *"It is not education but education of a certain kind that will save us."* And so, he reported several myths for this kind of education: First one is about the myth that, ignorance is a solvable problem. According to him, ignorance is not a solvable problem; it is rather an inescapable part of the human condition. The advance of knowledge always carried with it the advance of some form of ignorance. As for the case of chlorofluorocarbons (CFCs); *in 1929 the knowledge of what a substance like CFCs would do to the stratospheric ozone and climate stability was a piece of trivial ignorance as the compound had not yet been invented. But in 1930, after the compound discovered, what had been a piece of trivial ignorance became a critical life threatening gap in human understanding of the biosphere. Not until the early 1970's no one did ask "What does this substance to do what? In 1986 we discovered that CFCs had created a hole in the ozone over the South Pole the size of the lower 48 U.S. states; by the early 1990's*

CFCs had created a worldwide reduction of ozone. With the discovery of CFCs, knowledge increased, but like the circumference of an expanding circle, ignorance grew as well (Orr, 1994, p.9). Likewise, one can refer to the above mentioned history of London fog; it was 1257 when Queen Eleanor left Nottingham Castle because heavy coal smoke fouls the air, a series of "killer fogs" occurred in London in 1873, similar incidents happened in 1880, 1882, 1891, 1892 in London, in 1948 another "killer fog" in London caused 600 deaths, New York smog incident killed almost 260 in 1953 and another smog phenomenon in London caused 750 die in 1962 and 4000 people died in the worst of the London "killer fogs" in 1952, 7 centuries later then the first sign. What's more, the disaster caused by a pesticide factory caused unavoidable impacts on the people in 1984, although Rachel Carson's Silent Spring was seen as a turning point in environmental history because it opened a much stronger national dialogue about the relationship between people and nature in 1962. Besides, according to J.Russell (2007), in 2006, coal accounted for 25 percent of world primary energy supply. Due to its high carbon content, coal was responsible for approximately 40 percent of the carbon dioxide emissions from fossil fuels, despite supplying only 32 percent of fossil fuel energy. Management of this plentiful but heavily polluting energy resource has tremendous implications for human welfare, the health of ecosystems, and the stability of the global climate. World coal consumption reached a record 3,090 million tons of oil equivalent (Mtoe) in 2006, an increase of 4.5 percent over 2005. China led world coal use with 39 percent of the total. The United States followed with 18 percent. The European Union and India accounted for 10 percent and 8 percent, respectively.

Another myth Orr (1994) wrote is that, with enough knowledge and technology, we can "manage planet earth". According to Orr, however, what might be managed is us, human desires, economies, politics and communities. But, our attention is caught by those things that avoid the hard choices implied by politics, morality, ethics, and common sense. It makes far better sense to reshape ourselves to fit a finite planet than to attempt to reshape the planet to fit our infinite wants (Orr, 1994, p.9). The global trend on the use of materials, on the other hand, reveals that, we are, in the year 2011, still trying to reshape the planet: Yet, as Gardner (2010), global use of materials (the food, feed, forest products, metals, and minerals that constitute the foundation of modern economies) was up 2.7 percent in 2007, reminding that, materials use is a proxy indicator for environmental impact: the greater the tonnage of virgin materials extracted, processed, consumed, and disposed of, the greater an economy's environmental footprint.

One other myth Orr (1994) presented is related to knowledge, and by implication human goodness, is increasing. According to Orr, rapid increase in data, words and paper, that is an information explosion, should not be mistaken for an increase in knowledge and wisdom, which is not easy to measure. What can be said, as he states, is that some knowledge is increasing while other kinds of knowledge are being lost. As Lopez, B.(1989) says, *"[I am] forced to the realization that something strange, if not dangerous, is afoot. Year by year the number of people with firsthand experience in the land dwindles. Rural populations continue to shift to the cities…. In the wake of this loss of personal and local knowledge, the knowledge from which a real geography is derived, the knowledge on which a country must ultimately stand, has come something hard to define but I think sinister and unsettling."*

All in all, the message is that, we are becoming more ignorant of the things we must know to live well and sustainably on the Earth.

In an Environmental Impact Assessment study, for example, we ask public opinion on the investment. But, how public will assess the investment without having related evaluation skills on the causes and effects? Without having knowledge, positive attitudes, responsibility, skills on the environmental issue, public will act under the influence of economical and cultural pressures and cannot make sustainable choices. An investment even with a very efficient technology on environmental protection has no meaning if the people, who will use and be served by, do not share the responsibility, and responsibility comes with education. Similarly, that efficient technology on environmental protection has no meaning if manager, engineer, technician of the plant, governmental authority responsible for controlling the plant do not feel and share the responsibility. Didn't Chernobyl disaster happen because of such a reason, or what was the reason for the dispersion of the fatal gas in Bhopal? Or how items in the Kyoto Protocol will be effective if people do not understand and feel them? Therefore, education should be a part of any investment. Moreover, we don't need to look for investment proposals for educating people. We can create environmentally literate citizens through several means like integrating EE and ESD into the curriculum of preschools, high schools and universities, offering seminars for women, young, teachers, managers, decision makers, workers etc.

Because, helpless poor in Philippines who has to destroy their own values to survive children living in Africa, people of Bangladesh who do not have any idea about the threat of the global warming, citizens of the developed nations who consume to be happy; decision makers who sign world's biggest investments that discard environmental issues, managers who decide to build new highways instead of constructing new bicycle routes, teachers, students, politicians, women, ...Everybody shall take the responsibility.

Education for Sustainable Development (ESD) may be an answer: It was a quarter of a century ago, that education was described as the "greatest resource" for achieving a just and ecological society. Since then, a series of major international reports have emphasized the critical role education can play in the search for sustainable living. The Brundtland Report, (WCED 1987) argued that teachers had "a crucial role to play in helping to bring about the extensive social changes" (p. xiv) necessary for sustainable development. This message was reiterated by *Caring for the Earth* which identified education's vital role in ensuring that people learn, accept and live by the principle of living sustainably (IUCN, 2002).

> Sustainable living must be the new pattern for all levels: individuals, communities, nations and the world. To adopt the new pattern will require a significant change in the attitudes and practices of many people. We will need to ensure that education programs reflect the importance of an ethic for living sustainably.
> (IUCN, UNEP & WWF 1991 p. 5)

Unlike most education movements, the inception of ESD was not created by the education community. One major outside thrust for ESD came from international political and economic forums. From the time sustainable development was endorsed in the UN General Assembly in 1987, the parallel concept of education supporting sustainable development was being explored. From 1987 to 1992, the concept of sustainable development matured as committees discussed, negotiated, and wrote the 40 chapters of *Agenda 21*. The initial thoughts concerning ESD were captured in Chapter 36 of *Agenda 21*, "Promoting Education, Public Awareness, and Training" (UNESCO 1992).

> Education, including formal education, public awareness and training should be recognized as a process by which human beings and societies can reach their

fullest potential. Education is critical for promoting sustainable development and improving the capacity of the people to address environment and development issues. (*Agenda 21*, Chapter 36, p. 3)

3. Millennium goals and hopes for a sustainable air pollution solutions

Does it make a change if people know and feel deeply the difference they make - for their carbon footprint - by choosing to buy locally grown potatoes instead of imported ones?
Does it make a change if people know and deeply feel how an American meal – hamburger, fries and coke –pays for global warming?
Does it make a change if people can realise that we live the lives that "assigned" to us, and as we continue to do so, we'll continue to face with worse problems of air pollution?
It is our choice to make a change…
Former President of Ireland, Mary Robinson (Robinson, 2011) wrote about her visit to Bangladesh in The Hoffington Post: "Travelling by seaplane to Koyra, in the delta area of Bangladesh, was the equivalent of a journey some years into the future, when the devastating effects of climate change will be an accepted reality worldwide. We landed in an area still devastated by cyclone "Aila" which hit Bangladesh in 2009. A huge amount of once cultivated land was still under water, because of daily tidal fluctuations and the fact that some embankments had not been mended in the nearly two years since Aila… A memorable stop on my visit was to a local primary school run by BRAC (a development organization dedicated to alleviating poverty by empowering the poor to bring about change in their own lives). It was organized on the same principles as a BRAC school I had visited the day before in Korail slum, the largest slum in Dhaka. The schools have 30 plus pupils and one teacher, who teaches these children the five year curriculum in four years, .. In the school in Koyra the children enacted with great gusto -- and acting skills -- how climate change may happen. One of the taller boys acted as the tree which the others cut down, even though warned not to. The winds came, and the consequences were played out -- they all knew where the climate shelter was! As I watched with a grandmother's eye, it struck me that every primary school around the world should be beginning to bring home to children what we must all do to change our habits. Reminding that, Bangladesh is a least developed country (LDC) which has become the leading LDC negotiator on climate change issues. Its contribution to the problem of green house gas emissions is negligible, but the additional burden of climate change is already being felt. It is predicted by the officials in Dhaka that 20 million people may have to leave this region if the global temperature increases by more than 2° Celsius and sea levels rise as predicted. There is nothing theoretical about the climate change issue from this local viewpoint. The injustice of a poor LDC country having to bear huge additional costs from climate impacts it did not contribute to be self evident. Every school needs to be a "green school", so that children can educate their parents. For some it will be knowing where the nearest climate shelter is. For others -- in the developed world -- it will be learning to reuse, reduce, recycle, eat less meat, and travel by public transport, among other ideas".
One of the choices to make a change, eating less meat is investigated by Eshel and Martin (2006). Authors compared the energy consumption of animal- and plant-based diets and the range of energetic planetary footprints spanned by reasonable dietary choices. As a result it was demonstrated that the greenhouse gas emissions of various diets vary by as much as the

difference between owning an average sedan versus a sport-utility vehicle under typical driving conditions. Therefore, authors concluded with a brief review of the safety of plant-based diets, and find no reasons for concern. As Walsh (2008) wrote in his article titled "Meat : Making global warming worse", in 2008 the head of the U.N.'s Nobel Prize–winning Intergovernmental Panel on Climate Change, Pachauri advised people around the world to cut back on meat in order to combat climate change. Pachauri is absolutely right, considering the numbers. In a 2006 report, the U.N. Food and Agriculture Organization (FAO) concluded that worldwide livestock farming generates 18% of the planet's greenhouse gas emissions, by comparison, the entire world's cars, trains, planes and boats account for a combined 13% of greenhouse gas emissions. Much of livestock's contribution to global warming come from deforestation, as the growing demand for meat results in trees being cut down to make space for pasture or farmland to grow animal feed. Livestock takes up a lot of space, nearly one-third of the earth's entire landmass. And, there's manure, all that animal waste generates nitrous oxide, a greenhouse gas that has 296 times the warming effect of CO_2. And of course, there is cow flatulence: as cattle digest grass or grain, they produce methane gas, of which they expel up to 200 L a day (full). And as Tung (2010) reported, global meat production increased by 0.8 percent in 2009 to 281.5 million tons, a slowdown from the 2.4 percent growth rate of 2008. But, as he wrote, the increase continued the steady growth of the past decade. Since 2000, global meat production has risen by 20 percent. So, what if people have aware that, giving up average 176 lb. of meat a year is one of the greenest lifestyle changes one can make as an individual. One can drive a more fuel-efficient car, or install compact fluorescent light bulbs or improve insulation, really no green way to get meat — although organic, locally farmed beef or chicken is better than its factory-raised equivalents. If every American reduced meat consumption by just 20%, the greenhouse gas savings would be the same as if we all switched from a normal sedan to a hybrid Prius. Likewise, if a portion of French fries is preferred with meat one shall need to be sure if the potatoes are locally grown or not.

It's a lesser known fact that up to 1/3 of all greenhouse gas emissions (GHGs) can be attributed to the food we eat. Many people think of transportation of food as the big contributor to climate change. But, although transporting food over great distances does contribute to climate change, how food is produced matters just as much. In fact, growing and harvesting, heating and cooling, processing and packing, transportation and storage ALL contribute to the GHGs emitted by our food system. Focusing on the potatoes and a coke accompanying, yet, just simply try to imagine what it takes to consume a favourite American meal (apart from the meat); the potato was dug with a diesel-powered harvester and then trucked to a processing plant where it was dehydrated, sliced, and frozen. The freezing was done by a cooling unit containing hydrofluorocarbons, some of which escaped into the atmosphere and likely contributed to global climate change. The frozen fries were then trucked to a distribution centre, then on to a fast-food restaurant where they were stored in a freezer and then fried in corn oil heated by electricity generated by hydropower. The meal was served in a fast-food restaurant built on what once was originally forest, then farmland, and then converted to commercial/industrial uses as the city expanded. The ketchup in aluminium- foil packets came from Pittsburgh and was made from Florida tomatoes. The salt came from Louisiana River. The high-fructose corn syrup came from Iowa, as did the carbon dioxide used to produce the fizz, which is produced by fermenting corn. The caffeine came from a processing plant that makes decaffeinated coffee. The cola can was made from one-third recycled aluminium and two-thirds bauxite ore strip-mined in Australia. It came to Washington State on a Korean freighter, and was processed into aluminium using an amount of energy equivalent to a quart of gasoline. The energy came from some of the same dams mentioned earlier that have contributed to a 97 percent decrease in the salmon runs of the Columbia Basin. The cola came from a Seattle processing plant. It is made of 90 percent water from the Cedar River. The high-fructose corn syrup came from Iowa, as did the carbon dioxide used to produce the fizz, which is produced by fermenting corn (Ryan & Durning, 1997).

Knowing the theory of global warming, acid rains, photochemical smog is not enough to construct the links, knowledge should be accompanied by feeling the links deep inside and human being has this potential. ESD is the way to take out the potential and help individuals to feel how their presence is connected with nature, the way we arrange our lives in harmony with nature and to respect nature as well as respecting ourselves. The way each individual shape their living styles, on the other hand, is unique. ESD does not mean to outline the styles, but shows the human and environment interdependence considering environmental, economical, social aspects together. Therefore, having environmentally literate citizens promise to take a step in managing air pollution issues. Disinger & Roth (1992) provide a generally accepted 'definition' of environmental literacy pointed out that, Environmental literacy is essentially the capacity to perceive and interpret the relative health of environmental systems and take appropriate action to maintain, restore, or improve the health of those systems (p.2). They go on to explain that environmental literacy draws upon six major components: environmental sensitivity, knowledge, skills, attitudes and values, personal investment and responsibility, and active involvement (Disinger & Roth, 1992). This definition is in line with a growing literature that sees pro-environmental behaviour as a function of environmental literacy and the literature that views environmental behaviour as something that may be learned through increased environmental knowledge (Wilke 2005; Hines et al., 1986)

4. How can education help? How education can be integrated in managing air pollution?

> Education is how we live our lives and how you live with everything around you. Everything in existence teaches us something about life... everything around us educates, how we interact with the land, minerals, trees, sky, animals everything, even our thoughts. Our thoughts too can become a force, for we are in charge of them. (Profeit-Le Blanc 1996, p. 14)

The Intergovernmental Panel on Climate Change (IPCC, 2007) concluded in 2007 that, global warming is inevitable and that human activity is likely to be the main cause. However, according to a survey of the American public in the same year (ABC News 2007), while 33% cited climate change as the world's top environmental issue, and 84% thought it was probably happening today, only 41% of the American public believed that global warming was caused by human activity. Furthermore, while 86% believed global warming would become a serious environmental problem if not corrected, 63% thought it could be reduced, with 62% claiming they knew a moderate amount about global warming. However, reported that, only 18% of the US public agree that every time we use coal or oil or gas, we contribute to the greenhouse effect (Nisbet and Myers, 2007). What these findings imply is the importance of women, young, decision makers, teachers, ... learning about the greenhouse effect in order to understand the arguments and debates about the science of global warming and climate change to promote that is knowledgeable about global warming and climate change, and one which can assume informed responsibility for the management and policymaking decisions facing our planet (Brown 1992; Bybee 1993 quoted in Shepardson et.al. 2011).

Accordingly, ESD for air pollution issues have been covered in several researches all over the world and the major areas of focus are; implementations, cultural differences in attitudes, misconceptions, factors effecting satisfactory results, problems for an effective ESD. And although there are a number of recommendations made by researchers, the targets for ESD still have not been satisfied. But, it is reality that, although research in developing ESD is the task of the education society, increasing environmental literacy is the task for all sectors, including universities, governmental authorities, NGOs, private sector, etc. The following section, therefore, highlights the recent research, focusing on the results, on ESD implementations related to air pollution issues, global warming being the focus of almost all researchers, major challenges and the driving force.

Although air pollution has been a problem since 17th century (please refer to the 1st section of this chapter), it has just become more complex and difficult to manage and control, requiring a growing need for improvement in public understanding of environmental science and policy. Thus, comprehensive and meaningful education is decided to be a promising avenue for equipping members of society in identifying potential solutions to environmental problems in order to protect valuable natural resources. ESD can produce an environmentally literate citizenry able to actively address environmental challenges and problems (Hungerford and Peyton, 1976; UNESCO, 1980; Roth, 1992). Therefore, comprehensive environmental education may be an important mean for societies to meet the increasing need for improved public understanding of environmental issues, trade-offs and other alternatives. The 1972 United Nations-Stockholm Conference (UNEP, 1972) helped articulate a shared outlook and set of principles for inspiring and guiding efforts focused on helping the public learn to pre-serve and enhance healthcare and environment. In 1977, an international assembly of environmental educators developed a set of definitions and principles for environmental literacy and education at the Inter-governmental Conference on Environmental Education in Tbilisi (UNESCO, 1977). A decade later, Hines et al. (1986/87) pointed out that environmental education efforts must go beyond providing simplistic information and move towards providing: knowledge of complex environmental issues, specific knowledge about approaches for addressing such issues and decision-making skills. Hines et al. also called for efforts to change certain effective qualities (attitudes) that result in people caring about and paying more attention to environmental conditions. It seems axiomatic to observe that the current global environmental and natural resources conditions are worse today, the world over than in the past. Therefore, it is imperative that the goals of 21st century educational systems should include environmental education and the formation of an environmentally literate citizenry able to actively participate in solving environmental problems. Disinger and Roth (1992) provide a generally accepted 'definition' of environmental literacy and have pointed out that: Environmental literacy is essentially the capacity to perceive and interpret the relative health of environmental systems and take appropriate action to maintain, restore or improve the health of those systems (p. 2).

Individuals (adults and children) may change their environmental behaviour when their values, beliefs, and pro-environmental norms change (Dietz et al., 2005). Improved environmental education as well as increased environmental literacy may result in such changes. As Clair (2003) pointed out: Environmental literacy for adults means developing and participating in the social practices likely to change the way our societies think about and act upon ecological issues. Literacy is a powerful metaphor that contributes a great deal to thinking through the question of what each of us can contribute for a more just and sustainable way of life for the planetary community (p. 77).

Coyle (2005), in reviewing 10 years of NEETF/Roper research on environmental literacy in the United States, points out that creating more widespread environmental literacy depends on (1) bringing sound environmental education programming into the education realm, and (2) channelling public environmental literacy efforts to focus more on depth rather than accuracy. Thus, environmental literacy is distinct from simple awareness or personal conduct knowledge because of its depth of information and the actual skills (thinking and doing) imparted. Knowledge and attitudes are essential components of environmental literacy, especially if the goal of environmental education is to change behaviour. Individuals' environmental behaviours may change as a result of changes in their values, beliefs, and pro-environmental norms (Dietz et al., 2005).

A research on the environmental literacy of pre-service teachers in Turkey (Tuncer et.al, 2009) revealed that, the least percent (34%) of correct answer for environmental knowledge items were concerned about motor vehicles as the major contributor to carbon monoxide while more than 60% of respondents incorrectly identified factories and businesses as the major source of carbon monoxide. 77% of the teachers correctly answered the item that the ozone layer serves as a protective layer from cancer-causing sunlight. The most frequently answer received for the types of pollution "very concerned', on the other hand, was related to "indoor air pollution" (42%), "ozone depletion and global warming" (38%), while 42% of respondents indicated that they were very concerned about "ozone depletion and global warming," only 19% of them were 'very concerned' about automobile emission. Therefore, the conclusion was that, preservice teachers in this study either do not understand the cause - effect relationship between automobiles and global warming or that their fear of more restrictions on their automobiles motivated them to answer strategically. Besides, gender appeared to play a role in elucidating the variation in the two components of environmental literacy variables along with environmental attitude and uses. Female pre-service teachers tended to have more positive attitudes and have more responsible actions toward the environment than male pre-service teachers, which is in line with other studies (Alp et al., 2006; Berberoglu and Tosunoglu, 1995; Huang and Yore, 2003; Chu et al., 2007; Tikka et al., 2000; Yilmaz et al., 2004; Worsley and Skrzypiec, 1998; Zelezny et al., 2000). For example, according to Tikka et al. (2000), whereas males are more likely to emphasize mastering nature and taking benefits from natural resources, females obtain a more emotional attitude toward nature. Since females, as indicated by the authors, have traditionally been responsible for looking after the home and children, such behaviours could be perceived as a way of taking care of their offspring. Indeed parallel to this explanation, there are two theories, namely socialization-based theories and structural theories proposed in the literature to clarify the gender difference in environmental variables: Socialization-based theory posits (sets) that females are more likely than males to associate themselves with 'caregiver' roles. It is argued that this leadswomen to be more in tune with their locality and the world at large and, consequently, to turn their compassion toward the ecological environment. Women's close affinity with nature is viewed as a result of socialization due to cultural and social-structural forces rather than resulting from biological differences. Structural theories suggest that it is the gendered segmentation of the economy and workplace that frames the perspective of women and men toward the environment. It is argued that although women may be knowledgeable and accepting of the aims of economic growth, they are more prone than men to question the consequences of such growth. The reasoning behind this argument lies in the combination of women's role as caregivers for children and their role in the household, where they do most of the house work, in addition to working in the paid labor force. This role is in direct contrast to men's historical "breadwinner" role. (Weaver, 2002;p.83).

Referring to the above mentioned definition of environmental literacy, it seems to rest on an assumption that individuals have a competent level of environmental knowledge. It is concluded in one of the recent studies realised in Turkey, to determine environmental literacy of preservice teachers (Tuncer et.al, 2009), that, a majority of Turkish pre-service teachers do not possess enough knowledge to be classified as having an acceptable level of environmental knowledge (NEETF and Roper, 2005). Slightly less than half of the pre-service teachers of this study (49%) received a passing grade, based on the NEETF and Roper Starch grading scale. Yet, 66% of the students from Michigan State University [MSU] (Kaplowitz and Levine, 2005) reported to receiving a "passing grade" based on the same grading scale. Interestingly, in that same university, MSU College of Education students had one of the lowest mean correct scores compared with students from other MSU colleges such as agriculture and natural resources (8.84 of 11) and Natural Science (8.48 of 11). One possible explanation for the low level of passing grades for Turkish respondents' environmental knowledge, on the other hand, was explained be the absence of course works, relevant to environmental education in the current teacher education programs in Turkey. Of course, there may be demographic characteristics of students that relate to differences in academic level and environmental knowledge. Despite their low levels of environmental knowledge, Turkish respondents expressed positive attitudes toward the environment as well as high degree of concern about environmental problems. The respondents also expressed feelings of responsibility for environmental problems; expressed the view that environmental problems are one of the most important problems of their lives; and shared their feeling that they are comfortable with their background on environmental issues. Such results beg the question of what it would take to create a critical mass for increased environmental learning throughout the educational system.

In service elementary teachers' knowledge about air pollution in Turkey was the subject of another study (Tuzun et.al, 2008). The study pointed out specific results. One of the questions related to teachers' general knowledge about air pollution sources was that, "Which human activities contributed to air pollution?" The answers for this question were categorized as, individual, societal, and industrial contribution. Approximately 24% of the teachers reported exhaust gases as the main cause of the air pollution. Which, was not a surprising result for the authors, reminding that the study area is one of the crowded cities in Turkey with a number of cars has been increasing and traffic is getting more and more challenging every day. Use of sprays and deodorants was considered as the second main individual source of air pollution by the participants. The reason was explained in relation with not the content of textbooks but with the effect of media. Use of coal at home for heating was declared as the third most important individual source of air pollution by the teachers of the study. Teachers seemed to have knowledge about importance of improper use of other energy sources. Moreover, cigarette, forest fires, education, and trash were found as the other important sources of air pollution declared by the teachers. Contribution of community to air pollution was declared by Turkish teachers as exhaust gases, as the main source, unconscious use of energy especially at homes and forest fires. Besides, awareness toward environmental pollution, trash, and industrial emissions were also considered as the sources of the air pollution by the teachers. When industrial contribution was in consideration, more than 70% of the teachers were stated industrial emissions as the main contributor to air pollution. Whereas, they also indicated that lack of control mechanism is an important aspect for air pollution in Turkey. According to the teachers of this study, emissions from solid waste disposal areas also cause air pollution. Therefore, it was decided by the authors that, teachers participated the study have a considerable knowledge about the reasons of air pollution. However, teachers were asked further questions to get the nature of their knowledge. As a result, teachers reported that, global warming was caused mainly by carbon dioxide (44.3%), methane (28.4 %), chlorofluorocarbons (19.7 %) and ozone (10.4 %). But at the same time, 34.9 % of them declared CO as one of the gases causes global warming, while 18.1 % declared SO2 as so. Teachers declared CFCs (31.7%), spray deodorants (61.7%) and air conditioners and refrigerators (38 %) as the reasons for the ozone layer depletion problem. The interesting result at this point was that, although about 32 % of the teachers indicated CFCs as a reason for the ozone layer depletion, spray deodorants were indicated by 62 %. This result explained with the media effect as; teachers were more familiar with the term rather than the chemical formula. Because, ozone layer depletion has always coincided with the use of spray deodorants in media; especially by the phrase "ozone friendly" in the deodorant advertisements. Moreover, sulphur and nitrogen compounds were stated as the pollutants for acid rain problem by relatively higher percentages. It was inferred as a result that, teachers' knowledge about global warming, ozone layer and acid rain cannot go beyond just defining the concepts. Misconceptions of the teachers of this study about global warming and ozone layer were also obvious by their answers for the multiple choice question (*Which characteristic of the green house gases make them cause green house effect*?): Although 61% of the teachers answered this question correctly, as they absorb *the sunlight emitted from the Earth surface,* 15 % of them answered as they are *found in the upper layers of the atmosphere relative to other gases.* This result points out a common misconception observed in the literature. As was declared by Michail, Stamou and Stamou (2007), Cutter (2002) and Summers, Kruger, and Childs (2000), for example, teachers had confusion about green house effect and ozone layer depletion and they had misconception on these issues. In general terms, the location of ozone layer was confused with that of the greenhouse gases and stratospheric ozone with tropospheric ozone. Teachers in Turkey were also asked about renewable energy question, emphasizing that teachers need to teach renewable energy sources to allow their students to show appropriate actions to prevent air pollution. Just having knowledge about causes of the air pollution will not help them prevent air pollution, they can help prevent air pollution if they know the ways for sustainable use of the sources. As a result, majority of the teachers knew that wind, sun light, wave, and geothermal are renewable energy sources and some of them also indicated that biogas, hydrogen, and biodiesel as renewable energy sources. Similar research conducted with Greek, Australian, and English teachers had also revealed that they did not had enough understanding about air pollutants, green house gases and ozone layer (Cutter, 2002; Michail, Stamou, & Stamou, 2007; Summers et al., 2000).

Shepardson et. Al. (2011) derived five distinct mental models of the greenhouse effect from an inductive analysis of the content of the drawings and explanations of 225 students' from three different schools in the Midwest in the US. Based on the mental models identified it was apparent that students lacked a clear understanding of the greenhouse effect. At best 48% of the students realized that greenhouse gases, whatever they may be, cause the greenhouse effect and that the sun's energy is either 'trapped' by or 'bounced' back to the Earth by the greenhouse gas layer. On the other end of the spectrum, 29% of the students lacked an understanding of the greenhouse effect.

Investigation on the regional differences in 15 year old Turkish students' awareness, perception, optimism and responsibility development toward environmental pollution issues carried out, with 4942 fifteen year-old-students attending 160 schools across 78 provinces and 7 geographical regions (Teksoz, Tekkaya, Erbas, 2009). Results indicated that a minority of students, across seven regions reported as they aware of the increase of greenhouse gases in the atmosphere (9.7%). The percent of students who stated that they never heard about these issues were 26.5% in average. As far as the acid rain issue has been considered, 18.2% stated that they were familiar and 9.2% stated that they never heard about the acid rain issue. Furthermore, while 63.7% of the participants believed that they were familiar with the consequences of clearing forests for other land use, 3.9% declared that they never heard about the issue. The mean scores calculated for environmental perception, environmental awareness and responsibility development components show almost the same trend among the regions. Those for the regions in the West part of the country are higher than those for the Eastern regions. Whereas, mean values for the environmental optimism component show a different pattern between the regions: students living in the Eastern regions seem more optimistic than those in other regions. In the first place, the results of the study provided some evidence that the place where students live had an effect on their environmental awareness, concern, optimism and responsibility for sustainable development. For example, the most noticeable characteristic was that; although the students of the two of the least industrialized regions (Southeast Anatolia and East Anatolia) displayed lower awareness and concern toward environmental issues, they displayed highest degree of optimism concerning the development over the next 20 years of the problems associated with air pollution, clearing of forests for other land use. In fact, the results reflect the transcontinental feature of Turkey. Among the 7 geographical provinces, Marmara having students with comparably higher environmental concern, responsibility but low degree of optimism, distinguishes from the others with its being heavily advanced in industry, commerce, tourism and transportation because of its close location to Europe. Thus the children living in such circumstances are more aware of air pollution problems, are concerned about them and pessimistic about the future state of the problems. The significant feature of Aegean, which has the students with very high environmental awareness, concern, responsibility and comparably higher optimism, is that, most of the population and cities are concentrated on the coast line because of its convenience for sea transportation and tourism and it's also being both industrialized and agriculturalized. Students from one of the most important trading and tourism centre and the rapidly growing port, the Mediterranean region, on the other hand revealed high awareness and concern but lowest optimism toward environmental issues. The students from the Black Sea region, one of the most heavily forested regions with very rich fauna and flora, revealed a similar trend with the former regions. The students from plateau-like heartland of the country, Central Anatolia, revealed comparably higher perception, responsibility and optimism toward environmental issues are considered. The students showing a distinguished feature, lowest awareness, perception, responsibility but highest optimism toward environmental issues, in the current study come from the Eastern Anatolia, where the population and habitat not dense because of the harsh climate and high mountains and has the highest unemployment rate in Turkey and South-eastern Anatolia where a special atmosphere exists throughout, uniquely different from other parts of the country, thus, reflecting a specific life style over its land. Thus, referring the very well known phrase of the environmental studies, *think globally act locally* (UNEP, 1972), efforts to explain environmental perceptions and concern as a function of social structure and socio-demographic characteristics can be combined with the regional features of a country and such a relationship, if any, is valuable for strategy development for developing air pollution perceptions. Thus, as Matthews (1995) reviewed in his study, regional features and culture affect children's behaviour in large-scale environments and it follows that as the life worlds of children from different socio cultural backgrounds differ, the way in which children encounter place and make sense of their everyday worlds are also likely to be at variance. Therefore, it may be concluded that, "area of residence is a silent predictor of responsible environmental behaviour". Therefore, efforts for creating environmentally literate generations need to consider regional socio-economical features as well as the people's perceptions towards environmental issues. Such an evaluation will be very valuable leading the education specialists to establish a national strategy for and will help to make the strategy regional, as suggested in Chapter 36 of Agenda 21 (UNCED, 1992).

Efforts for developing EE and ESD as a tool for managing air pollution therefore continues under the light of the research, several examples of which are summarised above. Although there is are international conventions leading such efforts, regional, national and local strategies are needed to get more effective outputs. Efforts, although concentrated in teachers' education, shall diffuse to all areas, engineering being one of the vital one. Ashford (2004) pointed out the concern as follows: "Scholars and professionals committed to fostering sustainable development have urged a re-examination of the curriculum and the restructuring of research in engineering-focused institutions of higher learning. The focus is on engineering, more than on the natural and physical sciences or on social science, because the activities that drive the industrial state - the activities that implement scientific advance - are generally rooted in engineering. Moreover, engineers are known as 'problem solvers' and if economies are becoming unsustainable because of engineering, it is natural to ask whether engineering as an activity and as a profession can be re-directed toward achieving sustainable transformations. Of course, engineering cannot do it alone; scientific as well as social and legal changes must occur as well" (p.239).

Nevertheless, results of research investigating people awareness, attitudes, behaviour on the air pollution issues have a single common result that, most of the people from different countries with different socio-demographic features and life styles, aware of the natural resources, like air, are vulnerable resources and deserve conservation for the sake of supporting human life or just for the health of itself. Most of the people have an eco-centric approach to the natural sources. But beyond this point, when it comes to individual responsibilities, i.e. making changes in living styles for natural protection, such as preferring public transport instead of private cars or using less energy at home, the approach slides through anthropocentrism. Or, citizens both in the developing and developed world cannot make a relation between, for example the way we eat and the global warming, or the way we shop and acid rain, or the way we consume, we produce and ozone layer depletion, etc, which as a result makes it difficult to make a change. Education is a powerful tool to make people aware of such relations and get rid of the ignorance we have been carrying since 17th century.

Living sustainably depends on a duty to seek harmony with other people and with nature. The guiding rules are that people must share with each other and care for the Earth. Humanity must take no more from nature than nature can replenish. This in turn means adopting lifestyles and development paths that respect and work within nature's limits. It can be done without rejecting the many benefits that modern technology has brought, provided that technology also works within those limits. (IUCN, UNEP and WWF 1991, p. 8)

Knowledge, learning, information, and skilled intelligence are the new raw materials of international commerce and are today spreading throughout the world as vigorously as miracle drugs, synthetic fertilizers, and blue jeans earlier... Learning is the indispensable investment required for success in the "information age" we are now entering. (National Commission for Excellence in Education, 1983, p. 7)

5. References

ABC News (2007). Concern soars about global warming as world's top environmental threat. URL http://abcnews.go.com/images/US/1035a1Environment.pdf.

Alp, E., Ertepinar, H., Tekkaya, C., Yilmaz, A.,(2006). A study on children's environmental knowledge and attitudes: the effect of grade level and gender. *International Research in Geographical and Environmental Education* 15, 210–223.

Andersen, S.O. & Madhava Sarma, K. (2002). Protecting the Ozone Layer, *The United Nations History*, Sinclair,L (Ed.). Earthscan.

Berberoglu, G., Tosunoglu, C.(1995). Exploratory and confirmatory factor analyses of an environmental attitude scale (EAS) for Turkish university students. *The Journal of Environmental Education* 26, 40–44.

Ashford, N., A. (2004). Major challenges to engineering Education for Sustainable Development: What has to change to make it creative, effective and acceptable to the Established Disciplines, *International Journal of Sustainability in Higher Education* 5(3): 239-250.

Aydemir M. (2007). The investigation of teachers with respect to knowledge level on environmental concepts. Ms.Thesis. Middle East Technical University, Ankara.

Chu, H.E., Lee, E.A., Ko, H.R., Shin, D.H., Lee, M.N.,Min, B.M., Kang, K.H. (2007). Korean year 3 children's environmental literacy: a prerequisite for a Korean environmental education curriculum. *International Journal of Science Education* 29, 731–746.

Clair, R. (2003). Words for the world: creating critical environmental literacy for adults. *New Directions for Adult and Continuing Education* 99, 69–78.

Coyle, K.(2005). Environmental Literacy in America. The National Environmental Education Training Foundation, Washington, DC.

Cutter, A. (2002). The value of teachers' knowledge: Environmental education as a key study, ERIC DocumentReproduction Service No. ED 466456.Brelis, D., (1984). Time Magazine, December 17.

Dietz, T., Fitzgerald, A., Shwom, R. (2005). Environmental values. *Annual Review of Environment and Resources* 30, 335–372.Disinger, J. F., Roth, C. E. (1992). Environmental Literacy. ERIC/CSMEE Digest.URL http://eric.ed.gov/ERICDocs/data/ericdocs2/

Dove, J. (1996). Student teacher understanding of the greenhouse effect, ozone layer depletion and acid rain. *Environmental Education Research, 2(1)*, 89-100.Eshel, G., Martin, P.A. (2006). Diet, Energy and Global Warming, *Earth Interactions*, 10:1-17.

Gadgil, M. and Guha, R. (1992). This Fissured Land: An Ecological History of India. Berkeley: University of California in Orr, D., (1994) *Earth in Mind on Education, Environment and the Human Prospect.*

Gardner, G. (2010). Materials Use Up, *Vital Signs 2010*, Worldwatch Institute.Gayford, C. G. (2000). Biodiversity education: A teacher's perspective. *Environmental Education Research, 6:4*, 348–361.

Grove, R.H.(1995). Green Imperialism: Colonial Expansion, Tropical Island Edens and the Origins of Environmentalism, 1600-1860. Cambridge, England: Cambridge University Press, in http://www.radford.edu/~wkovarik/envhist/Envbiblio.html#G

Hines, J.E., Hungerford, H.R., Tomera, A.N.(1986/87). Analysis and synthesis of research in responsible environmental behavior: a meta-analysis. *Journal of Environmental Education* 18, 1–8.

Huang, H., Yore, L.(2003). A comparative study of Canadian and Taiwanese grade 5 children's environmental behaviours, attitudes, concerns, emotional dispositions, and knowledge. *International Journal of Science and Mathematics Education,* 1, 419–448.

Hungerford, H.R., Peyton, R.B.(1976). *Teaching Environmental Education.* Weston Walch, Portland, ME.

IUCN, The World Conservation Union (2002). Education and Sustainability Responding to the Global Challenge, Editors: Daniella Tilbury, Robert B. Stevenson, John Fien, Danie Schreuder.

IUCN, UNEP, WWF (1991). Caring for the Earth: A Strategy for Sustainable Living. IUCN, UNEP, WWF: Gland, Switzerland.

IUCN, UNEP, WWF (1991). Caring for the earth: a strategy for sustainable living. IUCN: Gland, Switzerland. in Education and Sustainability Responding to the Global Challenge (2002). Tilbury, D., Stevenson, R.B., Fien, J., Schreuder, D., (Eds.) IUCN – The World Conservation Union

IPCC (2007). Intergovernmental Panel on Climate Change., *Climate change 2007: The physical science basis.* Geneva ,IPCC.

Lopez, B. (1989), American Geographics, Orion , in Orr, D. (1994), *Earth in Mind on Education, Environment and the Human Prospect,* p.10.

Makra, L., Brimblecombe, P. (2004). Selections from the history of environmental pollution, with special attention to air pollution. Part 1, *J. Environment and Pollution*, 22:6, 641-656.

Matthews, H. (1995). Culture, environmental experience and environmental awareness: making sense of young Kenyan children's views of place. *The Geographical Journal, 161*, 285-295.

McNeill, J.R. (2001). *Something New Under the Sun. An environmental history of the twentiethcentury world,* W.W. Norton & Company.

Michail, S., Stamou, A. G. & Stamou, G. P. (2007). Greek primary school teachers' understanding of current environmental issues: An exploration of their environmental knowledge and images of nature. *Science Education, 91(2)*, 244-259.

National Commission for Excellence in Education (1983). *A Nation at Risk.* Washington, DC. in *Education and Sustainability Responding to the Global Challenge* (2002). Tilbury, D., Stevenson, R.B., Fien, J., Schreuder, D., (Eds.) IUCN - The World Conservation Union

Orr, D. W.(1994). *Earth in Mind, On Education, Environment and the Human Prospect,* Island Press.

Özdemir, N. ve Çobanoğlu, E. O. (2008). Türkiye'de nükleer santrallerin kurulması ve nükleer enerji Kullanımı konusundaki öğretmen adaylarının tutumları. *Hacettepe Üniversitesi Eğitim Fakültesi Dergisi, 34,* 218-232.

Profeit-LeBlanc, Louise (1995). "Transferring Wisdom through Storytelling" in Jickling, R. (ed.) *A Colloquium on Environment, Ethics and Education.* Whitehorse:Yukon. in *Education and Sustainability Responding to the Global Challenge* (2002). Tilbury, D., Stevenson, R.B., Fien, J., Schreuder, D., (Eds.) IUCN - The World Conservation Union

Robinson, M. (2011), Climate change: A Glimpse of the Future, The Huffington Post, March 4, 2011. URL http://www.huffingtonpost.com/mary-robinson/post_1734_b_823803.html).

Roth, C.E.(1992). Environmental Literacy: Its Roots, Evolution and Directions in the 1990s. ERIC/CSMEE Publications.

Russell, J. (2007), Coal Use Rises Dramatically Despite Impacts on Climate and Health World Watch Institute Vital Signs 2007.

Ryan, J.C. and Durning, A.T. (1997). *Stuff—The Secret Lives of Everyday Things,* Sightline Institute.

Schumacher, F. (1973). Small is Beautiful: Economics as if People Really Mattered. Abacus:London, in, *Education and Sustainability Responding to the Global Challenge* (2002). Tilbury, D., Stevenson, R.B., Fien, J., Schreuder, D., (Eds.) IUCN - The World Conservation Union

Shepardson, D.P., Soyoung, C., Niyogi, D.and Charusombat, U. (2011). Seventh grade students' mental models of the greenhouse effect, *Environmental Education Research,* 17:1, 1–17.

Summers, M., Kruger, C., Childs, A. & Mant, J. (2000). Primary school teachers' understanding of environmental issues: An interview study. *Environmental Education Research, 6(4),* 293 - 312.

Teksöz, G., C.Tekkaya, A.K.Erbaş (2009). "Geographical Regions as a Silent Predictor of Responsible Environmental Behaviour", *H.U. Journal of Education,* 36, 249-259.

Tikka, P.M., Kuitunen, T.M., Tynys, M.S.(2000). Effects of educational background on students' attitudes, activity levels, and knowledge concerning environment. *The Journal of Environmental Education* 31, 12–19.

Tuncer, G., C. Tekkaya, S. Sungur, J. Cakiroglu, H. Ertepinar, M. Kaplowitz (2009. "Assessing pre-service teachers' environmental literacy in Turkey as a mean to develop teacher education programs," *International Journal of Educational Development,* 29, 426-436.Tung, A. (2010), Vital signs 2010, World Watch Institute.

Tüzün, Ö., Y., G. Teksöz Tuncer, M. Aydemir (2008) "An investigation on the elementary teachers' knowledge about air pollution issues", *H. U. Journal of Education,*35, 374-385.

UNCED, (1992). United Nations Commission for Environment and Development *Agenda 21, the United Nations programme of action from Rio.* New York: United Nations.

UNEP (1972). *United Nations Conference on the Human Environment.* Retrieved from United Nations: URL http://www.unep.org/Documents.Multilingual/Default.asp?DocumentID=97

UNESCO, (1977). Final Report, Intergovernmental Conference on Environmental Education organized by Unesco in co-operation with UNEP, Tbilisi (USSR).

UNESCO, (1992). *United Nations Conference on Environment and Development: Agenda 21.* UNESCO, Switzerland.

UNESCO, (1977). In: International Conference on Environmental Education, Tbilisi, USSR.UNESCO, (1980). In: Environmental Education in the Light of the Tbilisi Conference, Paris, France.

UNEP, (1972). In: Declaration of the United Nations Conference on the Human Environment.

Walsh, B. (2008). Meat : Making global warming worse, Time Magazine, September 10, 2008. URL http://www.time.com/time/health/article/0,8599,1839995,00.html

Weart, S. (1988). *Nuclear Fear: A History of Images,* Harvard University Press.

Weaver, A. (2002). Determinants of environmental attitudes: a five-country comparison nternational Journal of Sociology 32, 77–108.

Wilke, R. (1995). Environmental literacy and the college curriculum. *EPA Journal 21,* 28-30.WCED, (1987). *Our Common Future,* Oxford University Press.

Worsley, A., Skrzypiec, G. (1998). Environmental attitudes of senior secondary school students in South Australia. Global *Environmental Change* 8, 209–255.

Yilmaz, O., Boone, W.J., Anderson, H.O.(2004). Views of elementary and middle school Turkish students toward environmental issues. *International Journal of Science Education* 26, 1527–1546.

Zelezny, L.C., Chua, P., Aldrich, C.(2000). Elaborating on gender differences in environmentalism. *Journal of Social Issues* 56, 443–457.

21

Do Developing or Developed Nations Pollute Air More? An Assessment of Health Consequences

Ramalingam Shanmugam and Attila Hertelendy
School of Health Administration, Texas State University, San Marcos, Texas USA

1. Introduction

Air pollution is a serious threat to the healthy life in society. Nations across the continents in the 21st century need to evaluate the air pollution data to make up fair and implementable global policies, which support healthy life. In this process of examining the data, the following relevant question could be answered. Do developing or developed nations contribute more air pollutions? As this chapter illustrates, the pollutions related data in the public domain WebPages (see [1] through [21] for details) hide the answer.

The developing nations aspire to improve their standard of life by advancing their industrial or agricultural productivities. The developed nations, on the contrary, tend to maintain their higher standard of living with the same or more industrial and agricultural productivities. When it comes to a discussion on who contribute the air pollutions, both the developed and developing nations argue and are willing to make a few policy changes but preach and expect the other side to do more to cut down the global air pollutions. How should this escalated conflict be resolved? What do the relevant data suggest? How a compromise between the developed and developing nations is plausible?

To answer this and related questions, this chapter examines and assesses existing global air pollutions data using statistical techniques. The findings as they are stated in this chapter are interesting and pave ways to formulate equitable global policies to reduce air pollutions and thereby increase the healthy life at large.

2. Motivation

Whether air pollutions are natural or human generated, the damages significantly affect the quality of life. Environmentalists warn that the human generated air pollutions are on the rise to an alarming level in the recent years. Such a rise results in tremendous health impacts. The illness and sufferings among the humans, animals, natural life and vegetations are on the steady increase. The land and water sources are also contaminated because of the air pollutions. A recovery from the damages due to the air pollutions requires enormous resources, efforts and money. Many health hazardous chemicals released by the human initiated activities in day-to-day life, in industrial and agricultural sectors are additional add on to the existing natural air pollutions due to phenomena like volcanoes or ozone leaks (see [23] through [28] for details on natural calamities). Some of these hazardous chemicals are absorbed by the air and stay there as permanent air contaminants. Not only the local communities where the air pollutions originate but also those reside in nearby or far away communities suffer as the air contaminants are blown through because of the wind. Consequently, many chronic diseases and illnesses increase everywhere in the world.

The World Health Organization (WHO) estimates over 2.4 million people die each year from the air pollutions. Among those, about 1.5 million deaths are due to the indoor air pollutions. About 310,000 Europeans died in year 2005 alone due to the air pollutions. The number of deaths due to the air pollutions is rising and it is more than the automobile accident deaths. The 1984 Bhopal's Union Carbide disaster caused more than 2,000 deaths and over 500,000 injuries. See [11] and [12] for historical details.

The health effects of the air pollutions include breathing difficulties, wheezing, coughing, pulmonary exacerbations, cancer, respiratory and cardiac failures. Consequently, the consumption of medications and the usage of emergency room visits increase. The hospitals and clinics are overwhelmed by the influx of patients for treatments.

The land and water sources are not immune from the air pollutions. When it rains, because of the air contaminants, the rain exhibits acidity. The acidity in rain causes water pollutions. Especially when the drinking water and arable land receive the acidity, the food chains are contaminated and the public health deteriorates. The land becomes less fertile and the crops are of less safe. The food sources are even depleted in some situations. On the health arena, the acid rain contains a high amount of carbon oxides (CO2), sulfur oxides (SO2) and nitrogen dioxides (NO2). The nations across all continents suffer even though they are not the originators of the air pollutions. The neighboring nations constantly complain to the polluting nations. For an example, Canada, in 1980s, complained to the USA that their air and water pollutions originated in USA and wanted the USA to make remedies. Such complaints were also the reasons for the US congress to legislate the US Clean Air Act of 1990. Similar complaints were lodged by some European nations resulted in the famous Helsinki Protocol Agreement of 1985 in which about 21 European nations committed to reduce their sulfates in the air. Much more, of course, need to be done to alert both the developed and developing nations to commit to drastic improvements for pure air. The findings in this chapter based on relevant data analysis are critical and educational for having healthy life.

Incidentally, drinking the distilled water does not remedy the problem of water contaminations. Why is it so? Though the drinking water contains no carbon dioxide (CO2) after the distillation, its average amount of phosphate (pH) is 5.6μ, which is still acidic. Note that an acceptable pH level in vinegar sold in the supermarkets is only 2.4μ on the average. Thus, the acidity in the rain needs to be taken seriously and corrected. However, the acidity in rain arises from the PM10, sulfuric, carbon or nitrogen oxides in the air.

The acid rain was first noticed in Manchester, England 1852. The carbon, nitrogen, PM10 and sulphate in the air are some important sources of acidity in the rain. The carbon dioxide and the sulfur dioxide (SO2) emanate from the auto emissions. The glacial ice and historical lakes attest to the fact, according to the environmentalists, that the volcanoes and environmental factors such as ozone leaks and deforestation contribute to the increased level of sulfur in the air. The coal and petroleum products contain more sulfur and hence, produce sulfur dioxide upon their combustion. The carbon monoxide and dioxide are due to the combustion of cooking or fossil fuels. The particulate matter (PM10) in the air is quite health hazardous (see [22] and [23] for details). The PM10 are tiny particles of solid or liquid in the gas. The ammonia and chlorofluorocarbons in the air are toxic and are so banned by the environmental protection agency (EPA). The principal and significant causes of the acid

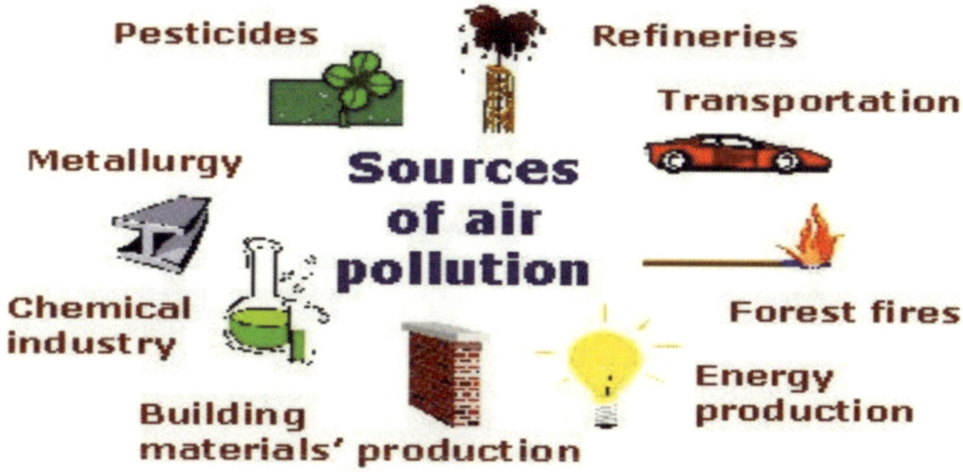

Fig. 1. The group of air pollutants

rain are therefore the chemical compounds emitted by the manufacturing industries or motor vehicles. See Shanmugam et al (2007) for details about the increase in stroke incidences due to the air pollutions. See Figure 1 below for the sources of air pollutions in general.

The developed nations pollute the air by their excessive use of fossil fuels and coal energies in manufacturing industries, oil refineries and the auto emissions in particular. The developing nations aspire to catch up with the developed nations' status and hence add up to the air pollutions by their increased efforts to use excessively the fossil fuels and coal energies in personal and public life. Their use of saltpeter, urea, ammonia (NH_3) and nitric oxides (NO_2) in agricultural activities do contribute to the air pollutions. The farming and agriculture activities in the developing nations are also sources of the air pollutions. Spreading fertilizers and pesticides from the airplanes causes the air pollutions with these solid particles. The soil surface erosions also generate large amounts of particulates with their diameter up to 2.5 μm (the so-called PM 2.5). The particulates are transported in the wind to near and far away communities.

To assess whether the developed and developing nations have parity when it comes to causing the air pollutions, we need to define first who the developed nations are really. How are the developed versus developing nations defined and distinguished? Many definitions are contentious. Could the *per capita income* be an identifier of a developed nation? There might be nations still in developing status though they have a higher per capita income. Could the level of *industrialization* be an indicator of a developed nation? Could the *gross domestic product* (GDP) be an indicator of the developed nations? The list of indicators goes on. The *human development index* (HDI) might be an appropriate and acceptable index according to the discussions in the United Nations, to classify a nation as developed or developing among the members of the United Nations.

If so, what is HDI? The HDI is an aggregate measure of life expectancy, literacy level, education rate, purchasing power and standard of living among the citizens in a nation. The two economists Sen and Haq created the concept and calculating formulas for the HDI during 1990. See [21] for details about the 2010 Human Development Report containing the calculation and interpretation of HDI. In this chapter, the HDI is involved to formulate a list of developed versus developing nations.

2. Do data distinguish developed and developing nations in air pollutions?

To answer this crucial question, an analysis of PM10, sulfuric, carbon or nitrogen oxides emitted to the air by the nations in all continents should be performed. To begin with, for comparisons sake, four groups of nations are formed and considered. All the nations in both the North and South American continental surroundings are grouped together for the analysis. The nations in Asia, New Zealand, Australia and Pacific Oceania are grouped together as part of Asia continental surroundings. The Russia, eastern and western European nations are grouped together as part of European continental surroundings. The Madagascar and all the nations in African continental surroundings are grouped together for the analysis. In all these four groups, there are developed and developing nations. How do these nations perform within the group and across the groups?

The median HDI in each of the four groups of nations is used to classify a nation as a developed or developing in the group. If a nation's HDI is more than the group's median HDI scores, it is a developed nation. See Tables 1 through 4 for the median HDI scores and the status of each nation in the group. Notice that the median HDI scores vary across groups. The median HDI scores are 0.403, 0.695, 0.61 and 0.81 among the 56 African, 51 American, 66 Asian and 54 European nations respectively. Most of the European nations are developed while many of the American nations are developing type. The African and Asian nations are evenly split. The percent of the developed nations are 50, 31, 56 and 61 among African, American, Asian and European continental nations respectively. Hence, no continent contains completely developed nations. A comprehensive, equitable and global air pollution policy is therefore necessary. All members of the United Nations should frame such a policy collectively.

Are the per capita *gross domestic productivities* (GDP) of the developed nations in these four groups close enough? The answer is no as they exhibit a quite heterogeneity according to the Box plots in the Figures 1 through 4 below. The European nations are more homogenous (with the exception of outlier nation Luxemburg). The African nations, American nations (with the exception of outlier nations USA and Canada) and Asian nations (with the exception of outlier nation Qatar) are quite heterogeneous in per capita GDP.

One can not avoid wondering whether the density of the population is a factor in for so much heterogenous per capita GDP or HDI. In fact, it is indeed a factor according to the Figures 5 through 8 which portray the density of the population in millions per square mile. The Asian (with the exception of outlier nations Hong Kong, Taiwan and Singapore) and European (with the exception of outlier nations Belgium, Ireland and Malta) nations have more dense population per square mile than in American (with the exception of outlier nations Bermuda and Barbados) or African nations (with the exception of outlier nations Mauritius, Rwanda and Comoros) in general.

However homogenous or heterogeneous are the nations within the groups, they differ significantly in their contributions to air pollutions. The Figures 9 through 12 portray their contributions to PM10 in the air. The European nations (with the exception of outlier nation Turkey) turn in consistently a low amount of PM10. The Asian nations (with the exception

Countries	HDI	Is it a developed nation?	Countries	HDI	Is it a developed nation?
Algeria	0.677	Yes	Madagascar	0.435	Yes
Angola	0.403	Yes	Malawi	0.385	No
Benin	0.435	Yes	Mali	0.309	No
Botswana	0.633	Yes	Mauritania	0.433	Yes
Burkina Faso	0.305	No	Mauritius	0.701	Yes
Burundi	0.282	No	Morocco	0.567	Yes
Cameroon	0.46	Yes	Mozambique	0.284	No
Cape Verde	0.534	Yes	Namibia	0.606	Yes
Central African Republic	0.315	No	Niger	0.261	No
Chad	0.295	No	Nigeria	0.423	Yes
Comoros	0.428	Yes	Reunion	0	No
Congo (Brazzaville)	0.489	Yes	Rwanda	0.385	No
Congo (Kinshasa)	0.239	No	Saint Helena	0	No
Cote d'Ivoire	0.397	No	Sao Tome	0.488	Yes
Djibouti	0.402	No	Senegal	0.411	Yes
Egypt	0.62	Yes	Seychelles	0	No
Equatorial Guinea	0.538	Yes	Sierra Leone	0.317	No
Eritrea	0.328	No	Somalia	0	No
Ethiopia	0.328	No	South Africa	0.597	Yes
Gabon	0.648	Yes	Sudan	0.379	No
Gambia, The	0.39	No	Swaziland	0.498	Yes
Ghana	0.467	Yes	Tanzania	0.398	No
Guinea	0.34	No	Togo	0.428	Yes
Guinea-Bissau	0.289	No	Tunisia	0.683	Yes
Kenya	0.47	Yes	Uganda	0.422	Yes
Lesotho	0.427	Yes	Western Sahara	0	No
Liberia	0.3	No	Zambia	0.395	No
Libya	0.755	Yes	Zimbabwe	0.14	No
Madagascar	0.435	Yes	Median HDI score	0.403	# developed nations=28

Table 1. Nations in Africa

Countries	HDI	Is it a developed nation?	Countries	HDI	Is it a developed nation?
Antarctica	0	No	Haiti	0.404	No
Antigua-Barbuda	0	No	Honduras	0.604	No
Argentina	0.775	Yes	Jamaica	0.688	No
Aruba	0	No	Martinique	0	No
Bahamas	0.784	Yes	Mexico	0.75	Yes
Barbados	0.788	Yes	Montserrat	0	No
Belize	0.694	No	Netherlands Antilles	0.29	No
Bermuda	0	No	Nicaragua	0.565	No
Bolivia	0.643	No	Panama	0.755	Yes
Brazil	0.699	Yes	Paraguay	0.64	No
Canada	0.888	Yes	Peru	0.723	Yes
Cayman Islands	0	No	Puerto Rico	0.79	Yes
Chile	0.783	Yes	Saint Kitts	0	No
Colombia	0.689	No	Saint Lucia	0	No
Costa Rica	0.725	Yes	Saint Pierre	0	No
Cuba	0	No	Saint Vincent	0	No
Dominica	0.663	No	Suriname	0.646	No
Dominican Republic	0.663	No	Trinidad-Tobago	0.736	Yes
Ecuador	0.695	Yes	Turks Caicos Islands	0	No
El Salvador	0.659	No	United States	0.902	Yes
Falkland Islands	0	No	Uruguay	0.765	Yes
French Guiana	0	No	Venezuela	0.696	Yes
Grenada	0	No	Virgin Islands, U.S.	0	No
Guadeloupe	0	No	Virgin Islands	0	No
Guatemala	0.56	No	Median HDI score	0.695	
Guyana	0.611	No	# developed nations	16	

Table 2. Nations in South and North America

Countries	HDI	Is it a developed nation?	Countries	HDI	Is it a developed nation?
Armenia	0.695	Yes	Macau	0	No
Afghanistan	0.349	No	Malaysia	0.744	Yes
American Samoa	0.414	No	Maldives	0.602	No
Australia	0.937	Yes	Mongolia	0.622	Yes
Azerbaijan	0.713	Yes	Nauru	0.414	No
Bahrain	0.801	Yes	Nepal	0.428	No
Bangladesh	0.469	No	New Caledonia	0.414	No
Bhutan	0.518	No	New Zealand	0.937	Yes
Brunei	0.805	Yes	Niue	0.414	No
Burma (Myanmar)	0.451	No	Oman	0.49	No
Cambodia	0.494	No	Pakistan	0.49	No
China	0.862	Yes	Papua New Guinea	0.431	No
Cook Islands	0.614	Yes	Philippines	0.638	Yes
Fiji	0.669	Yes	Qatar	0.803	Yes
French Polynesia	0.614	Yes	Samoa	0.49	No
Guam	0.614	Yes	Saudi Arabia	0.752	Yes
Hawaiian Trade Zone	0.902	Yes	Singapore	0.846	Yes
Hong Kong	0.863	Yes	Solomon Islands	0.494	No
India	0.519	No	Sri Lanka	0.658	Yes
Indonesia	0.6	No	Syria	0.589	No
Iran	0.702	Yes	Taiwan	0.869	Yes
Iraq	0	No	Tajikistan	0.58	No
Israel	0.872	Yes	Thailand	0.654	Yes
Japan	0.884	Yes	Timor-Leste	0.502	No
Jordan	0.681	Yes	Tonga	0.677	Yes
Kazakhstan	0.714	Yes	Turkmenistan	0.669	Yes
Kiribati	0.614	Yes	U.S. Pacific Islands	0.614	Yes
Korea, North	0	No	United Arab Emirates	0.815	Yes
Korea, South	0.877	Yes	Uzbekistan	0.617	Yes
Kuwait	0.771	Yes	Vanuatu	0.614	Yes
Kyrgyzstan	0.598	No	Vietnam	0.572	No
Laos	0.497	No	Wake Island	0.614	Yes
Lebanon	0	No	Yemen	0.439	No
#developed nations	37		Median HDI score	0.61	

Table 3. Nations in Asia, Australia, New Zealand and Pacific Islands

Countries	HDI	Is it a developed nation?	Countries	HDI	Is it a developed nation?
Albania	0.719	No	Latvia	0.769	No
Austria	0.851	Yes	Lithuania	0.783	No
Belarus	0.732	No	Luxembourg	0.852	Yes
Belgium	0.867	Yes	Macedonia	0.701	No
Bosnia-Herzegovina	0.71	No	Malta	0.815	Yes
Bulgaria	0.743	No	Moldova	0.623	No
Croatia	0.767	No	Montenegro	0.769	No
Cyprus	0.81	Yes	Netherlands	0.89	Yes
Czech	0.841	Yes	Norway	0.938	Yes
Denmark	0.866	Yes	Poland	0.795	No
Estonia	0.812	Yes	Portugal	0.795	No
Finland	0.871	Yes	Romania	0.767	No
France	0.872	Yes	Russia	0.719	No
Georgia	0.698	No	Serbia	0.735	No
Germany	0.885	Yes	Slovakia	0.818	Yes
Gibraltar	0	No	Slovenia	0.828	Yes
Greece	0.855	Yes	Spain	0.863	Yes
Greenland	0	No	Sweden	0.885	Yes
Hungary	0.805	No	Switzerland	0.874	Yes
Iceland	0.869	Yes	Turkey	0.683	No
Ireland	0.895	Yes	Ukraine	0.71	No
Italy	0.854	Yes	United Kingdom	0.849	Yes
#developed nations	33		Median HDI score	0.81	

Table 4. Nations in Europe

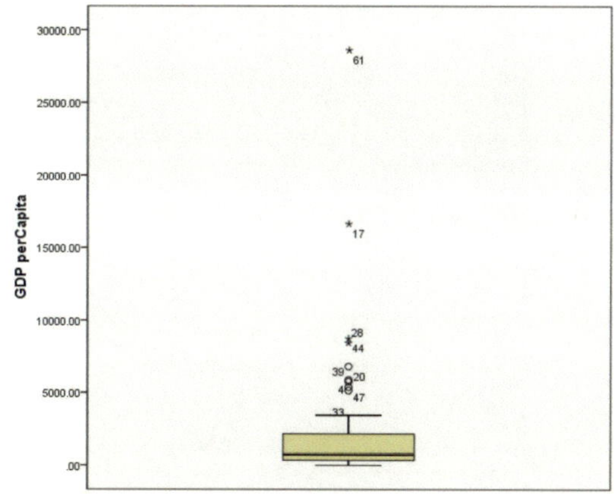

Fig. 1. GDP in Africa

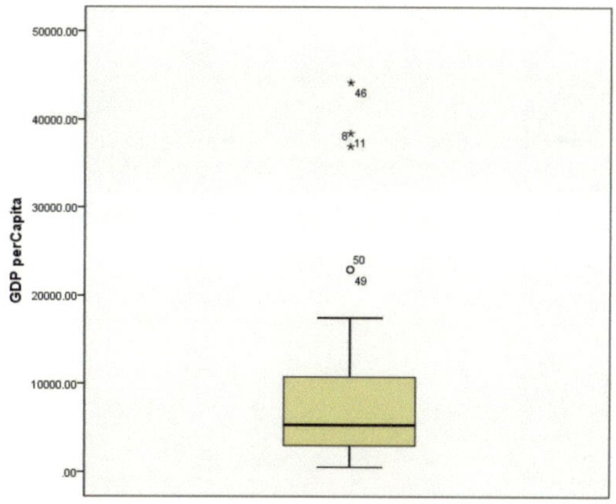

Fig. 2. GDP in America

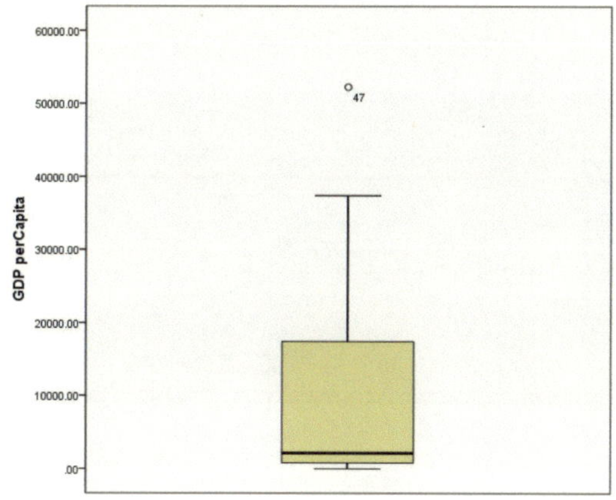

Fig. 3. GDP in Asia

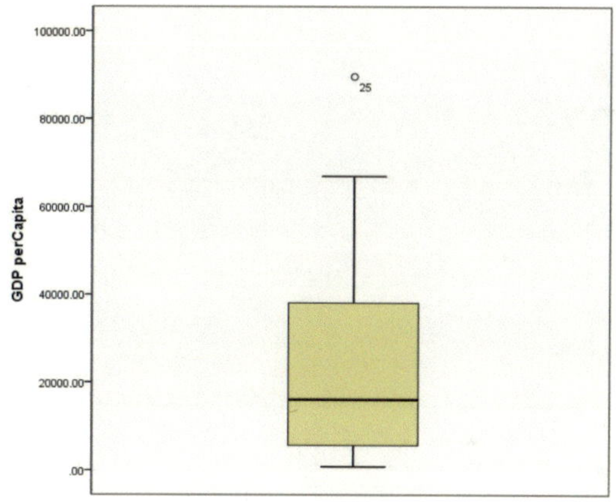

Fig. 4. GDP in Europe

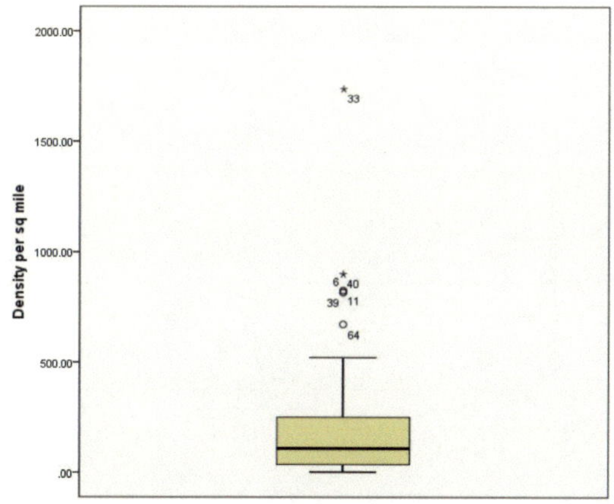

Fig. 5. Density/Square mile in Africa

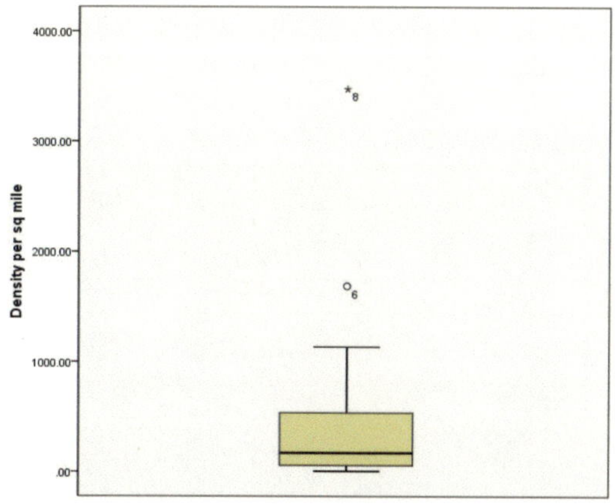

Fig. 6. Density/Square mile in America

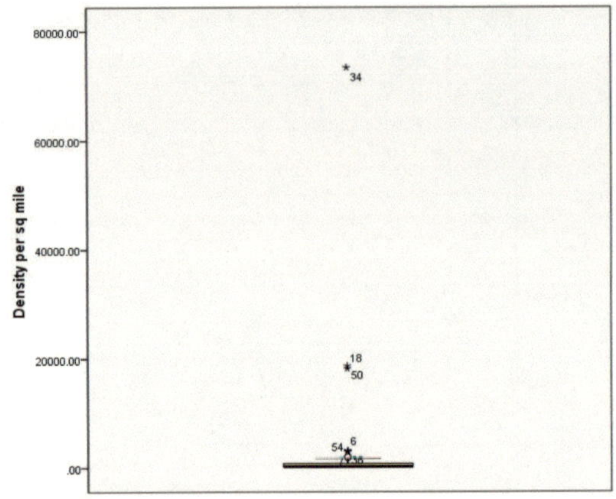

Fig. 7. Density/Square mile in Asia

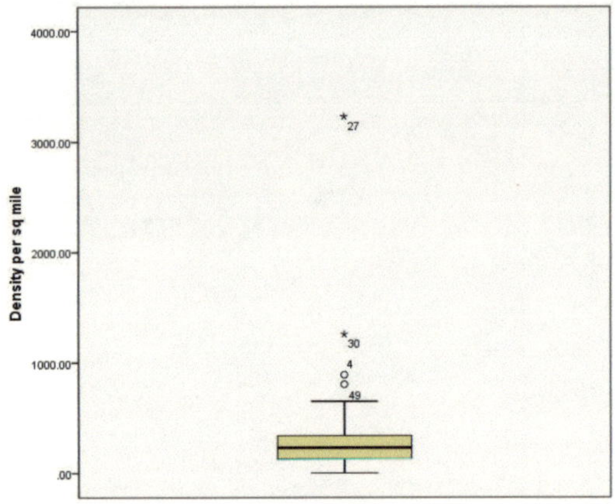

Fig. 8. Density/Square mile in Europe

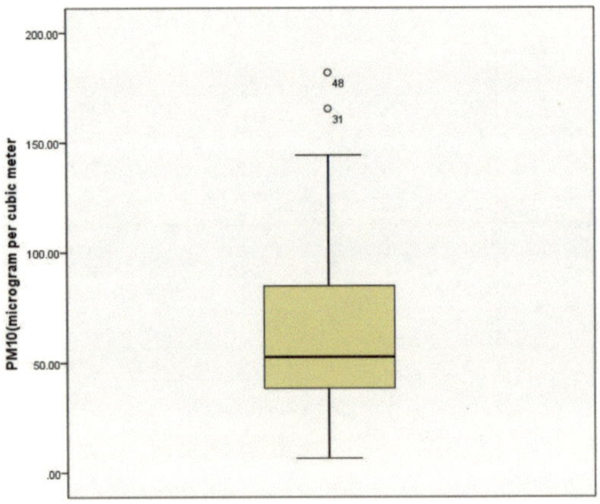

Fig. 9. PM10 in Africa

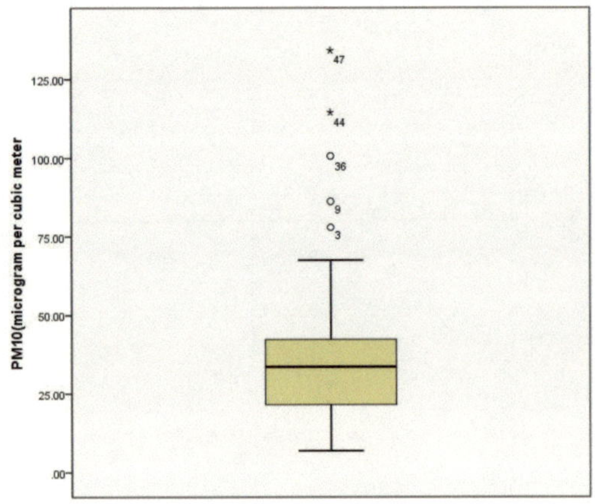

Fig.10. PM10 in America

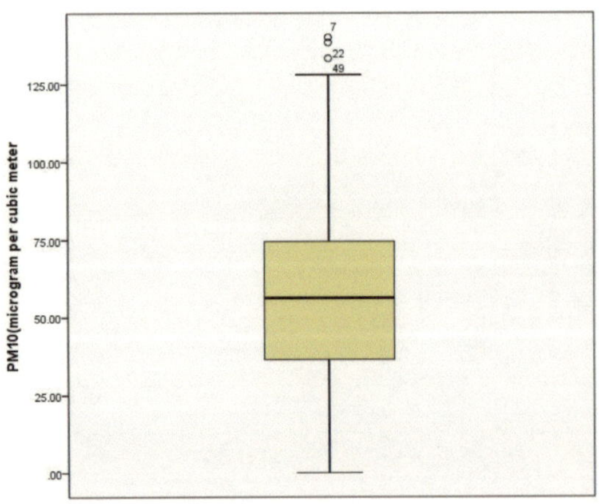

Fig. 11. PM10 in Asia

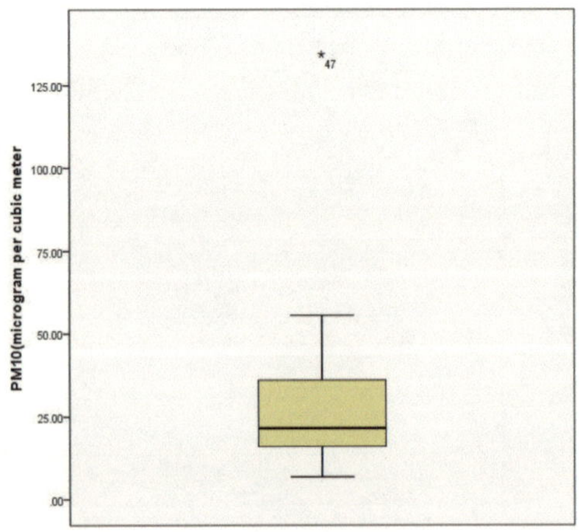

Fig. 12. PM10 in Europe

of outlier nations United Arab, Saudi Arabia, Bangladesh and Iraq) turn in a high amount of PM10. Though the median amount of PM10 is same in both the African and American nations, the nations within each group differ quite a bit. The African nations (with the exception of outlier nations Mali and Sudan) have much different PM10 while there are outlier nations in American continents like Paraguay, Trinidad and Uruguay. It appears that the presence of excessive PM10 in the air is due to industrial and agricultural sectors.

The contributions by the nations to the carbon dioxide (CO_2) in the air also differ significantly. The Figures 13 through 16 exhibit the CO_2 per capita in metric tons. The per capita CO_2 in Asia (with the exception of outlier nations Bahrain, Kuwait, Qatar and United Arab) and in Africa (with the exception of outlier nations Equatorial Guinea, Libya, Seychelles and South Africa) are low. The per capita CO_2 in America (with the exception of outlier nations USA, Aruba, Antilles and Trinidad) and in Europe (with the exception of outlier nations Belgium, France and Luxemburg) are high. It appears that the presence of excessive CO_2 in the air is due to industrial sectors.

The sulfur dioxide (SO_2) in metric tons are illustrated in Figures 17 through 20. The SO_2 is low in Asia (with the exception of outlier nations Tajikistan, Australia, China, South and

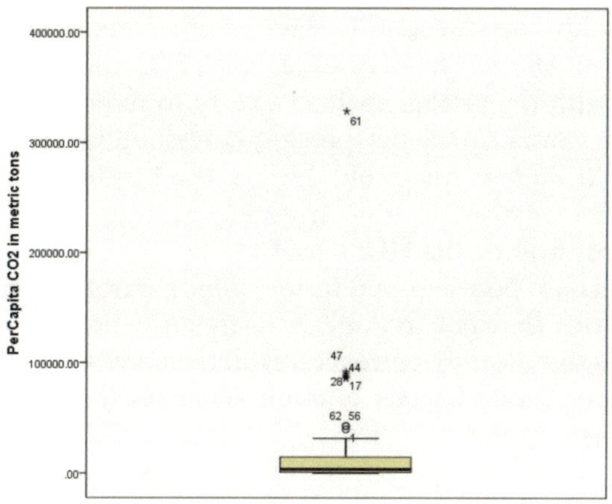

Fig. 13. Per capita CO2 in Africa

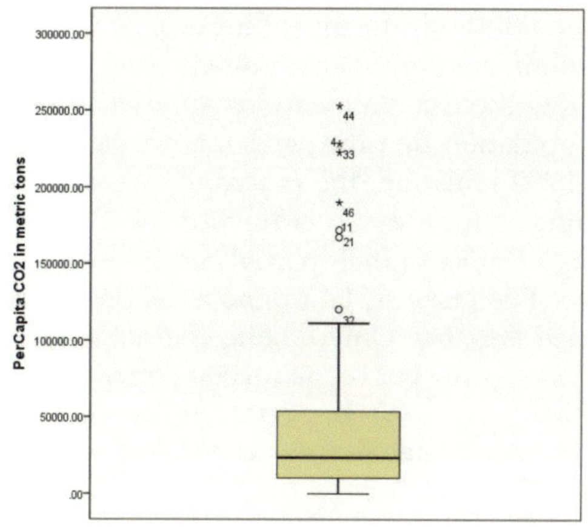

Fig. 14. Per capita CO2 in America

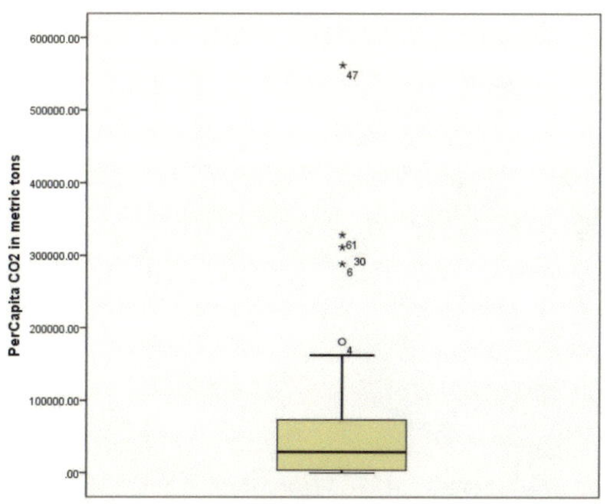

Fig. 15. Per capita CO2 in Asia

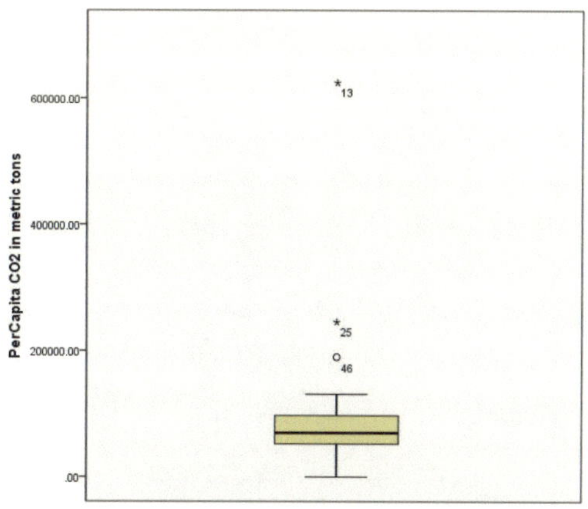

Fig. 16. Per capita CO2 in Europe

North Koreas) and in Americas (with the exception of outlier nations Jamaica, USA and Canada). The SO2 is more in Europe (with the exception of outlier nation Belgium). There are outlier nations in Africa (with the exception of outlier nations Egypt, Libya and South Africa) when it comes to the SO2 level. It appears that the presence of excessive SO2 in the air is due to energy sectors.
The nitric oxide (NO2) level in metric tons are displayed in Figures 21 through 24 for the nations in Africa, America, Asia and Europe. The NO2 is more in Africa (with the exception of the outlier nations like Congo, Ethiopia, Nigeria, Sudan and Tanzania) than in Asia (with the exception of the outlier nations like Cook Island and India). The NO2 in America (with the exception of the outlier nations like USA, Canada and Virgin island) and in Europe (with the exception of the outlier nations like Germany, Russia, Spain and United Kingdom) are low. It appears that the presence of excessive NO2 in the air is due to the agricultural sectors.

3. Clusters of nations based on the principal components

In the previous section, it was noticed that the developing and developed nations contribute differently to the air pollutions. A natural follow up question is then whether the nations within a group pollute more in a single contamination or in all of CO2, NO2, SO2 and PM10. How these contaminations are matching up with the factors such as acidity in the rain, the population density per square mile, the per capita GDP, the percent contribution to the global emission, the per capita expenditure in budget for public health, the incidences of lung cancer and cardiovascular deaths? An answer to this question resides on the significance of their partial correlations after controlling the HDI variable.
For this purpose, let r denote the partial correlation between two factors after controlling the HDI variable. Only among the nations in Africa (but not in American, Asian or European nations), the per capita carbon dioxide (CO2) is significantly correlated with the emission level (that is, r = 0.8). In Africa and in Asia, the per capita budget amount spent on the health services is significantly correlated with the per capita CO2 level (that are, r =0.9 and0.9 respectively). Interestingly, only in American nations, the partial correlation between the per capita CO2 and the PM10 is significant (that is, r = -0.9). Among the European nations, the per capita CO2 and the nitric oxide (NO2) are significantly correlated (that is, r = 0.7).

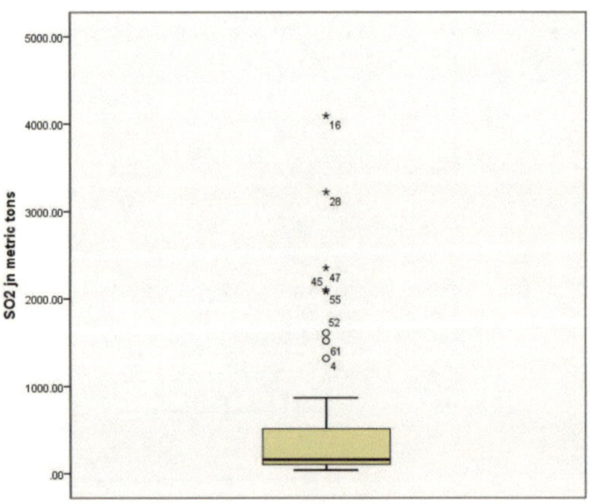

Fig. 17. SO2 in Africa

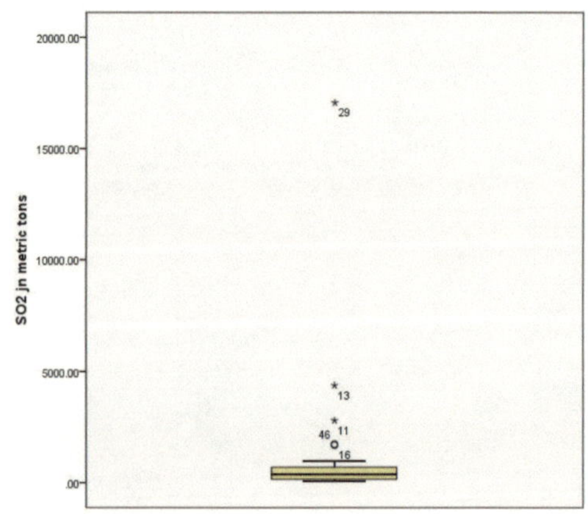

Fig. 18. SO2 in America

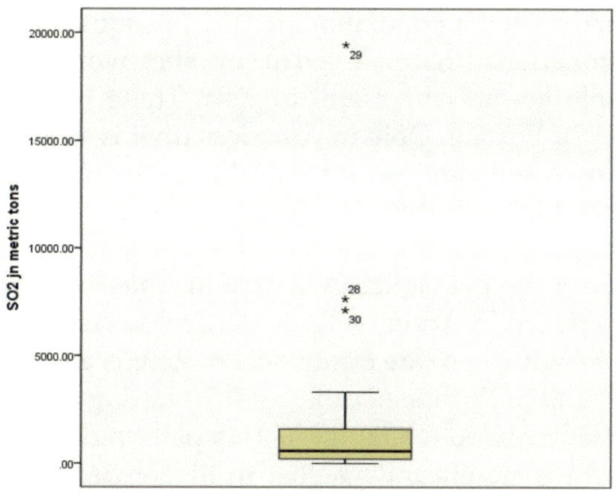

Fig. 19. SO2 in Asia

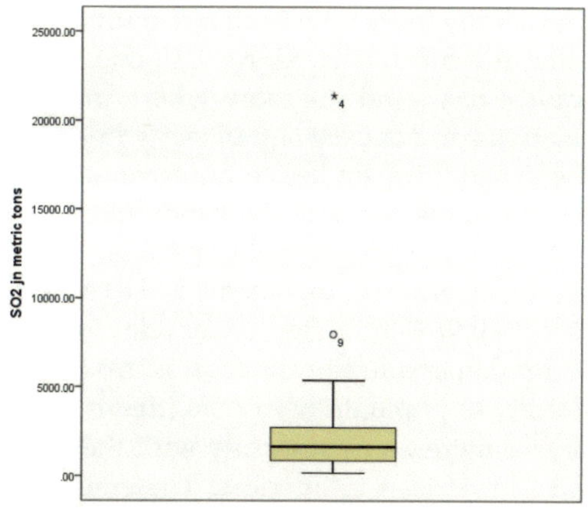

Fig. 20. SO2 in Europe

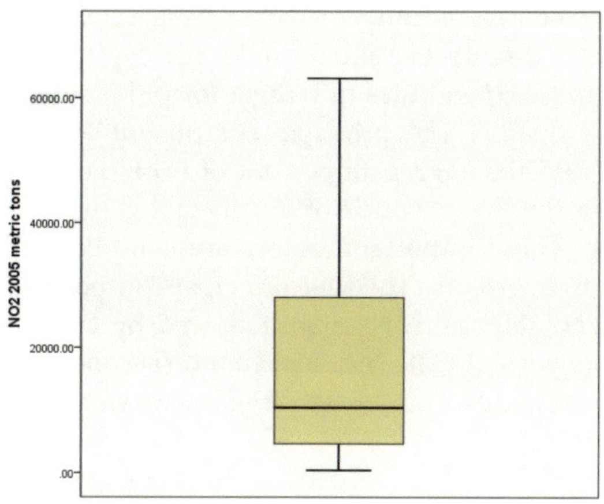

Fig. 21. NO2 in Africa

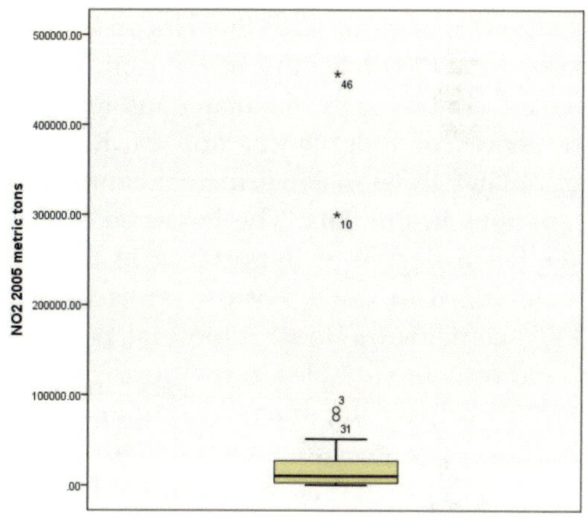

Fig. 22. NO2 in America

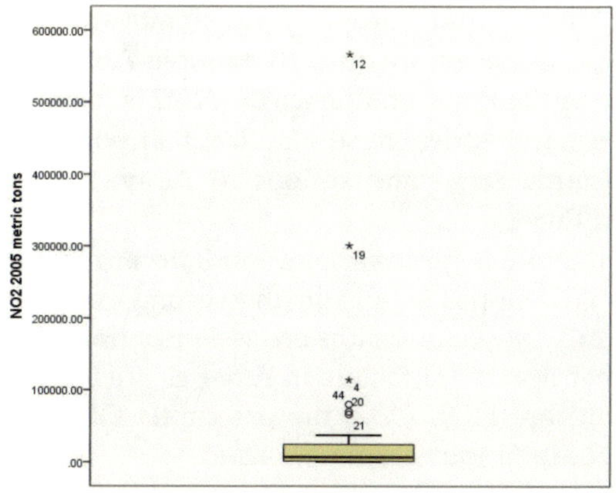

Fig. 23. NO2 in Asia

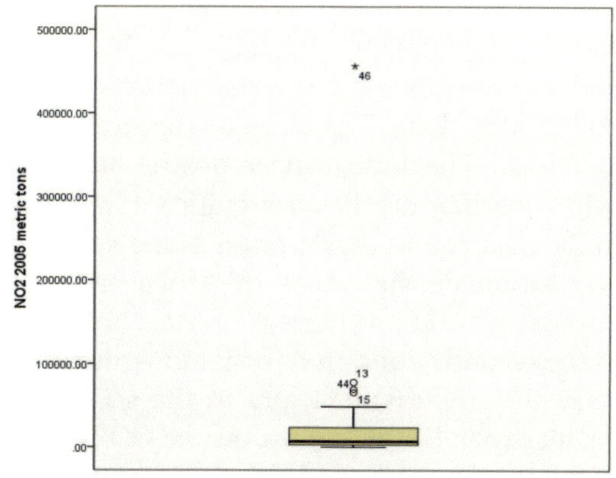

Fig. 24. NO2 in Europe

The acidity in the rain is significantly correlated with the population density per square mile only in African (that is, r = 0.9) and in Asian nations (that is, r = 0.6). In other words, the acidity in the rain is more where more people live in Africa and in Asia. These types of relations do not exist significantly in Americas or Europe. Only in Americas (that is, r = 0.9), the expenditure on health improvement has increased along with the acidity in the rain. The acidity in the rain is significantly more with the higher sulfur dioxide only in Asia (that is, r = 0.7) and in Europe (that is, r = 0.5).

Strangely enough, higher the PM10 means lower the per capita CO2 only in Americas (that is, r = - 0.9). Such a significant relation is not noticed in Africa, Asia or Europe. In Asia (that is, r = 0.5) and in Europe (that is, r = 0.5), more sulfur dioxide occurs when there is a higher density of population. In Asia (that is, r = 0.9) and in Europe (that is, r = 0.7), cardiovascular deaths increase significantly with the higher nitric oxide (NO2) level. This pattern does not exist in Africa or Americas. The emission level is significantly higher in higher density of nations only in Asia (that is, r = 0.9) but not in America, Africa or Europe.

No clear consistent pattern among the developed and developing nations either within or across the groups is emerging. It is, therefore, worthwhile to perform the principal component analysis with the data on CO2 from the coal burning, HDI, per capita GDP, CO2, NO2, SO2, PM10, acidity in the rain, population density per square mile, per capita GDP, percent contribution to global emission, per capita expenditure in budget for public health, incidences of lung cancer and cardiovascular deaths. The principal component analysis formulates three major principal components and the aggregate percent of explained total variations in the data. The larger coefficient in the Eigen vector of a principal component signifies a degree of importance of the factor. These important factors are compiled and summarized in Table 5 below for each of the four groups. The first principal component 1 (PC1) contains the most important pollution variables and these are followed by the next round of data variables in the principal component 2 (PC2). The final round (because only three principal components were chosen to begin with) of pollution variables are selected in the third principal component (PC3).

Notice that the coal burning is identified in the PC1 itself for all continents. The CO2 in the air is serious in Africa and Asia. The emissions cause major air pollution in all the continents. The GDP is quite diversified in Africa and Americas. Only in Asia, the health expenditure is not causing variation in the data. An implication is that the nations in Asia spend more for health. The nations in Africa, America and Europe spend significantly not enough amounts for the health. The SO2 and NO2 in the air are quite different in Africa and in Asia respectively. The Asian nations are more agricultural and hence the NO2 is major air pollution in there. The oil refineries and petroleum industries cause significant air pollutions in Africa. The lung cancer deaths are significantly large and serious in Americas. The cardiovascular deaths are significant in Asia and Europe.

In the next round of importance, the acidity in the rain is quite serious and it occurs in high population density areas in Africa only. The HDI appears only in the second round of importance and it is more so in Asia only. The cardiovascular deaths are seriously important in the second round in Africa and America. Other important factors in America are NO2. In Asia, the important factors in the second round are HDI, CO2, the per capita GDP and health expenditure. In Europe, the CO2 and NO2 are serious air pollutions.

In the third round, the HDI, PM10 and NO2 are important factors in Africa and only PM10 is important in America. The acidity, high population density areas and SO2 are important factors in Asia and only SO2 is an important factor in Europe so far as air pollutions. See Figures 25 through 28 for specifics.

Group	Principal Component 1	Principal Component 2	Principal Component 3	Explained variation
Africa	1. Coal burning 2. CO2 3. GDP 4. Health expenditure 5. SO2 6. Emission	1. Acidity 2. Population density 3. Cardiovascular death 4. Lung cancer	1. HDI 2. PM10 3. NO2	89%
America	1. Coal burning 2. HDI 3. Acidity 4. CO2 5. GDP 6. Health expenditure 7. Lung cancer 8. Emission	1. NO2 2. Cardiovascular deaths	1. PM10	72%
Asia	1. Coal burning 2. NO2 3. Cardiovascular deaths 4. Emission	1. HDI 2. CO2 3. GDP 4. Health expenditure	1. Acidity 2. Populati on density 3. SO2	77%
Europe	1. HDI 2. Health expenditure 3. Cardiovascular deaths 4. Emission	1. CO2 2. NO2	1. SO2	88%

Table 5. Principal factors in all four groups

4. Equitable global air pollution policy for better world health

In this section, the pollutions in both the developed versus developing nations are compared not only within each group but also across the groups. In developed African nations (see Figure 29), in developed America nations (see Figure 31), in developed Asian nations (see Table 33) and in developed European nations (see Figure 35), the per capita carbon dioxide, sulfur dioxide, carbon dioxide from the coal burning, nitric oxide, emissions, acidity and PM 10 are significant pollution variables. In developing African nations (see Figure 30), in developing American nations (see Figure 32), in developing Asian nations (see Figure 34) and in developing European nations (see Figure 36), the carbon dioxide from the coal burning, sulfur dioxide, PM 10, nitric oxide, emission and the carbon dioxide are the pollution variables. There is not much of difference between the developed and the developing nations in African and American continents. All nations are polluting the air significantly.

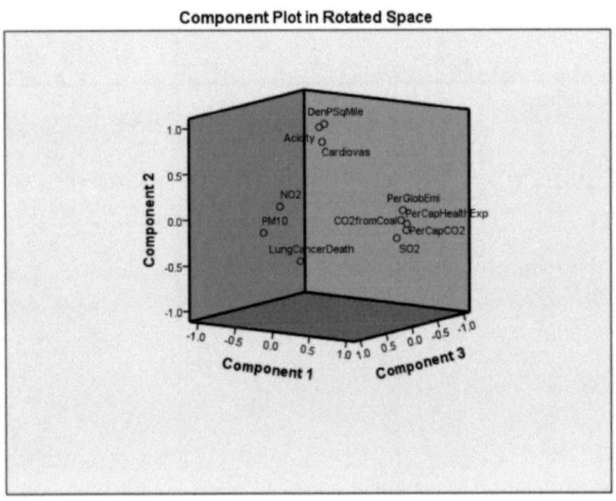

Fig. 25. Significant factors in Africa

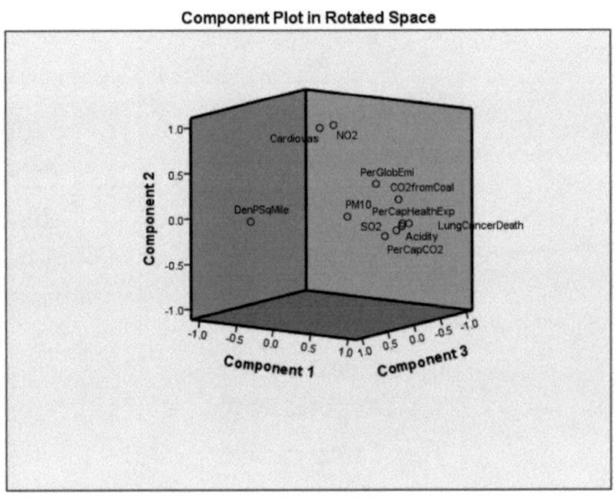

Fig. 26. Significant factors in America

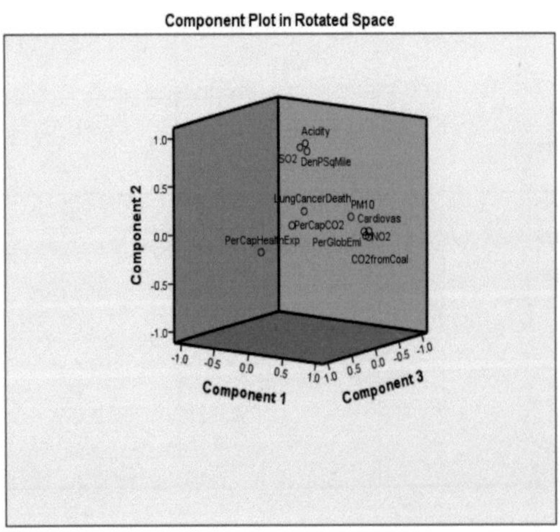

Fig. 27. Significant factors in Asia

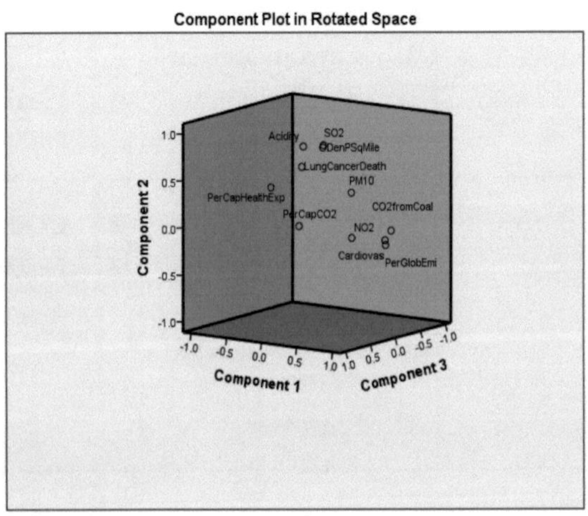

Fig. 28. Significant factors in Europe

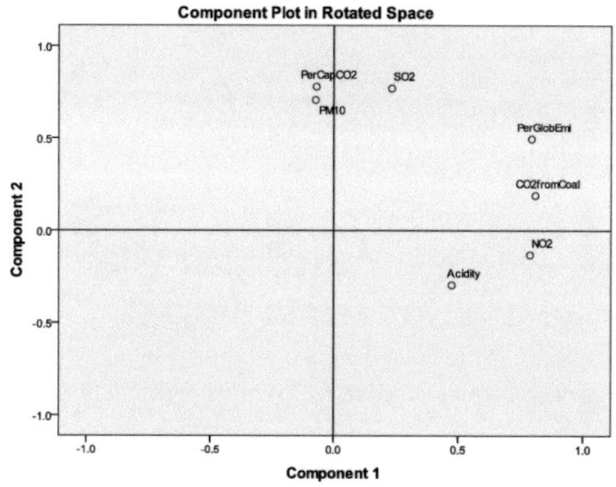

Fig. 29. In developed Africa

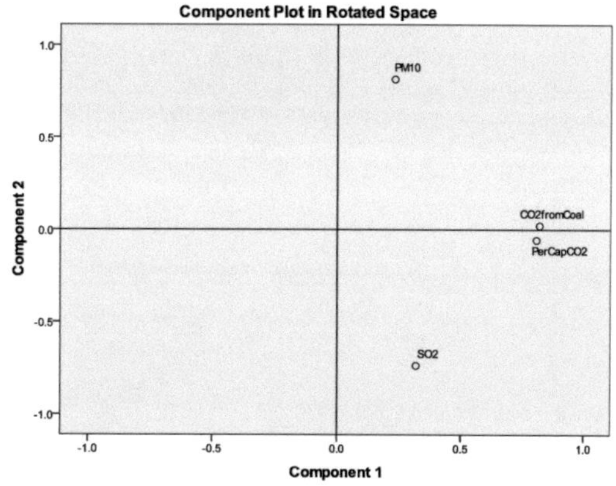

Fig. 30. In developing Africa

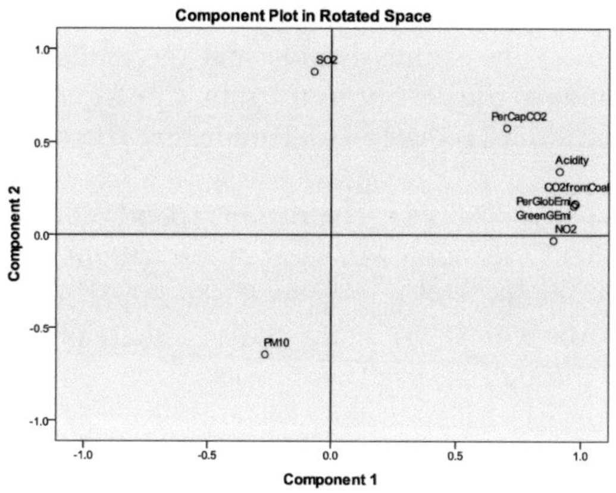

Fig. 31. In developed America

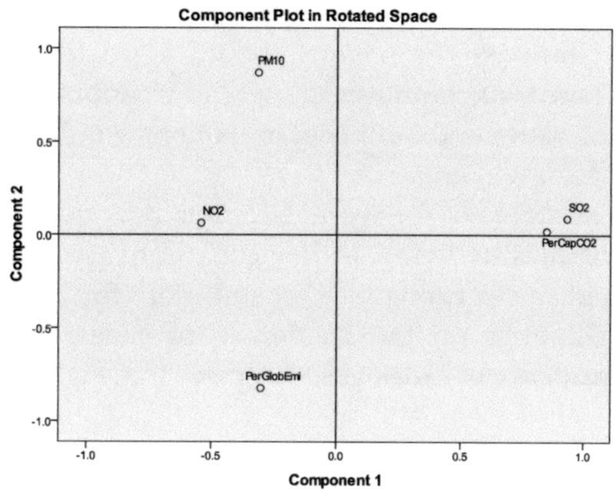

Fig. 32. In developing America

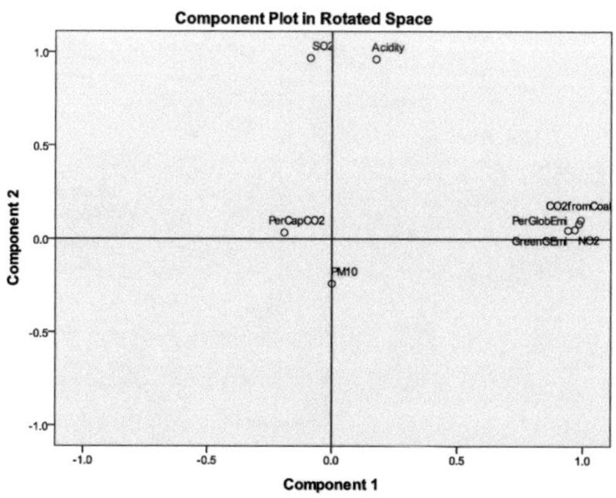

Fig. 33. In developed Asia

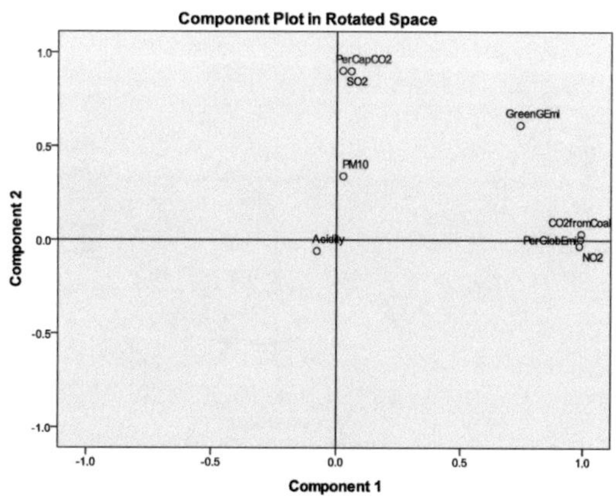

Fig. 34. In developing Asia

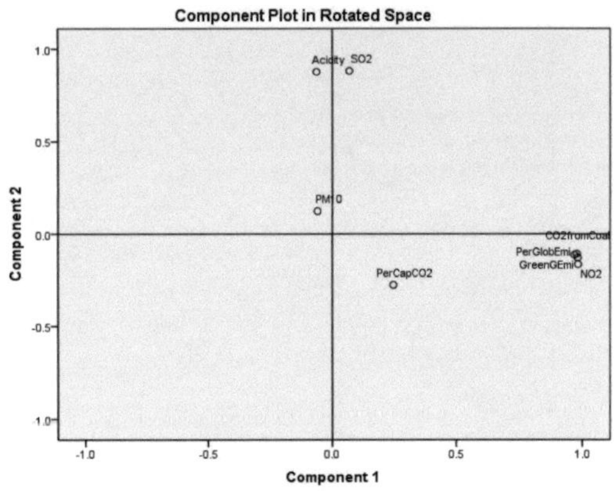

Fig. 35. In developed Europe

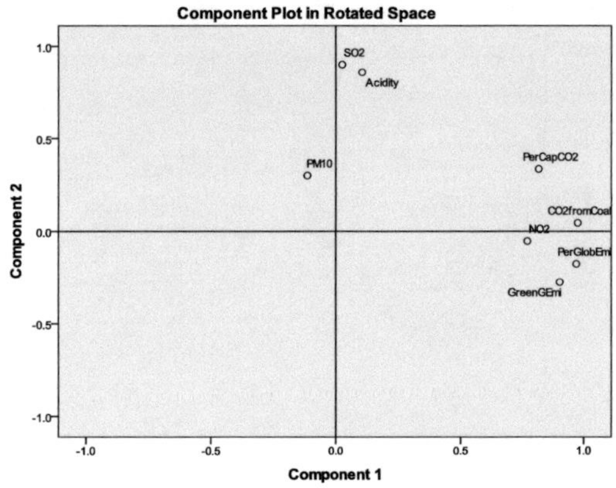

Fig. 36. In developing Europe

The nations in all continents should perhaps collectively create and implement a policy to restrict the coal burning as it produces significant amount of CO2 and PM 10 to the air pollutions. The nations should agree to minimize the sulfur dioxide and the emissions. These two commitments would undoubtedly reduce the acidity in the rain. Consequently, the water and land contaminations would significantly reduce in all continents. The world health will improve.

The status of the world health is now not in what is desirable. The nations spend more for the healthy living, if they can afford (see Figures 37 through 44). Most of the nations with higher per capita GDP spend more for healthy living. Notice in general, the impact of air pollutions on the incidences of cancer and tuberculosis is more positive than on the incidence of cardiovascular cases (see Figure 37 through 44) in all continents.

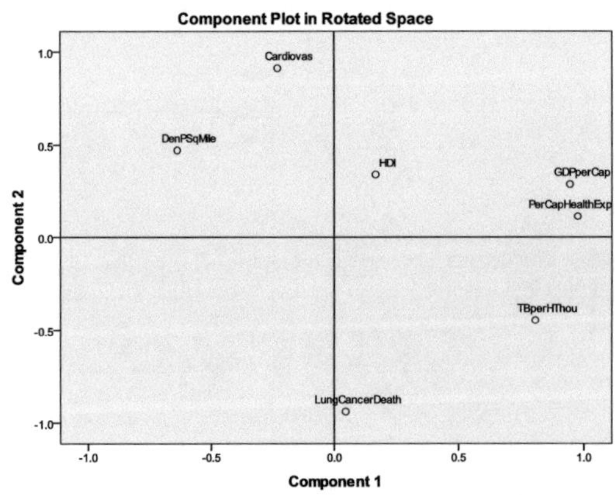

Fig. 37. Health in developed Africa

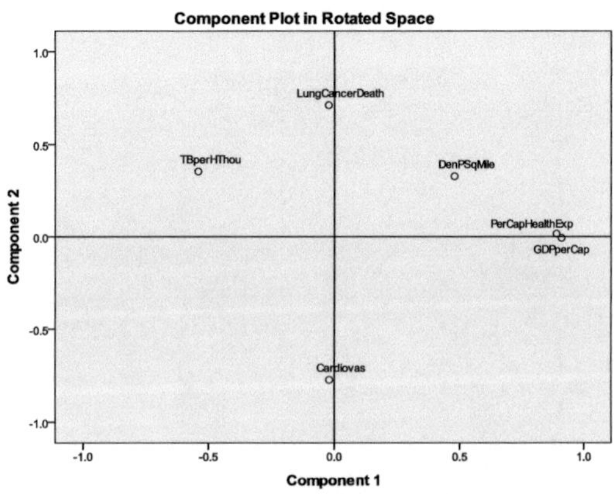

Fig. 38. Health in developing Africa

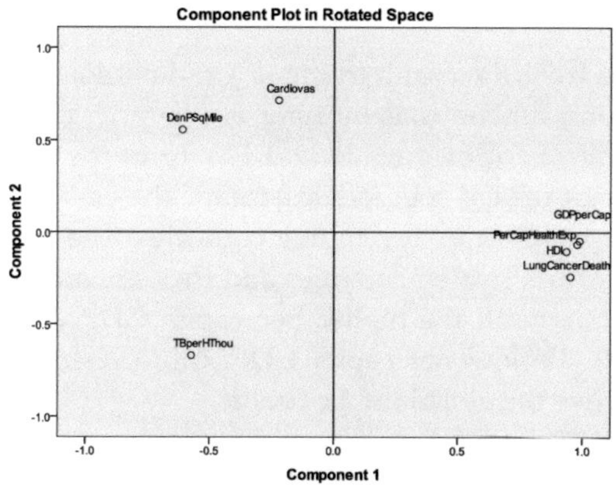

Fig. 39. Health in developed America

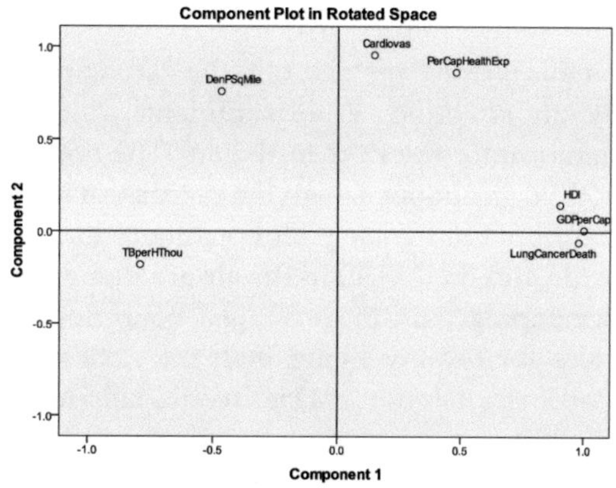

Fig. 40. Health in developing America

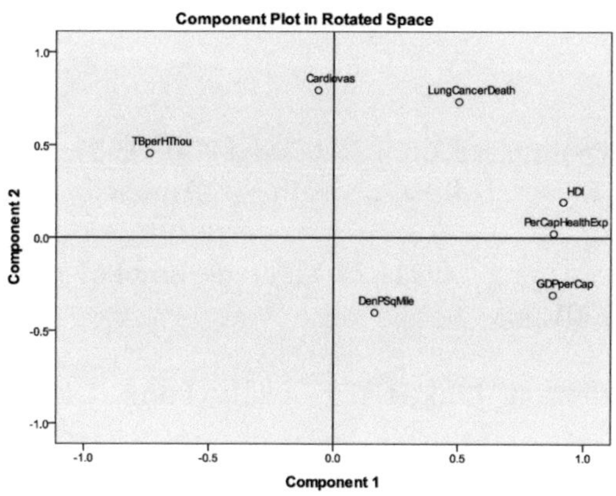

Fig. 41. Health in developed Asia

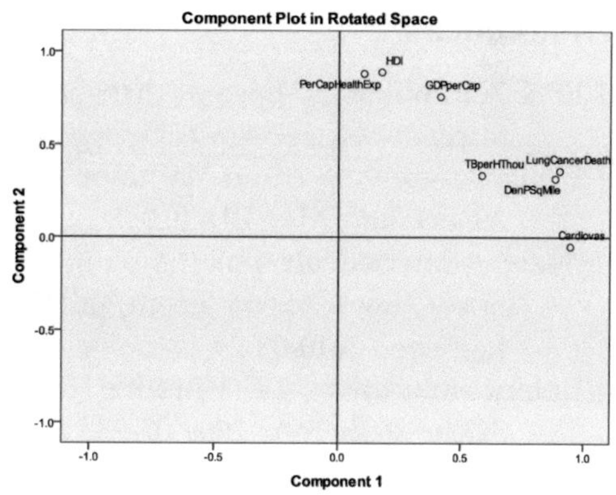

Fig. 42. Health in developing Asia

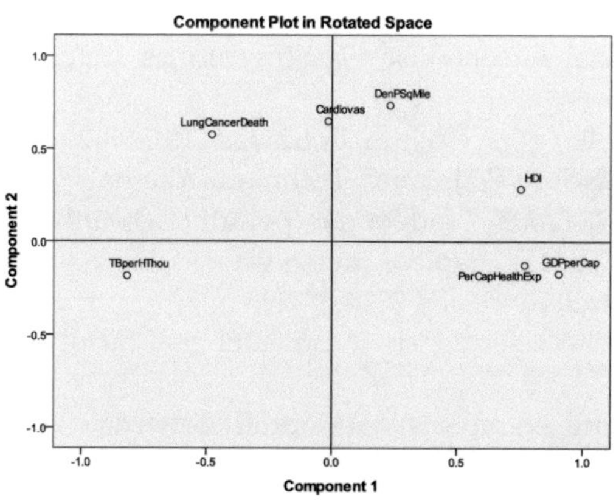

Fig. 43. Health in developed Europe

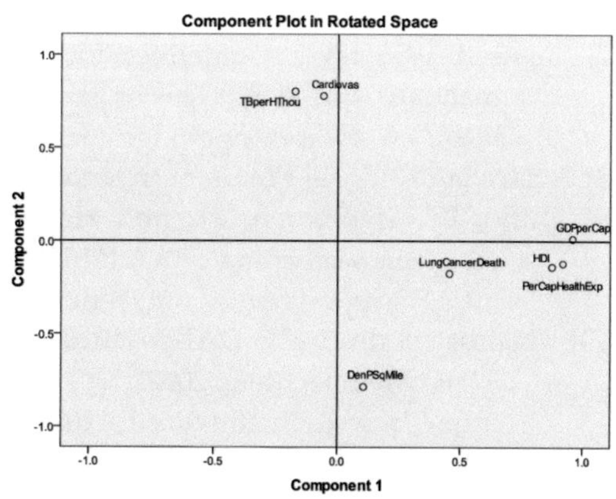

Fig. 44. Health in developing Europe

5. Conclusions

As this chapter pointed out, the carbon dioxide from the coal burning is a serious source of the air pollutions in all continents. A restriction on the coal burning is likely to reduce significantly the CO2 in the air. The abundance of sulfur dioxide and PM 10 in the air is noticed in nations where the petroleum industries related activities are more. The emissions contribute to the air pollutions more in developed than in developing countries. The nitric oxide and the PM 10 in the air are due to agriculture related activities and they are more in developing than in developed countries. Nations with the higher per capita GDP spend more for healthy living than the nations with the low per capita GDP. More restrictive global air pollution policies would help to improve the global public health.

6. Acknowledgement

This research work was supported by a Faculty Research Enhancement leave during the spring 2008 semester from the Texas State University, San Marcos, Texas.

7. References

[1] EPA: Air Pollutants".Epa.gov.2006-06-28 http://www.epa.gov/ebtpages/airairpollutants .html. Retrieved 2010-08-29.

[2] Evidence growing of air pollution's link to heart disease, death // American Heart Association. May 10, 2010

[3] "Newly detected air pollutant mimics damaging effects of cigarette smoke" (PDF). http://www.physorg.com/pdf138201201.pdf. Retrieved 2010-08-29.

[4] "Infant Inhalation Of Ultrafine Air Pollution Linked To Adult Lung Disease". Sciencedaily.com. 2009-07-23. http://www.sciencedaily.com/releases/2009/07/090722123751.htm. Retrieved 2010-08-29.

[5] "United Kingdom's emission factor database". Naei.org.uk. http://www.naei. org.uk/emissions/index.php. Retrieved 2010-08-29.

[6] European Environment Agency's 2005 Emission Inventory Guidebook

[7] "Revised 1996 IPCC Guidelines for National Greenhouse Gas Inventories (reference manual)". Ipcc-nggip.iges.or.jp. http://www.ipcc-nggip.iges.or.jp/public/gl/invs6.htm. Retrieved 2010-08-29.

[8] Australian National Pollutant Inventory Emissions Estimation Technique Manuals[*dead link*]

[9] "Duflo, E., Greenstone, M., and Hanna, R. (2008) "Indoor air pollution, health and economic well-being". "S.A.P.I.EN.S." "'1'" (1)". Sapiens.revues.org. http://sapiens.revues.org/index130.html. Retrieved 2010-08-29.

[10] "Estimated deaths & DALYs attributable to selected environmental risk factors, by WHO Member State, 2002". http://www.who.int/entity/quantifying_ehimpacts/nationprofilesebd.xls. Retrieved 2010-08-29.

[11] "Study links traffic pollution to thousands of deaths". *The Guardian* (London, UK: Guardian Media Group). 2008-04-15. http://www.guardian.co.uk/society/2008/apr/15/ health. Retrieved 2008-04-15.
[12] Simi Chakrabarti. "20th anniversary of world's worst industrial disaster". Australian Broadcasting Corporation. http://www.abc.net.au/worldtoday/content/2004/s1257352.htm.
[13] http://www.latimes.com/features/health/la-me-pollute13-2008nov13,0,5432723. story, http://www.sfgate.com/cgi-bin/article.cgi?f=/c/a/2008/11/13 /MNQP143CPV.DTL
[14] Christopher H. Goss, Stacey A. Newsom, Jonathan S. Schildcrout, Lianne Sheppard and Joel D. Kaufman (2004). "Effect of Ambient Air Pollution on Pulmonary Exacerbations and Lung Function in Cystic Fibrosis". *American Journal of Respiratory and Critical Care Medicine* 169 (7): 816–821. doi:10.1164/rccm.200306-779OC. PMID 14718248.
[15] Zoidis, John D. (1999). "The Impact of Air Pollution on COPD". *RT: for Decision Makers in Respiratory Care.* http://www.rtmagazine.com/issues/articles/1999-10_06.asp.
[16] Holland WW, Reid DD. The urban factor in chronic bronchitis. Lancet. 1965; I: 445-448.
[17] J. Sunyer (2001). "Urban air pollution and Chronic Obstructive Pulmonary disease: a review". *European Respiratory Journal* 17 (5): 1024–1033. doi:10.1183/09031936.01.17510240.PMID 11488305. http://erj.ersjournals.com/cgi/content/abstract/17/5/1024.
[18] "2005 BC Lung Association report on the valuation of health impacts from air quality in the Lower Fraser Valley airshed" (PDF). http://www.bc.lung.ca/pdf/ health_and_air_quality_2005.pdf. Retrieved 2010-08-29.
[19] "World Bank Statistics" http://siteresources.worldbank.org/DATASTATISTICS/Resources/table3_13.pdf. Retrieved 2010-08-29.
[20] International Carbon Dioxide Emissions and Carbon Intensity Energy Information Administration
[21] "2010 Human development Report". United Nations Development Programme. pp. 148–151. http://hdr.undp.org/en/media/HDR_2010_EN_Complete.pdf. Retrieved 4 November 2010.
[22] Shanmugam, R., Anand, S., & Burke, G. (2007). Mining air pollution Data to Learn Stroke Incidence, *Model Assisted Data Analysis, 2*, 53-58.
[23] World Health Organization 2010 Report at http://www.who.int/2010en/.
[24] The World Bank: Data Factoids at http://go.world.bank.org/.
[25] US Census Bureau. World Population Information at http://census.gov/world.html/.
[26] International Panel on Climate Change at http://www.ipcc.ch/

[27] International Agency for Research on Cancer at http://www.iarc.fr/
[28] US Environemental Protection Agency at http://www.epa.gov/
(Completed on Saturday, March 05, 2011)